THE OBJECTIVE IS
QUALITY

Cover Illustration: © Infinity – Fotolia.com

Michel Jaccard

THE OBJECTIVE IS QUALITY

Introduction to Quality, Performance and Sustainability Management Systems

Translated from the French by
Nadia Ljundberg

EPFL Press
A Swiss academic publisher distributed by CRC Press

This book is the revised and augmented edition of *Objective qualité*, published in French by the PPUR in 2002. The translation was made possible thanks to generous subsidies from Nestlé and from the laboratory LEM-EPFL of Professor Philippe Wieser, to whom the author and publisher express their great appreciation.

The International Organization for Standardization (www.iso.org) supported the project from its inception, and the publisher and author are thankful for the assistance and information they provided on the chapters related to ISO standards.

Taylor and Francis Group, LLC
6000 Broken Sound Parkway, NW, Suite 300,
Boca Raton, FL 33487

Distribution and Customer Service
orders@crcpress.com

www.crcpress.com

Library of Congress Cataloging-in-Publication Data
A catalog record for this book is available from the Library of Congress.

This book is published under the editorial direction of
Professor Philippe Wieser (EPFL).

The publisher and author express their thanks to the Ecole Polytechnique Fédérale de Lausanne (EPFL) for its generous support towards the publication of this book.

The EPFL Press is the English-language imprint of the Foundation of the Presses polytechniques et universitaires romandes (PPUR). The PPUR publishes mainly works of teaching and research of the Ecole polytechnique fédérale de Lausanne (EPFL), of universities and other institutions of higher education.

Presses polytechniques et universitaires romandes, EPFL – Rolex Learning Center,
Post office box 119, CH-1015 Lausanne, Switzerland
E-mail: ppur@epfl.ch
Phone: 021 / 693 21 30
Fax: 021 / 693 40 27

www.epflpress.org

ISBN 978-2-940222-65-0 (EPFL Press)
ISBN 978-1-466572-99-7 (CRC Press)

Printed in Italy

Our system of education is absolutely vicious: for 22 years we leave our young people in absolute ignorance of the inevitable conditions of the rest of their lives; every day, we force them to absorb new knowledge and to store it in a corner of their minds, giving them the false hope that it will be useful one day. As soon as they begin their professional lives, everything suddenly changes as if by some theatrical trick. It is no longer a case of absorbing, but rather of doing, of putting into action a very small number of skills acquired under the service of the employer.

After spending 22 years being served by teachers, it is necessary to serve one's boss. It is very difficult, after having been reduced for so long to the simple role of sponge, to put oneself in motion, and it is a marvel to see this transformation not taking place more slowly than it does, nor more painfully.

Frederick W. Taylor*

---

* Excerpt from a communication presented at the *Society for the Promotion of Engineering Education*, Proceedings, XVII, 1909 (pp. 79-92). Taylor was essentially an engineer by training, who climbed the corporate ladder from manual worker to workshop foreman, to finally become head engineer of his company, having received his diploma after taking night-school courses, which might explain the acerbic tone of the comments.

Table of contents

Foreword 19

Chapter 1
The Environment of Quality 23
1.1 Products, services and satisfaction of expectations 23
1.1.1 Distinction between products and services 23
1.1.2 A hierarchy of needs, Maslow's pyramid 24
1.2 Quality, definition 25
1.3 Quality and its place in the Marketing Mix 27
1.4 The quality cycle of a product 28
1.5 The portfolio of products (BCG matrix) and quality 29
1.6 Quality and stakeholders 31
1.7 Structure and functions of an organization 33
1.7.1 Representation by an organizational chart 33
1.7.2 Company functions and quality activities 34
1.7.3 Quality management and network of activity 35
1.7.4 Similarly managed organizational functions 36
1.8 Quality and performance of a company 36
1.9 Quality and Porter's value chain 37
1.9.1 Limits of representation by an organizational chart 37
1.9.2 Porter's diagram and the process approach 38
1.10 Optimization of a process 40
1.10.1 Optimization of processes: the algorithmic approach 41
1.10.2 The option *ex nihilo* 42
1.10.3 Optimization of the process: the classification approach 42
1.10.4 Optimization of the process: benchmarking 44
1.10.5 Reformulating processes: re-engineering 44
References 46

Chapter 2
The Approach to Quality: Concepts and Definitions 47
2.1 Quality: first approaches 48
2.1.1 The basics: smell and taste 48

2.1.2 Quality, an expectation 48
2.1.3 Quality: a moving target 48
2.1.4 Quality: a concept of an exchange economy 49
2.1.5 Two operational definitions of quality 49
2.2 Quality standard (or list of requirements) 50
2.2.1 Quality is associated with a measurement and a standard 50
2.2.2 The standard, an agreement between stakeholders 50
2.2.3 The standard, a foundation but not a blueprint for production ... 50
2.2.4 Applying Ishikawa's diagram to the Battle of the Somme 51
2.3 Standards of quality, poor quality, nonconformity and defects 55
2.4 Quality and innovation standard 56
2.5 Quality control 57
2.5.1 Sub-standard quality and costs 57
2.5.2 Above-standard quality and costs 58
2.5.3 Quality assurance 60
2.5.4 Quality management 60
2.5.6 Quality Management System (QMS) 61
2.6 Total Quality Management (TQM) 62
2.7 Domains of quality 62
2.8 Quality Function Deployment: The House of Quality 63
2.8.1 Several rooms, one entrance and one roof 63
2.8.2 Step one, defining subsystems 65
2.8.3 The central room 66
2.8.4 The roof 67
2.8.5 The southern room 68
2.8.6 The eastern room 68
2.8.7 Derived matrices 69
References 71

Chapter 3
History of Quality 73
3.1 Introduction 73
3.2 Quality has a history 73
3.2.1 The establishment of standards and measurements 74
3.2.2 Predominance of self sufficiency 74
3.2.3 Establishment of the first kingdoms and empires 74
3.2.4 Manufacturing standardized products 75
3.3 The Industrial Revolution and its effects 78
3.3.1 A decisive factor: the steam engine 78
3.3.2 The first managers were born 79
3.3.3 Standardization, the first industrial success in the US 81
3.3.4 A boost for American management: trains 81
3.3.5 The emergence of American industrial predominance 83
3.4 Generalized product inspection – the birth of the quality function 84
3.4.1 Management of engineers, Taylorism 84
3.4.2 From Taylorism to Fordism 85
3.5 Alfred Sloan, General Motors and the development of American management [3.18] 88
3.6 Quality control 90
3.6.1 Inspection overwhelmed by the quantities produced 90

3.6.2 Shewart and the Bell Telephone Laboratoires 90
3.6.3 World War II and the postwar period 91
3.6.4 The 50s and the emergence of Japan 92
3.7 The development of quality Assurance 93
3.7.1 Deming's SQC courses 93
3.7.2 The Deming prize portfolio 93
3.8 Quality management and the first steps toward Total Quality 94
3.8.1 Juran's seminars to Japanese middle and top management 94
3.8.2 From Quality Management to Total Quality Management (TQM)... 94
3.8.3 The birth of quality circles 95
3.9 The long route for the United States toward Total Quality 95
3.9.1 Certain blindness... 95
3.9.2 Failed initiatives of the 1960s and 70s 95
3.9.3 The third wave of SQC and initiatives of the 1980s 96
3.9.4 The rise of the 1990s and the Malcolm Bridge Quality Award... 99
3.9.5 Motorola and the Six Sigma approach 99
3.9.6 Japan in difficulty 100
3.10 The path taken by the EU: ISO 9001 100
3.10.1 American management introduced in the 1950s 100
3.10.2 The origin of ISO 9001: the United States once again 101
3.10.3 Consumer usage confiscated by perfidious Albion 101
3.10.4 A huge success... 101
3.10.5 followed by a diversification for the environment: ISO 14001 ... 102
3.10.6 triggering grimaces from the supporters of Total Quality... 102
3.11 Synopsis 103
References 104

Chapter 4
The ISO 9000 Family of Quality Management Systems 107
4.1 Evolution of ISO 9001 107
4.1.1 Taking criticism into account 107
4.1.2 New principles 107
4.1.3 Two objectives 109
4.1.4 Implementation through project management 109
4.2 ISO 9000: Eight principles 111
4.2.1 The basic eight principles of ISO 9001:2000, as well as ISO 9001:2008 111
4.2.2 ISO 9001 helps create a learning environment 114
4.3 The system approach in eight steps and four sections 115
4.4 ISO 9001: Emphasis on employees' skills 116
4.5 Documentary foundation 118
4.5.1 A pyramidal base 118
4.5.2 Structure of a quality manual 120
4.5.3 Contents of a procedure and other documents 121
4.5.4 Document management 122
4.6 Contents of ISO 9001:2008 123
4.6.1 Who develops ISO standards? And how? 123
4.6.2 Drafting of a standard and research activities: a large gap 123
4.6.3 Structure of the ISO 9001 standard 124
4.7 ISO 9004:2009 – beyond quality management 128

4.7.1 ISO 9004:2009, presentation ... 129
4.7.2 A model of maturity for lasting success ... 129
4.7.3 Content of the standard ... 130
4.7.4 The role of self-assessment ... 133
4.8 The ISO 9000 family ... 135
4.9 Standards derived from ISO 9001 ... 137
4.9.1 Specialized standards ... 137
4.9.2 Higher education and ISO 9001 ... 138
4.10 ISO 9001 and the role of audits ... 139
4.10.1 Types of audits ... 139
4.10.2 An audit and its procedures: ISO 19011:2002 ... 140
4.11 Certification and/or accreditation of the organization ... 143
4.11.1 Making the right choice ... 143
4.11.2 Steps towards certification of an organization ... 144
4.12 Post-ISO 9001 certification ... 145
4.12.1 Certification is... only the beginning of a quality approach! ... 145
4.12.2 The role of quality manager ... 147
4.13 Problem-solving after certification ... 149
4.14 Conclusions ... 153
References ... 154

Chapter 5

Quality Management in Laboratories ... 155
5.1 ISO/IEC 17025 is for who? ... 155
5.2 Measurements and tests: a process approach ... 156
5.3 GLPs of the OECD, the Directive 2004/10/CE and the Swiss OGLP ... 157
5.4 The key content of ISO/IEC 17025 ... 163
5.4.1 Management requirements (Plan) ... 163
5.4.2 Document control (Do): ... 164
5.4.3 Review of requests, tenders and contracts (Do): ... 165
5.4.4 Subcontracting of tests and calibrations (Do) ... 165
5.4.5 Control of nonconforming testing and/or calibration work (Act) . 166
5.4.6 Control of records (Do) ... 167
5.4.7 Internal audits (Check) ... 167
5.4.8 Reliability of tests (Do) ... 168
5.4.9 Tests and validation (Do) ... 169
5.4.10 Equipment (Do) ... 169
5.4.11 Measurement traceability (Do) ... 170
5.4.12 Planning of sampling procedures and handling of objects (Plan) 170
5.4.13 Monitoring the quality of test results and documentation of results ... 171
5.4.14 Summary ... 171
5.5 Quality and its management in leading research laboratories ... 172
5.5.1 QA within research, a disputed tripartite method ... 172
5.5.2 Resistance to quality management ... 172
5.5.3 An idealized vision of the researcher ... 173

5.5.4 A growing trend ... 173
5.5.5 AFNOR's first attempt ... 174
References ... 180

Chapter 6
Safety of the Food Supply Chain and ISO 22000 ... 181
6.1 Food safety and hygiene, a little history ... 181
6.1.1 Food safety, from antiquity to the Industrial Revolution ... 181
6.1.2 Reduced diseases in the West: a primary cause, hygiene ... 182
6.1.3 The introduction of food standards and regulations ... 183
6.1.4 FAO, WHO and Codex Alimentarius ... 183
6.1.5 HACCP enters the scene ... 185
6.1.6 Newly emerging risks ... 186
6.1.7 Reform of food legislation within the EU ... 187
6.1.8 ISO 22000 and private frames of reference ... 188
6.2 HACCP: content and implementation ... 189
6.2.1 Prerequisite programs, good manufacturing practices ... 189
6.2.2 HACCP: seven principles ... 192
6.2.3 Identification of hazards ... 192
6.2.4 Characteristics of CCP ... 193
6.2.5 HACCP decision tree ... 195
6.2.6 HACCP: 12 implementation steps ... 196
6.3 ISO 22000:2005 ... 197
6.3.1 An overview of the standard ... 197
6.3.2 ISO 22000: some salient points ... 203
6.3.3 HACCP, PRP, ISO 22000 and OPRP ... 208
6.3.4 Clause 7, the core of ISO 22000:2005 ... 210
6.4 Additional documents of the ISO 22000 family ... 211
6.5 First experiences ... 212
6.6 Conclusions ... 213
References ... 214

Chapter 7
The Off-shoots of ISO 9001 ... 215
7.1 Introduction ... 215
7.2 Risk management and ISO 31000:2009 ... 216
7.2.1 Introduction to risk management ... 216
7.2.2 ISO 31000:2009, Objectives, goals and expected benefits ... 219
7.2.3 ISO 31000:2009, principles of risk management ... 220
7.2.4 ISO 31000:2009, the organizational framework of risk management ... 220
7.2.5 ISO 31000:2009, Risk management process ... 222
7.2.6 Discussion ... 225
7.3 ISO/IEC 27005:2009 and the management of risks relating to information security ... 227
7.3.1 Threats to information systems ... 227
7.3.2 Elements involved in an information security system ... 227
7.3.3 Sensitive information and security criteria ... 228
7.3.4 IEC 27001:2005 and information security management systems. 228
7.3.5 ISO/IEC 27005:2011 ... 231

7.4 ISO 50001:2011, Energy management systems – Requirements with guidance for use 234
7.4.1 Introduction 234
7.4.2 Goals and objectives of ISO 50001:2011 234
7.4.3 Structure and content 235
7.5 Requirements of the energy management system (EMS) 236
7.5.1 General requirements 236
7.6 Discussion and conclusions 240
References 240

Chapter 8
Total Quality, Personalities who Stand Out 241
8.1 The triumvirate of the 1980s 241
8.1.1 Management according to Deming 241
8.1.2 Ishikawa and TQC (Total Quality Control) 245
8.1.3 Big Q and Little Q, the contribution of J.M. Juran 247
8.2 Total quality according to Shoji Shiba 248
8.2.1 From kaizen to hoshin 248
8.2.2 The WV model 249
8.2.3 Total quality by four management revolutions 249
8.2.4 BT, integration of strategy and forecasting in TQM 250
8.2.5 Other concepts of breakthrough management 252
8.2.6 Voice of the customer 254
8.3 Taguchi's quality engineering 259
8.3.1 A talented and somewhat isolated researcher 259
8.3.2 A new definition of quality – Loss of quality 260
8.3.3 The quadratic function of the loss of quality 260
8.3.4 Targeting the set-point value 262
8.3.5 Quality by design 263
References 266

Chapter 9
Total Quality Management: Awards 269
9.1 The Deming Prizes 269
9.1.1 The various types of prizes 269
9.1.2 The procedure for the Deming Application Prize 270
9.1.3 The expected benefits 272
9.1.4 Does the Deming Prize really increase performance? 272
9.2 The Malcolm Baldridge Award 273
9.2.1 The aim of the award is performance, not distinction 273
9.2.2 The frame of reference of the Malcolm Baldridge Award 274
9.2.3 Subdivision of criteria 277
9.2.4 Grading the subdivisions and sections of the prize; expectations 277
9.2.5 What to do to receive the Baldridge Award? 279
9.2.6 Differences between the Baldridge Award and the Deming Prize 280
9.2.7 Is the Baldridge Award in decline? 281
9.3 EU, the EFQM model and the EFQM 2013 Excellence Award 281
9.3.1 A foundation created by European companies 281
9.3.2 A definition of excellence and a call to 8 concepts 282
9.3.3 The EFQM Excellence Model: a system of 9 criteria 283

9.3.4 Itemization of the nine criteria 285
9.3.5 *RADAR* logic and performance evaluation 287
9.3.6 The stepwise path to EFQM excellence 289
9.4 Quality system: EFQM or ISO 9001? 291
9.4.1 A non-exclusive approach 291
9.4.2 From ISO 9001 certification to EFQM 292
References 293

Chapter 10
The Toyota Way and the Toyota Production System (TPS) 295
10.1 The Toyota phenomenon 295
10.1.1 A brief history of Toyota 295
10.1.2 Two key ideas of the Toyota Way 2001 299
10.2 Operational excellence as a strategic weapon 300
10.2.1 The 4P model [10.3] 300
10.2.2 The 14 principles of the Toyota approach: milestones towards excellence [10.4] 300
10.3 The Toyota Production System (TPS) 306
10.3.1 The two pillars of the Toyota production system 306
10.3.2 Heijunka 306
10.3.3 Just-in-time (JIT) production 307
10.3.4 Self-activation or autonomation 309
10.3.5 Eliminating waste and the 5S program 309
10.4 Training of staff at Toyota 311
10.5 Fordism and Toyotism 313
10.6 TPS, a model? 313
10.6.1 A model – adopted by the West – in search of meaning 313
10.6.2 Evolution and future of TPS 314
10.6.3 A bureaucratic risk? 314
10.6.4 Gray areas at Toyota? 314
10.6.5 Discussion and conclusions 316
10.7 Illustration: the financial situation proves TPS is right 316
10.8 A specific tool of TPS: the A3 Report 317
10.8.1 What is a problem for Toyota? 317
10.8.2 Toyota's A3 Report 317
10.8.3 Staff development at Toyota 318
10.8.4 The steps of A3 for solving problems 318
10.8.5 The content of the A3 report 320
10.8.6 Advantages of the A3 method for solving problems 321
10.8.7 Conclusions 321
10.9 Glossary of Toyotism 322
References 328

Chapter 11
Total Quality and the Six Sigma Approach 329
11.1 Introduction 329
11.1.1 History 329
11.1.2 Very visible savings 330
11.1.3 Six Sigma in brief 330
11.2 Is Six Sigma more efficient than other TQM approaches? 331

11.3 What does Six Sigma mean? 332
11.4 Concepts utilized by Six Sigma 333
11.4.1 Parameters of a distribution 333
11.4.2 CTQ: critical-to-quality value (SMART) 334
11.4.3 Capability 335
11.4.4 SIPOC 335
11.5 Before starting Six Sigma 336
11.5.1 Six Sigma, a road of variable width 336
11.5.2 Commitment of the Board to Six Sigma 337
11.5.3 Six Sigma, training of *Black Belts*, Green Belts and other players. 339
11.5.4 Employee training in general 342
11.6 How to carry out a Six Sigma project 342
11.6.1 Define 343
11.6.2 Measure 343
11.6.3 Analyze 344
11.6.4 Improve 345
11.6.5 Control 346
11.7 An avatar, Lean Six Sigma 346
11.7.1 What is Lean Six Sigma? 346
11.7.2 The supremacy of TPS (reminder) 347
11.7.3 The characteristics of TPS (reminder) 347
11.7.4 The design of Lean Management 348
11.7.5 Lean Six Sigma 350
11.8 The current focus on DFSS (*Design For Six Sigma*) 350
11.8.1 DFSS vs. DMAIC 350
11.8.2 Organizational aspects of DFSS 351
11.8.3 The stages of DFSS 352
11.8.4 Discussion 352
11.9 Conclusions 353
References 353

Chapter 12

Moral and Ethical Issues 355
12.1 Introduction 355
12.1.1 The presumption of acting rationally 355
12.1.2 The past: theoretical and practical ethics 356
12.1.3 Morals and ethics 356
12.1.4 Deontological, teleological and consequentialist ethics 357
12.1.5 Values, value systems and axiology 357
12.1.6 The ethical diamond of the expert citizen 357
12.1.7 Three adopted moral approaches 358
12.2 The practice of virtue 358
12.2.1 Passions and will, definitions 359
12.2.2 Against slavery and passions: virtues 360
12.2.3 What are the virtues for? 362
12.2.4 The example of Benjamin Franklin 363
12.2.5 Abandonment of the practice of the virtues in the 19th and 20th centuries 363
12.2.6 Renaissance of the practice of virtues 364
12.2.7 A universal taxonomy of virtues? 364

12.3 Kant, La Bruyère and the categorical imperative 364
12.3.1 What is Man? 365
12.3.2 Morality according to Kant 366
12.3.3 The categorical imperative and the resistance to totalitarianism 368
12.4. Ethics, morality and economy 368
12.4.2 Is capitalism moral? 369
12.4.3 The four orders 369
12.5 Utilitarianism 371
12.5.1 Definition 371
12.5.2 Rule-utilitarianism 371
12.5.3 Rule-utilitarianism and human rights of 1948 372
12.6 Summary: an ethical approach to a problem, an educational model... 372
12.7 A philanthropic application 372
Reference 374

Chapter 13
Deontology of Professions and Functions 377
13.1 Deontology 377
13.2 Deontology, the science of professional duty 378
13.2.1 The reason for deontology 378
13.2.1 First known deontological code 378
13.2.3 Deontology: neither morals nor ethics 379
13.2.4 Compliance between a deontological code and virtuous behavior 379
13.3 A pioneer, an example: the medical profession 380
13.3.1 A deontological code 380
13.3.2 Medical-ethical guidelines 381
13.3.3 An ethical think tank within SAMS: the CEC 381
13.4 Deontology of State functions [13.7] 382
13.4.1 Deontology of public functions 382
13.4.2 Deontology of public functions: cardinal principles 383
13.4.4 Values of a public administration 384
13.4.5 How do we deal with conflicts of value? 385
13.4.6 Value conflicts and whistleblowing 385
13.5 Deontological charters and codes: hazy communications 386
13.5.1 Codes or charters? 386
13.5.2 The proliferation of deontological codes: retreat of the State? 387
13.6 Deontology of engineering and technology 387
13.6.1 Ethics and deontology of scientists 387
13.6.2 Ethics and deontology of the engineer 388
13.6.3 A historical antecedent – the *Regius Manuscript* 388
13.6.4 Ethics of availability and effectiveness for engineers 388
13.6.5 Evolution of deontological charters for engineers 389
13.7 The FEANI Charter 389
13.8 Charters for software engineers 390
13.9 Business ethics in other professions 391
References 391

Chapter 14
Environmental Management
and the ISO 14000 family ... 393
14.1 Increasing number of environmental problems ... 393
14.1.1 The environment becomes a theme ... 393
14.1.2 The first shock ... 394
14.1.3 Chernobyl, a nuclear remake ... 395
14.1.4 Sustainability, an avatar of the results of the Club of Rome? ... 395
14.2 Environmental policy and sustainable development ... 396
14.3 Environmental policy and environmental management ... 396
14.3.1 What is the role of ISO 14001? ... 396
14.4 Private sector initiatives, and those of the UN and the EU ... 397
14.4.1 CERES ... 397
14.4.2 Business Charter for Sustainable Development ... 397
14.4.3 Global reporting and Global Compact ... 399
14.4.4 The EU and EMAS – Eco-audit (Smea II) ... 400
14.5 Benefits of an environmental management system ... 401
14.6 Presentation of ISO 14001:2004 ... 401
14.7 Other documents of the ISO 14000 family ... 403
14.8 The heart of ISO 14001: Clause 4 ... 405
14.8.1 Preparing the plan ... 405
14.8.2 Implementation ... 406
14.8.3 Measurement and monitoring ... 406
14.8.4 Identification and management of emergencies ... 406
14.9 EMS documentation ... 408
14.10 Moral obligations in ISO 14004? ... 409
14.11 Certification, recognition and the effects of ISO 14001 ... 409
14.12 Environmental audits and ISO 19011:2002 ... 410
14.13 Post-certification ... 411
14.14 Environmental communication and ISO 14063:2006 ... 411
14.14.1 ISO 14063 and internal communications ... 411
14.14.2 ISO 14063 and external communication ... 412
14.14.3 A proactive external communication for the Eco-audit ... 412
14.15 Conclusions: ISO 14001, advantages and ambiguities ... 413
References ... 414
Chapter 15
Social Responsibility of Organizations and ISO 26000 ... 415
15.1 Introduction ... 415
15.2 Corporate social responsibility (SR) and the Triple Bottom Line (TBL) ... 415
15.2.1 Corporate Social Responsibility and Sustainable Development 415
15.2.2 Ambiguous from the start? ... 416
15.2.3. The three pillars of social responsibility ... 417
15.2.4 Critics of social responsibility ... 417
15.2.5 An unfortunate omission of TBL: Corporate governance ... 418
15.2.6 Quality of corporate governance, OECD principles,
the Sarbanes-Oxley act and COSO2 ... 419
15.2.7 Governance and social responsibility ... 421
15.3 Corporate social responsibility (SR): what to choose? ... 421
15.3.1 A wealth of initiatives ... 421
15.3.2 Calling for meta-standards? ... 422

15.4 A pioneer, the French Standard AFNOR SD 21000 423
15.4.1 A standard and an application document 423
15.4.2 Strategic issues 424
15.4.3 The strategic approach 424
15.4.4 The deployment and implementation of an action plan 426
15.4.5 Self-assessment in the light of sustainable development (SD) .. 426
15.5 A promising start: ISO 26000 427
15.5.1 Foundations and functions 427
15.5.2 Comparisons with other basic texts 428
15.5.3 Scope of ISO 26000 429
15.5.4 Areas concerned 429
15.5.5 Type of standard 430
15.5.6 Structure and contents of ISO 26000 430
15.5.7 ISO 26000 in a nutshell 435
15.5.8 Key steps in implementing ISO 26000 436
15.5.9 Advantages of implementing ISO 26000 438
15.5.10 Towards integration? 438
15.6 Corporate charters on ethical principles – a marketing strategy? 441
15.6.1 Charter of ethics or social responsibility? 441
15.6.2 Three case studies: *The Body Shop*, *L'Oréal* and *Nivea* 441
15.6.3 A skillful and responsible policy, that of Nestlé 442
15.6.4 A paragon of ethics and sustainability (www.switcher.com) 445
15.7 Limits and opportunities of the market for virtue 445
15.7.1 The real impact is weak 445
15.7.2 Enhance collaboration between governments and organizations 446
15.7.3 The economy at a turning point? 446
References 447

Foreword

What is Quality?

Quality is a form of management (or leadership) that has a dual approach:

- *a call for and an approach toward excellence* (process, product, service, know-how, behaviors): the best product at the best price, with an optimal profit margin;
- *a commitment to comply to* legal requirements, standards and prescriptions (procedures, specifications, best practices, safety) of the product or service, intended to provide authorities, stakeholders and clients with complete confidence.

The world of quality applies the following principles:

- *work and manufacture by logical methodical and reproducible means*, taking achievements into account → intelligence, rationality and logic in the approach;
- *do what is communicated* → integrity, consistency, credibility;
- *communicate what is being done*→ transparency, openness, honesty;
- *keep track of what has been done* → responsibility, reliability, historical traceability.

The world of quality often uses the vocabulary and practice of virtues to develop views and methods. The ultimate mission is to facilitate exchange and mutually beneficial partnerships, especially on the international level, profit factors, but also globally, prosperity factors, well-being and peace worldwide (by interweaving national economies).

Why quality?

The globalization of the economy can no longer rely solely on the trust between trading partners; volatile activities, products, services, make relationships more ephemeral; intercontinental distances, cultural differences are often sources of misunderstanding. Faced with the globalization of exchange, companies establish systems of quality management that provide proof of their competitiveness and the reliability of their products, components or services. The best known is probably ISO 9001.

These quality management systems have another advantage: they make companies more resilient to ongoing changes - often unexpected, sometimes chaotic – that they may experience. They also make them more competitive by eliminating all of their process steps that do not provide added value (Six Sigma, Lean Management, Toyota Production System (TPS), all covered in this book).

The growing role of services in the global economy provides relevance to the ethical behavior of economic agents. The dominance of the economy and the role of companies (given the size of some) in the affairs of the world lead to the development, after that of quality and performance management systems, of management systems for the protection of the environment, and finally of systems for social responsibility management (sustainable development).

But how does one develop a vision? How does one plan, deploy, manage and account for all these actions? This is the role of social responsibility and related management systems, largely inspired by the ISO system. The ISO organization has edited an ad hoc norm on global sustainability, ISO 26000. Certain limitations will also be discussed in this book.

Quality in the academic world. What for?

The professional activities of a scientist or engineer working in the private and public sector will regularly be confronted with the world of quality. Some will even make it their profession. A course in quality, for which this book constitutes course material, will typically be a course giving the context for the trade.

A few years ago, the world of quality was separate from the academic world. However, this is in the past, because quality has entered academia, especially in higher education in Europe, by the ENQA[1] standard, closely associated with the Bologna process. It also blends into the lives of university laboratories:

- one cannot reasonably carry out research on technology management, or even develop new technologies, without implicitly and consistently referring to the world of quality;
- the world of quality knocks at the door of research laboratories: as proof, an extract from a document from the National Center of Scientific Research (CNRS) in 2003 [1]:

Quality becomes a concern of the CNRS (extract)

The introduction in research units of a "quality" approach is growing because of significant constraints in terms of efficacy and efficiency: the production and processing of information and knowledge aim to control the means of registration, reproduction, dissemination, and to determine the reliability and validity of the research process. This requires traceability procedures and methods used by the widespread deployment of a quality approach.

This is an essential element when the research unit provides a service, when designing a complex instrument with the involvement of various partners, or more generally when dealing with projects where the triptych of cost, quality, and timing is of importance. Quality is also imposed from an external point of view by "clients / beneficiaries / partners" (companies, scientific and technical communities, sponsors, guardianships, public and private industrial partners ...) that have long operated with quality requirements ... Generally speaking, quality becomes a national priority of the organization ...

[1] Cf. http://www.enqa.eu/

The world of quality seems ready to consolidate the activities of research laboratories and to strengthen the robustness of the results they produce. Who can complain?

The contents of the book

This book focuses primarily on a synthetic and comprehensive approach, and not on a detailed knowledge of standards, the respective implementation of statistical tools that can be controlled only in the intimacy of daily work within a management system. Few books offer an overview of quality and its tools, and this one intends to fill the gap.

Therefore, this book will not teach the reader to precisely establish procedures, instructions, flowcharts, forms, which is a typical activity for operational quality.

The emphasis is rather on the overview, the system's approach, the stakes and the role of quality in life and policies of an organization or company, but also on the world of quality spilling over onto other management systems, such as environmental protection, the sustainability of an organization and its social responsibility.

Moreover, the book provides additional perspectives, for example on:

- the strategy of a company, with breakthrough management according to *Shiba* (total quality);
- customer focus, mainly a sales and marketing activity, with CEM according to *Shiba*;
- quality engineering and experiment planning, with input from *Taguchi*;
- the functioning of a steering committee and the development of a project with the top-down *Six Sigma* approach;
- the role of a board of directors, in the module of social responsibility;

It also highlights the capping of American management methods and how they were boosted by contributions of the Japanese industrial organization.

Expected learning outcomes

Once the contents of the book have been learned, the reader will be able to:

- quote from memory, explain and use in a discussion, a work meeting, when writing a memo or a report, the concepts and basic vocabulary of the subject of quality;
- find the contributions of the history of quality in current management systems;
- situate the quality process in the functioning of an organization and assess the role of structures that are not included in this process;
- understand the content, as well as the issues of a product and process quality approach such as ISO;
- comprehend and review the general concepts of key quality management systems for companies, institutions, testing and analysis laboratories, and their interface with environmental management;
- argue with regard to the approach, the pros and cons of a system of quality management such as ISO, and distinguish it from a total quality approach;
- discover relationships that are maintained by quality and productivity in the total quality approach (Six Sigma, Toyota Production System, Lean Management);
- identify, describe and communicate the key factors of conduct of a company and distinguish them from elements of an individual moral approach;

- understand and assess the relationship between management quality systems such as ISO, professional ethics as well as corporate and social responsibility of organizations and companies.

However, the reader cannot hope to fully master the intricacies of the subject of quality with this book, which is merely an introduction and a solid initiation. The content of this work will allow the student to discretely resist – *honorably and as a upstanding citizen* – the onslaught of catchwords and automated statements used by corporate management .

This book differs from others by the fact that it is marked by the author's will to describe the overall subject of quality, to focus on basic concepts, and also provide (from an academic perspective) a critical look at quality and its history, its tools and management systems. Indeed, many books focus on a single system, or have a predominantly practical content, intended primarily for practitioners under pressure to set up a system (in a short amount of time).

Structure of the book

The book consists of 15 chapters that correspond to the content of a course of 4 ECTS credits in an Institute of Technology, keeping in mind that 4 ECTS credits approximately constitute a workload for the student equivalent to 90-120 periods, including classes. The document is mnemonically structured with:

- *a triangle of musts:* the environment, the basic concepts and the history of quality. These three chapters overlap and complement each other;
- *an octagon of management systems:* the quality management system (ISO 9000 family), the ISO 22000 management system for food safety, other management systems (ISO 31000, ISO 50000, ISO 27000), quality management in laboratories, total Quality (two approaches: significant personalities and awards), TPS, and the Six Sigma approach;
- *a golden or virtuous triangle* whose vertices are: ethics and moral principles, environmental management (ISO 14000 family), and corporate social responsibility (ISO 26000).

References

[1] MARIE-NOËLLE POGER, La fonction Qualité au CNRS, *Les cahiers de l'Observatoire des métiers* (Dec. 2003), p. 3, available at the URL: http://www.sg.cnrs.fr/drh/publi/documents/divers/metier-qualite.pdf

Chapter 1

The Environment of Quality

Key concepts: purchasing, expectation(s), need(s), Porter's value chain, product quality cycle, BCG matrix, logistics, market share, product portfolio, production, products, Maslow's pyramid, marketing, marketing mix, dead weight, question marks, competitive position, profit, expected quality, achieved quality, perceived quality, desired quality, research and development, satisfaction, service, stars, cash cows, sales.

1.1 Products, services and satisfaction of expectations

1.1.1 Distinction between products and services

It is customary to divide the services of a company, organization, institution or individual into two types :

- a tangible product (e.g., toothpaste, radio, car);
- a service that is consumed as it is produced (e.g., haircut).

But this distinction is artificial because a product is only purchased if it provides a service :

- toothpaste is purchased not only because it cleans your teeth, but also because it provides a feeling of well-being, gives good breath that is pleasing to the people close by, while protecting from dental caries that spoil a smile, and thus the power of seduction;
- a car is acquired for rapid transportation from one place to another, making it possible to vary pleasures and giving a taste of freedom (*via est vitae*, omnipresent);
- a radio can provide information and music, but can also satisfy a need for being connected and gaining knowledge (omniscient).

A ***product*** is therefore a form of ***service*** that is not fully consumed in a single use. Accordingly, gaming software, loaded on your PC from the internet, is not a service because it can be used repeatedly at no extra cost. Conversely, a railway ticket is not a product because it is only a physical representation of a service; in this case, transportation of a client.

This definition has the consequence that all services that meet the same implicit and explicit expectations of their beneficiaries are interchangeable. Here, one expectation is the price as paid as working hours according to the beneficiary's hourly wage (which represents a perceived sacrifice).

At the same level of satisfaction, should one then choose a product or a service? Again, we must analyze whether we really get the same level of satisfaction from these two objects; for the same price, the tangible product can satisfy our desire to own and consolidate external signs of wealth, both promoting social recognition and thereby well-being. In other situations, the tangible product costs valuable time or corresponds to an asset that provides no satisfaction, because social recognition can be acquired by other means (scientific reputation, army rank, normal or luxury car rental with or without driver, leasing, etc.).

There is a corollary to this assertion: the same tangible product can satisfy different expectations, which may lead to its diversification/differentiation (multiple products) implemented by marketing. Thus, in prehistoric times, obsidian shards were used as knives, arrowheads, ornaments and religious objects ...

1.1.2 A hierarchy of needs, Maslow's pyramid

Personal expectations arise when one or several desires are left unsatisfied. *Abraham Maslow's* work consists in classifying these needs into five or six levels (see Figure 1.1). This classification corresponds to the order of priority in which they appear to the individual; only when the needs of the first level are satisfied are we concerned with those of the next level. The motivations of people, their desires, can bloom only if the first three levels of needs (physiological, security and social needs) are largely satisfied. *Maslow* expressed this principle in 1943 [1.1], but it had already been discovered and put into practice by the Salvation Army[1] during the second half of the nineteenth century.

A service/product can be addressed towards a beneficiary to meet the needs/desires of one or more of the levels of the pyramid:

1. *Need (physiological) for the maintenance of life:* air, food, excretion and hygiene, maintaining body temperature, rest and sleep, bodily contact, sex.
2. *Need (psychological) for security, protection and property:* physical and psychological protection, employment, family and professional stability, to own things, to have places to oneself, and to have at least some power over external matters.
3. *Need (social) for love and belonging:* to be accepted – as we are – to give and receive love and affection, to have friends and a good communications network, to be recognized as being valuable, belonging to a group or hierarchy.
4. *Desire for self-esteem:* expressed by the desire for strength, achievement, merit, mastery and competence, for self-confidence to face others, for independence and freedom.
5. *Desire for self-realization:* to increase knowledge, develop values, do new things, create beauty, achieve the fulfillment of being a "natural" human being and of the ego. This

---

1 *William Booth*, the founder of the Salvation Army in 1865, who preached to the destitute during the English Industrial Revolution in Whitechapel, knew that before talking about religious faith to someone, one has to be in a position to offer decent living conditions on earth. Therefore his motto was: "Soup, soap, salvation". See the online book, *William Booth, founder of the Salvation Army* (1948), at the URL: http://www.regard.eu.org/Livres.8/William.Booth /index

Figure 1.1 The pyramid of needs and desires by *A. Maslow* (6-stage model).

desire must be balanced with the unmet needs of the other stages. The person must be in a favorable environment and be able to negotiate in stressful situations. According to *Maslow*, there are few people who reach the full development of this level.

Some people add a last stage:

6. *Spiritual aspirations*: development of the inner life, spaces for meditation, reflection, writing, asceticism. Temporarily achieving this level moderates the appetites and desires of the previous stages: proceed with caution according to the maxim: "Man is neither angel nor beast, and the misfortune is that he who would act the angel acts the beast" [1.2].

1.2 Quality, definition

The ***quality*** of a product or service is defined here as the level of satisfaction of its beneficiary (a customer in the market economy) in relation to the expectations expressed or implied in the different stages of *Maslow's* pyramid or any other frame of reference.

These expectations are listed in the form of specifications, standards and reference values that play a role in the world of quality.

For purchases of capital equipment, this frame of reference may be more objective specifications, even if the sales services of these companies know how to target the different levels of *Maslow's* pyramid in potential buyers.

The world of quality belongs to the world of management, defined as implementation of certain functions to obtain, to allocate, and to use human (personnel, supervision) and material (raw materials, energy, components, etc.) resources to accomplish a goal in line with the needs of a company (Figure 1.2, [1.3]). Management has two dimensions: the act of managing resources and the art of handling them (with patience, discipline, rigor, efficacy and efficiency). Quality management has interfaces with organizational theory [1.4].

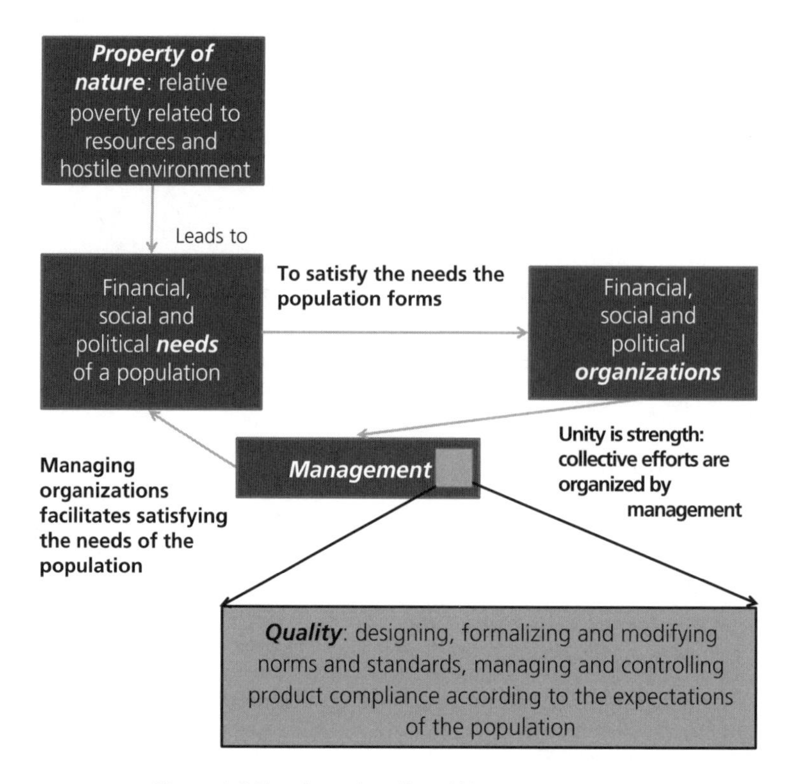

Figure 1.2 The place of quality within management.

The world of quality is therefore based on four basic assumptions of management, called myths or even beliefs yet often verified in economic reality [1.5]:

- *Assumption of rationality*, for which an action is essentially a choice that has consequences – the human mind finds an optimum choice among a variety of actions that are evaluated by comparing their future results, the rationality of certain previously set preferences or criteria.
- *Assumption of a hierarchy*, of which the basic idea is that the problems and actions can be decomposed into elements, which in turn consist of sub-problems and sub-elements, and so on. Consequently, the responsibility to perform a complex task can be delegated within a system of subordination where the highest level controls and integrates the solutions and actions of the lower levels. However, this assumption generates illusions about the real possibilities of control and accountability.
- *Assumption of the importance of the individual leader*, i.e., that the major developments in management are attributable primarily to the abilities and actions of a few exceptional heroes (*Henry Ford*, *Alfred Sloane*, etc.).
- *Assumption of historical efficiency*, which postulates that history will follow a path that leads to an equilibrium determined solely by the initial conditions, and produced by economic competition. This myth contains the idea of natural and fair competition where the best survive.

Although sensitivity to quality has been evident since ancient times, we attribute the birth of the quality function within management to *Henry Ford* and his invention of mass production in the early twentieth century.

1.3 Quality and its place in the Marketing Mix

How important is quality in business? One answer is provided by the *marketing mix*.

To reach the objectives of revenue from the sale of a product or service, one conceptual tool available to marketing divisions is the marketing mix. It is also called the 4P's of marketing (as described by *E. Jerome McCarthy* in 1960 [1.6] (see Figure 1.3):

- Price,
- Product,
- Place (in linear and distribution networks), or Positioning for services,
- Promotion.

In everyday activities, *price and promotion usually play the most essential roles*. The marketing mix gives marketing directors the means to verify that all elements of a product's definition are represented in a logical and transparent manner. The marketing mix is a generic tool for all marketing strategies. Some critics highlight the small financial implications of this model (margins, returns to shareholders), but it is a robust approach in the definition stage of a product.

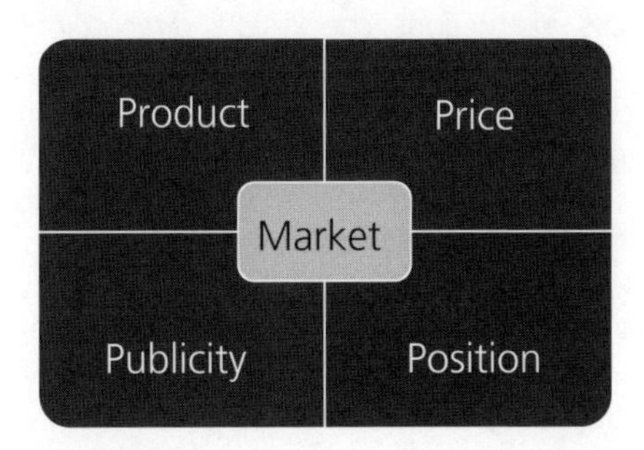

Figure 1.3 The marketing mix, or the 4 P's of marketing.

The ***marketing mix*** means optimal use of the levers used in marketing:

- product or service (definition of its functionality and intrinsic quality); position in the existing portfolio;
- price (and any variation following distribution channels);
- deployment of sales forces;
- definition of the services associated with the product;
- how to use a brand policy;
- definition of distribution channels → transport and logistics;
- promotion policy, communications, advertising and advice to consumers, lobbying.

Optimization of the marketing mix also involves taking control of different variables: financial (such as return on shareholder investments), legal, technical, temporal (products for summer, Christmas or current season), human (social, demographic, cultural values), political, competitive, consumer and environmentally friendly.

These variables affect the definition of the quality of a product, its standard and its specifications (including packaging). However, quality is only one element in the issues associated with a product.

Quality is one of several business functions, such as manufacturing, research and development and marketing. Research and development are closely related to the quality approach, which begins in the early stages of design.

1.4 The quality cycle of a product

The role of *marketing* is to find and identify client expectations, define and produce a service according to these expectations, and to the role of *R and D* as well as *manufacturing*. These relative roles make it possible to define four types of quality in *a quality cycle of a product* [1.7]:

- The *expected quality*, which is a summary of the expectations of the beneficiary, developed by the marketing team. It can, for example, be determined through surveys, group meetings or market analysis.
- The *desired quality*, which takes the form of a prototype (virtual; specifications) developed by the R&D team, determined under the assurance of quality, subjected to quality control, and protected from external changes and fluctuations by quality management. It involves site manuals, quality standards defined by the organization, operating procedures, instructions, checklists and recordings.
- The *obtained quality* after manufacturing and logistic operations (packaging, transport, storage, placing on shelves) have been performed. This objective and quantifiable measure of quality is what is actually delivered to points of sale, for instance.
- The *perceived quality* is what the beneficiary feels with his five senses (and a sixth sense: the mind and memory, which can create associations of ideas) when using the service for the first time. For a tangible product, functionality and ergonomics, ease of use, appearance, harmony, and material selection are essential ingredients.

Figure 1.4 shows that the difference in beneficiary satisfaction, which indicates the degree of quality achieved, is the sum of the achievements of R&D, production, marketing and communication teams. To this we add a third factor that is not included in the figure: the various achievements of marketing in the search for explicit and implicit beneficiary expectations.

The ideal scenario is when the expected quality, the obtained quality and the perceived quality are identical or largely overlap. *Control of the perceived quality is key to the sustainability of a product or even the company.*

If the perceived quality satisfies one or several of the higher stages of *Maslow's* pyramid, the intangible part of the product is very important, and the cost to the customer increases because the price is part of the perceived value, in the same way as an advertisement refers to its use by movie or sports stars. Thus, the price of major-brand perfumes (Chanel, etc.) is irrelevant to the costs of design, development, manufacturing and delivery, for two reasons:

- the price is part of the positioning: everything that is prestigious is expensive and vice versa;
- the costs of design, advertising, and brand positioning are high.

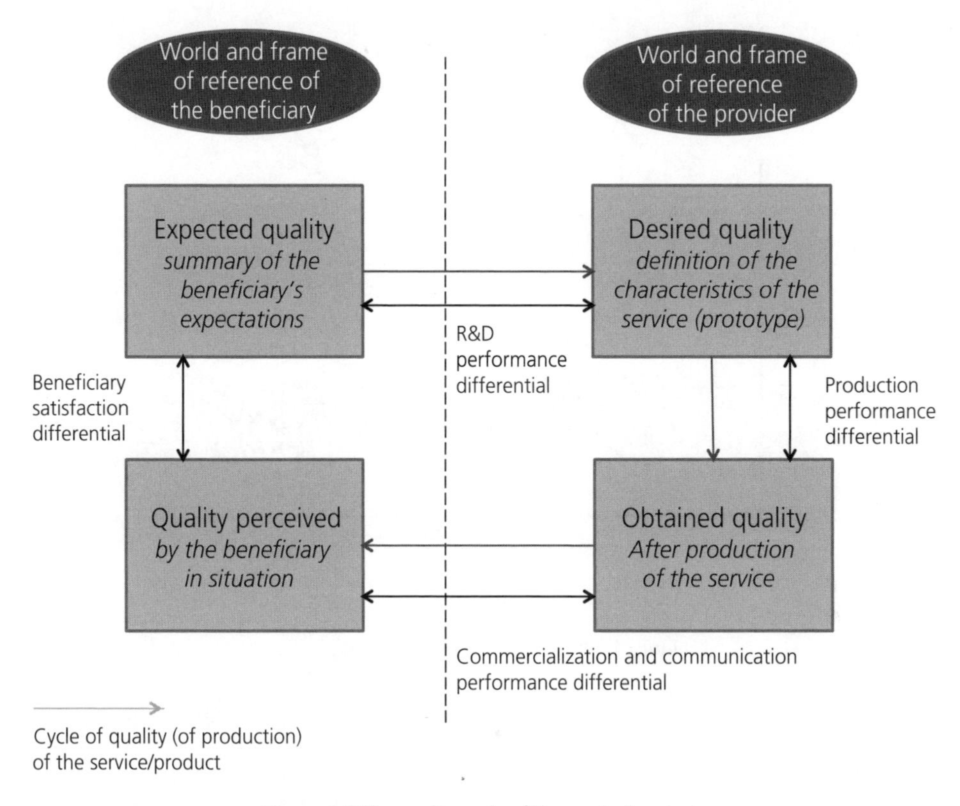

Figure 1.4 The quality cycle of the service/product.

1.5 The portfolio of products (BCG matrix) and quality

Each product/service coexists with others, thus forming the portfolio of services of the organization or company that produces them. The portfolio is dynamic and must therefore be managed, and the most commonly used tool is the BCG matrix (Figure 1.5) named from the company *Boston Consulting Group* who introduced it 30 years ago (the 1970s). It places products according to a 2x2 matrix (given the pressure of time, managers rarely exceed this level of complexity ...):

- *x-axis*: the market share in % relative to that of the market leader of this segment;
- *y-axis*: the annual growth rate of the market segment in %.

This diagram shows four families that correspond to separate commitments to quality:

- *Dead weight* corresponds to a product that has a low market share and a low annual growth rate. There is normally little hope for it, especially if this type of product is getting old. The most common decision when it comes to this product type is to get rid of it, or to develop it in a niche where it can pretend to be a star if the product has been launched to the general indifference of the company (the pet project of the Director, etc.)..

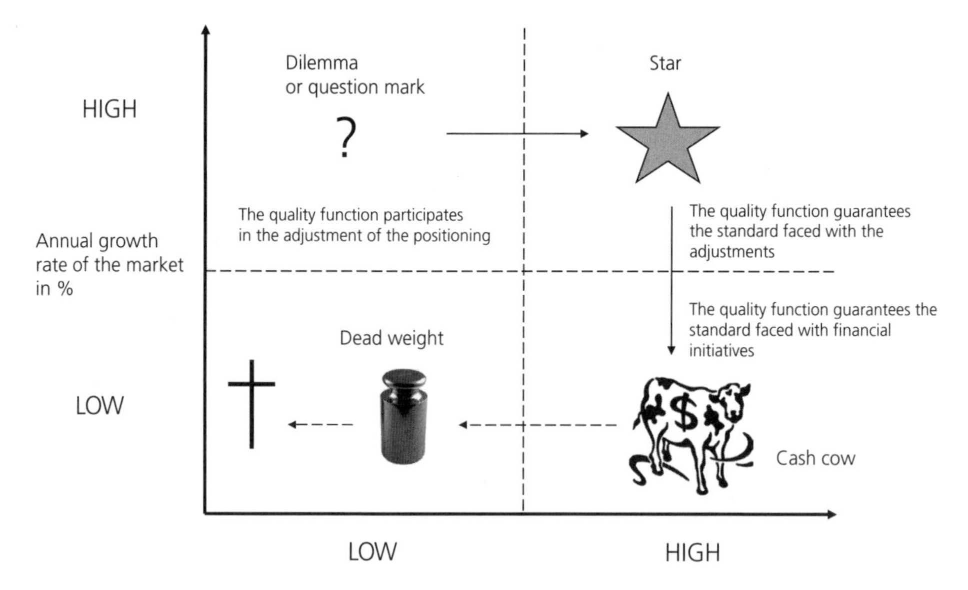

Figure 1.5 BCG matrix of the product portfolio.

- *Question marks or dilemmas* are in an unstable situation: the market share is low, while the annual rate of market growth is strong. This may be due to improper positioning, an incorrect marketing mix, or a late entry into an already filled market space. Dilemmas become stars or dead weight. *The quality function is involved in adjusting the positioning (improvements, changes) of the product.*
- *Stars* are products of a rapidly growing market segment and for which the company is well positioned. Stars generate large profits; this is the best position in the portfolio. Marketing extends this star status with new releases of the same product under various conditions of the marketing mix (packaging, retail change, etc.). This operation is called ***diversification***. *The quality function guarantees the standard against these adaptations.* Over time, there is a high chance of seeing the star reach the status of a *cash cow*, unless the market is in constant innovation, in which case they disappear in full glory.
- *Cash cows* are products that have a large market share, but in a market with low growth, thus subject to strong competitive pressure. Margins are generally lower than for stars and, to obtain improvement, savings must be sought in manufacturing, purchasing and finance functions (less costly technologies, increasing returns and hourly capacities, job cuts, lower-priced raw materials and packaging). *The quality function guarantees the standard against all these often unexpected initiatives.* Cash cows are particularly sensitive to substitution products, but also to changes in modes of financial calculation of the profitability of the product and to the accuracy of the assessment (watch out for creative accounting by management control units who have become profit centers…).

Analysis of the portfolio is performed by plotting each type of product in the matrix as a circle whose area represents the relative annual sales. This helps to gauge the health of the portfolio, and thus the prospects for sales revenues and margins for the coming years.

1.6 Quality and stakeholders

The concept of *stakeholder* is antagonist to that of *shareholder*. ***Stakeholders*** are thus forces that do not necessarily follow the goals of the shareholders, but who have a say in the objectives and operation of the company. Here are two definitions:

- *Freeman* (1984) [1.8]: an individual or a group of individuals who can influence or be influenced by the company.
- *Clarkson* (1995) [1.9]: a person or a group running a risk by having invested some form of human or financial capital in a firm.

The concept of stakeholders is widely used in the world of quality, but does not seem adequate when referring to social responsibility [1.10]. Figure 1.6 illustrates the types of stakeholders in a company.

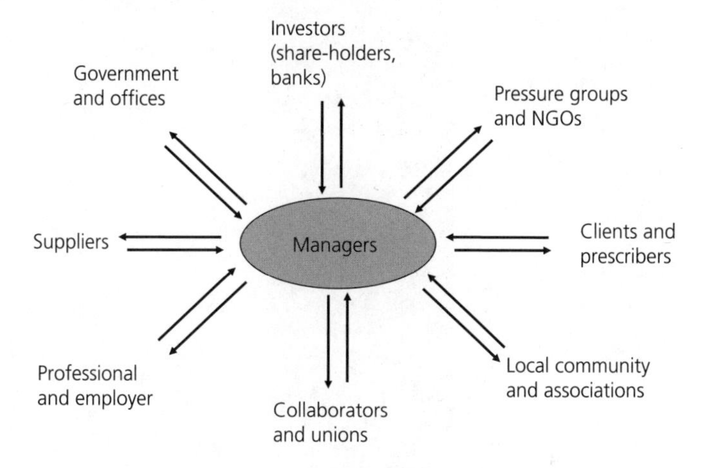

Figure 1.6 The stakeholders in a company.

More specifically, the stakeholders of the quality of the product outside the company are:

- the client, often the consumer;
- the purchasing department of the distributor (sometimes having the most weight);
- those who influence or prescibe, e.g., the media, or for drugs, doctors;
- the authorities of the country (services for certification, standards, ordinances and laws);
- NGOs, if the product carries ethical claims;
- consumer groups;
- suppliers of raw materials, components and equipment;
- investors (but who in principle manifest themselves through the product and profit strategies endorsed by the board supposed to represent them).

Stakeholders have expectations that may be contradictory (Figure 1.7).

Managing and meeting the expectations of stakeholders is a fundamental element of managing the social responsibility of organizations (Chapter 13). There exist stakeholders who have no voice, but who may in the future have rights and people willing to defend them, including members of the animal – and possibly plant – kingdoms. Taking this into account could be a key element of sustainable development of our society.

Managers of the organization
- the organization's history, values;
- organizational structure;
- management system (quality, environment, social responsibility, risks, etc.);
- financial performance;
- knowledge of the competition;
- mission, strategy and goal.

Employees of the organization
- salary, bonus, compensation, reward;
- social benefits (health, retirement), vacation;
- assistance to families, single parents, employees having difficulties;
- absenteeism and turnover;
- career plan and job security;
- continuing education;
- organization of leisure activities, sports facilities;
- health and safety at work;
- job atmosphere, internal communication, hierarchical relations, participation in decisions;
- socially advantageous profile of the organization.

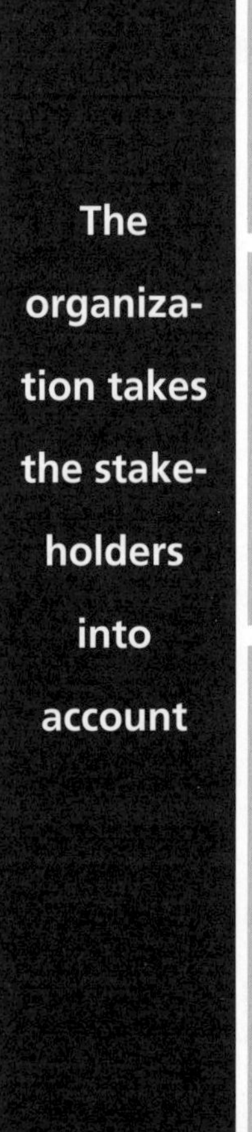

Shareholders
- return on invested capital;
- future perspective and sustainability of the company;
- claims processing and communication;
- salary policy especially for the managers;
- shareholders rights, participation in strategic decisions, in the board of directors, etc.

Customers
- general policy;
- price policy;
- quality of the product/service;
- service related to the product (after-sales, etc.);
- image of the organization;
- claims processing.

Suppliers
- purchasing policy (price of the delivered components);
- solvency, durability of the organization;
- aptitude for partnerships and collaboration;
- information on strategies to facilitate forecasting;
- profile (pioneer, follower);
- prestige of the organization;
- capacity to resolve conflict.

Public authorities
- committed to being socially responsible;
- employment and salary policy;
- health and safety at work;
- taxes and fees;
- donations and philanthropic actions;
- relations with the public community
- image of the organization;
- claims processing.

Figure 1.7 Stakeholders. Examples of segmentation and expectations.

1.7 Structure and functions of an organization

1.7.1 Representation by an organizational chart

An organization, especially a company, is a system (i.e., a set of elements) in constant interaction. These elements are usually organized in a hierarchical structure of separate functions (Figure 1.8). In Figure 1.8 and those following, the quality function is associated with the R&D function, as is often the case in significantly technique-oriented SMEs (small to medium-sized enterprises). As part of a large company or a policy of total quality, the connection is direct to the management, because the quality policy covers all the organization's functions.

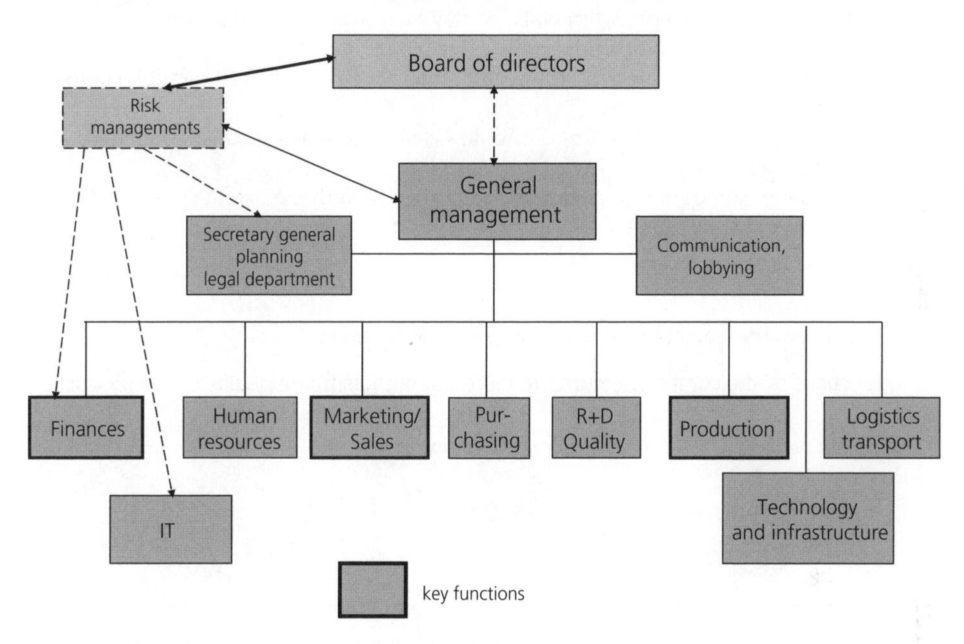

Figure 1.8 The functions of an organization operating in a market (organizational chart).

The key functions of an organization are:

- finance (management and control), to which IT management is often attached,
- marketing/sales, managing the strategic product portfolio and contacts with customers and advisors,
- manufacturing, assembling or finishing the product to be delivered by logistics and transport.

This distribution, typically for a product that requires little know-how, can be modified:

- if the product is surrounded with an aura of prestige, institutional communication is clearly important;
- if the product has a short-life and/or high-tech niche, the R&D/quality function is paramount;
- if manufacturing is limited to an assembly of components, the purchasing function is then essential, such as quality management of the suppliers;
- if the product requires the management of high-tech facilities, the technical function will be highlighted;

- if the product is sold in a highly regulated environment in which disputes are common, the general administration and the legal function will be heavily involved;
- if the added value depends mainly on the intellectual activity of employees, human resources should be given close attention by senior management. Indeed, the company would be weakened and lose its competitiveness if the best elements should leave. This risk is largest for an organization at the forefront of the information society and knowledge-based economy.

The company management is accountable to a board of directors representing the shareholders and investors. The board may have a risk manager who has access to information (especially financial information) which will not stay exclusively within management. Quality management of a board of directors is briefly discussed in Chapter 13.

All functions may not be present. In this case, they are then managed through contracts with providers:

- the firm Lacoste[2], for instance, distributes licenses to manufacturers, boutiques and some chain stores,
- many SMEs do not have a personnel department and thus leave their financial management to an ad hoc service.

1.7.2 Company functions and quality activities

Some functions of the company contribute more than others to the quality of the product.

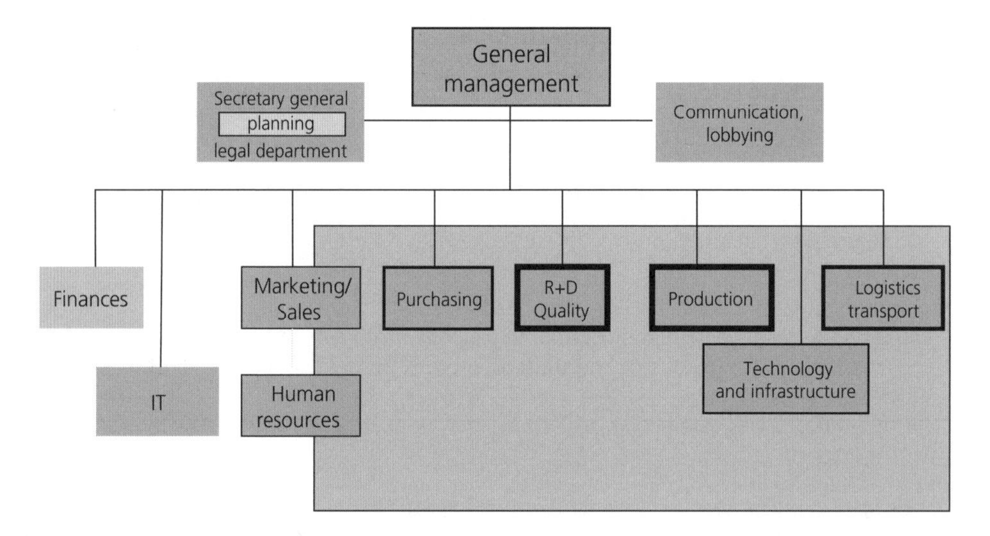

Figure 1.9 The functions significantly affected by quality management.

These include in descending order (Figure 1.9): the systemic management of quality, R&D (based on the importance of already paying attention to quality at the design stage), manufacturing, logistics and transport, purchasing, marketing and sales, technology and

[2] Prestigious apparel company, URL: http://www.lacoste.com

infrastructure. The board of directors should also be included in this list, but experience shows that some of them delegate this concern to lower levels.

The finance, IT and corporate communications functions are only marginally affected, except in the context of total quality, since large sums of money can be spent on communication and advertising. These functions may sometimes resist quality management but, since the company's financial performance is linked to its quality performance, finance and quality management must be able to get along.

1.7.3 Quality management and network of activity

Quality management requires a matrix organization, a networking of different services as shown in Figure 1.10.

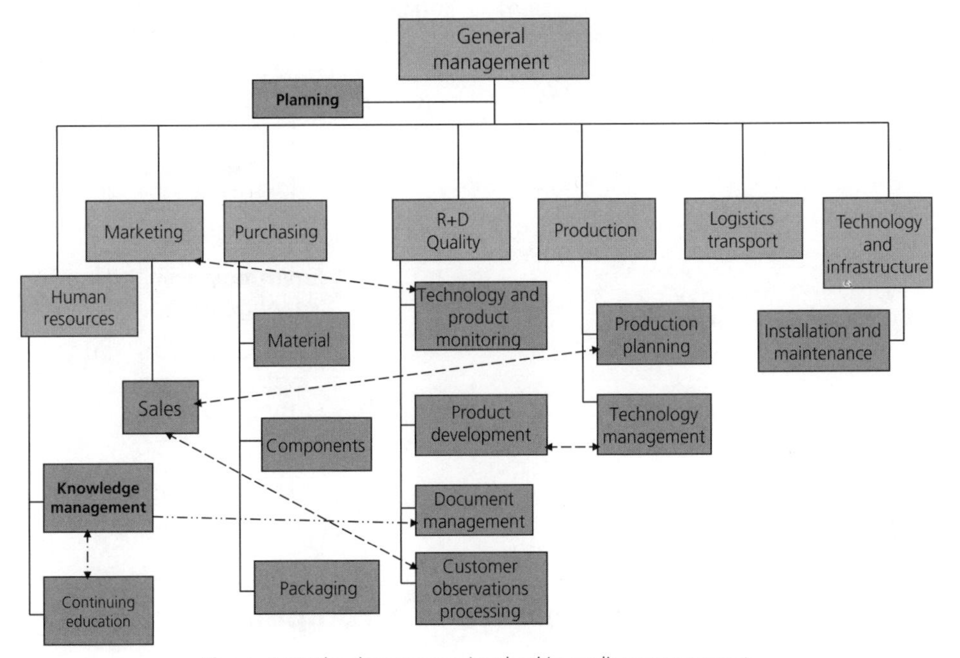

Figure 1.10 The departments involved in quality management.

Some examples follow:

- The quantities delivered are the result of arrangements between the sales department and the production (failure to deliver = poor quality).
- The dynamics of market shares is influenced by the synergy between marketing and the department in charge of monitoring products and new technologies within R and D.
- The management of staff competence is closely related to their continuing education. These two factors of human resources must be taken into account when documenting the quality system of the company (staff competence is critical).
- The sales department works with the service processing client observations, which records claims (poor quality) or suggestions for improvement.

1.7.4 Similarly managed organizational functions

Finally, some business activities that do not affect product quality but do influence the reputation of the organization are handled by management systems similar to that of quality. These include safety, environmental management and built infrastructure (Figure 1.11).

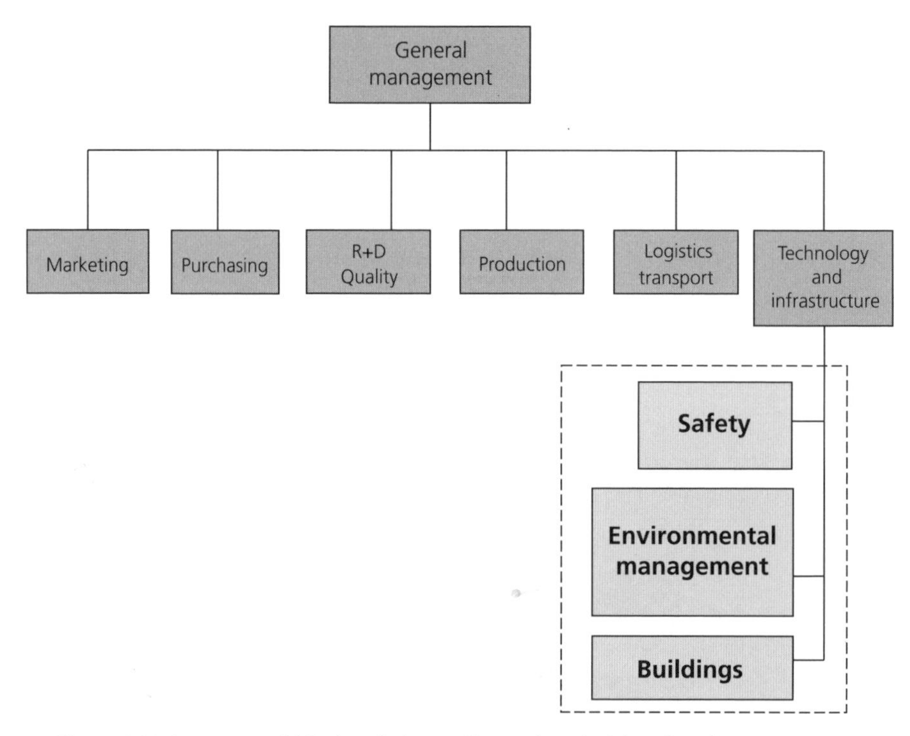

Figure 1.11 Company activities handled according to the principles of quality management.

1.8 Quality and performance of a company

What is the role of quality in the competitiveness and performance of a company, i.e., in its ability to occupy a leading position in a sustainable way? In fact, it has a major role. To show this, we must develop [1.11] Figure 1.8 into Figure 1.12.

If we consider the segments of areas of expected and perceived quality (customer viewpoint) as well as desired and obtained quality (management and operation viewpoint), Figure 1.12 shows that quality works at two levels to increase revenues by growth in sales and margins:

- by maintaining a high capacity for differentiation, in order to have a high-performing, reliable product, and a leading position in the market (generating capital gains);
- by having significant ability to control costs and changes (productivity), which improves the profit margins of the product.

These two factors are key to the competitive position of the company, but the procedural and standardized quality function often only covers issues from a management and

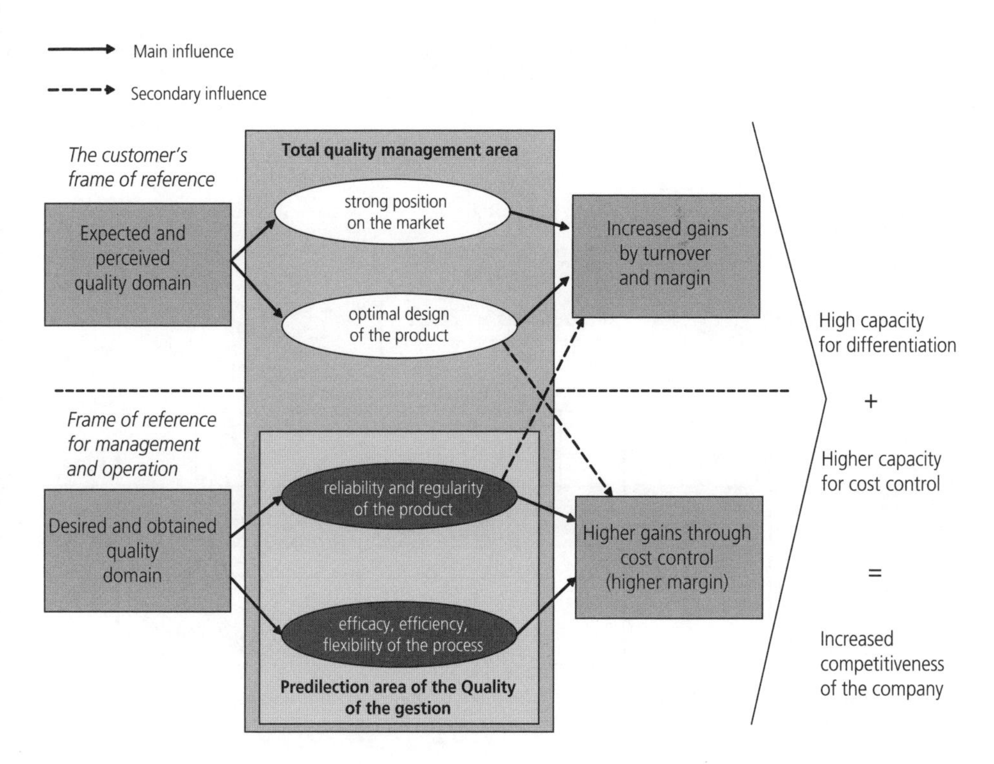

Figure 1.12 An organization's quality and position in relation to the competition.

operational viewpoint. Only a transition to the management of both areas, also known as total quality management – commitment to excellence – is able to provide all the desired impact and support.

1.9 Quality and Porter's value chain

1.9.1 Limits of representation by an organizational chart

The limits of representation by an organizational chart appeared in the late 1970s, when the pace of innovation started to accelerate. The various functions of the company represented in its management struggled to work efficiently as a team. In fact, most activities providing added value required a chain of actions transversal to several departments. The hierarchical relationship as defined by the organizational chart leads to conflicts, and the creation of ivory towers (Figure 1.13). The emergence of opposing factions can develop into a system of clientelism and can increase power struggles.

There is an even more worrying issue: *the client, the object of attention of the organization, is absent.*

Finally, a chart representation does not lend itself to optimization of production support activities (formerly of marginal cost), which now represent a significant portion of the costs. The same applies to services, where the production of tangible goods is close to zero.

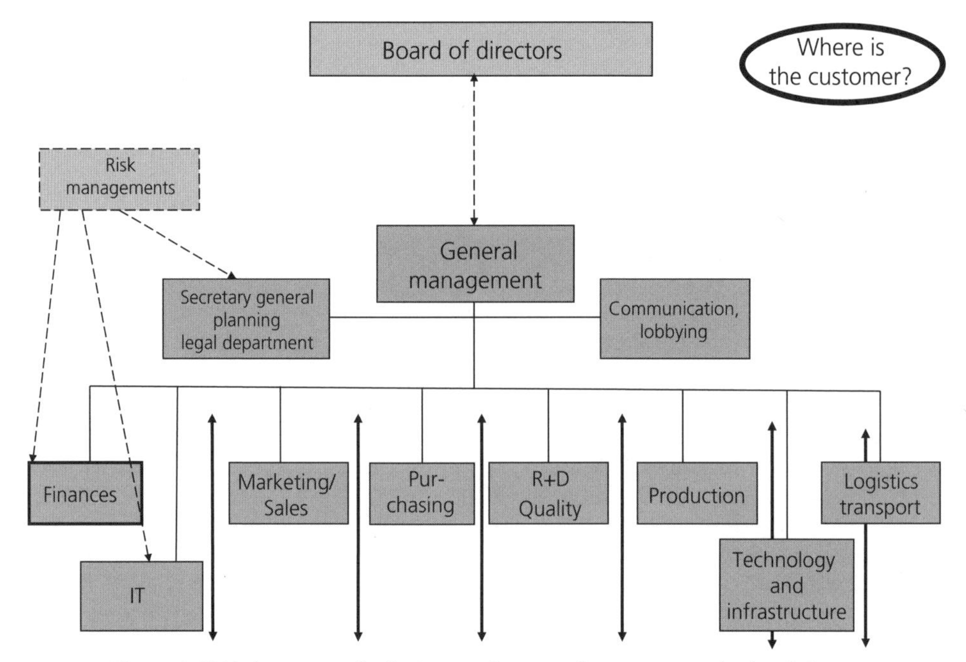

Figure 1.13 Limits an organization's operation according to an organizational chart.

1.9.2 Porter's diagram and the process approach

How do we view the activity of an organization that combines high-performance services with or without the production of tangible products? We need a representation that is compatible for both activities, which are juxtaposed or complementary. This is the process approach.

Services and the production of tangible goods work by means of implementation ***processes***, defined as a *set of interrelated or interacting activities transforming inputs into outputs.* These items are either material objects or information, or both. In everyday language, we talk about *business processes* when the activity is that of a service (where human activity is predominant) and *production processes* when we refer to tangible goods.

For the role of quality in the company functions and its management, it is convenient to refer to *Porter's* diagram [1.12] (Figure1.14), which distinguishes between main and support processes.

The main process performs the principle steps in the production of product. Therefore, it mainly involves marketing, Q/R&D, purchasing, manufacturing, logistics, sales and services (after sales). These sub-processes of the main process have support processes such as strategic and financial management, human resource management, infrastructure and information technology management, etc.

Historically, the quality process had its starts in manufacturing, since most other business functions were at an embryonic stage at that time (marketing flourished in the second half of the twentieth century). *The areas of focus for quality are R&D, purchasing, manufacturing, and logistics.* The quality management system according to ISO 9001 has these four functions as strong points. However, there is another management system that directs all functions of the company towards excellence: total quality management.

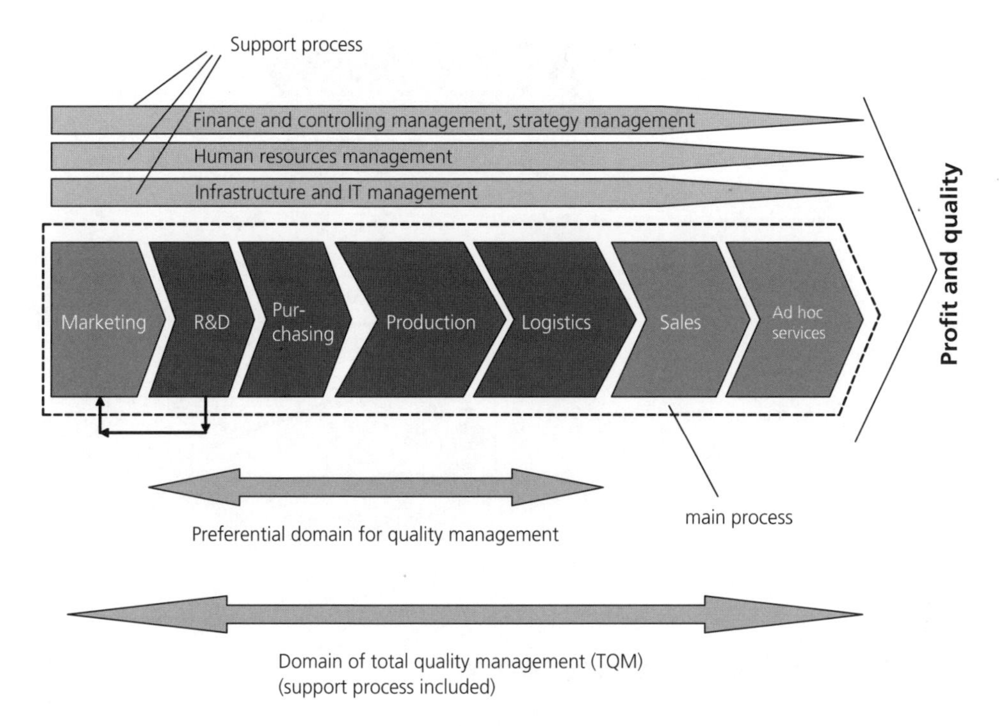

Figure 1.14 Porter's value chain for Quality.

Each company creates value, but also quality, which is closely linked to the proper working of the main and support processes: the control of quality, not to mention weak links, largely (with the usual quality management) or completely (with total quality management) contributes to the competitive performance of the organization.

In the book *Business Process Improvement*, H.J. Harrington describes the benefits of a process approach and notes that a process has several features that, once optimized, make it possible to improve the quality [1.13]:

- ***flow***, the essence of the process, which converts the inputs into outputs;
- ***effectiveness***, which represents the degree of meeting the expectations of internal or external customers;
- ***efficiency***, the degree of judicious use of resources to produce the output;
- ***hourly capacity*** of the process (number of units of output per unit time) or the duration of the cycle;
- ***cost*** of the entire process.

The key values of these characteristics reflect the adequacy of the process. They are often used as ***performance indicators*** of the process. Efficiency is a particular performance indicator, as it only really takes on meaning once the effectiveness of a process is reached.

Very often, the processes themselves are divided into mini-processes, which in turn can be represented by micro-processes. The finest division may be the sum of activities of a service or an employee. The last working mode corresponds to a procedure or even an instruction (Figure 1.15).

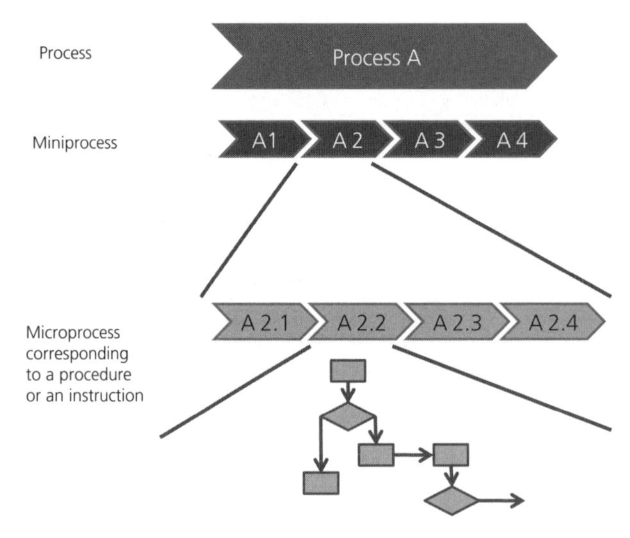

Figure 1.15 Decomposition of a process.

1.10 Optimization of a process

The breaking down of a business process can make it possible to analyze its characteristics and especially to strengthen those related to the production of the service or product. Over time, even a previously optimized process becomes complicated and unproductive. Figure 1.16 shows an algorithm for value analysis and simplification of an activity [1.14].

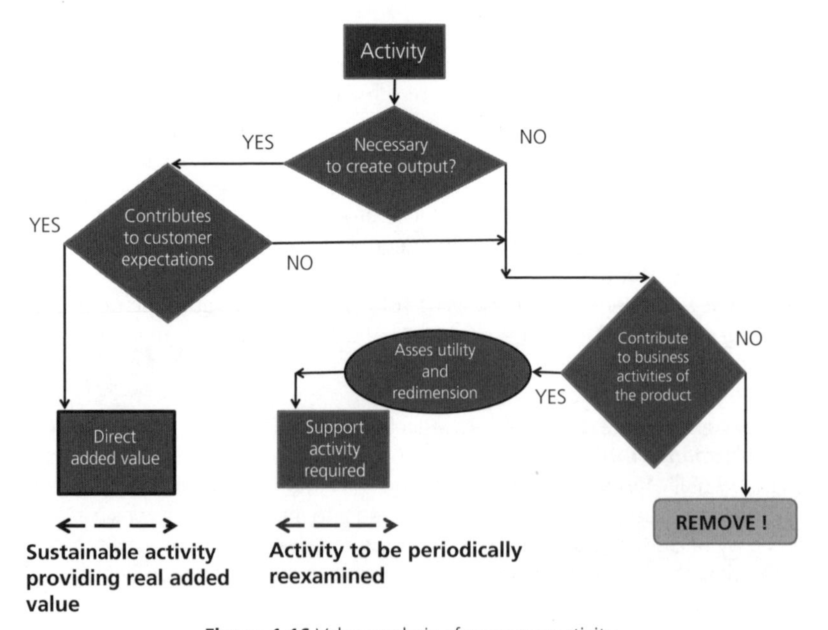

Figure 1.16 Value analysis of a process activity.

The definition of the key processes of the company is a major step in the optimization of performance and quality.

As a result of a company's existence, there are processes in place, in part or in whole, but they are rarely optimized. Process definition, and especially their efficiency, can be questioned after changes in company policy, a change in the product portfolio, a change of mission, or as the result of a policy acquisition and merger. Its relevance must be regularly tested. Once identified, the processes are analyzed, regularly revised and systematically improved.

1.10.1 Optimization of processes: the algorithmic approach

How should we start? Algorithms have been proposed, such as the one by Harrington [1.15], comprising five stages [1.16]:

1. *Preparation: assuring success through leadership, understanding and involvement*:
   a) set up a structure for improvement and communication, and train employees;
   b) formulate customer expectations and complaints, customer service and sales organization and/or objectives set by management, and communicate them to employees;
   c) determine the process categories that require improvement.

2. *Comprehension: understanding all aspects of the process under scrutiny*:
   a) analyze all aspects;
   b) develop an overview (purpose, scope) and consider the interfaces (system approach) as well as the chain of service providers/internal customers;
   c) formalize the current operation by an ad hoc diagram;
   d) determine the metrics to assess the activity and the results expected.
   e) reformulate 1c, if required.

3. *Rationalization: improving the effectiveness, efficiency and adaptability of the process*:
   a) identify opportunities for improvement;
   b) eliminate activities not providing added-value to the product;
   c) optimize the number and role of interfaces in the process operation;
   d) simplify the process and compare the results with the classification approach.

4. *Measures and controls: implementing a system of process control for continual improvement*:
   a) develop internal measures and targets;
   b) periodically audit the process;
   c) establish metrics of measuring the costs of poor quality.

5. *Continuous improvement: developing a sub-process for continual improvement*:
   a) describe the process;
   b) identify, analyze and eliminate the problems;
   c) assess the impact of change on the client (internal or external);
   d) apply *benchmarking* to the process.

To summarize, the objective of the approach is to redefine the existing processes by:

- compacting them;
- clearly delineating their outline;
- providing a logical structure with regard to flow;

- defining their interfaces according to a provider/customer (internal and external) logic;
- integrating them into a perspective (i.e., by considering their mutual interactions).

1.10.2 The option *ex nihilo*

Sometimes the organization (although currently rare) does not have a process approach when the analysis is started. Reduced to trial and error, one proceeds as follows:

- identify products and services, i.e., the output of the business;
- identify customers and suppliers;
- identify the sequence of activities starting from the research and formulation of the clients' expectations through delivery of products and services to the client;
- arrange the activities sequentially in a single row oriented toward a common goal, i.e., *shaping the process*;
- assess whether reformulation of the hierarchical structure is necessary;
- conduct an initial summary, revise it, review it with the people involved and, if necessary, carry out a re-iteration.

1.10.3 Optimization of the process: the classification approach

The algorithmic approach to process optimization, even if done successfully, does not mean that the choice is sure to be a good one. Having models of good practice beforehand makes it possible to correct or to approach the optimization in a more secure manner. The process approach has been significantly developed since its systematization by *Porter* [1.17]. There is now a generic approach to the definition of processes, including the production of specialized literature by branches of services and products, as well as the expertise – sometimes overused and/or expensive – of consultants.

As an example, a robust approach and classification is provided by the APQC[3] model (organization similar to the American Society for Quality) called *Process Classification Framework*[SM] or PCF. The 2007 version of PCF[4] defines 12 categories of key processes (Figure 1.17) divided into main and operational processes, management processes and support processes.

Each process category is subdivided into process areas, which are in turn segmented into process units consisting of a list of activities in the form of procedures, instructions or competences[5].

EXAMPLE:

- *Process category*: 4. Logistics, deliver products and services.
- *Area or process group*: 4.1 Plan the acquisition of necessary resources.
- *Process unit or element*: 4.1.1 Manage the supply of products and services.

---

[3] The *American Productivity and Quality Center* (APQC), created in 1977, and protagonist of the implementation of the Baldridge Total Quality Award (since 1985). The standard can be downloaded at the URL: http://www.apqc.org

[4] APQC is not the only organization to have a classification. That of PWC (PricewaterhouseCoopers) is divided into 13 processes and puts more emphasis on market research. It is available at: http://www.globalbestpractices.pwc.com (accessed November 2007).

[5] Can be downloaded at the URL: http://www.apqc.org

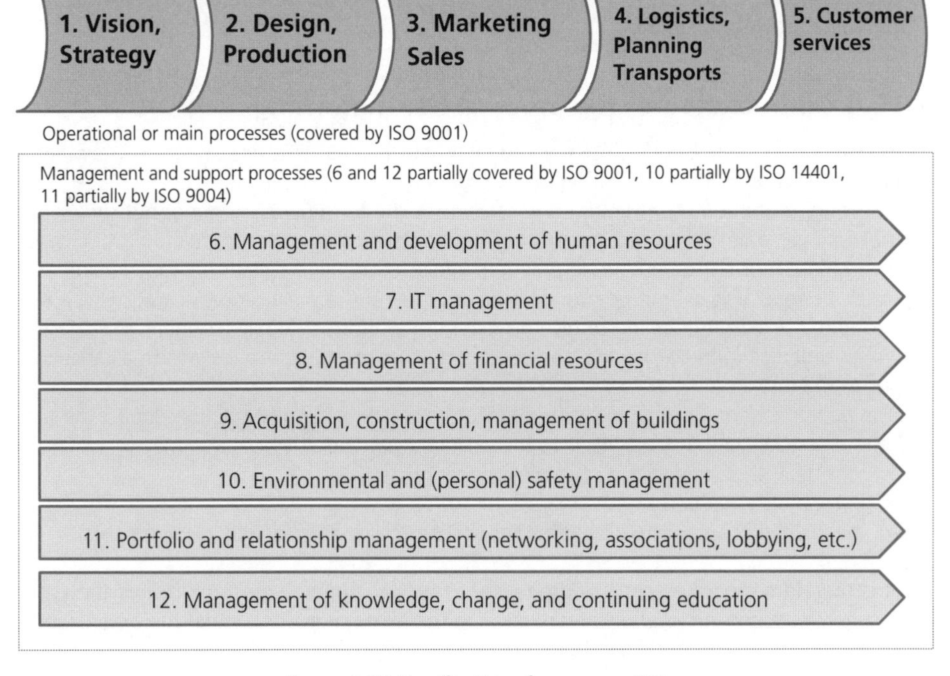

Figure 1.17 Classification of processes: PCF.

- *Activity*: 4.1.1.1 Define estimates of weekly or monthly requirements (algorithm defined by a procedure or in a simpler case by an instruction).
- *Activity*: 4.1.1.2 Work with clients (implemented by competence).

The classification is tedious, but can in a short time make it possible to:

- define the categories of the company's processes and their importance;
- identify and select the process areas affecting quality, and limit the scope of the system;
- verify and/or decide whether the process units are in place, whether all activities have been identified and listed in the form of procedures, employee competence or instructions.

Not all of the process categories of the PCF approach are covered by ISO 9001 (Chapter 4):

- management of information technology (No. 7) follows from the application of ISO 27001;
- financial management (No. 8) is covered by the internal control standard COSO or equivalent;
- environmental and safety management (No. 10) is covered by the ISO 14000 family (see Chapter 5) and OHSAS 18001;
- management of public and external relations (No. 11) is covered by ISO 9004, the standard for social responsibility ISO 26000, or the principle of sustainable management TBL (*Triple Bottom Line*, Chapter 13).

Classification is often used to optimize management and support processes that are not considered strategic or essential. Indeed, as this is in the public domain, the competitive advantage it provides is limited.

1.10.4 Optimization of the process: benchmarking

Benchmarking is one method of obtaining quality, which consists of acquiring the knowledge and expertise of another organization.

If this appropriation is unfriendly, one rarely uses the term *benchmarking*, but rather the neutral (and versatile!) term "economic intelligence", which ranges from reading trade magazines to more morally reprehensible practices (indiscretion, etc.). Since the added value of a product can, to a large extent, be found in the activities of the main process category, it is not easy to find a competitor willing to give his expertise without strategic compensation. This is where an organization's bargaining power as well as its prospects for long-term alliance will be put to the test.

The *benchmarking* approach is different from a simple copy/paste action due to the fact that the organization that lends itself to this mutually beneficial transaction may:

- not be in the same sector of services or products;
- have a different corporate culture;
- be located in a country with different cost structures (and quality requirements).

It is therefore crucial to analyze the reasons for the improved performance of the company suitable for *benchmarking* – and relate it to the existing process within the organization – before carrying it out. This done, here are the key steps of *benchmarking* (*reverse engineering*):

1. define the process for which best practices are sought;
2. attempt to find organizations with this excellence;
3. negotiate terms of agreement that are mutually beneficial;
4. establish a specification for the analysis and key objectives;
5. determine the performance gap between the two processes;
6. identify improvements to be implemented in order to eliminate this gap;
7. define the objectives for the implementation of the new conditions;
8. implement them;
9. control the implementation by analyzing the new performance;
10. add the optimized process to the continual improvement process.

1.10.5 Reformulating processes: re-engineering

Optimizing processes through a series of improvements may not be enough if:

- the competition threatens the organization's market share;
- the company wants to resize its operations or significantly change its policy;
- profit margins become eroded and the dividend yield, or the value of the stock, is no longer attractive to investors; they might then sell their shares to a competitor who is preparing a hostile takeover, putting the company under the control of a predator ready to take its market share and then break it up;
- the leading circles of the company wish to increase the value of the stock to obtain favorable loan terms from banks;

- senior management wishes to issue new shares and seeks to make the company more attractive;
- product quality stagnates or even declines, despite improvements to the process.

This type of situation, which can be experienced as a crisis, requires strong and long-term action from the company's management, which should involve a complete redefinition of the organization's processes. From an optimization perspective, we move to a drastic change in the organization, a quantum leap with regard to management in a complex environment (Figure 1.18) which must be carefully analyzed.

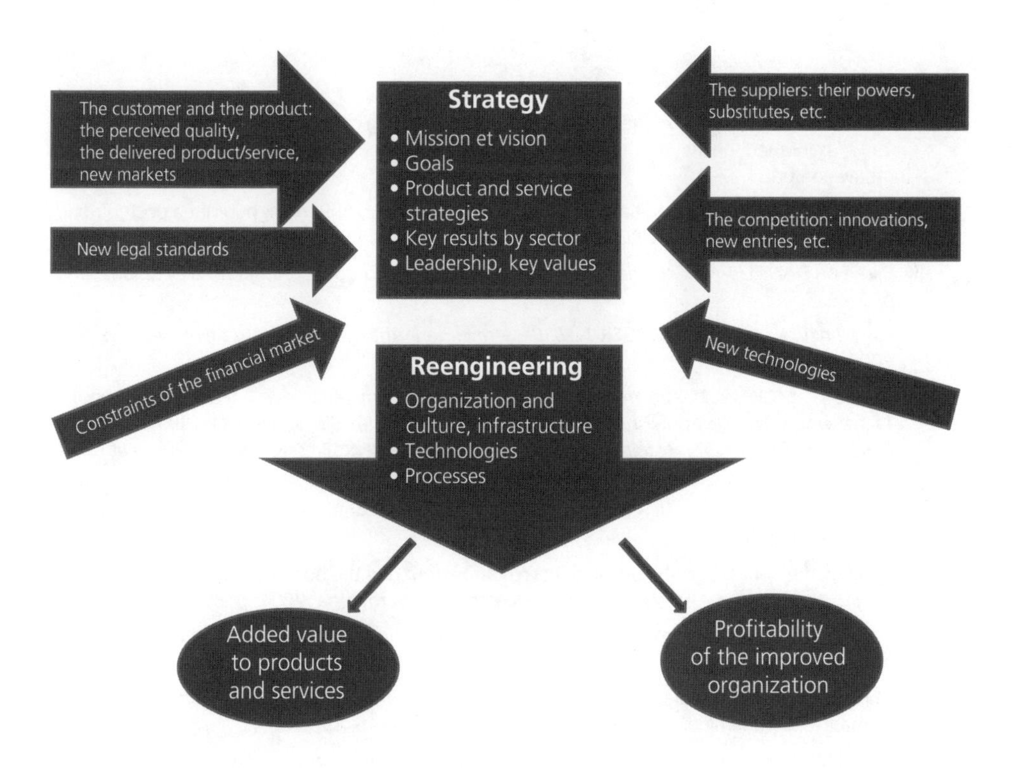

Figure 1.18 The re-engineering environment.

This search for a new mode of operation often calls for international consultants, which reinforces the neutrality from the management viewpoint, while increasing the visibility of the internal operation as well as its prestige. This is the reason why *re-engineering* is not the only aspect of the quality function that does not make use of human resources, logistics and expertise for it to be carried out. To the *re-engineering* of a company it is possible to attach the strong measures of Total Quality, e.g., *Six Sigma* projects, coupled with the *Lean Management* approach.

Re-engineering is a risky activity that is not very visible on the outside. It may not attract the attention of management involved in public or commercial relations, or in strategic business processes with customers. The action of the board of directors is thus fundamental.

Management will sometimes be tempted to delegate the task of *re-engineering* to senior "fall guys" (or those that some factions want to set aside). Just by existing, these work groups will have little support and will test the saying that "friends come and go, but enemies accumulate" because actions such as preparing layoffs and suspensions give rise to long-lasting hatred.

Many *re-engineering* actions do not have their expected success. Although hailed by the stock market and board of directors, such operations are greeted with suspicion and anxiety by staff. While productivity can be increased by playing with the stress levels of staff, this has a rather adverse effect on quality and attempts to increase it within the organization.

However, in cases where the very survival of the organization or the company is at stake, management often has no choice but to proceed with this reformulation.

References

[1.1] ABRAHAM H. MASLOW, A Theory of Human Motivation, *Psychological Review*, vol. 50, 370-396, 1943; published online at the URL: http://www.advancedhiring.com/docs/theory_of_human_motivation.pdf.

[1.2] BLAISE PASCAL, *Pensées*, édition Port Royal, chap. 7, n° 12, 1669, book published posthumously and available online at the URL: http://fr.wikisource.org/wiki/Blaise_Pascal.

[1.3] Based on DANIEL A. WREN, *The Evolution of Management Thought*, ed. John Wiley & Sons,1994, p. 9.

[1.4] For a brief overview, see, ROGER AÏM, *L'essentiel de la théorie des organisations*, ed. Lextenso, 2009.

[1.5] JAMES G. MARCH, *Quatre mythes du management*, Conférence, May 1998, Paris School of Management-Full Text published in *Gérer et Comprendre*, n° 57, September 1999; taken from CHRISTIAN THUDEROZ, *Histoire et Sociologie du Management*, Presses polytechniques et universitaires romandes, 2006, pp. 260-263.

[1.6] E. JEROME McCARTHY, *Basic Marketing: A Managerial Approach*, ed. Irwin Homewood Il, 1st ed.1960, 2001.

[1.7] This scheme, which is in the public domain, can be found in the book by DOMINIQUE SIEGEL, *Le diagnostic stratégique et la gestion de la Qualité*, ed. L'Harmattan, 2004, p. 85.

[1.8] R. EDWARD FREEMAN, *Strategic Management: A Stakeholder Approach,* ed. Pitman Publishing, 1984.

[1.9] M.B.E. CLARKSON, "A stakeholder framework for analyzing and evaluating corporate social performance". *Academy of Management Review*, *20*(1), 92-116, 1995.

[1.10] MICHEL CAPRON and FRANÇOISE QUAIREL-LANOIZELÉE, *La responsabilité sociale d'entreprise*, ed. La Découverte, 2007.

[1.11] Figure 7.1 is based on Figs. 7 (p. 80) and 26 (p. 223) of the book by D. SIEGEL, op. cit.

[1.12] The diagram in Figure 1.12 is based on Figure 2.2 *The generic value chain*, p. 37 of the book by MICHAEL PORTER, *Competitive Advantage*, The Free Press, 1985.

[1.13] H. JAMES HARRINGTON, *Business Process Improvement: The Breakthrough Strategy for Total Quality, Productivity, and Competitiveness*, McGraw Hill Inc., NY, 1991, p. 114 and seq.

[1.14] Based on Figure 6.1, of the book by H. J. HARRINGTON, ibid., p. 141.

[1.15] H. J. HARRINGTON, ibid.

[1.16] Based on the reformulation of MICHEL WEIL, *Le Management de la Qualité*, ed. La Découverte, 2001, p. 95.

[1.17] MICHAEL E. PORTER, *Competitive Advantage – Creating and Sustaining Superior Performance*, The Free Press, 1985, p. 33 and seq.

Chapter 2

The Approach to Quality: Concepts and Definitions

Key concepts: quality assurance (internal or external), implicit or explicit client expectations, quality control, criteria of quality, Ishikawa diagram, quality deployment standard, the house of quality, mastering quality, nonconformity, poor quality, stakeholders, quality planning, quality, total quality, re-engineering, Deming cycle, sub-standard quality (and costs), specification, quality standard, above-standard quality (and costs), quality system, zero defects.

A first approach to the world of quality requires an understanding of 20 essential concepts. These are valid for all production sectors (tangible products and services):

- Quality: definitions, marketing and production
- Quality standard, criteria and substituted criteria of quality
- Inspection
- Non-standard quality: standard and nonconformity
- Non-standard quality: sub-standard quality and above-standard quality
- Quality control
- Mastering quality
- Quality assurance
- Quality management
- Systems for quality management
- Quality planning
- Total quality
- Deployment of quality
- The House of Quality

It is vital to first understand these concepts before entering into the formal world of quality.

2.1 Quality: first approaches

2.1.1 The basics: smell and taste

The instinctive and intuitive basics when it comes to quality reside in the senses, especially smell and taste. Their stimulus, sometimes perceived as unpleasant, lead human beings to spontaneously reject spoiled food. The emotional response to this stimulus is disgust, which can cause spontaneous vomiting.

One feature of the olfactory system is the wide range of substances that can be identified. However, it is accompanied by a low discrimination when it comes to intensity. This binary character is noted as it sets a fundamental characteristic of quality: conformity or nonconformity, acceptance or rejection.

It is not surprising that the first approaches to quality involved the selection, processing, and preservation of food, and the associated fraud. The most common have been the dilution of milk, the addition of mineral powder in flour, the replacement of consumable alcohol with methanol that causes blindness, the addition of mineral oil to vegetable oils, selling minced meat that includes pork or horse to ethnic groups with one or more food taboos, the addition of common herbs in preparations of fine herbs and spices, wine adulteration with beef blood, etc.

2.1.2 Quality, an expectation

More sophisticated perceptions of objects or tools are grafted onto this instinctive basis, in relation to more or less formalized expectations. Thus, it is expected that a garment purchased, even at a modest price, does not tear, that it does not shrink after washing, and that its color does not change after prolonged exposure to sunlight. Detergent is also expected to have a pleasant smell when you open the box. Some expectations are obvious, others more difficult to define.

Quality can therefore attempt to describe minimum expectations (threshold effect) below which the object is not chosen, used, exchanged or purchased. It can also mean a degree of excellence. Thus, a work (of art, in particular) of great beauty has – from the start – a high degree of quality. Of course, many of the attributes of beauty depend on the state of mind and cultural values of the man or woman who perceives the work. Regardless of whether an object is crafted by hand or produced industrially in numerous copies, its essential characteristics are defined by the expectations of the customers or consumers.

2.1.3 Quality: a moving target

The quality criteria of an artifact or a product[1] of nature are not static; they evolve over time. The Wright brothers' first plane can in no way be compared to the latest Airbus in terms of comfort, reliability, flight distance, security and maneuverability. Generally, consumer expectations increase over time.

[1] According to ISO 9000, a product is the result of a process. ISO 9000 distinguishes between four generic product categories:

- services (e.g., transport);
- software (e.g., dictionary);
- material products (e.g., mechanical motor parts);
- products from continuous processes (e.g., lubricants).

However, this law is not always true. Beyond a certain level of sophistication, expectations can be reversed, leading to the development of a simpler or more robust product in the portfolio of offers. Mobile phones for seniors are an example. Another is the stereo: in the 1970s, the average degree of sophistication of a stereo was well above the current one (filters, speakers, etc.). Today, simplicity of operation is the primary expectation.

2.1.4 Quality: a concept of an exchange economy

The concept of quality[2] is associated with an exchange economy, i.e., the production of consumer goods and equipment (individual, group or public), but also with services or a combination of services and goods – in short, with what is known as a business. Quality in the creation of a work of art – whether unique or produced in a very limited number of copies – and the qualities attributed to such a piece of art will not be addressed in this book.

2.1.5 Two operational definitions of quality

2.1.5.1 Quality of an object of exchange

In the context of economic exchange, a first concept of quality is based on the expectations of the customer or partner, whether expressed or implied. In a strict exchange economy, customer expectations are met if the customer and not the product returns after the sale. This statement is the marketing approach of the definition of quality, whose development ultimately leads to total quality.

2.1.5.2 Quality of a production batch

Another concept of the quality of a product is linked to mass production. Mass production has several steps or processes, leading to a degree of variation within the produced batch. This amplitude must be controlled in order to reduce the spread – if the population according to the criterion of measure is Gaussian, this is known as the variance – so that the characteristics of all products in the batch are within the tolerance area of the client's expectations. If this degree of variation is reached, the quality will be high because the number of returns and claims will be low. Statistical analysis is one of the tools of quality, and it is particularly used in quality control. The most complete development of this approach is the system of quality management.

A lost customer reports his dissatisfaction to many families, which is less often the case for a satisfied customer (remember the instinctive basics of quality). The goal is to strive for a claim rate close to zero, or in other words to promote a policy known as Zero Defect.

In the area of services, the situation is more insidious: in view of the intangibility of the product, customers rarely complain, thus depriving companies of any warning signs.

---

[2] According to ISO 9000, Quality is defined as the ability of a set of inherent characteristics to fulfill requirements. A characteristic is a distinctive feature (physical, sensory, behavioral, temporal, ergonomic, functional). A requirement is a conveyed need or expectation, usually expressed, implicit or statutory.

2.2 Quality standard (or list of requirements)

2.2.1 Quality is associated with a measurement and a standard

The two definitions mentioned above are related to the measurement of quality criteria. This is either associated with real (measurement of a physical variable, etc.), natural (microbiological or organoleptic evaluation, for example) or binary results (presence or absence of odor, color specification...).

The criteria (or requirements) that define the quality of a product or service are united in the form of a standard. If the number of criteria in the product standard is limited, the term "in-house requirements" is often used. The term "specification" is for more complex products.

2.2.2 The standard, an agreement between stakeholders

The quality standard of a product specifies the expectations of the customer and the consumer, but these expectations alone generally do not define the profile of the product. The standard brings together the expectations of all the stakeholders and acts as a facilitator in the international politics of trade. In addition, the standard incorporates the requirements of the shareholder or client: profitability, efficiency of investment.

Stakeholders can be:

- governments (safety, hygiene and health – e.g., for cosmetics – environmental protection);
- manufacturers' associations (policies with regard to safety, accidents, hygiene, health, generic definition or expectations specific to the entire category, but also market protection);
- standardization bodies;
- associations of suppliers of raw materials, semi-manufactured products and components used in the composition of the product;
- wholesalers and distributors (size, weight, features);
- consumer associations (expected features);
- NGOs concerned about the civic responsibility of companies and consumers (protection of fauna and flora, sustainable development, protection of workers, etc.);
- Board of Directors and shareholders (reputation, profitability).

There are numerous norms or standards of product quality. Many of them are international and are mainly managed by ISO (*International Organization for Standardization*) and, in the EU, by CEN (*European Committee for Standardization*) or by ad hoc Directives.

Manufacturers adhere to international quality standards of the products they manufacture, and integrate their own specifications, especially those associated with the product's marketing mix.

2.2.3 The standard, a foundation but not a blueprint for production

The quality standard of a product represents the common expectations of all stakeholders. However, it does not specify the conditions for production nor systematically the raw materials or semi-finished products used. Thus, a car can be built by craftsmen (Rolls-Royce) or can be industrially constructed (Toyota) with the same generic standard. A warplane of World

War II may have been manufactured from wood (the very efficient twin-engine attack plane, the Mosquito) or from metal (the American fighter aircraft, Tempest).

Some criteria are translated into more technical specifications (e.g., impact resistance) to be part of a production process. This is where the genius of the manufacturers, their know-how obtained by long experience of developing products and technologies, comes into play. Some of this knowledge is in the public domain, in other cases it is protected by a set of patents, and sometimes it consists of jealously guarded innovations that competitors seek.

The quality standard must be transcribed as an internal specification, placed at or near the production center. This document specifies the quality criteria of the components, raw materials, semi-finished products and mode of assembly (Figure 2.1). These specifications or derived quality standards will be used during the design phase to adjust or determine the tools, processes and operating conditions, as well as the mode of production assembly. Often, a criterion of quality of a product cannot be easily measured, such as the comfort of a car driver's seat. Going from the "comfort" quality criterion to a technical one (size and shape of the seat, sense of touch of the seat cover, softness and elasticity of the padding) is a sort of substitution, and this is the reason why these parameters are sometimes called substituted or derived quality criteria. Thus, the power of a car and its maximum speed, which can be customer expectations, are related to derived quality criteria such as the engine size, weight, etc.

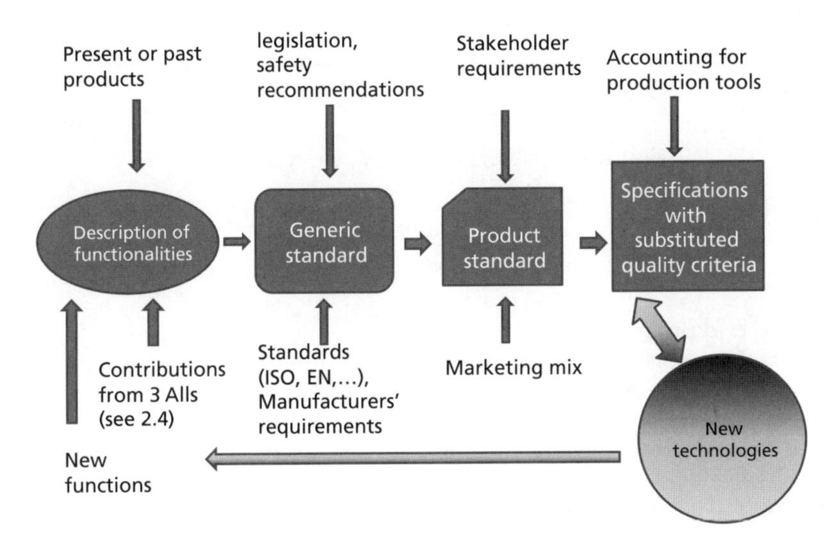

Figure 2.1 Quality standard: steps in a specification.

This concept of substitution is used to analyze the complexity of a process and the effect of its variables (derived quality criteria) on a product quality criterion. This is *Ishikawa's* fishbone diagram, also known as the 6M chart (man power, machine, measurement, material, method, management), or the pattern of cause and effect [2.1] (Figure 2.2).

2.2.4 Applying Ishikawa's diagram to the Battle of the Somme

It may be interesting to apply *Ishikawa's* diagram to a non-industrial – military – activity, more specifically to the dramatic example of the Battle of the Somme (British Front) of World War I. Things hadn't got off to such a bad start, but ...

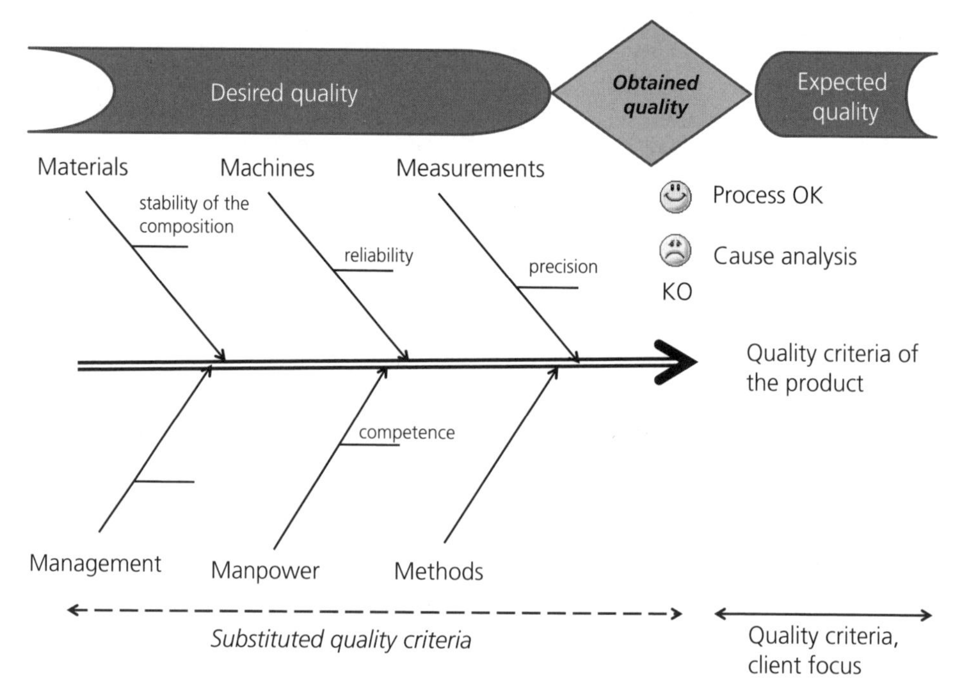

Figure 2.2 Ishikawa's cause and effect diagram.

Background and preparation

The Battle of the Somme (department of northern France in the region of Picardie) went on from July 1 1916 to the end of November 1916 during World War I. It was started in order to relieve the hard-hit French troops at Verdun. The British and French troops attacked and attempted to break through the fortified German defense lines north of the Somme on a north-south line of 45 km.

The back of the trenches had been transformed into a giant warehouse with complete transport logistics: roads, railway and aviation. The Allies had temporarily (until September) complete control of the airspace. The artillery, including large caliber guns (of 380 to 400 mm) on railway tracks, achieved peaks of destructive power. The artillery preparation, originally scheduled for five days, started on June 24 with both ranging and barrage shots. It intensified from June 26 with a general and continuous bombardment of German lines. In one week, the British artillery, comprising 1500 guns, fired 1,750,000 shells.

In addition, 17 tunnels had been dug near the enemy lines, where they had amassed tons of explosives: 21 tons in each of the three largest tunnels. Explosions were planned shortly before the attack of the British infantry.

On July 1, in the morning, the final Allied bombing began. From 6:25 am, the artillery fire reached a rate of 3,500 rounds per minute, producing a sound so intense that it was heard in England. At 7:30 am, at the whistle, the British infantry crossed the parapets, their bayonets fixed, and attacked the enemy trenches, persuaded by their officers that the German troops were annihilated and that the attack would be almost a "walk in the park".

The results of the first day

The first day of the Battle of the Somme holds the sad record of being the bloodiest day for the British army, with 60,000 victims (the number of injured survivors will never be released) of which 20,000 were killed. At 10 pm, the first orders to stop the attack were given, but these orders were not carried out by all units of the army until July 3.

Indeed, the Germans had a well-organized front, occupying elevated positions almost everywhere. It was structured into three parts:

- a first position that was well supported, with front line trenches as support and backup, but also with a succession of deep shelters with relatively modern comfort.
- a second line hosting the artillery;
- further back, a second position, nearly as well defended as the first.

Further back, there were woods and villages connected by narrow passages, to form a third and even fourth line of defense, all hard-surfaced and fortified thanks to a rocky terrain that could be easily cut and that hardened when dried.

End of the battle

It was the rain, snow and sleet that ended the battle. In late November, due to an insignificant advance, the British decided to stop the Battle of the Somme, with a balance sheet of 630,000 dead and wounded allies. After five months of fierce fighting, the Allies had penetrated only about 13 km within a 35-km front. The estimated losses to the Germans were 660 000 men.

The Battle of the Somme was one of the bloodiest battles and a major confrontation of World War I. Despite the heavy losses of his army, British general *Douglas Haig* was promoted major in 1917 and knighted after the war.

The cause of the disaster

Artillery (machine) and ammunition (material)

- The ammunition for the artillery was hastily manufactured in the UK by makeshift personnel with inadequate training, because the workers had gone to the front. The percentage of defective ammunition was very high.
- Of the 12,000 tons of ammunition used, two-thirds were fragmentation shells that had an effect on people, but no impact on the bunkers and shelters of the German soldiers, or on the barbed wire and trenches. Only 900 tons of adequate ammunition were fired.
- 1,450 British guns were used, but only 450 could be counted as heavy artillery and only 34 had a high caliber.

Troops (man power) and environment (Mother Nature)

- The infantry carried 32 kg of equipment, considerably hampering its mobility.
- In 1916, the British army in France lacked experience, because its professional part, six divisions, had been eliminated. The majority of the soldiers was composed of volunteers from the Territorial Force with little experience and a new army (Kitchener's) that hadn't been trained.
- In the trenches, the soldiers lived in miserable conditions under constant deprivation. Exhausted soldiers got little sleep in makeshift shelters. During rainy days, they found themselves knee-deep in mud.
- Disease proliferated. Fleas and lice caused the soldiers to get trench fever. The constant

immersion in cold water gave the soldiers "trench foot", which could sometimes lead to amputation.

- The diet – monotonous, consisting of stewed or canned beef, bland jam, hard biscuits and sweet tea – made the soldiers especially susceptible to a variety of skin diseases.
- The land where the trenches were dug was also dangerous. Fertilized with manure for generations, it was home to bacteria that contaminated wounds and caused gas gangrene, a disease that was often fatal in an era without antibiotics.

Officers (management) and communication (method)

- The officers had been promoted rapidly and lacked both training and experience. *Haig* himself had obtained lightning-fast promotion.
- Prior experience had shown that similar shelling had been unsuccessful, but the General Staff did not seem to have this information, or did not consider it.
- Orders had been given to the troops to advance slowly, upright and uncovered.
- Communications were completely inadequate and the General Staff was largely ignorant of the conditions and progress of the battle.
- *Haig* administered the huge British army in France with competence, but regardless of casualties. He shared with most of his subordinates a faith in the power of artillery bombardments, and an unlimited willingness to continue the offensive.
- *Haig* was eager for recognition, anxious to join the aristocracy, very concerned about royal intrigues. He often wrote private letters to King George V. He liked to be served champagne in his castle, which was his headquarters, while his men, most of them common people, barely survived in the unsanitary trenches.

Since the bombing had had only little effect, the German trenches were left interspersed with bunkers with heavy weapons and machine guns sweeping the open ground between the lines of the trenches: the British forces advancing there were eliminated. The barbed wire slowed the soldiers, forcing them to crawl, nailing them in place, or driving them out into the areas of machine gun fire.

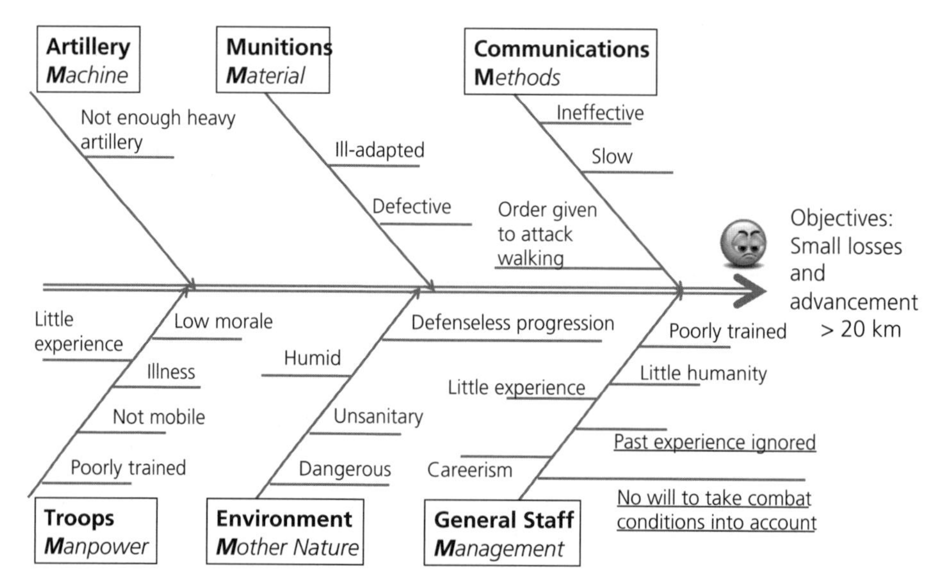

Figure 2.3 An *Ishikawa* diagram of the Battle of the Somme.

Remarks

Based on this report, the establishment of an Ishikawa diagram (Figure 2.3) provides answers to several questions:

- What are the three leading causes for the significant losses of the British army? Inadequate and defective ammunition, past experience ignored by the General Staff, a lack of heavy artillery.
- What is/are the error(s) that contributed to 80% of the losses? Past experience ignored by the General Staff.
- What are the errors attributable to the troops? Their only error was lack of experience.
- What are the errors directly attributable to the General Staff? Its lack of professionalism in the consideration of conditions and battle management, including information of past experience and a blind belief in the power of technology.
- And attributable to the British government? Its desire, although worthy, to hastily relieve pressure from the French and allied troops, and the fact of assigning ultimate responsibility for the operation to a careerist officer without sufficient experience.

Conclusions

The losses of the Battle of the Somme almost completely destroyed an army and practically put it out of combat (not to mention political problems in light of the democratic system):

A) In fact, 80% of the losses could have been avoided if the munitions had not been defective and inadequate, if the heavy artillery was truly present, and if the General Staff had shown more attention with regard to the conditions of combat and included previous experience in their plan of attack. These are 4 causes. The Ishikawa diagram mentions 21 causes, of which 4 result in 80% of the losses. *This is the Pareto or 20/80 law that states that in the search for causes, 20% or less of the causes produce 80% or more of the effects*. It applies to the Battle of the Somme. One can visualize the *Pareto* law in the form of a diagram (Figure 2.4).

B) The errors of the General Staff alone caused huge losses. Had they been more attentive, they could have stopped the attacks earlier, or even deferred them, and paid more attention to effective communication. This illustrates another law, *that the primary responsibility when it comes to sub-standard quality is management.* This is an assertion made in the absence of controversy and trade-union influence!

C) These results led to the strengthening of the BSI (British Standards Institution) in order to have better methods to produce quality ammunition, but probably not to a more efficient education of the General Staff. This illustrates another trend, *the fact of focusing on technical solutions to minimize errors due to management and its processes.*

2.3 Standards of quality, poor quality, nonconformity and defects

Poor quality includes nonconformity and defects. In the world of quality, nonconformity and defects are not synonymous, although defects are nonconformities. Indeed, it may be that the product meets the criteria of the quality standard, but not the standard comprising the substituted quality criteria (specifications).

Thus, a car model may be equipped with tires other than those that were originally planned (breakdown in the delivery within an in-time production process), the product is nonconforming, but the customer is satisfied because the product was delivered on time

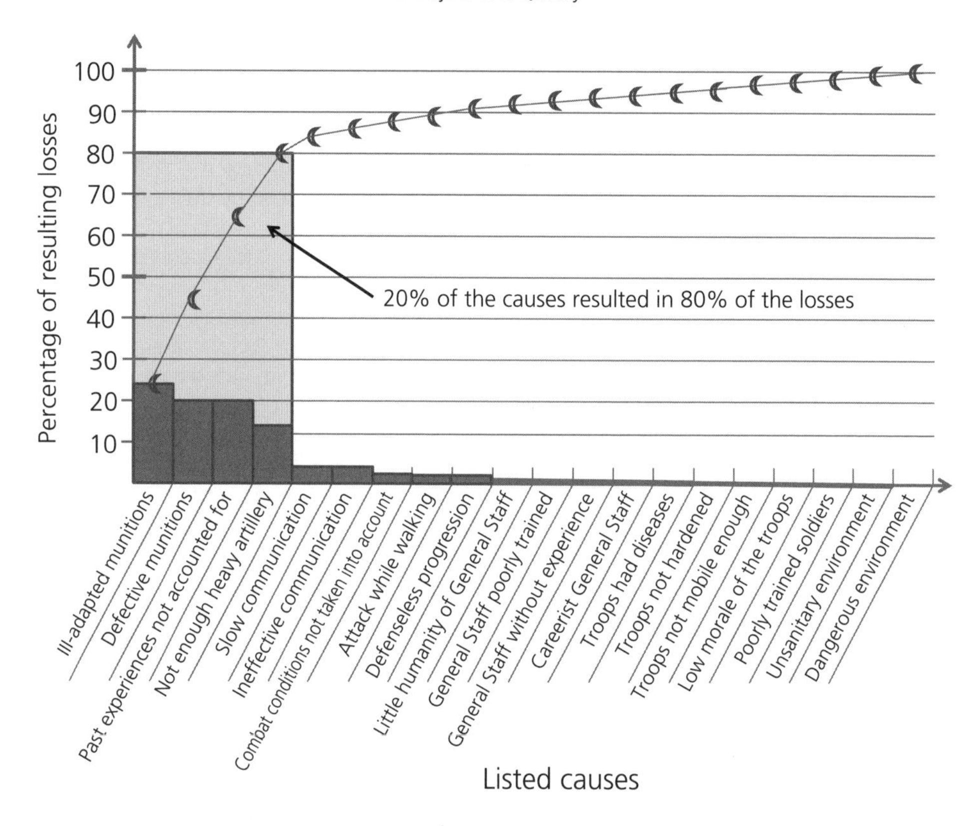

Figure 2.4 (Hypothetical) Pareto diagram of the losses of the Battle of the Somme.

and in line with expectations. Similarly, the PC you use can have a more powerful hard drive, instead of a standard one, perhaps because the standard hard drives that were delivered to the assembly station were defective or, simply due to a mistake in the assembly process (error in the reduction of stock). The product does not present defects, but it is nonconforming.

How one treats nonconformity is part of the quality policy of a product. Nonconformity detection and corrective actions are accompanied by a risk assessment. The batch produced must be released or permanently blocked by a special procedure.

2.4 Quality and innovation standard

Many products have seen their standards evolve over time through regular revisions and the integration of knowledge of increasing complexity. But how can one create an "ex nihilo" standard when it is not a case of product development but a true innovation?

A standard is rarely started ex nihilo, even if the product is a true innovation because, after satisfying basic needs (lodging, food, clothing, etc.), the desires of man since the dawn of humanity can be expressed in a conceptual, almost magical, triangle: all-powerful (including seduction), all-present and all-knowing.

Thus, the standards for motor cars were inspired by some of the criteria for carriages, the hackney coach and train wagons; mobile phones have features from landlines, walkie-talkies of the 1970s and the mobile stations of the US Army in Vietnam; personal computers maintained some features of large computers and typewriters.

In the most general case, the pioneer and market leader (which is unfortunately not always the same company) set the outline of the standard, although not yet recognized as such. If the product is distributed on a large scale, the need for a specific international standard emerges. It is set up in stages:

1. selection by the market leader (or group of manufacturers) of the standardization organization;
2. proposition of the standard to the agency;
3. acceptance (or rejection) of the proposal by the standardization organization;
4. the standardization organization proposes a first working version of the standard;
5. the standardization organization provides this version to the stakeholders and requires comments and proposed amendments;
6. the modified standard is published as a new standard;
7. the standardization organization makes periodic revisions of the standard.

Does this mean that all quality criteria will be taken into account? The answer is unclear. For example, the risks that mobile devices pose to the brains of users have long been ignored by manufacturers. It is not certain that this file is closed. It could reappear in a future conflict between manufacturers on the one hand, and NGOs and consumer associations on the other.

However, although the standard of customer requirements and stakeholder expectations can be largely the sum of expectations of products of the past; this is not the case for the content of the specifications, which may include numerous innovations, whether of a technological or other nature.

2.5 Quality control[3]

2.5.1 Sub-standard quality and costs

The term sub-standard quality is used for a batch of a product with quality criteria below the defined standard. Sub-standard quality is poor quality. While it is easy to understand the disadvantages of sub-standard quality in terms of customer satisfaction, it is more difficult to assess the damage associated with it. However, it includes damage that is:

- Measurable in terms of cost:
  - urgent product development to achieve conformity;
  - cost of production (complete or partial) for nonconforming products;
  - cost of processing claims;
  - cost of product recall, potential sorting and destruction;
  - cost of treatment of nonconforming product (rework);

[3] According to ISO 9000, quality control is the part of quality management focused on fulfilling quality requirements.

  - loss of sales;
  - loss of market share;
  - temporary or permanent decline in equity values of the company stock, etc.
- Evaluated in terms of loss of image and trust:
  - with consumers and communities;
  - in the media;
  - with competitors/industrial partners;
  - with national and transnational legislative and political authorities;
  - with distributors and prescribers, such as NGOs and consumer associations;
  - with key suppliers, who until this point reserved their new innovations for you.

The loss of image and trust is more insidious than the loss of hard cash because the consequences, sometimes global, are diffuse and difficult to identify at first sight. This is why the majority of multinational companies attempt, in the case of consumer products, to attain zero defects.

2.5.2 Above-standard quality and costs

2.5.2.1 The best is the enemy of the good

The term above-standard quality is used when a more detailed standard – or one with a stricter amplitude of variations of acceptable quality criteria – is followed, and when this standard does not bring greater satisfaction/customer loyalty (and therefore does not lead to increased sales!).

Above-standard quality is poor quality. It has its costs, although their impact is more limited than sub-standard quality:

- higher design and development costs;
- sophisticated production equipment, even for buildings;
- more expensive supplies and raw materials;
- use of a more qualified staff than necessary;
- higher packing and transport costs if the product is more fragile.

Production costs and logistics will be higher than for the competition, and the price of the product may be affected. If the price is set by the competition, the profit margin is lower and the durability of the product line, as well as that of the company, can be threatened.

2.5.2.2 Costs of above-standard quality, marketing and promotion (new product)

Although this risk is particularly sensitive to production units regarded as profit centers in a multinational company or those manufacturing products with low profit margins, it can be put in perspective by balancing the costs of marketing and promoting the launch of a new product.

The costs of marketing and promotion can be higher than production costs directly related to the introduction of a new product (awareness awakening). Sub-standard quality cancels the impact of this expenditure and hampers sales and, in the case of consumer products, denies the product a valuable place on retailers' shelves, which is instead given to the competition. It is understandable why taking the risk of above-standard quality seems more appropriate than that of sub-standard quality, especially for new products. This risk-taking must prevail in a tight market facing a competition that monitors any mistakes.

2.5.2.3 Risk management for sub – and above-standard quality

Quality control begins during the design process[4] of a product and is supported by the company's quality policy[5].

Panels, market analysis and intelligence

The risks of above – or sub-standard quality are kept under control by market analysis, the use of panels of prescribers and/or customers, business intelligence from the sales front, the search for strategic information in professional and international environments, and from prescribers. These activities help to refine or recognize quality standard, in short, to highlight the expected level of Quality. The effectiveness of such tools in the case of a real innovation is limited by the absence of a frame of reference.

For a family of products, it is the quality standard itself that can be modulated with the support of a whole group of manufacturers (e.g., minimum content of cocoa fat in chocolate).

Inspection and quality control

One way to obtain the desired quality of a product (i.e., quality control[6]) is to inspect the batches (inspection method) or to sample them (by a simple survey or by using statistical methods of quality control to define a representative sample):

- raw materials or components at the beginning of the production process;
- at critical points during production or assembly;
- batches of finished product.

Inspection criteria are the quality criteria contained in the design specification of the product. For some, all items in the batch are inspected, but this activity is generally not possible for reasons of cost, storage, or because the inspection may destroy the product. The inspection is then performed on a representative sample, or by survey. However, this is not true for the aerospace industry, where special procedures are applied.

Quality control is effective in detecting systematic errors, but is ineffective if the poor quality is sporadic. It is thus impossible to claim zero defects. As for the inspection (product after product), it is so expensive for consumer products that it is never used. However, it can be implemented if poor quality is detected.

---

[4] According to ISO 9000, the design process is a set of processes that transform requirements into specified characteristics or into the specification of a product, process or system. Specifications correspond to a document stating the requirements.

[5] According to ISO 9000, quality policy is the overall intentions and directions of an organization related to quality as formally expressed by top management.

[6] According to ISO 9000, control is an evaluation of conformity by observation and trial, accompanied if necessary, by measurements, testing or calibration.

2.5.3 Quality assurance

Quality assurance[7] focuses on the entire process (supplier relations, purchasing, production, storage, logistics and delivery) and places it under control to ensure it attains the standard. Usually, this is achieved by careful analysis of risks at all stages, followed by quality control at critical points in the process. The process is subjected to an analysis of variations of the critical parameters to determine their tolerances and obtain a consistent product.

This work requires the writing of procedures, technical control sheets and guides, and a study of measurement systems. The purpose of quality assurance is to secure fluctuations at a level determined by the standard. It is not in itself a process of quality improvement.

Quality control of a finished product is, in the context of quality assurance, some evidence that the objectives of quality assurance have been reached. Quality assurance and quality control come at a cost. They are optimized according to the ***KISS*** principle (*Keep It Simple, Stupid*; do what is necessary and only what is necessary).

We speak of ***internal quality assurance*** for procedures, processes and guidelines of the company or institution, and of ***external quality assurance*** for the same elements when external experts inspect/audit the company for accreditation (from customers or authorities) or certification (issuing a certificate of conformity to established standards).

2.5.4 Quality management

The quality of a product or service is a moving target, given the competitive pressures and the different stakeholders.

In addition, changes to raw materials, packaging and legislation are regularly implemented (change of suppliers, new requirements from authorities, use of other raw materials or components). Quality and quality assurance are thus constantly changing. Methods that contribute to the validation and adjustment of quality assurance are part of quality management[8]. Quality management [2.2] will therefore:

- **P**lan: formulate the objectives and processes necessary to obtain a consistent product.
- **D**o: implement the standard and the processes as well as means of control.
- **C**heck: monitor and evaluate the results and determine corrective actions to be consistent with conformity.
- **A**ct: implement corrective actions to improve the performance of services and products.

Quality management is in line with continual improvement of services and tangible products: it describes a cycle also called the Deming cycle (Figure 2.5). However, quality management may also involve the control of changes that do not increase the quality of the product (change of suppliers, materials, or legal constraints without affecting the intrinsic quality of the product).

---

[7] According to ISO 9000, quality assurance is the part of quality management focused on providing confidence that quality requirements will be fulfilled.

[8] According to ISO 9000, quality management is coordinated activities to direct and control an organization with regard to quality. An organization is a group of people and facilities with responsibilities, authorities and relationships. Examples: company, corporation, firm, enterprise, institution, charity, self-employed person, association, or parts or combination thereof.

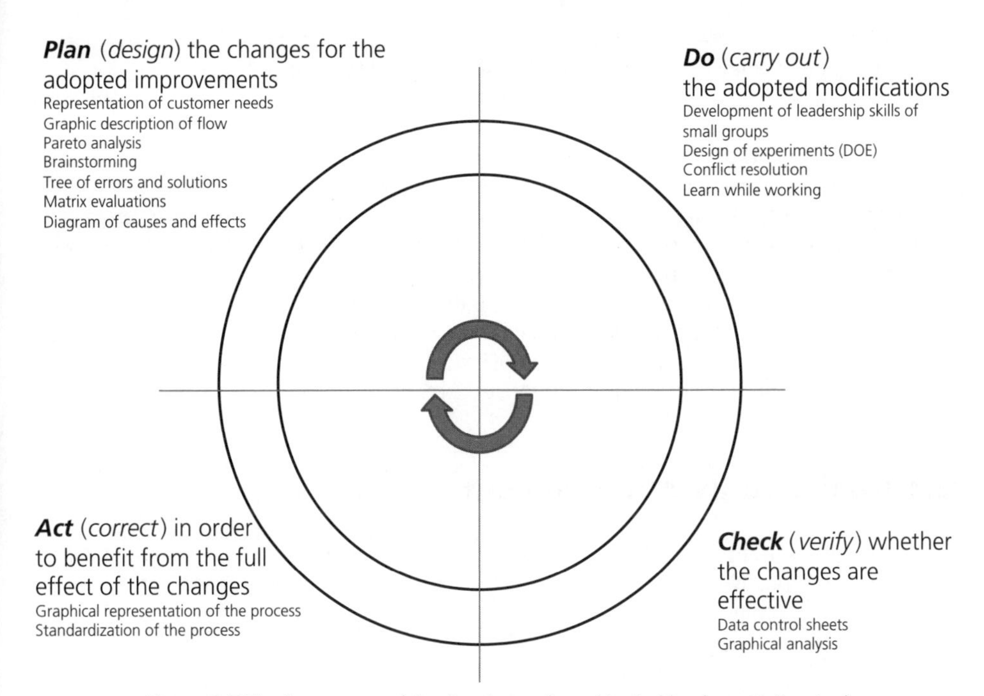

Figure 2.5 The four stages of the Deming cycle and typical implementation tools.

Quality management based on the Deming cycle presupposes a line of products beyond the initial phase of innovation. In addition, continual improvement can reach a plateau, as any process is accompanied by a minimum poor quality below which it is impossible to go. Therefore, in a program to improve quality, we may be in a quantum leap phase that disrupts the structure of the company by calling for a new organization or new technology. The Deming cycle, a tool widely used and recognized worldwide, is not the panacea for all challenges related to quality.

Sometimes it is not possible to keep the same type of product. In other cases, lower prices from the competition give rise to a complete overhaul of the production process. Organizational changes and technological modifications thus disrupt the logic of continual improvement. This disruption also occurs with the re-engineering of a process. Such an authoritarian (top down) change process is often associated with waves of dismissal, and in view of the resistance it generates, has had a very variable success rate.

There is sometimes no alternative but to close a company when faced with the marketing of new products. A concrete case is that of the producers of ice blocks in the Vallée de Joux (Switzerland) for consumers and restaurants. Their destiny was doomed when refrigerators and freezers came onto the market. Increasing the quality of the product (more transparent ice blocks) would only have delayed its obsolescence.

2.5.6 Quality Management System (QMS)

Descriptions of manufacturing processes, procedures, record sheets for product control and release, quality management and quality assurance documents and their links to the strategy and policy of the Senior Management should be logically ordered. This structure is called a Quality Management System or QMS.

QMS plays a vital role in case of disputes with authorities, customers and consumers. Thanks to its transparency and the traceability of the products it generates, it administers the proof that good manufacturing practices (broadly defined), approved by the stakeholders, are in constant use.

This emphasis on producing evidence instead of prioritizing quality improvement was a major criticism of the most widespread quality management system, i.e., ISO 9001. The administrative (and mental) burden related to this system gave rise to detractors, who accused it of representing too much bureaucracy. For these critics, ISO 9001 represented only the "small Q" and not the "big Q" (Total Quality).

Revision of ISO 9001 in 2000 took account of these criticisms. Finally, ISO 9004 has made it possible to include stakeholders in the quality system, and ISO 14001 involves environmental management. Now the standard ISO 26000 encompasses aspects of social responsibility (environment, society, economy and corporate governance).

2.6 Total Quality Management (TQM)

Total quality is a set of principles and organized methods of global strategy, aimed at mobilizing the entire company to achieve better customer and stakeholder satisfaction at the lowest cost. It aims for excellence.

It concerns:

- all functions of the company or institution;
- all activities of the company or institution;
- all employees regardless of their hierarchical level (Board included);
- all customer/supplier relationships in the company;
- all quality improvements: resolution of existing problems and prevention;
- the entire product life cycle, from design to destruction (including recycling);
- all relations with suppliers, subcontractors, partners, networks;
- all existing and potential markets (vigilance, monitoring, business intelligence).

Total Quality Management (TQM) has its origins in Japan, but its name in Japan has long been Total Quality Control. This country already started laying the foundations of Total Quality in the 1950s.

The effort required for the implementation of total quality is not always perceived by the management of western companies as sufficiently useful and profitable. The same goes for the boards of directors, the members of which rarely fully acknowledge the total quality approach. In these companies, the approach to quality by quality systems, a mentality resembling financial control (with which the boards are more familiar), is therefore preferred.

2.7 Domains of quality

The progression of a quality approach makes it possible to determine domains that also correspond to the historical stages of quality, for which the ultimate step is the social responsibility of a company, which includes financial perspectives. That is why, in the Boolean diagram of the domains (Figure 2.6), total quality management (very product-driven) is distinguished from the management of social responsibility. The new version of the standard for the American Malcolm Baldridge Award prize for total quality does, however, detail this difference (see Chapter 8).

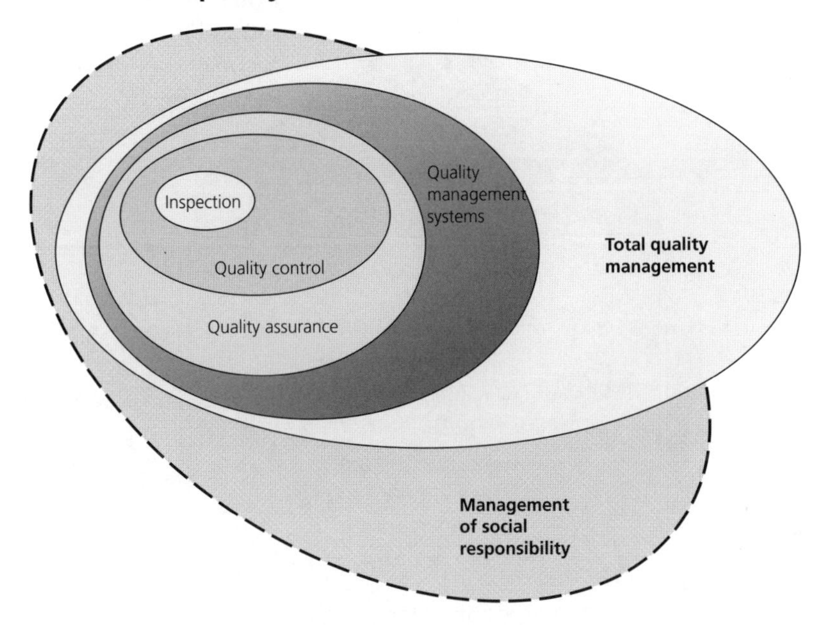

Figure 2.6 Domains of quality.

2.8 Quality Function Deployment: The House of Quality

2.8.1 Several rooms, one entrance and one roof

In his writings, the American guru of quality *Juran* says that the quality function has to be planned. What then are the instruments that enable quality to be deployed?

This process, Quality Function Deployment (QFD), is often cited as an element of TQM. The process of quality function deployment begins during product development.

The foundations of quality function deployment were laid by *Yoji Akao* of Japan in 1966 when the Japanese industry began to apply this concept. A decisive book by this author on the subject was published in Japan in 1978. However, the approach was only systematized and completed in 1990. The American promoter of QFD in 1984 was *Don Clausing*, who later taught at MIT.

According to the authors, when the developed methods are applied with skill and discernment, they make it possible to divide the product development time by a factor of two or three, while providing better guarantees that the product meets stakeholder expectations. Quality function deployment, to paraphrase *Juran*, is not for amateurs. Rather, it must be rooted in the everyday life of the company.

Quality function deployment is a complex process [2.3] that uses a document prepared by an interdisciplinary team, which includes all functions of the company. The document:

- defines as inputs the expectations of customers and stakeholders (quality criteria: QC), and
- links them to the proposed technical solutions or substituted quality criteria (SQC).

The former are provided by the marketing and sales departments as market analysis and diversification of a portfolio, and the second can be deduced from an *Ishikawa* diagram

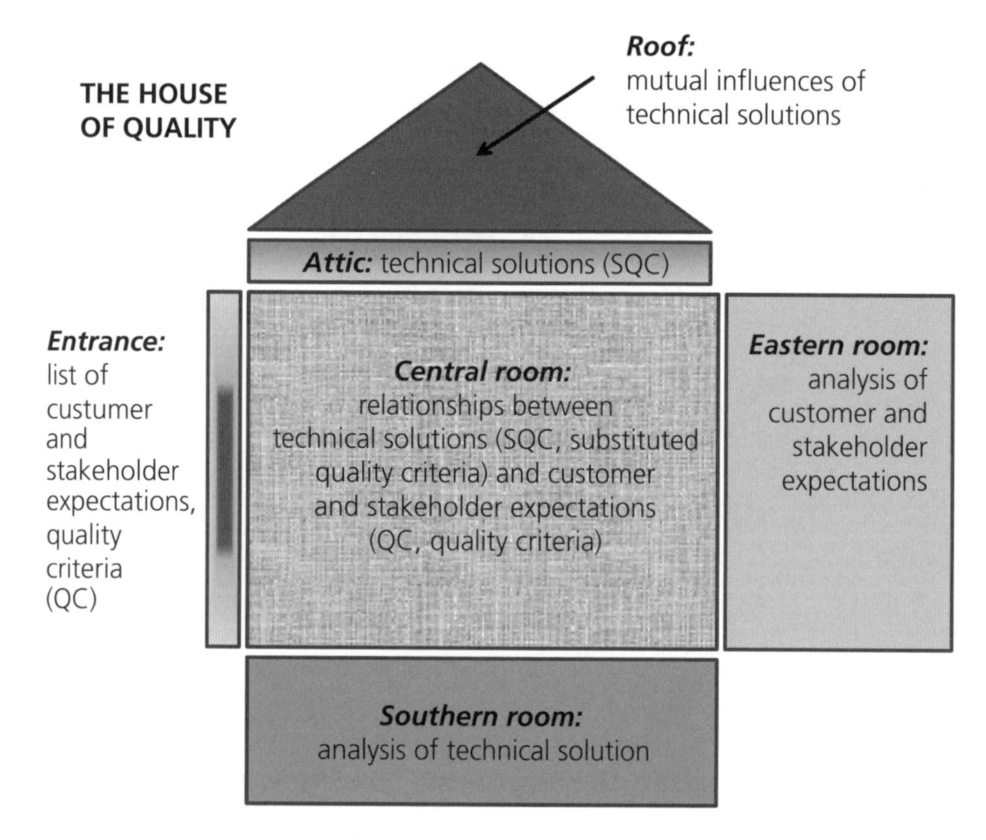

Figure 2.7 The House of Quality and its rooms.

developed by the functions R+D, production and technical support: the most common approach for the deployment of quality, integrating the many documents produced during the process, is the **House of Quality** [2.4] (Figure 2.7).

This analysis table of the quality criteria and the substituted quality criteria contains the following elements:

- *an entrance (the What?), the list of quality criteria (QC)*, corresponding to the expectations of customers and other stakeholders. When it comes to total quality, *Shiba's* algorithm is also used (see Chapter 7);
- *an attic (the How?)*, containing a list of technical solutions to expectations of the customers and stakeholders, *the substituted quality criteria (SQC)*;
- *a roof*, an analysis grid specifying the mutual influences of possible technical solutions and their level of importance; these may or may not be reciprocal;
- *a central room*, which details the relationship between the expectations of the customers and stakeholders and the proposed technical solutions (one can imagine that a technical solution may have an impact, either good or bad, upon the satisfaction of

expectations other than the one(s) targeted in the first place;

- *an eastern room*, a grid for thorough analysis and synthesis of customer expectations;
- *a southern room*, a grid for thorough analysis and synthesis of technical solutions.

As a consensual and comprehensive tool for internal stakeholders within the organization or company, the completion of a House of Quality is a receipt that customer expectations are properly considered, interpreted and integrated into the product. But the House of Quality is not a blueprint for production, since it does not give any indication of the manufacturing process. Nevertheless, it adjusts the technologies used, and may even introduce new ones (new features, "quantum" leap of performance, for instance). Moreover, the House of Quality should address a specific and sole segment of the relevant market. Some critics have stressed that the House of Quality does not give sufficient weight to cost factors, which often define the price of the service or product, and limit the services required by the marketing and sales departments. The House of Quality is a less common instrument than the *Pareto* and *Ishikawa* charts.

2.8.2 Step one, defining subsystems

Before constructing a House of Quality, the first representation of the product involves its division into subsystems and parts, and giving these descriptions (Figure 2.8). Some of them (the majority) can be borrowed from a previous product or the competition (copy marketing known as *Marketing "Me too"*).

This diagram should be kept in mind when constructing the House of Quality. If required, one may even construct several, which cater to every example or subsystem of the product or service (provided that there is no mutual influence between different subsystems). In this way, for software sold over the counter, it is possible to differentiate between the packaging, cover graphics, the user manual, the physical disk and the software itself.

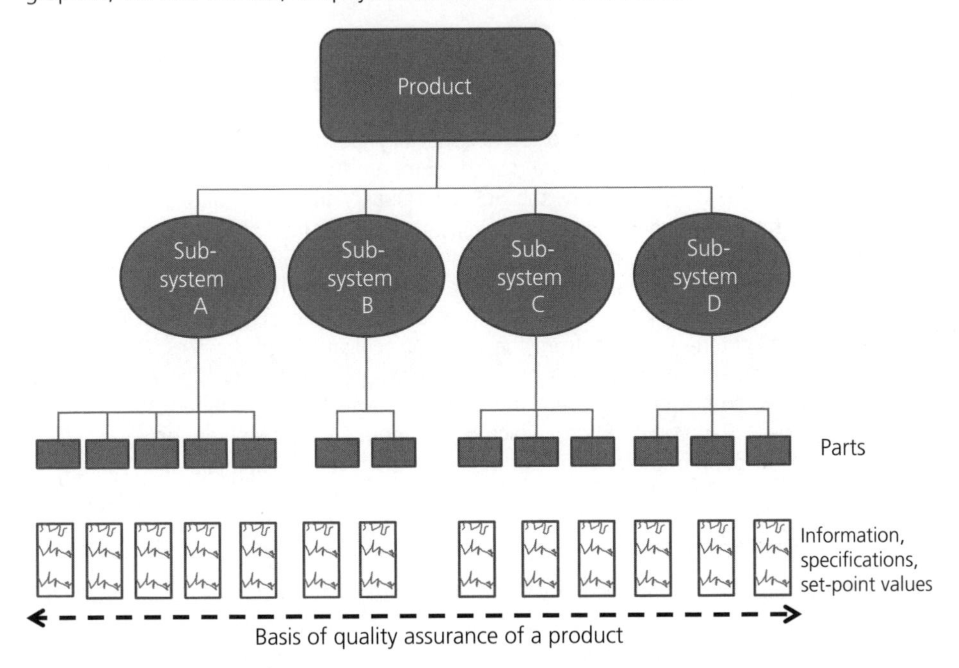

Figure 2.8 Division of a product into subsystems and parts.

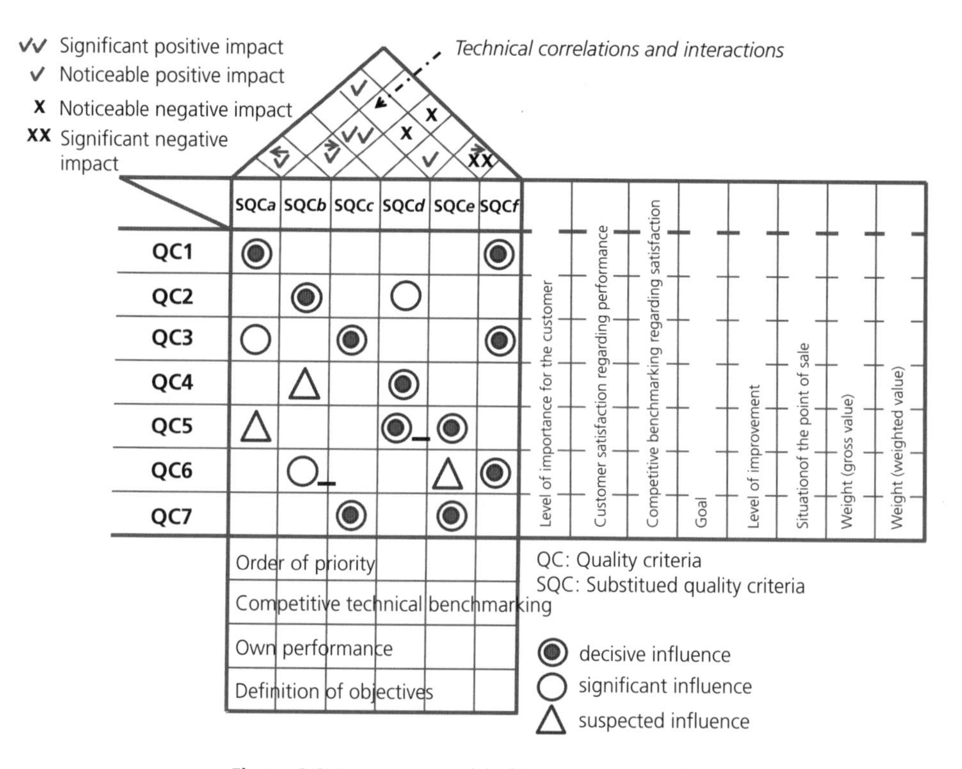

Figure 2.9 A common model of the House of Quality.

The subsystems, their components or parts, especially the information, specifications and set-points (operating conditions) associated with them, are the basis of quality assurance of a product. The information provided by the House of Quality adjusts or complements them.

Thus, in the new car model, the safety specifications of the fuel tank do not necessarily appear in the House of Quality, because they may have been taken from a previous model and do not correspond to any specific customer expectation, although it is clear that the buyer expects the gas tank not to explode in case of an accident. On the other hand, the substituted quality criteria resulting from the safety specifications of the tank are included in the information, plans and set values of the product description, which can be a lengthy document. This would not be the case if the customer expectations included, for example, the ability to drive 800 km on one tank, which would require a new tank design.

A detailed description of the elements of the House of Quality is exemplified in Figure 2.9.

2.8.3 The central room

This room describes the relationship between quality criteria and the proposed technical solutions. It also includes an assessment of the strength of these relationships.

It is intuitive to match a customer expectation with an appropriate technical solution, but this approach is only the beginning of product design: in fact, a technical solution can affect many expectations, and this in a decisive (symbol: dotted circle, point value of 9) or significant (symbol: circle, point value of 3) way. In the case where this interaction is suspected but not proven, the symbol used will be that of a triangle (point value of 3), which will be replaced once a series of tests has been performed.

It is also possible that the technical solution negatively affects another expectation (the dotted circle, the circle or the triangle are then followed by the sign – "minus"). In this case, which is not favorable, it must be ensured that the impact of all technical solutions on the expectation remains positive, otherwise another technical modification must be sought.

2.8.4 The roof

This part, also known as the technical analysis section, correlates the relationships and interdependencies of these technical solutions. This influence can be positive (symbols used: strong interaction: **vv**, significant interaction: **v**) or negative (symbols used: strong interaction: **xx**; significant interaction: **x**). However, this relationship is not necessarily reciprocal. For example, the air-conditioning system of a vehicle affects its weight, but the reverse is not true. That is why, when the influence is not mutual, the sign is topped with an arrow that specifies the direction of the relationship (Figure 2.9).

One advantage of the analysis of the roof is that it identifies the interactions that the team members belonging to the purchasing department, technical support, R and D or production, must implement to solve the problem. It highlights the interdependence of the departments. For this, it is possible to use a matrix representation specifying the degree of involvement of each unit (Figure 2.10).

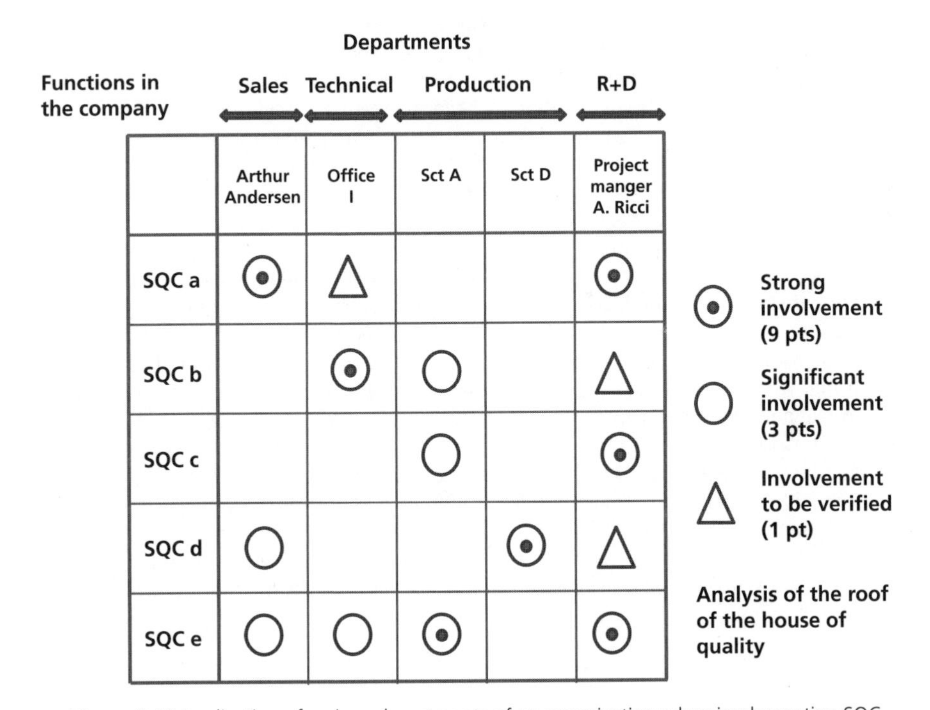

Figure 2.10 Implication of various departments of an organization when implementing SQC.

2.8.5 The southern room

The southern room includes four lines that deepen and clarify the relevance of the proposed technical solutions:

- *The order of priority*: this calculation is relatively complex. It consists of multiplying the value of the interaction of an SQC (9,3 or 1) with each QC by the weighted value of the weight of each expectation (last vertical column of the eastern room, see Section 4.5). The greater the result, the better the substituted quality criteria are considered to satisfy the customer expectations.
- *Technical competitive benchmarking*[9]: do the competition's products have identical or similar technical solutions? If not, are there others? If so, are they efficient? This line summarizes the information collected.
- *Own performance*: here is listed the current state of the product and service, especially after comparing with competing products.
- *Definition of objectives*: in this line we find the set-points of the substituted quality criteria, after analysis of the technical features of the competition. These values can also be determined by planning experiments based on quality engineering by *Taguchi* (Chapter 8: Total quality, the leading personalities).

2.8.6 The eastern room

The eastern room consists of eight lines that deepen the relevance of customer expectations reflected in the House of Quality:

- *The degree of importance for the customer*: this line, resulting from the results of market analysis is obtained by asking a sample of clients to fill out a questionnaire to determine the importance of their expectations (e.g., on a scale from 1 to 5).
- *The performance of customer satisfaction*: indicates the current level of satisfaction with the existing product for all customer expectations, for instance on a scale from 1 (very dissatisfied) to 5 (very satisfied). The idea is to take the observed average satisfaction (following a survey), with if possible the standard deviation of the distribution, possibly a "box plot" (see the Chapter on Total Quality with the Six Sigma Approach) if the distribution is not normal, or with the statistical confidence interval.
- *Competitive benchmarking of customer satisfaction*: making it possible to identify how competing products are positioned in relation to customer expectations (are the expectations satisfied and, if so, to what degree?). How are they better or worse than the existing product? Figure 2.11 shows one possible representation, which may appear vertically in the table, also stemming from the analysis of questionnaires:
- *The purpose*: the customer satisfaction that the team hopes to attain with the new version of the product and service. Marketing and sales always seek a satisfied customer with regard to all expectations (i.e., 5), but technical and financial requirements (quality at all costs, but not at any price) can modulate these requirements.
- *The proportion of improvement*: the relationship between current and targeted satisfaction of a customer with regard to expectations of a company's product or service. This proportion of improvement is sometimes juxtaposed with an adjacent column that assesses the degree of difficulty to obtain this improvement. Indeed, it is generally much easier to move from a rating of 1 (angry customer) to 3 (neutral customer), than to go from 4 (satisfied customer) to 5 (very satisfied customer), even though the

9 Benchmarking is a process of observation and analysis of practices used by competitors or industries that may have modes of function that are reusable by the company: typically, a comparative analysis of performance of the products or services.

proportion of improvement is much lower in the second case than in the first ...

- *The factor relating to the point of sale*: this is a weighting that assesses whether the expectations can be insupport of a promotional action, or a weighting argument at the point of sale (weighting, an arbitrary factor, generally varies between 1.0 and 1.5).
- *The final weight of each expectation* (gross value), which is the product of the factors below:

  (degree of importance for the customer) x (proportion of improvement) x (factor relating to the point of sale)

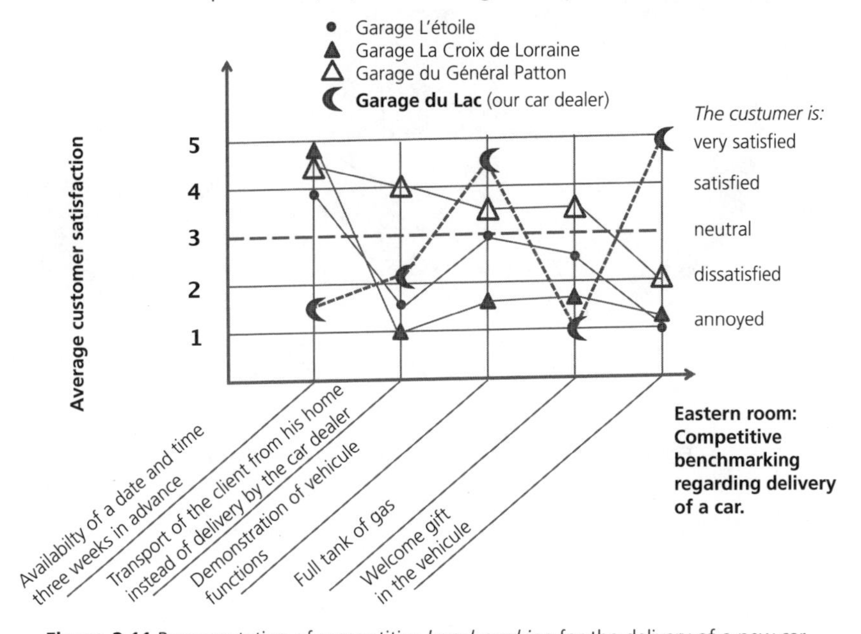

Figure 2.11 Representation of competitive *benchmarking* for the delivery of a new car.

- *The weight of each expectation (weighted value)*: this is a gross value, transformed into relative ones (the considered weight divided by the sum of weights of all expectations). This value describes a hierarchy of customer expectations and can be focused on those that appear to be essential. It is also used to calculate the order of priority, an important value in the southern room.

2.8.7 Derived matrices

The House of Quality can be translated into three other matrices by successive substitution operations. One starts from the House of Quality, cascading down to the component and its operating conditions, using the model shown in Figure 2.12. This process ensures that all technical solutions related to quality are implemented in the production: from the product to the manufacturing or purchasing of components, because the construction of the House of Quality is only the beginning of the design of the product and focuses on customer expectations, even if it is a key step.

The "Hows" become the "Whats" of the following matrix and the priorities of the "Hows" then become the priorities of the "Whats". Each matrix has only one row and one column: on the top and bottom, and to the left and right of the central room. It is possible, if it makes

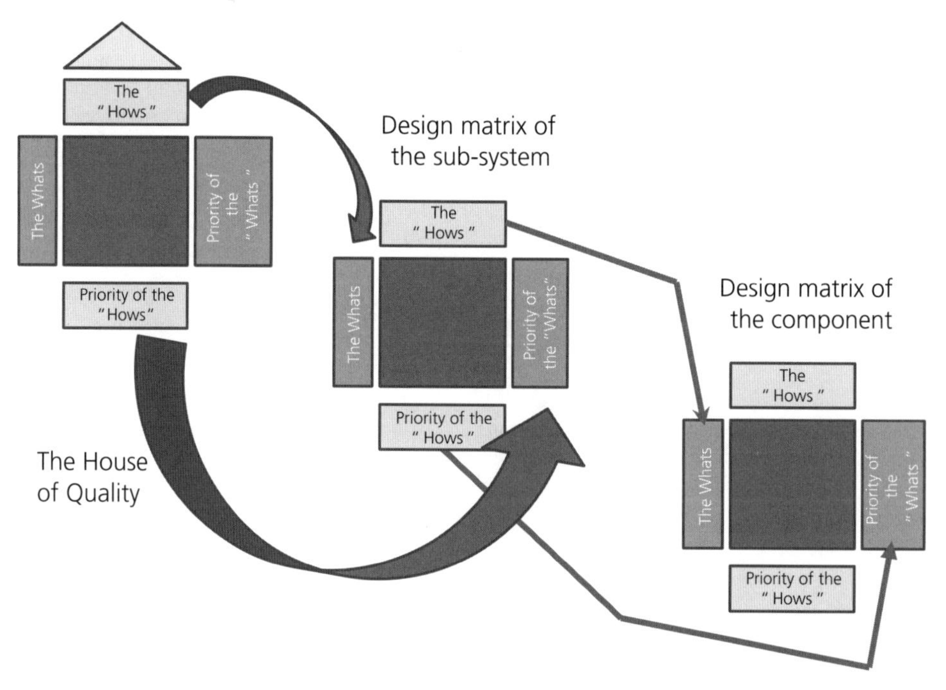

Figure 2.12 The House of Quality and two derived matrices.

sense, to add a roof. In this way, you can get a House of Quality and up to three derived matrices (Figure 2.13).

Matrix	What	How
House of quality	Client expectations	Technical solutions, CQS
Design matrix of the sub-system	Technical solutions, CQS	Characteristics of the components
Design matrix of the component	Characteristics of the components	Parameters of the process
Design matrix of the process	Parameters of the process	Operating conditions of the manufacturing

Figure 2.13 The House of Quality and three derived matrices.

For many quality experts, however, these derived matrices are not used and quality deployment ends with the completion of the House of Quality.

References

[2.1] According to the diagram III-5, p. 63, in the book by KAORU ISHIKAWA, *What is Total Quality Control – The Japanese Way*, ed. Prentice Hall, 1985.

[2.2] K. ISHIKAWA, ibid., diagram III-4, p. 59.

[2.3] YOJI AKAO, *Quality Function Deployment – Integrating Customer Requirements into Product Design*, Productivity Press, 1990.

[2.4] LOUIS COHEN, *Quality Function Deployment – How to Make QFD Work for You,* Engineering Process Improvement Series, Addison-Wesley Publishing Company, 1995.

Chapter 3

History of Quality

3.1 Introduction

Quality can be defined in many ways; it can be succinctly defined as: *Fitness for use* or also *Fitness of use*.

Its meaning and deployment are conditioned by trade, the economy, and exchange policies with regard to goods, raw materials and services. Even before the advent of writing, these exchanges were often performed over long distances.

EXAMPLES:

- Obsidian blades were produced from the sixth millennium BC in several production centers in Europe (Sardinia), but were distributed throughout the Mediterranean [3.1]. As obsidian is brittle, knives could be made extremely sharp so that even during the 20th century scalpels with obsidian blades were used for surgery. The production of obsidian blades required specification of the raw material, a manufacturing technique involving several delicate tools, and a minimum annual production of blades by the cutter/peddler to maintain a level of expertise.
- From the Bronze Age, Mediterranean people imported amber. In the burial chamber of Tutankhamen (Ancient Egypt), objects were found made of amber from the Baltic Sea.
- Over several millennia, tea and silk from Asia were brought west on the famous Silk Road, passing through Samarkand.

3.2 Quality has a history

The implementation of quality has always been in phase with the large-scale production of consumer goods, equipment (including infrastructure) and services.

Its development resembles that of physics: every growth stage includes each previous growth phase without making it completely obsolete (e.g., Newtonian mechanics is a "special case" of the special theory of relativity). Similarly, quality control procedures include product and inspection standards from earlier times in history.

To understand the evolution of quality, certain historical stages should be mentioned (bottom-up approach):

a) the establishment of standards and valid measurements as references for a community: weight, length, area, volume;
b) the manufacture of standardized products by blueprint, sizes and standards (norms) for a community of customers or for the purposes of logistics or power;
c) product inspection – the quality function is born;
d) quality control;
e) quality assurance and associated tools (capability, reliability, etc. ...);
f) total quality (total quality control, Deming prize, Baldridge prize, EFQM diplomas and prizes);
g) quality management systems such as ISO 9001;
h) development of QA systems that apply to the environment (ISO 14001);
i) development of management systems (security, IT) inspired by the ISO 9001 approach;
j) integration of professional deontology in quality assurance (under the influence of service development);
k) integration in the quality system of ethics or the social responsibility of a company;
l) quality as a policy in a private or public company, covering its strategic and operational management.

3.2.1 The establishment of standards and measurements

(Valid as standards for a community.)

3.2.2 Predominance of self sufficiency

Even though trade has existed since the dawn of time, it was reduced by the low level of consumption in communities, living self sufficiently and mostly practicing barter, but also by poor roads and the dangers of communication routes.

The introduction of currency brought flexibility, but coins were often scarce, as they depended on the supply of gold. Paper money was put on the market by the Law Bank in the second half of the 16th century (even though the first paper money in China dates back to the 10th century). Moreover, the generalization of paper money in the form of bills really began in the early 19th century, and therefore accompanied the industrial revolution.

Since families depended mainly on agriculture, they made their own clothes, cooking and washing utensils. Furniture was scarce, sometimes made by the users themselves. Many homes (shacks) were built by the farmers themselves. The largest farms were built with a carpenter. People went to the local blacksmith when pots and rare metal objects were needed. The place of trade was therefore mainly the local market: customers and suppliers were known, so the few products and their faults were quickly identified by the seller and the buyer in a context of general misery.

3.2.3 Establishment of the first kingdoms and empires

Agrarian kingdoms and empires introduced standards and measures, in parallel with a communication system, generally roads, aimed at ensuring safe passage.

Before the desire to trade, it was the quest for power (collecting "fair" taxes, repressing fraud during these operations, having strategic needs for reliable and rapid transport of troops) that hastened the establishment of measurement units (value, length, area, volume, weight, time), of standardization (such as the width of the axles of carts to facilitate road construction) and of communication systems (isn't the internet the best example yet?). Commercial trade followed.

Systems of weights and measures probably existed in the third millennium BC, in Mesopotamia, Egypt and the Indus Valley. Length units were generally related to key dimensions of the human body (length of the foot, thumb; the ell as a unit of measurement was the distance from one end to the other of both arms extended, etc.). Other measurements, such as the extent of land, referred to a work unit (such as the surface that a farmer could plow in a day, about twenty acres – a unit of measurement that was still in use in France under the *Ancien Regime*). As these measurements varied depending on the size of the individual (or his labor), a finer standardization was necessary. This was implemented by the great empires.

History records no sharp upswing in the standardization of weights and measures, as Germany, for example, had different systems even in the 19th century. The reform of the English system dates back to the early 20th century. Worse, although the MKSA system is in force worldwide, the flight altitudes of airliners are still given in Anglo-Saxon feet!

The first large-scale organizations

As an early outstanding example, the Chinese empire had a bureaucratic system probably as early as the first millennium BC. However, it was not until the second century BC that the system of hiring (and promotion) based on merit (initiated during the Sung dynasty in 962 BC) became generalized, under the influence of *Confucian* thought. This system will long survive despite the pressures and intrigues of the existing hierarchies and the force of corruption. However, civil servants were hired exclusively based on their (academic) results during entrance examinations: Chinese leaders quickly noticed that this was perhaps not the best way to guarantee future managerial talent. Roman and Egyptian empires also used sophisticated hierarchical systems.

Yet it should be stressed that the Chinese bureaucracy was at odds with Confucian thought: it used a system of rewards and punishment to guide the work of its civil servants toward good. *Confucius*, on the other hand, advocated the development of virtues and moral exhortation to obtain cooperation. This conflict between the two schools, sometimes called formalist and humanist, is still ongoing, particularly in the field of quality.

3.2.4 Manufacturing standardized products

(For a community of customers or for purposes of logistics or the desire for power and control.)

Ancient and Medieval times and the Renaissance[1]

Since antiquity, and particularly during the Roman Empire, architects, craftsmen, teachers and even doctors received no social consideration – they were often slaves or freedmen, such as the Stoic philosopher *Epictetus*.

The concept of geometry in ancient Greece was used to a greater extent in politics than in technology [3.2] – which was considered as the work of slaves. This situation was not

1 A study could be devoted to China, which early recognized the importance of standardization for common objects (crockery, porcelain, fabrics) as part of its trade policy.

conducive to the research of new technologies unless they gave an immediate military or political advantage (as in the case of *Archimedes*).

This continued in feudal society, but also for a long time among the Arabs. Thus, in the seventh century, an Arab patient simultaneously consulted a doctor and an astrologer. The latter was closer to the eternal values of the Cosmos, so had precedence over the physician, who was a man of contingent things, at the mercy of the mutability of the elements. The doctor could only act during the times prescribed by the astrologer [3.3], which was fatal for many patients.

Before the industrial revolution, the need for standardization of products was shown in the following areas:

- construction of vessels;
- construction of roads and fortifications;
- construction of prestigious art buildings or religious structures;
- manufacture of weapons, particularly firearms.

During the reign of Louis XIV, initiatives were launched:

- *Jean-Baptiste Colbert* (1619-1683) was Controller-General of Finances in France from 1665 to 1683. His role under the leadership of Louis XIV was to give economic and financial independence to France. *Colbert* focused on the economy through protectionist and interventionist government policies designed to increase internal wealth. To achieve this goal, he increased exports from France, pursuing excellence in manufactured goods through the implementation of drastic manufacturing processes and stringent regulations, by introducing quality control of all manufactured goods, and by creating royal factories. In 1670 he created a body of inspectors able to impose sanctions. *Colbert* registered in law standards for the quality of many articles, and any deviation from this standard was punished by public exposure of the offender and the destruction of the goods. If this lack of quality was repeated three times, the craftsman was pilloried. The virtues of this system remain controversial, as some historians believe that its rigidity hindered all creativity and the manufacture of inferior products that would have found buyers at a lower price. Defining and ensuring a quality standard cannot easily be the task of the state. *Colbert*, however, created more than 400 factories, which are considered as the beginnings of modern industry. Under *Colbert*, engineering of major works was also born.

- *Sébastien Le Prestre de Vauban* (1633-1707) was a military engineer, commissioner general of fortifications and a predictor, by his thoughts and work, of the Age of Enlightenment. He surrounded France with a belt of fortresses (approximately 100), a buffer that made the country invincible. The innovative architecture of these fortresses meant that they were not obsolete until the end of the 18th century. The aim was not to make the fortresses impregnable, but to save time by forcing the attacker to use ten times more men than those of the besieged. *Vauban* was of a systematic and practical mind: by performing overall timing of the earthworks included in the fortifications, he paved the way for the rationalization of work that provided a comprehensive approach to a construction site as a whole, and led him to attempt to unify and standardize construction techniques.

In many cases, such as the Arsenal of the Republic of Venice in its glory days from 1450 to 1670, which regularly employed 2000 people, a first division of labor was noticeable, as well as a hierarchical organization. However, key craftsmen (and since the Renaissance, engineers) were also used. These groups were divided into two classes: masters and apprentices. The masters' knowledge was imparted to the apprentices sparingly, and much was kept secret, either at an individual level, or within a professional association, also known as a guild. A breach of confidentiality could result in severe penalties.

The role of guilds and craftsmen (companionship, corporatism)

Learning was reserved for the sons of craftsmen or rich peasants, because the wages given by the supervisors were meager and the training long (7 years or more according to the *Regius Manuscript* which can be consulted still). This must be compared to the expected lifetime that barely exceeded 35 years. The guild was a privilege of the masters. Guilds did not promote competition among craftsmen: the size of their business was often regulated and the price differences between products or services were small. The craftsmen grouped together according to neighborhood.

To consider the guilds as corporatist and servile structures compared with nobility would be inaccurate: for example, the 1566 statutes of the dyers of Bourges in France indicate that after an apprenticeship, the apprentice was knighted [3.4]. The social position of a dyer is not that of a base craftsman. Here we have a mixed art including both mind and skill, and more mind than skill [3.5]. Thus, belonging to a guild meant taking pride in one's trade and committing oneself to excellence (which is also one of the bases of the culture of total quality).

Although some changes occurred locally, the productive function of Europe and the United States was based exclusively on the knowledge of craftsmen up to the late 18th century. A craftsman made the quality of his work known by means of a mark (stone cutter, manufacturer of paper, china, kitchen utensils and knives, for example) registered by the regional guild, often recognized by the state, and for which the nobility could provide a guarantee.

This mark of quality was one of the first components of product traceability. It was relative, because one of the first successes of Swiss watchmakers in the 18th century was that of counterfeiting. They were selling watches and clocks under English names [3.6] – the market leaders at that time. The term "Swiss made" still had its best days ahead of it (as did the protection of premium Swiss brands[2]).

The degree of power held by a guild varied: it was very significant if the product was considered vital (especially weapons and boats). The guild could issue standards, but the craftsman was nevertheless allowed a certain flexibility of implementation. The quality of the product was entirely the responsibility of the craftsman. Any noncompliance gave rise to penalties – death by immersion in a river for the baker who sold bread that weighed less than the standard in Germany, for example.

In terms of design, plans or models showed only the basics and were therefore approximate. Much was left to large-scale tests, which introduced errors, delays and costs. The estimation of costs and delivery time was difficult, so traditional well-tried methods were preferred and risk-taking careful. The system of guilds and crafts was not very innovative: some discoveries certainly disappeared with the death of their inventors. Competition existed between city guilds, but this competition concerned only easily transportable products, given the poor condition of the transport network.

The guild system persists in current professional associations, especially the liberal professions. It is also the basis of regulated professions.

2 History does not reveal whether the United Kingdom had established criminal procedures for citizens bringing these Swiss counterfeits within its borders...

Constructions of the Middle Ages

One of the major innovations of the Middle Ages was cathedrals, which were built as power symbols of the Church, the image of society, but also as centers of attraction, places for networking and exchange in a city with a bishop. After Mass, the transactions could begin. Guilds and merchants financed in part the construction of cathedrals and, in return, wealthy donors had their faces shown on a prominent statue of a saint [3.7].

Many types of stone items did not require mathematical knowledge for their manufacture. The object could vary in shape or volume. The guilds were weakly hierarchical. However for cathedrals and castles, the solidity of the walls essentially depended on the strict size of the blocks in the form of parallelepipeds. This required knowledge of elementary geometry, but also a study of available quarries and specific methods of extraction. Standardization was welcome because, after a demolition, the blocks were reused (a nightmare for archaeologists when attempting dating) [3.8].

An organization was thus in place. Representations in the form of scale plans were drawn on the walls of the building to facilitate vertical communication. The time it took to build cathedrals, however, shows that the number of workers on each site was small, simplifying their management. The architect often played the role of foreman. It is therefore not surprising that the first preserved code of conduct and organization of a guild is a masonic charter, the *Regius Manuscript* (1390) [3.9].

In 15 articles, the *Regius Manuscript* limits the apprenticeship to a maximum of seven years. It recommends the practice of the seven sciences, calls for honesty and loyalty, and regulates the conditions of exclusion from the guild. It can be considered as the first model code of conduct of a trade association.

3.3 The Industrial Revolution and its effects

3.3.1 A decisive factor: the steam engine

The Industrial Revolution began in the UK in 1770 and was primarily an energy revolution: the replacement of wood and charcoal – the latter was vital for the production of iron[3] – with mined coal (the patent was filed in 1701). The widespread construction of steam engines was the second step. This invention dates back to antiquity, to the aleolipile, a spherical pot with a tight lid from which rose two hollow arms causing a steam jet effect and allowing the sphere to rotate. It was attributed to *Heron of Alexandria*, during the second century AD, and was long considered a mere curio.

Its first application was implemented by *Thomas Newcomen* in 1711, and involved pumping water that infiltrated the tin mines of Devon (UK). This was an atmospheric steam engine as it used air pressure to move the piston. Its principle was based on the condensation of steam (injected into the piston chamber) by a cold water source (Figure 3.1). This machine was, however, very energy-consuming, because the piston chamber had to be heated and cooled for each cycle. These machines were almost exclusively used in mining, in regions where fuel prices were low enough.

But it was *James Watt*, whose profession was the manufacture of scientific instruments, who introduced several improvements as of 1765. He invented a new machine, also atmospheric, by juxtaposing a cooled condenser to the piston chamber, which in turn was heated by steam. The energy savings were such that the steam requirements were reduced by over 60%.

[3] It is impossible to melt iron at the temperature obtained with a wood fire.

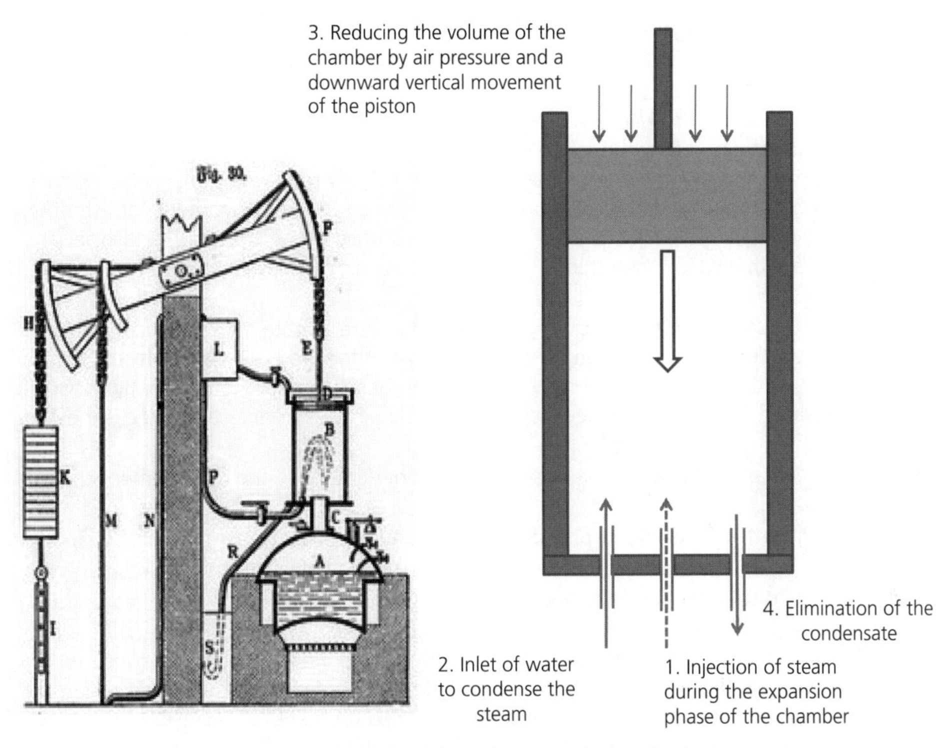

Figure 3.1 Newcomen's atmospheric steam engine and its 4-phase cycle.

Until 1782, steam engines were, however, used exclusively for pumping water and ventilating air for metallurgy. In that year, *Watt* included several enhancements, transforming the to-and-fro action of the piston of the machine into a circular motion. This innovation led to the mechanization of many other activities: elevators and lifts for coal and ore mines, steamboats and locomotives (patented by *James Watt* as early as 1784), power supplies to mills, breweries, and various machines (for the textile industry or the lathes of various craftsmen, etc.).

The first Industrial Revolution, dating back to 1780, did not take place *ex nihilo*. It was preceded in 1730 by the Agricultural Revolution, which increased productivity, leaving farmers and laborers without work. They formed the first proletariat. The Industrial Revolution was one of techniques and mechanization, leading to the emergence of scientific discoveries in the world of production. Man was gradually replaced by machine, and the work was aided by a previously unknown energy, originating from coal, provided in workshops where craftsmen worked side by side because of the cost of energy production.

3.3.2 The first managers were born

The energy supplied by the steam engine lowered production costs and the sales prices of manufactured goods, leading to an increased market size. Market needs accelerated the demand for labor, machinery and large-scale steady production. This development accentuated the demand for capital, but also that of people willing to manage and coordinate resources (human capital, machinery, raw materials, transportation, real estate partners, etc.).

The position of manager was born, even if the first were primarily entrepreneurs with a stake in the company. They had to face several difficulties connected with the first attempts to organize factories:

- *The recruitment of skilled workers*: the majority of available human resources were agricultural workers without qualifications, who were torn, not without difficulty or pain, from their rural communities and families, and forced to comply with schedules and often monotonous factory tasks, and to the constant requirement for attention to their work. Consequently, the work force tended to be unstable, antisocial and often apathetic. To make things worse, the craftsman guilds tended to boycott factory work.
- *Training in the workplace*, as very few workers were literate, or knew how to draw. There was no standardization of tasks and most of the workers were instructed orally by colleagues who did not know much more than they did. Consequently, workers specialized in certain tasks because of lack of skills; the efficiency and effectiveness of the whole were therefore limited.
- *Discipline and motivation*, because factory work demanded the development of new qualities:
  - regularity of work as opposed to the fluctuations of farming;
  - accuracy and standardization rather than the individuality of the craftsman with regard to design and method.

  These habits required getting used to and absenteeism was high, especially during Christmas. The workers often worked intensely for periods but then disappeared for a while once having received their pay. Often, penalties were put in place to curb absenteeism. It was hard to reward workers based on merit by increasing their salary, because of the belief, dating back to pre-industrial times, that argued that the best way to lose a worker was to pay him better, since the immediate consequence would be his idleness.
- *The search for competent managers*: when the size of the company grew, the contractor had to find people to replace or assist him. Often, these would be illiterate workers who had been promoted because of their technical skills. They often accepted this position, which was slightly better paid (salaries were rarely commensurate with responsibilities and achievements, but with social class), as it made it possible for them to hire their wives and family members. Entrepreneurs also hired family members, assuming their integrity, or clerks, tellers or bank employees in the hope that their knowledge of a business could be transposed. Training was non-existent.

The misery[4] of the working classes created by the Industrial Revolution had an impact on global politics: the advent of Marxism and later Marxist-Leninism. This had disastrous consequences in the two countries that at the time had an imperial system: Russia, as of 1917, and China (and its satellites), as of the end of World War II. Not to mention the indirect influence of the USSR in the advent of fascism in Italy, Germany, Spain and Portugal before the war[5].

The Industrial Revolution was not insensitive to quality, but efforts were largely invested in capital equipment, whereas innovation was more technical (new processes, adjustment of operating conditions using analytical methods, new materials and formulations, patents, etc.).

---

4 Have you ever wondered why Jack the Ripper ravaged Whitechapel and why Sherlock Holmes carried out some of his investigations there?

5 Proponents of ultra-liberalism, sometimes not inclined to respect human rights, minimized this impact.

3.3.3 Standardization, the first industrial success in the US

The first rational division of labor to increase productivity appeared with the Industrial Revolution, although it was not widespread. Factories saw the light of day. As early as 1776, *Adam Smith* in *The Wealth of Nations*, noted that the production of pins included 18 steps, each of which could be assigned to a semi-skilled worker. This would lead to significant gains compared to those obtained by production using craftsmen.

For over 70 years, the industry of Great Britain was the undisputed leader in the development of the economy of Christendom. This was mainly due to its textile industry, which was also protected for a long time. We recall that the British Empire prohibited Indian craftsmen to spin fabric so that the Indian population was forced to buy clothes manufactured in Britain with raw materials from their own country.

The industrialization of the United States followed that of Britain but, in about 1836, the British still enjoyed better organization and management of factories. This was because the UK had introduced a series of barriers to limit the dissemination of knowledge: workers did not have permission to emigrate to the United States and it was prohibited to export production machines. However, living standards rose in the United States: life expectancy increased from 34.5 to 38.7 years from 1789 to 1855.

In 1851, on the occasion of the World's Fair in London, the American industrial genius showed its teeth [3.10]. Visitors could discover the pickproof locks by *Alfred Hobbs*, the loom by *Isaac Singer*, the revolver by *Samuel Colt* and the mechanical reaper by *Cyrus McCormick*. Not only were these products superior, but all parts were carefully standardized: the parts were interchangeable, a worker could therefore draw a piece from the stockpile at random for assembly. This breakthrough, called the "American system of production", was largely the result of an initiative by the US military at the armory in Springfield (Massachusetts) in 1820. The first machine made of standardized components, however, emerged in the United States in 1793 (a cotton ginning machine), and the concept of interchangeable parts can be attributed to *Gutenberg*, around 1640, with the invention of mechanical movable type printing.

But standardization had limits, given the poor qualities of the methods used. A new profession was born, the adjuster, a resourceful and skilled worker who overcame assembly problems as he went along. This entailed additional costs. In the late 19th century, France was the market leader in the manufacture of automobiles, using a process from an emerging technology; automobiles were still produced to order in small quantities and at elevated costs. Each car was a prototype and its components were developed by craftsmen in their workshops, brought to the place of assembly and adjusted.

3.3.4 A boost for American management: trains

The development of the rail network in the United States began in 1830, and it became widespread at the start of the second half of the 19th century. Its development was associated with the introduction of the telegraph by *Samuel F.B. Morse*. The complexity of its operations, the network, the size of its infrastructure, and the geographic areas in which the railway companies were operating, were such that their size and number of employees, as well as the capital requirements, could not be compared to American manufacturing industries, which rarely had more than 250 employees. These companies also faced a high number of accidents. In fact, as of 1850, the second industrial revolution was underway. Unlike the first, which used coal and the steam engine, and was focused on manufacturing, especially cotton, the second revolution developed steel, electricity, rail networks and chemistry.

It was *Daniel McCallum* (Figure 3.2), a Scottish emigrant, who gave a new dimension to the organization of the *New York* and *Erie Railroad Company*. He was appointed General

Superintendent there in 1854. *McCallum* advocated that good management must be based on [3.11]:

- good discipline;
- detailed descriptions of the duties of the employees;
- regular, frequent and accurate performance reports;
- payment and promotion on merit;
- clear lines of authority between the superiors and their employees (he was one of the first to design and use an organizational chart);
- improved self-discipline.

Figure 3.2 General Superintendent *Daniel McCallum.*

His management principles were the following:

- appropriate division of responsibilities;
- sufficient authority given to each employee to carry out the duties entrusted to him;
- rapid communication of incidents in order to address them promptly (intensive use of the telegraph);
- optimization of reporting so as not to drown the line managers with too much or unnecessary information;
- adoption of an information system that not only allowed the hierarchy to rapidly detect incidents, but also to identify those responsible.

The function of each employee was accurately described and was illustrated by a uniform or a distinctive sign. The employee was also instructed on what he was not entitled to do.

The *McCallum* system was not well received by company employees, but his methods spread widely across the railway companies. Several young ambitious men used them to learn the basics of management: one of them was *Andrew Carnegie*, who was hired by the *Pennsylvania Railroad*. At the age of 24, he managed to quadruple the traffic, double the mileage and adjust the costs in such a manner that his division was the most efficient of the whole company. However, he gave up a promotion and left the railways in 1865 to become an entrepreneur.

3.3.5 The emergence of American industrial predominance

Andrew Carnegie was to become the tycoon of the steel industry, accumulating a considerable fortune and giving American philanthropy one of its best chapters, if not its *lettres de noblesse*. In the last decade of his life, he donated his entire fortune and built libraries; he financed many educational institutes and universities, one of which would become the Carnegie-Mellon University of Technology in Pittsburgh.

Cast iron is sensitive to impurities and its properties can vary significantly (poor quality was responsible for several bridge accidents). This is not the case with steel, the manufacture of which was developed in Britain by *Sir Henry Bessemer* in 1855. *Carnegie* quickly saw the advantage that steel could provide and decided to focus exclusively on its production.

Carnegie was the first to use accounting to assess and reduce costs (while his competitors did not assess their situations at all). He developed vertical integration, simplified operations by eliminating the middleman, increased produced volumes and developed an impressive sales force. A careful integration of all operations (from extraction in the mine to sales) made it possible to accelerate the production process.

When *Carnegie* began his business, one ton of iron rails cost 100 USD. When he sold his business in 1900, one ton of steel for rails cost 12 USD [3.12]. In 1868, Britain produced 110,000 tons of steel and the United States barely 8,500. In 1889, twenty years later, the US was the leading producer of steel, and in 1902, the volume produced by them exceeded 9 million tons, while Britain was left in second place with 1.8 million tons. The American industry gradually started to show its supremacy (synopsis, Figure 3.3)

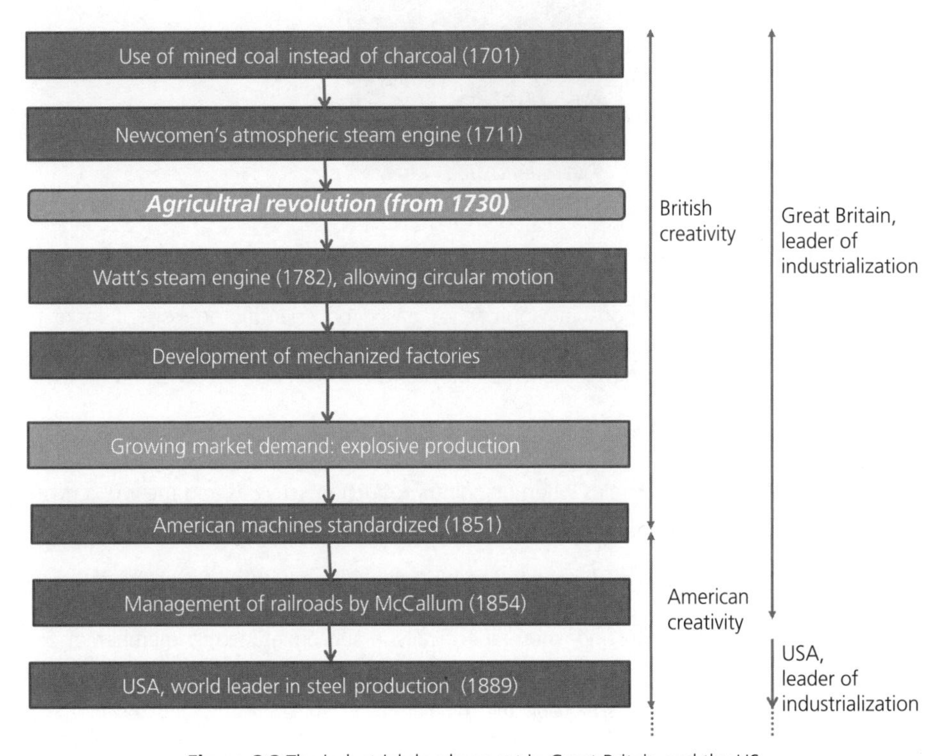

Figure 3.3 The industrial development in Great Britain and the US.

3.4 Generalized product inspection – the birth of the quality function

3.4.1 Management of engineers, Taylorism

Given the rapid development of the infrastructure and size of companies employing engineers, US engineers frequently became managers during the second half of the 19th century. The first professional engineering organizations in the United States date back to 1852 (civil engineers). Management would (very gradually) become one of their concerns, especially for mechanical engineers, whose association was born in 1880.

The emergence of quality as a distinct function of production occurred soon after the publication of the book *Principles of Scientific Management* by *Frederick Winslow Taylor* (engineer) (1911). These principles were applied with worldwide success by *Henry Ford* (also an engineer). The book's basic concepts[6] involve the division of labor and capacity measurement.

Figure 3.4 F.W. Taylor.

Taylor defined four principles of scientific management in opposition to the production logic of the 19th century:

- replace empirical methods by approaches based on scientific analysis of tasks to accomplish; this affects both the organization and the work tasks to be performed;
- scientifically train and develop each employee (called worker) instead of having them "learn by doing";
- cooperate with the workers in order for the scientific methods to be implemented;
- divide the work between managers and workers, so that the managers apply scientific management principles when planning production activities and the workers are dedicated to their productive tasks.

Taylorism, or line work, was denounced as inhuman (along with its counterpart, Soviet Stakhanovism, in Marxist countries) since it reduces the employee to the status of an unskilled laborer, a proletarian and, more specifically, a sub-proletarian, a member of the

[6] The first publication by *Taylor* is probably the communication, *A Piece-Rate system, Being a Step Toward Partial Solution of the Labor Problem,* that appeared in the Transactions (No. 16) of the Association of American Mechanical Engineers in 1895.

Lumpenproletariat (an expression of *Karl Marx* and *Engels* that defines a status close to the current working poor who work on call). Moreover, the employee's work is reduced to a monotonous and mechanical activity. Nevertheless, this method of organization paved the way for mass production at low prices for all classes. In the 1980s and beyond, much of these activities were automated.

The customer quality approach of Taylorism is that of engineers, decoupled from the sales function. This results in ***production marketing***, that is to say for a unique product. Hence the famous phrase allegedly uttered by *Henry Ford*: "The client may have the color of the car he wants, provided that it is black" [3.13]. (The reasons for choosing the color black are still under discussion: the lowest paint price or the effect of Puritan values cultivated by *H. Ford*). In Fordism, Taylorism is associated with another energy revolution, that of oil.

Production marketing is often outdated, but continues to exist for products whose price and availability are the main values of appeal, or in the case of an innovation or unique selling point, without the possibility of market comparison (budget brands, retailers, discounters such as the German chain stores *Aldi* and *Lidl*, first cell phones). Curiously, some "budget" products are appreciated by the consumer (very low costs of marketing and promotion, composition and product packaging to the point, good functionality).

For *Taylor*, as an engineer, the aim of management is not only to reward shareholders. It should provide maximum prosperity for the employer and employees. Salary is not the only prosperity; *Taylor* actually believed that it includes the professional development of the worker to the height of his abilities. Scientific management will make it possible to preserve the identity, originality and personal initiative of the worker, while working harmoniously with colleagues and under the benevolent control of its leaders [3.14]. *Taylor* can thus be seen as the father of paternalistic socialism.

3.4.2 From Taylorism to Fordism

Fordism is born

This utopian vision was embodied by *Henry Ford* at the age of 50. We call it Fordism [3.15]. *Ford* created the functions of manager and worker by giving the latter tasks divided into small parts, and so drastically increased productivity. We sometimes talk about the appearance of a third industrial revolution with Fordism, based on the use of oil, the development of roads, and the production of motor vehicles.

Figure 3.5 *Henry Ford.*

In 1913, the first assembly line appeared in the Ford factory, and in 1915 *Ford* drastically increased the workers' wages to 5 USD per day (an incredible amount at that time), expecting gains in productivity and market growth to support this increase. This wage of 5 USD per day was a success with the working classes and made *Ford* a hero of the American nation, revered and widely listened to.

In the reality of Fordism, workers' labor consists of the tiresome repetition of simple tasks. There is thus a multiplication of functions within the factory. An assembly line brings the car to the construction workers, whose movements are restricted to a minimum – replenishment of components for example.

Within this framework, the quality function of Fordism is to detect defects in finished products or components. This is carried out by officials, especially foremen, who inspect production lines. Quantity and quality appear to be linked by the work organization and the use of standards, templates and measurements, but the reality of the organization is different: an inspection department or office appears as a separate function of production, while being fully integrated into the company. *In this way the quality function was born in the United States in the first quarter of the 20th century.*

One of the reasons for the success of Fordism is that it accommodated poorly trained and even disabled staff. In his memoirs, published in 1922, *Ford* notes [3.16]:

a) that 43% of jobs in his factory needed only 1 day of training, 36% required 1 day to 1 week, 6% required 1 to 2 weeks, 14% required 1 month to 1 year, and 1% 1 to 6 years. With the simplicity of the assigned tasks, Fordism ruined apprenticeship training. The current poor professional level of those Americans who have not benefited from higher education is evidence of this fact.

b) That out of the 7,800 jobs in the factory, only 950 required men with above-average physical strength, 3,340 required a normal force, and the rest of the work could easily be accomplished by pre-teens and women.

c) That during his last census, his workforce counted 9,100 employees who did not conform to standard, including 1,000 patients with tuberculosis (assigned to tasks outside the factory in special halls since they were contagious), 120 one-handed people, 4 blind and 500 one-eyed people, 37 deaf and dumb persons, 60 epileptics, and 234 one-legged people. *Ford* also stated that hospitalized employees were able, once their beds properly fitted and provided that they could sit, to work as well (or better) as a healthy employee. Fortunately, this last observation never left the testing stage! ... For *Ford*, all these people could work – one just had to find the right work for them to do with their disability. This was an improvement[7], because insurance did not exist at that time, and a disabled worker found himself unemployed and was an expense to his family... if the family could support him.

The Ford factory functioned satisfactorily even though the majority of workers did not speak English. Indeed, many of them were first generation immigrants. But the T-Model was robust and could be repaired with simple tools by anyone who had knowledge of mechanics.

Thanks to *Ford*, who gave birth to the quality function, the United States had at its disposal organizational and industrial engineering leadership for about 50 years. However, the birth of inspection departments occurred against the will of production managers, who did not want to be deprived of that jurisdiction. Also, if a faulty product arrived on the market,

---

[7] On construction sites of the Middle Ages, according to some sources, the following rule reigned: being blind, lame or hunchback resulted in an exclusion from the construction site.

the inspection department was responsible. Quality thus became the responsibility of solely the inspection department.

But *Henry Ford* and his Board allied the desire for absolute control, domination and perfect management accepted by all, with the paternalistic socialism of Taylor. It was in this atmosphere that the quality function was born. The inspectors had counterparts outside the factory. Indeed, 150 intelligence officers, gathered in the social department of the company, prepared regular reports on the private life of each employee, especially their morals. Any deviation was deemed synonymous with dismissal. *Unpredictable and autocratic, perfectionist, compulsively orderly, having received a puritanical upbringing and praising work to the highest moral values – was this Ford's state of mind when introducing quality?*

In 1916, the Ford company produced more than 400,000 vehicles. In 1918, half the cars produced in the United States came from the Ford plants.

But in 1919, the "light of Detroit" dimmed and relations between workers and management became tense. The Russian Revolution worried America: the values of Fordism were used against the revolutionary ideology that was supposedly booming. The trade union movements were strongly repressed by private militia (an understatement). The honeymoon lasted less than 4 years.

Taylorism and Fordism, which both secularized and applied values of Puritanism in the working world of the industrial revolution, have definitely made their mark on management science. They have this in common with the viewpoint of microeconomics, without sharing all the values. It is important to bear this in mind when dealing with the world of quality and its ethics.

And afterwards?

Ford created a vertical integration that was beneficial to him as long as he had more or less a monopoly: in 1915, he decided to have his own steel mills; in 1922, he produced his own electricity; and in 1923, his own ships carried his ore and his cars on the Great Lakes. But the tide turned: at the end of 1927, *Ford* rushed out Model A, forced to admit that the Model T was outdated.

And then 18 years later, *Henry Ford II* or *Jr.* took over the company at the age of just 25. It had been transformed into a foundation to avoid shareholder returns, because it no longer showed profit as of some time. After a generation of mismanagement, the company lost 10 million USD per month. Accounting was primitive, there was no coordination between purchasing, production and sales. For years, the financial reports were kept secret even within the company, so as not to tarnish its image and cause an inquiry to be held. The company had no research and development program.

Worse, the only son of *H. Ford, Etsel,* died suddenly and panic blew through the Detroit establishment. A project was launched, asking the US government to lend enough money to *Studebaker*, a manufacturer whose annual production volume was 16% of that of Ford, in order to buy and manage the company.

However, it was *Henry Ford Jr.,* the grandson of the pioneer, who at only 25 masterfully turned the family business around, much to the surprise of the high society of Detroit. *Henry Ford Jr.* took his inspiration for the task from the structure and policy of another group that had become very successful: *General Motors*.

In the early 1920s, *Ford* had come to possess two-thirds of the automobile market. Fifteen years later, at the dawn of World War II, this market share had fallen to 20%. What had happened [3.17]? It is now widely agreed that this decrease was due to the systematic, conscious and deliberate decision by *Ford to lead a billion dollar business without any responsible senior managers*. From the outset, *Ford* considered his senior executives exclusively as assistants: they should do only what he ordered[8]. He refused to share the responsibilities of management with anyone. Yet, it has been proved that the fundamental problems of hierarchy, structure and business conduct within a company must primarily be resolved by senior management. This point, which is now a part of Total Quality, was lost on *Ford*.

Ford's secret police were not only watching the workers but also the managers, and informed him of any attempt on their part to make a decision. If they gave the impression of having decided on an activity or management responsibility on their own initiative, they were fired.

The result was:

- that there were only very few competent managers at Ford, when Ford Jr. took over. Those who were competent, and had not been dismissed, left the company as soon as the New Deal of President Roosevelt had shown its effects;
- that the man with the most power after Ford was Harry Bennett, the head of the secret police, but he had neither the skills nor the experience to assume a management function.

The overall effect of Fordism was thus to maximize obedience from employees, thus resulting in a lack of responsibility that would handicap the US automobile industry for a long time to come.

3.5 Alfred Sloan, General Motors and the development of American management [3.18]

Compared to *Henry Ford*, who was an engineer having received his education through an apprenticeship, *Alfred P. Sloan* (1875-1966) was a graduate of MIT and remained at the head of *General Motors* (GM) for over 30 years.

Figure 3.6 Alfred Sloan.

In 1923, when *Sloan* took his position as president of GM, the automobile market gave its first signs of running out of steam. Manufacturers had to face the growing market for

8 This behavior went even further: Ford changed the senior foremen every two to three years so that they would not become "authorities".

used vehicles. *Sloan* took several initiatives:

- he encouraged dealers to take old models against the purchase of new cars;
- he developed a national credit system that encouraged new buyers;
- he multiplied product lines; he sought to increase the amount of money a customer spent on cars;
- he pushed for increased consumption by making cars old-fashioned by changing their style (introduction of an annual model) – this strategy is still known as planned obsolescence. He hired Hollywood specialists who designed extravagant car models for movie stars. The result was a heavy and spacious design that would have to be corrected in the late 1960s.

But *Sloan* was also an outstanding organizer who introduced flexible mass production because even if the GM models were different, they contained common components, often produced by subcontractors. He also organized GM through a dual mode that was to go down in history: reflections and strategic decisions regarding company goals and resource allocations were taken centrally by management and finance committees, while tactical decisions on the current use of resources were the result of GM's operating divisions. When he left in 1957, *Sloan* was replaced by a succession of financiers who concentrated on short-term economic goals for the company. This was the end of the dominance of engineers as managers. This fact is often cited to explain the collapse of US manufacturers vs. their Japanese counterparts in the late 70s (Toyota executives were engineers).

Sloan did not give up Taylorism, but he increased the volume and pace of work, which caused discontent among the workers. He tried to curb this discontent, unlike Ford, by developing an unmatched information system in the company. In the 1970s, 75% of jobs in the automobile industry involved very few skills, to be compared with 10% in the rest of the US economy. The absenteeism rate was above 10%.

In the 1930s to 40s, *Sloan* was one of the instigators of American management, whose vision and methods would be extended, at first through consultants such as *Peter F. Drucker*[9], and then by famous business schools which would start to expand in the late 1960s throughout post-war Europe. As was the case with the birth of quality with *Ford*, the impetus provided by *Sloan* would give a competitive advantage to the US economy until the late 1960s.

In his memoirs [3.19] published in 1963, *Sloan* devotes Chapter 14 to scientific and technical personnel: the quality function was never mentioned by name. *Sloan* did not innovate in this area. This is probably why the quality function did not attract the attention of Sloan's followers and why it was not included in the curricula of business schools.

Finally, even though *Sloan* founded much of modern management, he hardly considered management a science but rather put emphasis on the profession of manager.

---

9 *Peter F. Drucker* remained a very revered management consultant until his death: he published two key works to describe the model that Sloan had put in place: *Concept of the Corporation* (1949) and *The effective executive* (1966). He is considered one of the founding fathers of modern management theories; in the preface of *Alfred P. Sloan's* memoirs, *My years with General Motors* (1963), he acknowledges his debt to the president of GM.

3.6 Quality control

3.6.1 Inspection overwhelmed by the quantities produced

Taylorism, as well as Fordism, became a worldwide success and had an impact on the economy similar to that of the first industrial revolution. However, it was also a victim of its own success:

- the quantities produced increased significantly;
- the products themselves became more complex: the number of components increased and the defects were not always apparent. With increasing complexity, repairs were not as easy as with the Model T;
- and the product could be partially or even completely destroyed during the inspection phase.

Inspection of each finished product or component when entering the production process was no longer possible, the costs were prohibitive. What to do? The answer came once again from the United States: replace the inspection by control; but its introduction was slow.

What distinguishes control from inspection is the methodology – a control of control:

- a generalization of laboratory analysis methods, to detect any hidden defects, and especially
- the use of statistics, including that of sampling; to issue a certificate of conformity for a batch of finished products, semi-manufactured goods or raw materials by examining only a portion of the batch.

3.6.2 Shewart and the Bell Telephone Laboratoires

This approach began in 1926 in the *Bell Telephone Laboratories*. In order to improve the production quality of the *Hawthorne Works* factory of the company *Western Electrics*, a work group proposed the implementation of certain tools:

- a control chart developed by Walter Andrew Shewhart (Figure 3.7), a physics graduate from Berkeley (viewing quality fluctuations of a parameter over time) (Figure 3.8);
- the use of statistical theory to set the size of control samples;
- a method for assessing the quality of the phones at the end of production.

Figure 3.7 Walter Andrew Shewhart.

Of these three tools, sampling was the most used. However, statistical quality control (SQC) did not go beyond the borders of the *Bell* company. An interesting question is why SQC was introduced in the communications branch: indeed, mathematical techniques were

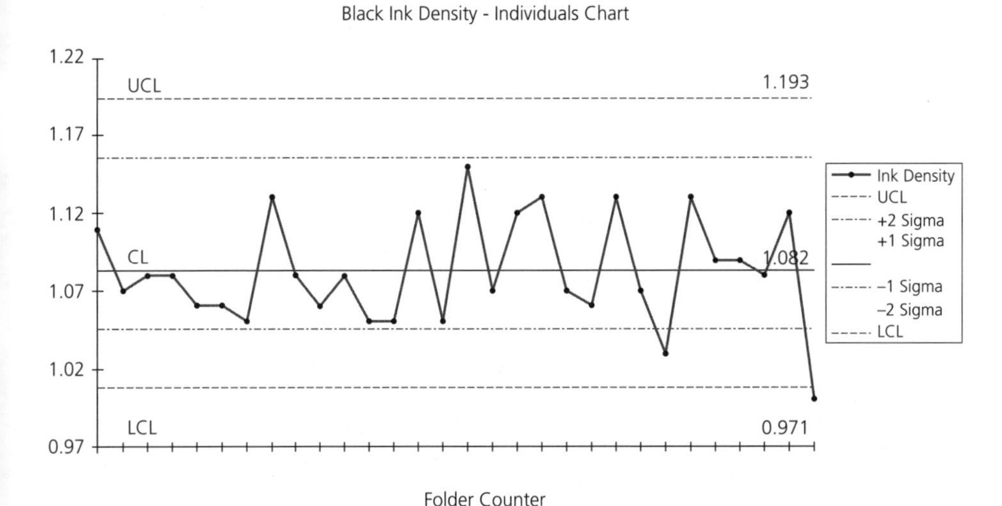

Figure 3.8 Shewhart's control chart.

already applied (signal processing), but in the case of *Bell*, it was the company itself that was affected by the problems and costs of poor quality (and not the clients). This increased the internal motivation of the management.

3.6.3 World War II and the postwar period

A change occurred when the United States entered World War II in 1941. Production quantities were significant and had to be promptly delivered to the front, limiting the time for inspection and requiring human resource skills that were hard to find. The US military used specialists from *Bell Telephones*, who adapted the methods and published a new table, the *MIL-STD 105*, which was incorporated into purchase contracts of the US military. However, as the price paid by the US military to the best suppliers was based exclusively on the accuracy of the date of delivery and production quantities (and not on the quality), some doubts still remained on the optimal use of this table.

The implementation of *MIL-STD 105* tables had a positive impact:

- training programs for specialists provided by the scientists at Bell Telephones to military suppliers, and
- the creation of a new position within the company: the quality control engineer.

Among the instructors was a physicist, *W. Edward Deming*, who would later play a key role.

Many engineers went beyond the use of tables and prepared procedures, carried out audits, sought sources of poor quality in industrial processes, etc. The early stages of quality assurance began to see the light of day. Some experts gathered together in associations, which eventually coalesced to give the ASQC (American Society for Quality Control).

In order to differentiate this task from inspection, large companies created a quality control department that included the services of inspection and statistical quality control, headed by a quality control manager, who deferred to the production manager of the company. The management of quality and quality assurance was under creation. Again, product quality was the sole responsibility of the quality manager, and quality training was provided only to employees of that department.

To make matters worse, each functional unit of a production center generated its own poor quality and transmitted it to the next manufacturing unit, with no concern for the needs of its internal and external customers. This mindset was called the *Throw it Over the Wall* (TOW) syndrome. The result was edifying: in the most spectacular period of statistical quality control (SQC), the *Hawthorne* plant employed 40,000 people, of which... 5,200 worked in the quality department.

This situation did not improve after the war, especially for products aimed at civilians. In view of the large quantities to be delivered over very short times, only the quantity and time mattered. Often, direct clients never filed any claims for fear of not receiving the delivery. Companies issued a guarantee for each product, which decreased customer frustration and made it possible to treat the detection of poor quality after purchase. In this case, the client was the one responsible for quality control of the finished product ...

Finally, some companies used the poor quality of their products to generate lucrative revenues based on costly after-sales services and the delivery of spare parts (e.g., *Xerox* copiers). Unknowingly, European and US productions were at the verge of being overtaken by competitors who were more creative, but also more respectful of their customers.

Such a situation lasted in the United States and Europe until the early 1980s. The US would lose its leadership in quality without even realizing it; consumerism put an end to the tolerance of consumers, and their associations ended acceptance of poor quality and product defects. A demanding society of consumption took the place of a society of shortages avid for basic necessities. Legislation was being introduced: it was soon the manufacturer who had to be able to prove to the authorities that the product was free of defects if the consumers filed claims.

3.6.4 The 50s and the emergence of Japan

The end of World War II left a helpless and battered Japan; three million inhabitants had tuberculosis (the main cause of death), inflation was chronic, the food supply hazardous, industry ruined, and road conditions were worrying. Economic development now had to be done without the geographic expansion of the Empire of the Rising Sun. But Japan had few resources, and thus had to rebuild its secondary sector. Its economy was very weak and completely protected. Moreover, given the development of the Cold War with the USSR and the proclamation of Communist China, the occupying power, the United States, would benefit by having an ally with a prosperous economy in the region. What would have happened if communism had taken root in Japan?

A first step was to rebuild the faulty[10] telephone network, both for the Japanese and for the occupying forces. They called in specialists from *Bell Telephones*, who inspected all the production plants of electrical and telephone equipment from 1946 to 1950. Their judgment was clear: despite the formation in 1942 of the Japanese Association of Management, the quality standard was inadequate, confirming the reputation of Japanese products having poor quality and being sold cheaply before the war (illustration of the motto: cheap is always too expensive). The situation was serious, given the low amount of natural resources in Japan.

The two main weaknesses were:

- Japanese members of top management could not imagine that it was their responsibility to increase product quality and to train their employees;

[10] The Japanese were without telephones 50 days per year ...

- management tools when it came to financial control, technical (process) control and quality control (SQC) were underdeveloped.

Therefore, the communication section of the staff of *General MacArthur* decided to introduce training seminars. The Japanese were urged by the occupying power to participate and could not refuse, but they were gradually impressed by the enthusiasm, selflessness and professionalism of the American specialists. In May 1949, the Japanese (mainly professors at the University) began to teach their colleagues SQC by translating the English texts into Japanese.

3.7 The development of Quality Assurance

3.7.1 Deming's SQC courses

Japan would benefit from the lessons of two American heavyweights in quality assurance, who had talked to deaf ears in the United States: *W. Edward Deming*, then a university professor and a specialist in SQC, and *J.M. Juran*, an expert and consultant in quality management. Both had worked at *Bell Laboratories* in *Hawthorne*. Their courses were complementary, given in 1950 and 1954 respectively, and this partly explains the success of the implementation of quality in Japan. The work of *Deming* and *Juran* had a global impact: they were the two undisputed leaders in quality until the late 20th century.

Deming came to Japan in 1950 to teach SQC to quality specialists. He only gave a brief presentation to top management but its success was considerable.

The introduction of the early stages of quality assurance and SQC in the production centers, however, was not easy. In 1955, a specialized journal reported that only 34% of industries for machinery and electrical components and 25% of chemical industries had succeeded in introducing SQC. However, there was also some success: for example, in 1949 the Japanese scientist *Masashi Asao* had successfully produced a drug for tuberculosis. Its production was very delicate, and SQC was used to control the manufacturing process. Moreover, the launching of a new product was usually accompanied by the introduction of SQC on the production line: the policy consisted in taking small steps forward.

3.7.2 The Deming prize portfolio

One reason for Japanese success was the development in 1951 of the *Deming Prize* portfolio. These prizes rewarded success and breakthroughs in quality (and *Masashi Asao* was among the first to win):

- *Deming Prize* for an individual contribution in the fields of SQC education and services (1951);
- *Deming Application Prize* for the exceptional implementation of SQC by a company (1951);
- *Deming Application Prize* for a small business (1958);
- *Deming Application Prize* for a division within a large company (1966);
- Japan Quality Control Medal (1970);
- Distinction of Excellence award in quality control of a company (1973).

In comparison, the US had only put in place individual prizes or those allocated to products. The introduction of the *Deming Prize* portfolio produced in Japan an intense emulation of specialists, departments and companies that progressively started incorporating quality assurance.

3.8 Quality management and the first steps toward Total Quality

3.8.1 Juran's seminars to Japanese middle and top management

In 1951, *Juran* published what would remain a classic: *Quality Control Handbook* [3.20]. This book was soon translated into Japanese with the title *Total Quality Control*. It sparked the second wave of knowledge transfer in quality to Japan.

Even though *Deming's* SQC courses had certainly convinced specialists, engineers and scientists, the middle and top management – devoted to increasing margins and market shares – remained skeptical about its implementation. Japan's situation was worrying: imports in 1953 were only 46% and exports only 34% of articles as compared to the prewar period.

At the request of the President of the JUSE (*Japanese Union of Scientists and Engineers*), *Juran* visited numerous production centers and gave seminars in Tokyo and Osaka in 1954. He noted that the camera manufacturers *Nikon* and *Toshiba* already had a level of quality equal to or higher than the US. These remarks, never listened to by American managers, only provoked annoyed shrugs and amusement.

Juran noted that although SQC education was useful, it had to be completed by quality management tools (diagnosis systems for improvements, resolution of problems, the use of Pareto charts, etc.) for quality engineers and heads of production divisions. Over 300 managers of such units took these courses.

Juran explained the importance of their responsibilities to top management (over 100), when it came to:

- establishing a quality policy, with clear objectives;
- the choice of the definition of quality;
- the organization chart for the implementation of quality;
- the definition of measuring the level of quality achieved, as compared to that desired;
- the implementation of remedial actions to achieve the desired quality.

3.8.2 From Quality Management to Total Quality Management (TQM)

This seminar for senior managers was the first step in the transition from classical Quality Management to Total Quality Management, but without a holistic view of quality in the company being formalized. *Juran* was impressed by the active involvement of top management in these seminars. He had not previously had such a large audience. And this would not happen again outside Japan.

After *Juran's* courses, the JUSE took over and began training in Japanese of middle (1955) and top management (1957). In 1990, these courses reached 1,500 middle and 300 top managers. JUSE also prepared radio courses on quality control and standardization for the staff and management of production lines.

In 1958, Japanese executives organized a trip to the United States where they visited several factories, and their diagnosis was gratifying to Japan but rather worrying for the United States: quality status in the US was only marginally higher than in Japan. Again, this information, when it reached the West, was ignored. In their benchmarking, the Japanese noted the existence of the ASQC (*American Society for Quality Control*) and the application of quality to services (*United Airlines* and *Bank of America*). These elements were introduced in Japan in the 1970s.

3.8.3 The birth of quality circles

From 1959 to 1960, a first economic boom saw exports and imports exceed the prewar level. Manufacturers of low-value products, such as clothing, gradually gave place to production centers for transistors, cameras and machines. The first trade liberalization was decided in Japan in 1960, under pressure from the United States, five days before *Juran's* second program in Japan. Local Japanese companies quickly became aware of the risk if their quality was not superior to that of companies exporting goods to Japan. Soon, the slogan was "*Ride out Liberalization with Quality Control.*"

In 1966, in his third and last trip to Japan, Juran noted the operation of quality circles (production staff developing their own programs to improve quality), established in 1962. This was the end of Taylorism and the separation between the quality function and production.

When *Juran* was in Europe a few months later for a congress of the European Organization for Quality Control, he gave a presentation about the importance of this tool and predicted that Japan was quickly becoming a leader in quality control. This information had no impact in Europe (or in the United States). These two regions would find themselves more than 20 years behind in terms of quality, efficiency and productivity in the secondary sector.

Gradually, a company emerged in Japan that would become an example of Total Quality. *Toyota*, with its production model of small and medium-sized series, would revolutionize the industrial organization. Its system, TPS (*Toyota Production System*) or Toyotism would allow the company to become one of the world's largest car manufacturers. But this complex model would be difficult to copy. It nevertheless gave rise to American Lean Management, included in the *Six Sigma* approach under the term *Lean Six Sigma*.

3.9 The long route for the United States toward Total Quality

3.9.1 Certain blindness...

During the 1960s to 70s, American managers became aware of the growing competitiveness of Japan. However, they only rarely assigned this to the approach of quality and production efficiency:

- Few journalists, scientists and engineers read Japanese, and moreover, the Japanese remained discreet.
- Top management did not include quality specialists and only rarely engineers (and if so, they came from the production – long frustrated by the quality function – or sales departments).
- Those teaching quality were rarely university professors.
- The Throw it Over the Wall syndrome endured.
- Managers were fixed on the progression of sales and higher margins. They thus lacked sensitivity when it came to quality and attributed the success of Japanese products only to the competitiveness of their prices, Japanese "collectivism", and the low wages of the Japanese working force.

3.9.2 Failed initiatives of the 1960s and 70s

Because of this information gap, the initial reactions of American managers were:

- to block imports through political lobbying in Washington (protectionism); and
- to move production facilities abroad in order to lower production costs (relocation).

These measures were without any major effect. In 1960, the first Japanese car was exported to the United States. It was a few years after the Korean War, during which Japan manufactured trucks for the US military. Already in 1975, the quality of Japanese cars was at least equivalent to that of American ones. In 2007, *Toyota* surpassed *General Motors* in the number of vehicles produced. 30 years after 1975, the quality standard and the robustness of Japanese cars remains superior. In 2009, the three largest US car manufacturers showed an extremely dysfunctional economic situation.

Faced with Japanese quality, however, a reaction appeared. But American managers were not trained in quality and thus delegated reforms to consultants or internal experts. Unfortunately, although many generally proven and validated tools and methods gave some positive results, a global overview was not taken into account and the causes of poor quality were not identified: the largest of the initiatives had very little effect.

Caught in a fix-it-now mentality, middle and top managers were unwilling to invest in an area that, in the United States, was not considered the starting point for a successful career (somewhat like medical, scientific and technical professions in the Roman Empire?).

3.9.3 The third wave of SQC and initiatives of the 1980s

In the early 1980s, the decline of American industries became a political issue across the nation (massive layoffs, businesses closing, falling stock prices). Companies took the problem very seriously.

3.9.3.1 Exhortation: a zero-defect policy [3.21]

Managers attributed the first responsibility for poor quality to the employees and not to the shortcomings of technological and managerial processes (it was easier...). They persuaded them to work better and to renounce their alleged negligence. *Do it right the first time* was one of the slogans. Its effect was limited, although it did help to educate employees that quantity was no longer the only item on the agenda.

3.9.3.2 Call for statistical techniques: SPC

Given the ineffectiveness of the exhortation, employees were at last educated in quality, including the use of a quality control tool that complemented the SQC: SPC (Statistic Process Control, an extension of the control diagram developed by the physicist *Walter Andrew Shewhart*). This method was new and effective, but many managers and consultants thought they had found a panacea. As the causes of poor quality still had not been identified, the employees were in the situation of a doctor with a drug but who was unable to diagnose the patient. Many contributors therefore had no opportunity to put SPC into practice.

Moreover, nothing had fundamentally changed the level of commitment and training of top management with regard to quality, thus weakening even further the poor leadership in this area.

3.9.3.3 1982: The *In Search of Excellence* smokescreen

In 1982, in the United States and Europe the book *In Search of Excellence, Lessons from America's Best-Run Companies* [3.22] by *Tom Peters* flew off the shelves; the author claimed to have identified the best practices of the most successful American companies, capable, so it was thought, to show the Japanese that the Americans also had competitive knowledge.

However, this know-how was not directed simply toward quality. It had another goal, more prestigious, but very ill-defined: *excellence*, of course, which would ensure a bright future instead of the very drab "collectivist" quality management. Centralization was seen as the cause of the difficulties of American companies. More initiatives should be given to the front: *Think globally, act locally*, was a slogan among others. Unfortunately, the Japanese with their circles of quality were already able to triumph over the consultants' proposals.

Many companies were impressed and applied these good practices, but without carrying out preliminary studies, always with a *fix it now* mentality. Two years later, the situation was different: the book was published after an empirical study, without a methodology or an investigation plan. It had been conducted from 1979 to 1980 by an employee of *MacKinsey* in San Francisco. Worse, five years after the study, in 1984, most of the companies mentioned in the book were going through serious financial crises. Nonetheless, *MacKinsey* was not worried, because at that time, international consultancy companies had the aura and respectability of a guru!

3.9.3.4 1982: Deming steps up to the plate

In 1982, Deming published a landmark book, *Out of the Crisis* [3.23], which is still read today[11]. He recalled that pressuring employees, or even using methods of intimidation or exhortation, to produce more does not make them work harder. He claims instead that if the increase in productivity is to be sustainable, it has to be accompanied by an increase in quality. He puts into perspective 14 points for a successful transformation of the American management system: some of the recommended reforms are iconoclastic, such as the elimination of quotas and management by objectives, the rehabilitation of pride in work well done, the focus on trust between customers and sellers, suppliers and producers. Fordism was put to death.

3.9.3.5 1986: Juran's trilogy

In 1986, *Juran* introduced a trilogy of quality [3.24], based on three approaches: *Quality Planning, Quality Control, Quality Improvement*. The trilogy takes into account the concept of the Deming cycle (PDCA). *Juran* also published the fourth edition of a reference book, the "bible" of quality, *Juran's Quality Control Handbook* (1988) [3.25].

Quality planning follows a universal sequence of steps:

1. Identify target customers and markets.
2. Discover the hidden or undetected needs of the customer.
3. Translate these needs into product or service requirements that can fulfill the expectations (new standards or specifications, etc.).
4. Design a service or product that exceeds customer requirements.
5. Develop a process that can provide the appropriate product or service in the most effective and efficient manner.
6. Transfer the prototype or design to the organization and operations.

Quality Control uses a feedback control loop through corrective actions. The elements of the feedback loop are:

1. the subject of control, product or process features that must be brought under control;
2. a detector that measures the features that have to be controlled;
3. an arbitration unit that receives the measurement values;
4. a standard or set value of the feature or variable (in measurement units);

[11] See, for instance, the rave reviews from readers on Amazon.com.

5. if the standard is not reached, the arbitration unit activates an operator to adjust the process to bring it back into the tolerance range of the setpoint; the operator can be a task, an employee, a supervisor, or a feedback mechanism.

Control is a form of improvement: an undesirable value is corrected by returning to the initial value, but it's not a breakthrough, a quantum leap in quality improvement (Figure 3.9).

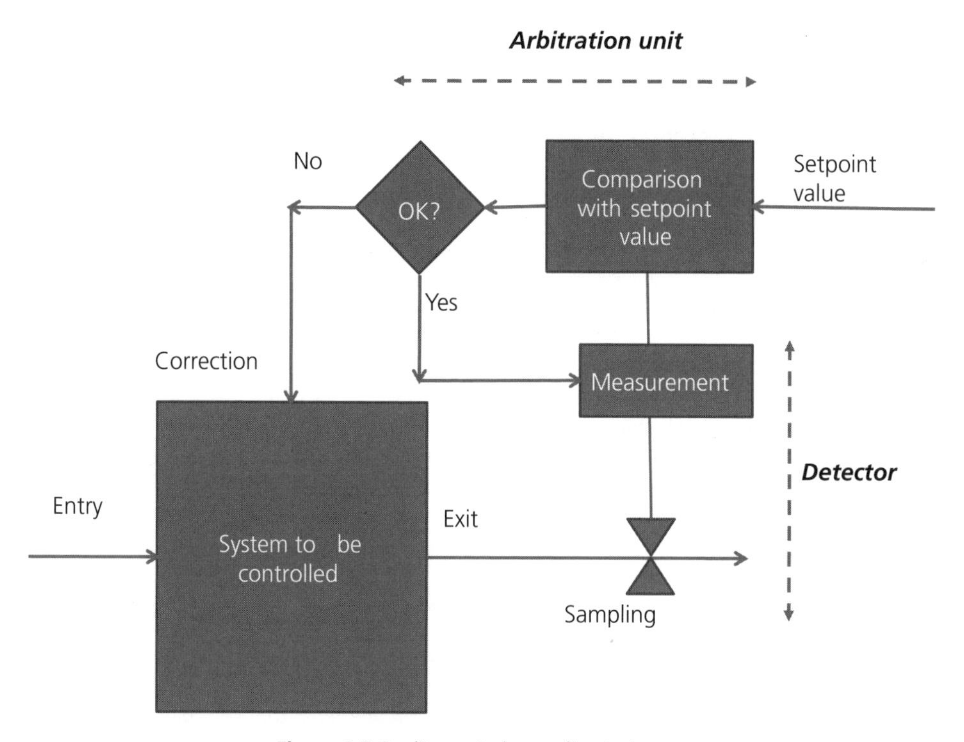

Figure 3.9 Quality control according to Juran.

On the contrary, ***Quality Improvement*** is obtained by breakthroughs or quantum leaps to create beneficial and previously unknown changes, by improving the current standard of organization: a quantum leap in quality is implemented by following a universal sequence of actions:

1. identify a problem in the company, the poor quality of a product or service that affects its performance;
2. establish a project;
3. measure and analyze the current process in order to have precise knowledge of its basic performance;
4. generate and test hypotheses about the causes of the poor performance;
5. determine the root causes of the poor performance and provide evidence;
6. design and develop remedial measures, process changes that eliminate or bring under control the causes of the poor performance;

7. establish new control procedures to prevent recurrence of the defect and to make the new standards permanent;
8. face resistance to change within the organization;
9. reproduce the results and define a new project.

The concept of quality improvement has been deepened by one of the gurus of Total Quality, *Shiba* from Japan (Chapter 6).

3.9.4 The rise of the 1990s and the Malcolm Bridge Quality Award

In the late 1980s, however, a handful of American companies had taken on the challenge – they had reached a world-leading standard of quality. A Total Quality Award was established in 1987: the *Malcolm Bridge National Quality Award*. Some of them had won this distinction and were willing to share their experience with other American companies that were interested and thus to act as role models (*Motorola, IBM, Xerox, Milliken*, etc.).

3.9.5 Motorola and the Six Sigma approach

The first company to win the *Malcolm Bridge National Quality Award* was *Motorola* (1988). The quality system in place was an eclectic combination of ideas and methods of old and new SQC or SMC. *Motorola* agreed to disclose this method under the name Six Sigma. The Six Sigma approach (in honor of *Shewhart* and *Bell Laboratories*) targeted less than three noncompliances in 1 million products or components and was approved by the American Society for Quality. Many companies, noting the success of *Motorola*, adopted this approach. It was estimated in 2003 that 100 billion USD of savings were obtained by the adoption of this method. *The Six Sigma approach is still the most applauded global standard of quality in the United States*, but it is not easy to implement and a slogan that characterizes it is: *No pain, no gain*.

At the same time, *Juran* and *Deming*, who had been very little in demand until then, progressively were raised to the status of gurus. However, some criticized them for having provided tools to the Japanese without having diligently taught Americans.

Emulation began. Although the strategies used were many and varied, often related to the product and the market, certain constants appeared:

- allowing the customer, who is ultimately a stakeholder, to set the quality standard;
- middle and top managers accepting responsibility for the quality function; this function cannot be delegated;
- including quality in strategies and business plans (Strategic Quality Planning);
- considering quality in all management functions, not only production (Total Quality); companies providing services will also be affected by this reform;
- directly relating quality management to business processes; a manager should be appointed for each process.
- considering new methods for assessing quality, and implementing them in a logical way, this time on a business level: customer satisfaction, analysis of the quality of competitive products, etc.;
- terminating Taylorism and restoring responsibility and initiative to employees, thus making them responsible for the control (union resistance possible);

- entering a new task in the job specifications of all employees: increased quality, and taking this into account in performance analysis;
- using the Malcolm Bridge Award as a standard for the measurement of quality, even if the company that implements it does not participate in competitions;
- carrying out benchmarking: adopting, in a friendly manner, efficient processes of non-competing companies, ready to exchange know-how in process management.

3.9.6 Japan in difficulty

There are very few points in the list above that have not been met by the Japanese, except for services. This had a serious impact on Japan in the 1990s:

- Japanese banks realized that many national companies borrowing money were not in the situation to pay back – was this due to a QA problem in the banks[12]?
- the financial and stock market situation caused a real estate crash;
- vertical integration of Japanese companies was tainted by cronyism and corruption;
- the Japanese had not liberalized the services market early enough; it remained national, and had fallen behind and was not very competitive, while the development of service activities in the global economy was booming.

Concerning services, Japan was not in the leading group, and even less so when it came to quality. Its exclusive focus on applied research had probably made the country miss out on some technological innovations, especially with regard to computers, communications (often seen as the foundation for a fourth industrial revolution), the internet and life sciences.

Moreover, at equal competitiveness, Japan invested more than the US economy. Finally, the relocation of industrial activities to China and the rest of Asia deprived the country of some income. As an expert and master of the secondary sector, Japan saw its specialty coveted by other Asian countries (China, Korea, India, and soon Vietnam). Since Japan was a closed culture, it struggled to implement human relations and international mobility, the key to the expansion of services. This situation still persists.

3.10 The path taken by the EU: ISO 9001

3.10.1 American management introduced in the 1950s

Although it was introduced in the 1920s, mass production according to Fordism wasn't really established in Europe until after World War II (national barriers created a delaying effect). Management methods based on the ideas of *Sloan* were introduced at the start of the 1960s, being popularized with the publication of the book *The American Challenge* [3.26] and the creation of the first European business schools.

One of the success factors was the introduction of the *Marshall Plan* in 1947 to get the European economy back on its feet. This assistance plan, primarily economic, was accompanied by the dissemination of intensive American practices in the economy and agriculture, and study trips by European leaders to the United States to gain knowledge of industrial methods used, etc. The result: the differential with regard to comfort and consumption for

[12] Given the collapse of the banking sector in 2008, quality and risk management should be the agenda of directorates and boards of banks (and not the reinstatement of bonuses).

an average person in the two continents, which was considerable in the 1950s, had already been significantly reduced in the early 1970s.

With the success of the *Deming Prize* in Japan and the *Malcolm Bridge National Quality Award* in the United States, Europe logically should have taken the same path. This was almost the case. A similar standard does exist: the Award of Excellence from the *European Foundation for Quality Management* (EFQM), established in 1989 by the CEOs of major European airline. It had notable success.

However, this was not the standard that was chosen by the majority. Why not?

3.10.2 The origin of ISO 9001: the United States once again

Admittedly, the first ISO 9000 quality system saw the light of day in 1987, the same year the *Malcolm Baldridge Award* was founded, but this is not the right track.

It all started in the Department of Defense of the United States (DoD), which, in 1959, defined a quality assurance program for its suppliers, the MIL-Q-9858. This standard was adopted by NASA in 1960 and revised by the DoD in 1963. It then became the MIL-Q-1958 A. This program was then generalized in all of NATO under the name AQAP (*Allied Quality Assurance Publications*).

3.10.3 Consumer usage confiscated by perfidious Albion

In 1970, a Quality Assurance Office of the Ministry of Defence (the DQAB) of the UK was created to give its opinion on new QA procedures to be applied by the Procurement Division. The Ministry of Defense adopted the AQAP, modified them in order for the focus to be on design and, with support from industry, published them as the Defense standards DEF/STAN 05-8.

The Director General of DQAB, the rear-admiral *D.G. Spickernell*, left the navy in 1975 to head the technical division of the *British Standards Institution* (BSI). When BSI created a committee for a quality system for businesses, *Spickernell* advised its members to closely follow the defense standards. The committee's work led to the standard BS 5750, approved and published in 1979, which was the precursor of ISO 9000. Meanwhile, other standards of the same type were being developed in Canada (Can 2229) and Switzerland (SN 029100). The risk of proliferation of standards was real.

3.10.4 A huge success...

BSI then proposed that ISO begin work on international standards relating to quality assurance (QA) and quality management (QM). The Technical Committee ISO/TC 176 was established in 1979. It started preparing the ISO 9000 standards in 1980 and published them in 1987. The same year, BSI and ASQC (*American Society for Quality Control*) adopted ISO 9001, in 1992 NATO integrated ISO as a standard quality system and in 1994 it was the turn of the Department of Defense and the Federal Department of Commerce of the United States. The second version of the ISO standards appeared in 1994. Since then, they have been reviewed every five years.

But the greatest success was in the European Union: the EU demanded that all medium – or high-risk fabrication of devices or products must be supervised by a certified quality management system. In this context, the European system was based on product compliance through the application of management systems including product-specific requirements. ISO 9001 is widely used. Companies are audited and a certificate of compliance is issued. Moreover, the ISO system is modular and a quality system of an institution can be built using other norms or standards from the ISO portfolio (or other standard writers), to adapt the content of the standard to the market and product situation.

In the United States, the US administration in charge of the food and pharmaceutical sectors (*Food and Drug Administration*) also relies on the structure of ISO 9001 to impose its own requirements for quality systems.

3.10.5 ... followed by a diversification for the environment: ISO 14001

In 1992, given the success of ISO 9001 and the development of standards for the protection of the environment (of which a British standard, once again, dates back to 1992: BS 7750), ISO considered the timely introduction of a standard:

- with a managerial approach to the environment similar to that of ISO 9001;
- increasing the competence of organizations when it came to achieving and measuring improvements in environmental protection;
- and as a result facilitating trade and lifting tariff barriers (consumer confidence);

ISO 14001 was published in 1996.

3.10.6 ... triggering grimaces from the supporters of Total Quality...

The worldwide success (except in Japan) of ISO 9001 sparked criticism from the supporters of Total Quality and the fathers of quality, *Deming* and *Juran*. This justified criticism is re-echoed by their followers. Thus, in 2004, *Juran* wrote:

> This system lacks essential elements needed to achieve worldwide quality: the personal commitment of the leadership at the highest level in favor of quality, quality management training of the hierarchy, recording quality objectives in the business plan, maintaining a high rate of continuous quality improvement, participation and motivation, assigning quality responsibilities to all employees. All this is very incomplete. The risk is great that this standard will lead to lower quality [3.27].

Moreover, it was estimated in 1996 that the contents of ISO 9001 only corresponded to between 30% and 40% of the requirements of the *Malcolm Bridge National Quality Award* [3.28].

ISO 9001 version 2000 and ISO 9004

This criticism was heard, and the 2000 revision of ISO 9001 put emphasis on the process approach, the inclusion of the quality strategy in the fundamental documents of the company, continual improvement of quality and the responsibility of managers (Figure 3.10). The criticism regarding the training of employees, however, seems to persist. As for the actual commitment of management, it was indeed very variable. With ISO 9004, the standard now includes stakeholders. The revision of ISO 9001 in 2008 resulted in only very minor changes.

The *Malcolm Bridge National Quality Award* still remains a worldwide standard of excellence in Total Quality. However it tends to be replaced by a broader vision of ISO 9001 and ISO 9004, and by the EFQM model in Europe.

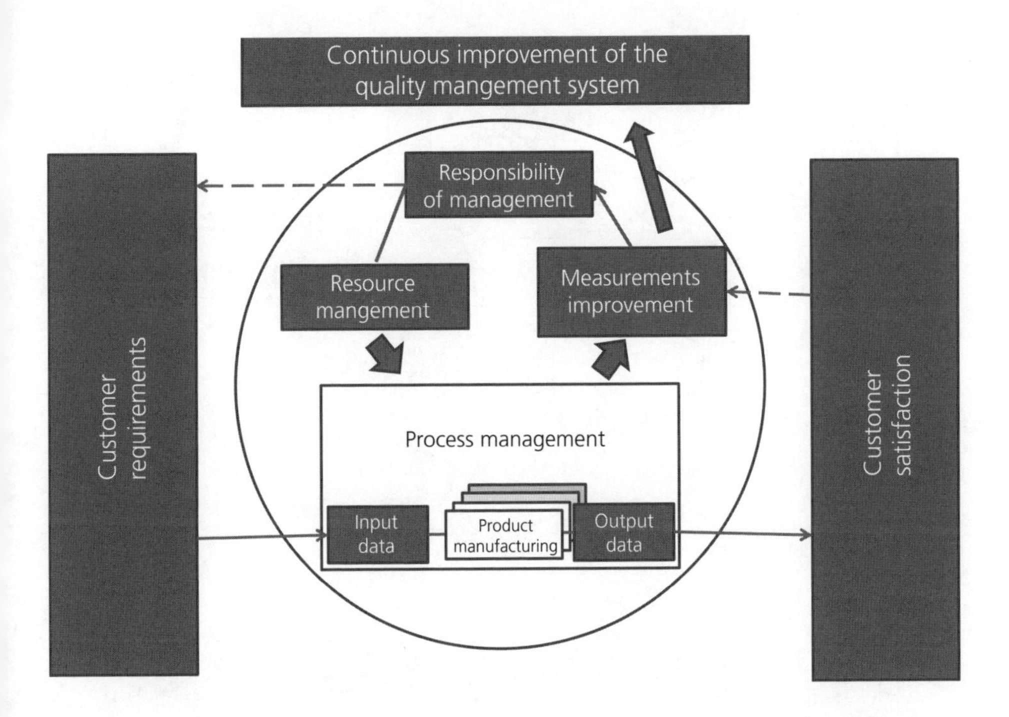

Figure 3.10 Block diagram of ISO 9001 (rev. 1999).

ISO has a weak point: the quality of certification does not depend only on the company, but also on the certifying agency that has an external quality assurance. However, ISO does not have direct relationships with certification agencies. Who audits those auditing? The answer differs from country to country. This is why many companies and many countries are auditing their suppliers (companies and institutions) in addition to their certification.

A new ISO standard on Social Responsibility

An international ISO committee has been working on guidelines to assist organizations and companies to operate in a socially responsible fashion. ISO 26000, *Guidance on social responsibility*, was published in 2010.

3.11 Synopsis

The history of quality is illustrated in Figure 3.11.

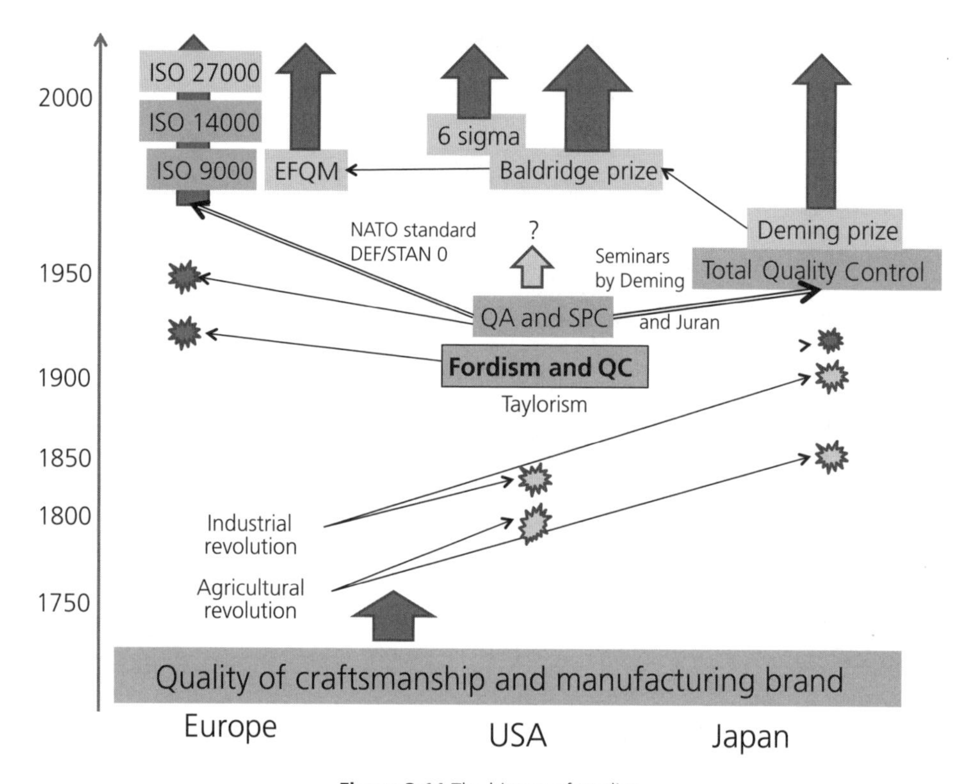

Figure 3.11 The history of quality.

References

[3.1] See the book by LAURENT JACQUES COSTA, *L'obsidienne, un témoignage d'échanges,* en Méditerranée préhistorique, Errance, 2007.

[3.2] JEAN-PIERRE VERNANT, *Les origines de la pensée grecque*, 1962, rev. 2007, Chap. VI et seq.

[3.3] FELIX KLEIN-FRANKE, *Iatromathematics in Islam*, ed. Olms (1984). He refers to the muslim doctor Avicenne, a universal genius (Ibn Sina 980-1037), who started separating astrologers and medecine …

[3.4] *Ordonnance sur la franchise de la teinture de la ville et des faubourgs de Bourges (…) donnés en cette notre ville de Bourges en mois de février en l'an de grâce mille cinq cents soixante six*, copy from June 15 1575, transcribed in E. TOUBEAU DE MAISONNEUVE, *Les anciennes corporations ouvrières à Bourgs* (…),Pigelet and Tardy, Bourges,1881, pp 100-106.

[3.5] Elements provided by Y. BOYER, *Histoire des corporations et confréries d'arts et métiers de la ville de Bourges*, IInd volume, Librairie A. Picard, Paris (circa 1890), p. 287.

[3.6] JOSEPH M. JURAN, *A History of Managing for Quality*, ASQ Quality Press, 1995, Chap 10, p. 349.

[3.7] on the importance of laymen and their power struggle with the clergy, see DOMINIQUE IOGNA-PRAT, *La Maison-Dieu – Une histoire monumentale de l'Eglise au Moyen-Âge,* Seuil, 2006, p. 539 et seq.

[3.8] JEAN-CLAUDE BESSAC et al., *La construction – la pierre*, Coll. «Archéologiques», Errance (1999).

[3.9] RENÉ DEZ, R., *REGIUS –Manuscript (1390)*, ed. Guy Tredaniel, 1985.

[3.10] D.A. WREN, *The Evolution of Management Thought*, Wiley and Sons, 1994, p. 73.

[3.11] Ibid., p. 77.

[3.12] Ibid., p. 87.
[3.13] HENRY FORD, *My Life and Work*, BN Publishing, 2008, Chap. IV, *The secret of manufacturing and serving*, p. 62.
[3.14] FREDERICK WINSLOW TAYLOR, *The Principles of Scientific Management (1911),* Dover Publications, 1998, p. 74.
[3.15] DAMIEN AMBLARD, *Le fascisme américain et le fordisme*, Berg international, 2007, p. 449 et seq.
[3.16] H. FORD, op.cit. Chap. VI *Machines and Men*.
[3.17] PETER DRUCKER, *La pratique de direction des entreprises*, Edition d'organisation (1957), *L'histoire de Ford*, p. 119-127, reproduced in CHRISTIAN THUDEROZ, *Histoire et sociologie du management,* ed. PPUR, 2006, p. 110–118.
[3.18] See also the book by BERTRAND HERIARD DUBREUIL, *Imaginaire technique et éthique sociale – essai sur le métier d'ingénieur,* De Boeck Université, 1997, p. 71 et seq.
[3.19] ALFRED SLOAN, *My Years with General Motors*, Broadway Business, 1990.
[3.20] JOSEPH M. JURAN, *Juran's Quality Control Handbook*, McGraw-Hill, 1988.
[3.21] PHILIP B. CROSBY, *Quality is free,* Signet, 1980.
[3.22] THOMAS J. PETERS, *In Search of Excellence Lessons from America's Best-Run Companies,* Warner Books, 1982.
[3.23] W. EWARDS DEMING, *Out of the Crisis,* MIT Press, 1982; already in 1991, the book was in its 13th edition.
[3.24] JOSEPH M. JURAN, The Quality Trilogy, *Quality Progress*, Vol. 9, No.8, p. 19-24, 1986.
[3.25] JOSEPH M. JURAN, op. cit., 1988.
[3.26] JEAN-JACQUES SERVAN-SCHREIBER, *Le défi américain,* Denoël, 1967.
[3.27] JOSEPH M. JURAN, op.cit., 1995, p. 595.
[3.28] WYLLY. A. SUSSLAND, *Le Manager, la Qualité et les normes ISO*, PPUR, 1996, p. 115.

Chapter 4

The ISO 9000 Family of Quality Management Systems

Key concepts: FMECA, auditors, self-assessment, factual approach to decision making, audits (customer, supplier, internal, for certification), certification, audit findings, management commitment, records, forms, document management, eight principles, involvement of staff, instructions, leadership, management by system approach, quality manual, certification body, customer orientation, sustainable performance, quality plan, quality policy, audit evidence, procedures, four sectors, mutually beneficial relationships with suppliers, management review, quality management system

4.1 Evolution of ISO 9001

4.1.1 Taking criticism into account

Earlier versions of the standard (1987, 1994, see Chapter 3) were criticized for:

- the administrative burden of the system, i.e., lots of "red tape";
- the abundance of instructions: writing procedures and instructions is not a solution to all organizational issues;
- the optimal *design* for the organization's activity focuses exclusively on procedures;
- the minor role of staff involvement and training;
- content and structure being derived from the needs of industry and therefore not suited to service-providing companies;
- looking at quality management from the perspective of compliance is not equivalent to improving it.

4.1.2 New principles

ISO 9001:2000, as well as the revised edition of 2008 [4.1], is based on:

- eight management principles;
- three orientations, i.e., customer orientation, process orientation and that of continual improvement;
- a documentary base;

- a system approach, and enhancement of staff skills and participation, which are essential and innovative.

The standard incorporates elements of total quality and refocuses the company's business on its objectives and the value of its activities from a customer-centered perspective.

ISO 9001 is undoubtedly a global success. This is demonstrated by the numbers below obtained from the survey conducted by ISO in 2011, even though they show a slight decline in the number of certifications issued. The total number of certificates issued in 2011 was 10182 in Switzerland, 49540 in Germany and 328213 in China [4.2].

ISO 9001:2008 Certifications issued [1].

Worldwide results	December 2003	December 2004	December 2005	December 2008	December 2011
Total	497 919	660 132	773 867	980 322	1 111 698
Annual growth	330 795	162 213	113 735	28 836	- 6812
Number of nations and economies	149	154	161	176	180

ISO 9001 will be treated in depth, because its generic structure and its "philosophy" have been echoed in many quality system standards (ISO and others) [4.3]. This introductory book gives an overview, and Figure 4.1 lists the various headings. However, it is the full content of ISO 9001 that is the reference.

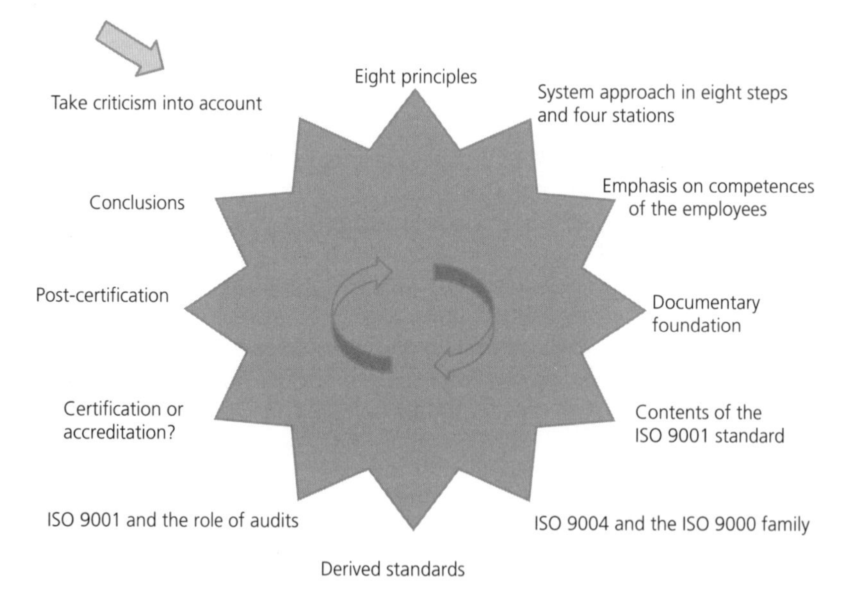

Figure 4.1 The headings of Chapter 4.

1 Main data can be downloaded at: http://www.iso.org

4.1.3 Two objectives

ISO 9001 specifies the basics of a quality management system for an organization that

a) needs to prove its ability to provide consistent and regular production, as well as a product that meets the requirements of customers and stakeholders (legislators, etc.), and
b) wants to steadily increase customer satisfaction through the effective application of the system, including continual improvement of this system and the assurance that the product conforms to customer and stakeholder (legislators, etc.) requirements.

The implementation of ISO 9001 is not always suitable for very small businesses (restaurants, craftsmen) for which a more specific quality system, a simple code of ethics, or even a quality manual inspired by the ISO standard may be more suitable. However, the ISO handbook *ISO 9001 for small business is one of its most successful publications, apart from standards, ever!* A recently ePub edition is the third edition of the handbook, which was first published in 1996. It is one of the most successful handbooks ever produced by ISO – national editions having been published by numerous ISO member countries including Bulgaria, Denmark, Estonia, Finland, Hong Kong, Hungary, India, Japan, Republic of Korea, Norway, Slovenia, Spain, South Africa, Sweden and Uruguay. In Switzerland, for example, one of the largest internationally active Swiss certification bodies (SQS), found that 60 % of delivered certificates (more than 12 000 in total) were issued to companies employing fewer than 30 people, and about 15 % were issued to companies with less than 5 employees.

4.1.4 Implementation through project management

The participation of top management is a major role when it comes to demonstrating that the implementation of a quality management system (QMS) is not an additional administrative burden, but rather a commitment to the excellence and sustainability of the company. The formal side of a QMS, often inconspicuous and mostly unproductive, barely motivates managers or CEOs coming from more exciting functions such as marketing, sales, or communications.

The implementation of ISO 9001 mobilizes all corporate functions, even the financial department; its successful implementation is the result of teamwork (process definition, writing and validation of procedures, instructions, specifications, records). Its implementation goes hand-in-hand with a project management approach.

The *PMBOK Guide – A Guide to Project Management (Fourth Edition),* [4.4] from the PMI[2] affirms that project management involves controlling a group of interacting processes according to the sequence below:

- Launch a process group
- Plan group activities
- Implement activities
- Monitor and control activities
- Stop the activities of a process group when the project is completed.

2 Project Management Institute.

The main activities considered are seven:

1. Management of the execution of the project
2. Definition and implementation of quality assurance
3. Recruitment, selection and integration of the group of active contributors to the project
4. Development of the group of employees
5. Management of the group of employees
6. Management of the expectations of stakeholders (internal and external), including customers
7. Management of acquisitions, purchases and contracts (with the suppliers of materials or services).

To implement quality assurance, the standard PMI introduces, not surprisingly, three activities of a character that is sometimes sequential, sometimes interactive, taking into account the need for customer satisfaction, prevention of nonconformities, continual improvement, management commitment and responsibility:

- planning for quality, which is the process of identifying requirements for quality and criteria for the project and/or product, and documenting how the project will adapt to these requirements;
- establishing quality assurance, i.e., an audit process of the quality requirements and quality control results, to ensure that the defined criteria and operational definitions are used; and
- quality control activities, i.e., using an indicator acquisition process that is the metric of quality defined in the planning and analysis stage to assess project performance and propose possible improvements

For each step, the *Guide* defines the inputs (which are the outputs of the previous activity), instruments and tools, as well as outputs (which form the input of the next activity).

A quality system makes it possible to respect the wishes and expectations of customers, both in terms of product quality and delivery time. It also enables adjustment of the project plan and deadlines, review of the deliverables, changes in the organization and team, adjustment of financial resources and contracts with external and internal providers, as well as providing stakeholders with targeted information at a suitable time.

ISO 10006:2003, *Quality management systems – Guidelines for quality management in projects*, gives advice on the implementation of a QMS, but should not be seen as a guide to project management [4.5]. Its aim is to create and maintain quality in a project by implementing a systematic process that ensures that:

- the implicit and explicit needs of the customer are understood and taken into account;
- the needs of the stakeholders are understood and evaluated;
- the quality policy of the organization is integrated into project management.

The implementation of the standard can be accelerated by hiring a knowledgeable consultant to increase the visibility of the operation in-house. Finding the right person can be made easier by the use of ISO 10019:2005, *Guidelines for the selection of consultants in quality management systems and for using their services.*

4.2 ISO 9000: Eight principles

4.2.1 The basic eight principles of ISO 9001:2000, as well as ISO 9001:2008

ISO 9000 cites eight principles that form the essence and philosophy behind ISO 9001 and enable a quality management system to deliver total quality management:

1. ***Customer focus* (*Plan*):** *Organizations depend on their customers and therefore should understand current and future customer needs, should meet customer requirements and strive to exceed customer expectations.*
   - *Main benefits:*
     - Increased revenue and market share are obtained through flexible and fast responses to market opportunities.
     - Increased efficacy and efficiency in the use of the organization's resources increases customer satisfaction.
     - Improved customer loyalty leads to repeat business.
     - The company's response to the strict requirements (no above-standard quality nor sub-standard quality) are enhanced and lead to better use of resources.
   - *Aspects that may arise:*
     - Identify and understand customers' needs and expectations, whether explicit or implicit.
     - Ensure that the organization's objectives are in tune with the customers' needs and expectations.
     - Describe the customers' needs and expectations throughout the organization.
     - Measure customer satisfaction and act on the results.
     - Systematically manage customer relationships.
     - In attempting to achieve customer satisfaction, ensure a balanced approach to other stakeholders (including owners, employees, suppliers, financiers, local communities and society as a whole).

2. ***Leadership* (*Plan*):** *Leaders establish unity of purpose and direction of the organization. They should create and maintain an internal environment in which people can become fully involved in achieving the organization's objectives.*
   - *Main benefits:*
     - The goals and objectives of the organization are understood by the employees and this motivates them.
     - Activities are evaluated, aligned and implemented in a unified way.
     - Errors in communication between levels of an organization are minimized.
   - *Aspects that may arise:*
     - Take into account the needs of all stakeholders, including customers, employees, suppliers, financiers, local communities and society as a whole.
     - Establish a clear vision of the future of the organization.
     - Set goals and achievable targets.
     - Create and sustain shared values and behavioral patterns based on fairness and ethics at all levels of the organization.
     - Establish trust and eliminate fear.
     - Provide the staff with the resources and training required, as well as the freedom to act responsibly.
     - Inspire, encourage and recognize people's contributions.

3. ***Involvement of people* (Do):** *People at all levels are the essence of an organization and their full involvement enables their abilities to be used for the organization's benefit.*
   - *Main benefits:*
     - Employees are motivated, committed and involved in the organization.
     - Innovation and creativity achieve the objectives of the organization.
     - Employees are accountable for their own performances.
     - Employees are eager to participate and contribute to continual improvement.
   - *Aspects that may arise:*
     - The staff understands the importance of its contribution and its role in the organization.
     - The employees identify constraints to their performance.
     - They agree to be accountable and to fulfill their responsibilities when it comes to solving problems.
     - Employees evaluate their performances against goals and objectives.
     - They actively seek opportunities to enhance their competence, knowledge and experience.
     - Employees freely share knowledge and experience.
     - They openly discuss problems and issues.

4. ***Process approach* (Do):** *A desired result is achieved more efficiently when activities and related resources are managed as a process.*
   - *Main benefits:*
     - Lower costs and shorter cycle times are achieved through effective use of resources.
     - Results are improved, consistent and predictable.
     - Added value is looked for in all essential components of the organization.
     - There is a focus on improvement opportunities and ranking by priority.
     - Greater responsiveness to external changes is achieved by better communication between the organizations' components (interrelationships).
   - *Aspects that may arise:*
     - The activities necessary to achieve the desired result are defined by analyzing the process input and output data.
     - Clear responsibilities are established for managing key activities.
     - The potential of key activities is analyzed and measured.
     - The interfaces of key activities within and between different functions of the organization are identified.
     - Factors – especially resources, methods and materials – that will improve key activities of the organization are brought into focus.
     - Risks, consequences and impacts of activities on customers, suppliers and other stakeholders are evaluated.

5. ***System approach to management* (Do):** *Identifying, understanding and managing interrelated processes as a system contributes to the organization's effectiveness and efficiency in achieving its objectives.*
   - *Main benefits:*
     - Integration and alignment of processes will best achieve the desired results.
     - Effort can be focused on key processes.
     - Interested parties (stakeholders) will have confidence in the consistency, effectiveness and efficiency of the organization.

- *Aspects that may arise:*
  - The system can be structured to achieve the organization's objectives as effectively and efficiently as possible.
  - Interdependencies between system processes are understood.
  - Structured approaches harmonize and integrate processes.
  - Better understanding of the roles and responsibilities necessary for achieving common objectives will reduce cross-functional blockage.
  - An understanding of organizational capabilities can establish resource constraints prior to action.
  - The way particular activities should be operated can be targeted and defined within a system.
  - The system can be continually improved through measurement and evaluation.

6. ***Continual improvement* (*Act*):** *Continual improvement of the organization's overall performance should be a permanent objective of the organization.*
   - *Main benefits:*
     - Improved organizational capabilities lead to competitive advantages.
     - Improvement activities are aligned at all levels with regard to the organization's strategic objectives.
     - Flexibility and the ability to react quickly to opportunities are achieved. It is easier to adapt to external change (technology, markets, regulations, etc.).
   - *Aspects that may arise:*
     - Use a consistent approach across the organization for continual improvement of its performance.
     - Provide staff training methods and tools for continual improvement.
     - Continual improvement of products, processes and systems should be an objective for every individual in the organization.
     - Set goals to guide continual improvement and provide measurement data to monitor this.
     - Recognize and acknowledge improvements.

7. ***Factual approach to decision making* (Check):** *Effective decisions are based on the analysis of data and information.*
   - *Main benefits:*
     - Informed decisions are made.
     - It is easier to demonstrate the effectiveness of past decisions through reference to factual records.
     - The ability to review, challenge and change opinions and decisions is increased.
   - *Aspects that may arise:*
     - Ensure that data and information are sufficiently accurate and reliable.
     - Make data accessible to those who need it.
     - Analyze data and information using valid methods.
     - Make decisions and take actions based on factual analysis, balanced with experience and intuition.

8. ***Mutually beneficial supplier relationships* (Do):** *An organization and its suppliers are interdependent and a mutually beneficial relationship enhances the ability of both to create value.*
   - *Main benefits:*
     - It is easier to create value for both parties.

Flexibility and fast responses to market changes or customer needs and expectations are enhanced.
Costs and resources are optimized.

- *Aspects that may arise*:
  Establish relationships that balance short-term gains and long-term considerations.
  Pool expertise and resources with partners.
  Identify and select key suppliers.
  Make clear and open communications.
  Share information and future plans.
  Establish joint development and improvement actions.
  Inspire, encourage and recognize improvements and achievements by suppliers.

The main benefits and the aspects that can arise from these principles are not trivial:

- their order is logical;
- they also provide criteria for judging, during internal or external audits, the effectiveness of the implementation and the advantages that can arise.

If, in the course of implementing a QMS, the staff does not know about these eight principles and certain aspects and benefits arising from them, the causes of this must be analyzed and corrective actions introduced. There are significant risks if the quality system is perceived only as an additional legislative burden binding employee autonomy and stifling creativity/business competitiveness.

The eight principles are recognized worldwide by those concerned with management in general and quality in particular. They represent the basis for the lasting success of any organization or company providing a service or product. They are, in fact, taught in all business schools and executive education programs. Many failures in public institutions and private economies are due to the fact that some of these principles are not respected (urgent decisions, partial investigations of a project, power struggles, demotivated staff, etc.).

4.2.2 ISO 9001 helps create a learning environment

Compared with previous editions, the current version of ISO 9001 is less litigious and instead focuses significantly on process management and its interfaces. The management of a process and its interrelations with other processes is not just a matter of describing a procedure, and the fact of having written procedures does not create the basis for process management.

This view is consistent with difficulties encountered by some companies, which are most often related to poor quality in the management of processes and their interfaces rather than procedures of more or less good quality.

For example, a customer-supplier relationship is defined by a contract, i.e., a set of requirements, expectations and needs (not only "technical") as well as a set of "game rules" to follow. The relationship is rarely defined by procedures to be followed.

Therefore, the standard places great emphasis on planning and validation of approaches to meet needs and variables. A quality management system is a strategic decision of the organization. The design and implementation of a QMS reflect varying needs, particular objectives, products provided, processes used, as well as the size and structure of the organization.

In this sense, the idea is to define the basic rules for planning activities as well as the key stages of validation. It is evident that the arrangements must be closely related to the level of competence of the staff, and this is why the standard emphasizes training and employee skills, resulting in a significant reduction in basic procedures and instructions. The definition of processes and the selection of only those that are important to the quality of the product or service will determine the best scope for the QMS. These are the bases for developing the organization as if it were learning [4.6] (Figure 4.2). The standard corresponds here to the views of *Deming* and *Juran*.

Another fundamental principle of quality management is the prevention of nonconformities. However, the level of prevention necessary is always related to the risk level (this is a general principle that does not apply exclusively to a management system). For this reason, systematic and methodological risk management is gaining importance in all areas (financial, legal, technical, etc.). It is very useful to assess the importance of the process and its requirements when it comes to descriptions, training and any other preventive measures according to an established risk management approach, where the "product" corresponds to an updated map of risks to be controlled. The current standard (ISO 9001:2008) only addresses this issue superficially so ISO 31000:2009, *Risk management*, is the reference for this area.

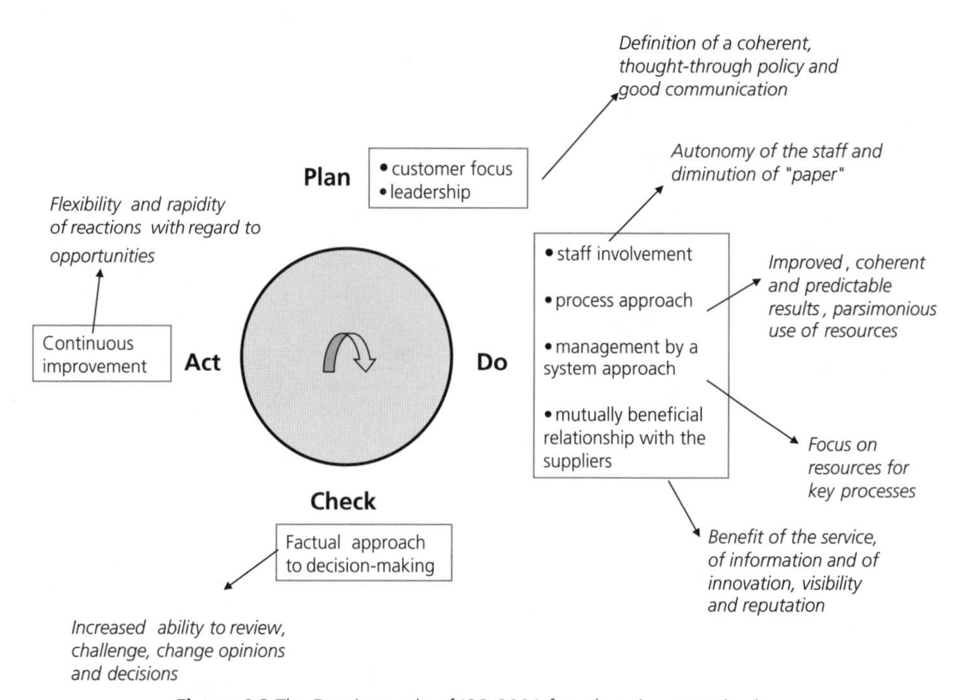

Figure 4.2 The Deming cycle of ISO 9001 for a learning organization.

4.3 The system approach in eight steps and four sections

The system approach to implementing quality management is based on the following steps:

- determine customer and stakeholder expectations and needs;
- determine the quality policy and quality objectives of the organization, taking into account the expectations and requirements of internal and external stakeholders;

- identify processes and responsibilities relating to quality objectives;
- assess the resources required to obtain quality as a result of the policy;
- develop methods that gauge the efficiency and effectiveness of the process;
- apply measurement methods to obtain the first results;
- describe and implement the methods, procedures and potential equipment to prevent nonconformities and eliminate their cause(s);
- establish a process of continual improvement of the quality management system, including reviews of the progress and the effectiveness of the measures taken.

These steps are conducted over four sectors: management responsibility, determination of resources, product realization, and measurement, analysis and improvement. *These four sectors strongly influence the process of continual product improvement* (Figure 4.3).

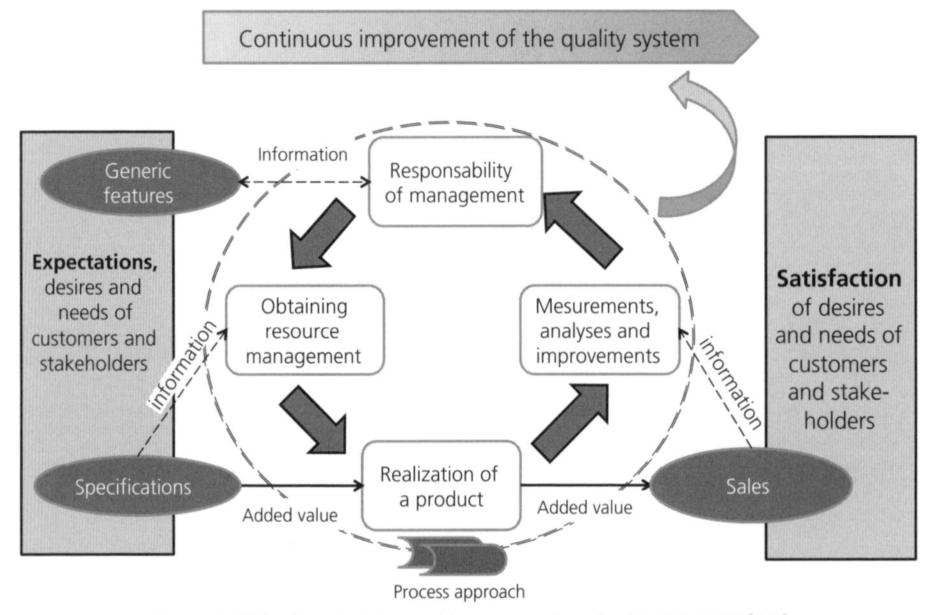

Figure 4.3 The four decision-making sectors described in ISO 9001 [4.7].

4.4 ISO 9001: Emphasis on employees' skills

The ISO approach incorporates the limitations of ensuring conformity (essential when it comes to tangible products and the scope of operations) and the requirements for total quality. The latter play a key role in all sectors and at all levels of any organization. The level of compliance and ad hoc behavior resulting from employee skills should be evaluated for each process (Figure 4.4).

In general, employee skills can easily be grouped into three categories:

- professional skills or trades, which refer to technical knowledge relating to the job (know-how);
- ethical skills, a form of ethical and moral knowledge related to carrying out the trade; and
- generic skills, a form of knowledge and know-how related to trade competencies that are not specific to a certain profession (communication, teamwork, basic principles of management, use of basic software packages, etc.).

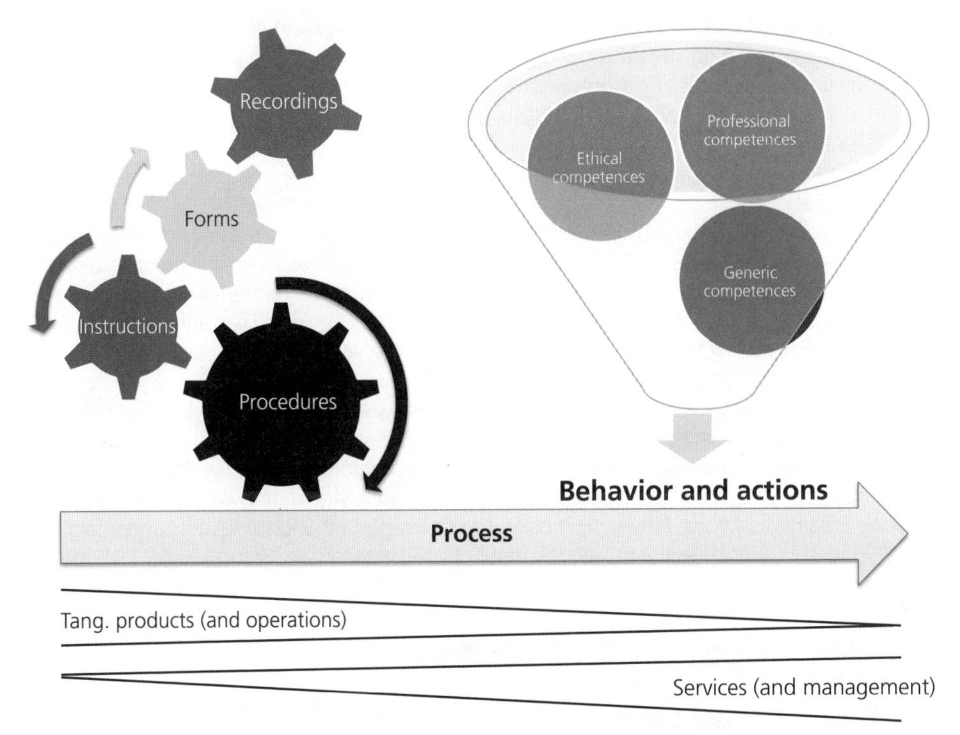

Figure 4.4 Procedures, instructions, competences and processes.

The application of ISO 9001 to service companies often leads to the importance of competences over procedures and instructions. This is true for public services, such as hospitals, clinics, health centers and medical institutions. On top of the three skills listed above, their employees must have an additional quality: a sincere and genuine compassion for patients and residents. If this is insufficient, the application of a quality system, with its underlying formalities, can at best help to gain certification and at worst reduce the quality of service to the patients [4.8].

ISO 9001 focuses on the development and sustainability of staff competence[3] at all levels. It not only stimulates training within the company, but also "learning by doing", and finally an increase in capabilities, interpersonal skills, knowledge and communication skills, which require individual or team training.

The implementation of ISO 9001 therefore requires significant skills in human resources, a department that has more detailed specifications than those of a pay office, or which plays the role of "axe man" during dismissals, lay-offs or "bad cop" during the recruitment of personnel. The management of human resources is the key in any organization concerned with information and knowledge, where services, human contact, and negotiations are essential.

Once again, the *Deming cycle* is a valuable tool in order to:

- define the competencies needed in a sector allocated to a process;
- identify existing skills in the sector or team; look for weaknesses;
- identify specific training needs, as well as teaching methods that promote the necessary learning;

[3] ***Competence*** is deep knowledge that is recognized, and which confers the right to judge, decide, advocate, direct and act in certain matters.

- implement the training of, and search for, targeted personnel;
- formalize existing skills through an ad-hoc function sheet, and verify that all responsibilities for a process or procedure have been established;
- at the annual review of an employee, check the status of how well the skills are mastered;
- start the cycle again during internal audits, or the development of new products or markets.

The development of employee competences and the fact of mastering them reduce the number of procedures and instructions. However, employee skills can quickly become outdated; continuing education supports the development and maintenance of capabilities. For this reason, any training undertaken is recorded in the *Life Long Learning*[4] project of the EU, which takes into account the aging of the workforce within the EU, and was formalized by the Copenhagen process (2002).

A survey of employee skills can, when it comes to sensitive and critical activities, be accompanied by a risk management approach according to FMECA[5] (see section 4.13). This approach [4.9] helps to identify what training actions should be prioritized in order to keep within an acceptable range the risk of serious malfunctions due to an organization's staff. It may well be that after such an analysis some functions, previously classified as less important and sometimes underpaid, will be adjusted (e.g., personnel controlling the input of raw materials and components).

4.5 Documentary foundation

4.5.1 A pyramidal base

In any organization there are procedures[6], instructions[7], activities[8], forms[9] and records[10]. With this system, know-how is durable and is not exclusively found in the sometimes failing memory of employees, who may be assigned to other tasks, be transferred, retire or change jobs. These documents facilitate the order, transparency and structure of the organization's functions. If they are analyzed and formalized, it is possible to create collective and organizational effectiveness and to increase the acceptance of change and learning. They must be applied and updated regularly and kept separate from other information or training, for which random updating can be safely carried out.

---

[4] Cf. the presentation at: http://ec.europa.eu/education/policies/lll/lll_fr.html

[5] ISO 9001 is among ISO's most well-known standards. Like all ISO standards, it is the result of international, expert consensus. The 2008 version addresses the concept of risk but there is no obligation to perform a risk analysis before defining the process or the need for procedures, for instance.

[6] A ***procedure*** is a system of unitary or elementary actions, activities or operations connected in series or parallel, in order to achieve a common goal (e.g., a procedure for handling complaints). It is often represented as a flowchart. A freeware for writing procedures in the form of flowcharts can be found at http://www.logigramme.fr/.

[7] ***Instructions*** relate to an activity or several activities. They are intended for the employees, and recommend a sequential list of actions oriented towards a goal. They are formal (starting and stopping of equipment, etc.).

[8] An ***activity*** is a step in a procedure in which an elementary action is executed.

[9] A ***form*** is a specific document that must be respected, on which questions are printed and opposite which a person must register his or her answers (registration form).

[10] A ***record*** is a written notation (on paper or computerized) of information, signals and various phenomena (recording a temperature, but also the registering of entries in a building, etc.).

The existence of documents is not in itself evidence of effectiveness and efficiency (or the Austro-Hungarian empire would have been efficient…), nor of resilience in a turbulent market. The organization still needs to systemize its working methods and put them in hierarchical order, without losing sight of the fact that they are primarily to satisfy the customers or beneficiaries of the service – who sacrifice some of their resources to obtain this. This initiative also assures the shareholders, owners and investors that resources are used sparingly and thoughtfully.

From the perspective of logic, subsidiary, and hierarchy of actions (ranging from strategy to the employee's activity), the documents of a quality management system have a hierarchical structure and a pyramidal shape (Figure 4.5).

From top to bottom:

- the quality manual, which is the only public document, is a "showcase" for customers regarding the quality of the organization, and contains management commitments as well as general information;
- the processes;
- the procedures;
- employee skills in the form of job descriptions;
- instructions and procedural guidelines;
- forms and records; a form becomes a record when the expected data is written down.

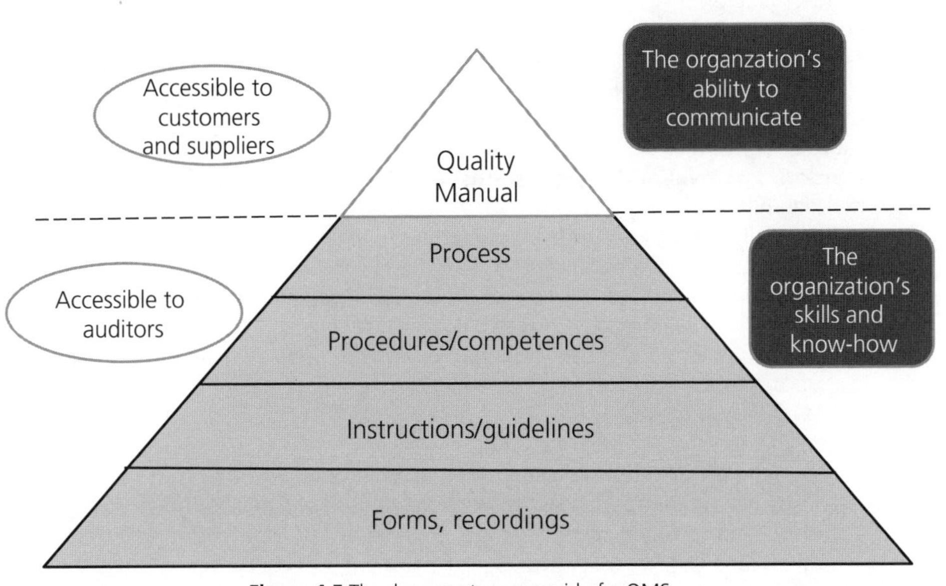

Figure 4.5 The documentary pyramid of a QMS.

These levels are connected to each other by a system of numbering and references.

Process or procedure? Processes and procedures may describe the same object: however, a process highlights the outcomes and resources to be used, whereas a procedure focuses on how to accomplish these activities. These two approaches are not mutually exclusive.

Some cases require an ad-hoc document, a ***quality plan***[11], which treats the possible dilemmas that may occur when applying a procedure to a particular case.

An important category when it comes to forms and records is the certificate of conformity of the product, which may be required by the purchaser. It specifies that the batch produced meets the specifications and further attests to the liability of the company.

4.5.2 Structure of a quality manual

Of all the documents, the quality manual is the basis of the entire process. For small businesses, all quality documentation may be found in this manual. Therefore, it is all that is required in order for the quality management of the organization to comply with the standard. The quality manual should provide information about the philosophy and methods used by the organization to meet its policy and objectives. **ISO/TR 10013:2001,** *Guidelines for quality management system documentation*, specifies the required contents:

- **title**;
- **subject:** e.g., *This quality manual (hereinafter abbreviated QMM) describes the quality policy and quality management system (hereafter abbreviated QMS) implemented by MECANCO SA in order to achieve the objectives. The QMM of MECANCO SA is the result of reflection and work of all employees to achieve all the objectives of improving their performances. This manual was developed to ensure optimal products and services, as well as an ethical performance of the branch. It is part of the quality policy of MECANCO SA;*
- **scope of application**: e.g., *The QMS applies to all services provided within MECANCO SA, except for the gas station owned by the company. The entire QMS is implemented as of September 11, 1998. It has been revised and adapted to meet the requirements of ISO 9001:2000.*
- **table of contents**;
- **revision, approval and review status,** date of issue, explanation of any changes (tracking history);
- **quality policy and (measurable!) objectives:** e.g., *Management and all employees of MECANCO SA have established a QMS based on ISO 9001:2008 to ensure the constant improvement of services, of the performance and of the organization of MECANCO SA. The objectives of the company policy are to:*
  - *satisfy the needs and expectations of the clients through a proper identification of specific needs;*
  - *continually improve:*
    - *the operation of MECANCO SA;*
    - *manufacturing technologies and control methods;*
    - *the level of performance, maintenance and use of equipment;*
  - *adapt manufacturing practices to the advancement of knowledge and technological evolution;*
  - *satisfy MECANCO SA employees through continuing training and knowledge transfer as well as a structured and open communication policy;*
  - *prevent the occurrence of potential adverse events by adopting a good process control;*
  - *use internal and external resources at MECANCO SA effectively and efficiently;*
  - *ensure the sustainability of MECANCO SA.*

[11] The exact definition is: document specifying which procedures and associated resources shall be applied by whom and when for a project, product, process or contract. (ISO 9000)

The quality manual should then show how these objectives will be measured and how often.

- ***Organization responsibility, authority***: hierarchical structure, organization chart and list of responsibilities, who decides what, functions, their duties, responsibilities and competences.
- ***Reference documents***: list of all relevant documents completing the quality manual (standards and legal acts, the group's internal standards, branch recommendations, etc.).
- ***Description of the quality management system***: ***presentation of the implementation of the QMS***, description and list of processes related to the quality standard and their order, tables and analyses of their interactions, procedures or list of the latter.
- ***Appendices***: all documentation that judiciously supplements the various entries in the quality manual (e.g., company brochure communicating the quality policy, etc.)

4.5.3 Contents of a procedure and other documents

A procedure can contain text (particularly as lists), flow charts[12], and tables. It may also contain references to instructions used for the procedure. Instructions describe the sequential activities of a person, while a procedure is more complex, and refers to the execution of a task that can take on several functions in the company.

The contents of a procedure can be formalized as:

- ***title***;
- ***goal***;
- ***scope of application***;
- ***responsibility and authority,*** functions and people involved, decision makers;
- ***description of activities:*** who does what, for whom, by whom, why, when and how, with what resources and materials, from the input to the output, by listing the services/functions/responsible persons, control steps, and actions;
- ***records*** associated with the procedure;
- ***appendices,*** *if any*: conversion tables, graphs, diagrams, etc.;
- ***revision, approval and review status,*** date of issue, explanation of any changes (tracking history).

The content of ***instructions*** is similar to that of a procedure, although simpler and written in a way that is adapted to the employees.

Forms comprise a title, identification number, date and revision level and must be listed in a procedure or instructions, but more rarely in the quality manual. Everything must conform to the contents of clause 4.2.3 "Control of documents" of ISO 9001:2008.

The ***quality plan*** is part of the quality management system and may include procedures, instructions and forms. It must demonstrate how it is applied to a particular situation, and identify and document how the organization will implement the conditions that are specific to a project, product, or particular contract. The writing of a quality plan is specified in ISO 10005:2005, *Quality management systems – Guidelines for quality plans*.

---

[12] A freeware for creating flowcharts from a process perspective can be found at: http://www.logigramme.fr/ (in French) or http://www.yworks.com/en/products_yed_about.html (in English).

Specifications are documents containing requirements.

Finally, the literature often refers to ***external documents*** (laws, regulations, charters, best practices, codes, technical norms, guidelines, standards, maintenance manuals, etc.). These should be referenced at their rightful place in the documentary system.

4.5.4 Document management

Document management is the subject of a specific ISO document, ISO/TR 10013:2001, *Guidelines for quality management system documentation.*

This is a six-step process, and the detailed degree of completion is a critical success factor in the implementation of ISO 9001:

1. ***Identification of documents***:
    a. ***Referencing***: This step is crucial:
        i. Identification of the type; quality manual (QM), procedures (PRO), instructions or procedural guidelines (MO), forms and records (FO), work sheets and staff expertise (CC), quality plan (QP).
        ii. Identification of the process to which the document refers, e.g., sales.
        iii. Identification of the document by means of the date of the last revision (tracking history).
        iv. Identification of the issuer (department, unit, person).
    b. ***Pagination***;
    c. ***Title***;
    d. ***Logo*** (of the company).
2. ***Drafting the document***: this is generally done by a drafting committee.
3. ***Validation of the document (visa)***: this includes the topics:
    a. written by:
    b. validated by:
    c. approved by (if necessary): *the review may be conducted by a single person*.
    d. authorized by:
4. ***Dissemination of the document or making it available***: this step is facilitated by IT management, by establishing functions sheets (describing competences), and by assigning responsibilities to the relevant people.
5. ***Modifying the documents***: a modification must be justified to maintain the history. Steps 1–4 are performed again in full.
6. ***Archiving and deleting documents***: here, to delete documents means to remove them from circulation and not to destroy them forever. Archiving is critical and compulsory as it enables, in the case of a dispute or proven nonconformity (even months after placing the product on the market), to present elements in addition to the certificates of compliance. Archiving can be divided into:
    a. identification;
    b. collection;
    c. indexing;
    d. access;
    e. classification;
    f. storage;
    g. conservation.

The current software packages for document management, intranet systems development and search algorithms such as *Google Desktop* greatly facilitate these operations, which were previously budget- and time-consuming.

4.6 Contents of ISO 9001:2008

4.6.1 Who develops ISO standards? And how?

The contents of ISO 9001 were decided by an international committee working in an atmosphere of consensus. For ISO 9001, this was ISO's Technical Committee (TC) 176[13] (ISO/TC 176), Sub-committee 2, bringing together partners from the private sector and a broad representation of stakeholders. ISO/TC 176, like other ISO committees associated with other standards, has three activities:

- the gathering of broad conceptual knowledge and a catalog of state-of-the-art good management practices;
- the precise, careful and structured formalization of this knowledge according to the most generic form possible (Anglo-Saxon heritage);
- editing and translation into several languages.

ISO/TC 176 also has information on interpretations of the chapters of the standard[14] (a gloss on the "officially recognized" text of ISO 9001).

The drafting of standards is also based on:

- *ISO/IEC Directives – Part 1: Procedures for the technical work* (http://www.iso.org)
- *ISO/IEC Directives – Part 2: Rules for the structure and drafting of International Standards* (http://www.iso.org)
- *ISO/IEC Directives, Procedures for the technical work of ISO/IEC JTC 1* (http://www.iso.org)
- *ISO/IEC Directives, Supplement – Procedures specific to ISO* (http://www.iso.org)
- *ISO/IEC Directives, Supplement – Procedures specific to IEC* (http://www.iec.org)

Writing an ISO standard is similar to writing a patent. Both summarize the state of an art at a certain time, but standards widely disseminate good practices, and are therefore in the public domain (the cost of purchasing a standard is a few hundred Swiss francs). The rate of a consultant is certainly higher, but it is true that understanding a standard is a very demanding exercise that first requires comprehension, and then communication, in order to be transcribed…

4.6.2 Drafting of a standard and research activities: a large gap

Writing a standard is fundamentally different from usual research activities in that it is the distillation of the sum of proven experiences, accepted by many experts, then drafted in a generic, comprehensive and concise fashion. There is a huge amount of knowledge within the enormous catalog of ISO standards. This is valuable for the global business economy, and technological and analytical activities worldwide, but also for the transfer of knowledge and expertise to emerging countries.

Colleges and universities do not play a leading role in the establishment of ISO or EN standards, even if technological university professors often take part in expert committees. The different approach to academic research explains why universities have so far been little involved in this activity, despite its obvious impact on management, particularly that of technology.

[13] Information available at: http://www.bsi.org.uk/iso-tc176-sc2

[14] Available at: http://www.tc176.org/Interpre.asp

4.6.3 Structure of the ISO 9001 standard

The structure of the table of contents of ISO 9001 is often repeated in other similar standards. It is frequently adopted from Chapter 5 onwards when establishing a quality manual – for an organization of regular size and/or delivering complex products/services.

The topics in the table of contents below that have not been previously treated will be briefly commented on. However, only the detailed content of the standard can be considered binding.

The key chapters of the standard are the four sectors of the system approach:

5. Management responsibility
6. Resource management
7. Product realization
8. Measurement, analysis and improvement.

Table of contents

0 Introduction

The introduction states that the adoption of a quality management system should be a strategic decision of an organization. Its design and implementation are influenced by:

a) its organizational environment, changes in that environment and the risks associated with it;
b) the varying needs of the organization;
c) its particular objectives;
d) the products it provides;
e) the processes it employs;
f) its size and organizational structure.

It is stated that the aim of the standard is not to unify the structure of quality management systems. It is possible to implement ISO 9001 and ISO 14001 together:

a) in order to understand and take into account the needs of internal and external stakeholders;
b) when it is required to consider processes in terms of added value;
c) to obtain results in terms of performance, efficiency and effectiveness, and processes;
d) for continuous process improvement based on measurable data.

1 Scope

This item appears at the beginning of each ISO standard and clearly defines the topic, the aspects covered and the limits of applicability of the document or part thereof. It must not include requirements.

1.1 General – *for* ISO 9011, *the contents can be found in Section 4.3*

1.2 Application – *the generic aspect of the standard is highlighted*

2 Normative references

This element is not found in all standards. It must be included if there are documents cited in the standard that are indispensable for its application. Dated references contain the year of publication. Here, ISO 9001 only lists the ISO 9000:2005 standard (see Chap. 4.8).

3 Terms and definitions

This optional element, found in many standards, provides the definitions necessary for understanding several terms used in the document. For ISO 9001, the most important are found in the form of footnotes in this module.

4 Management quality system

4.1 General requirements – *they are found at the beginning of Chapter 4.3 of this module.*

4.2 Documentation requirements – *these are summarized in section 4.5 of this chapter.*

5 Management responsibility

5.1 Management commitment – *Remember that the role of general management is to show to the employees its commitment for the development and implementation of the quality management system, but also for improving its effectiveness and efficiency by:*

a) *communicating to the entire organization the importance of satisfying the needs of customers and other stakeholders;*
b) *establishing the quality policy;*
c) *ensuring that quality objectives are defined;*
d) *conducting management reviews of the system;*
e) *allocating and distributing resources.*

5.2 Customer focus – *General management ensure that the customers' needs are identified and addressed, with a view to continual improvement of the services offered to them.*

5.3 Quality policy – *It is the management's responsibility to ensure that the quality policy is tailored to the goals of the organization, that it includes a commitment to comply with its requirements and to continually improve the effectiveness and efficiency of the management system, to establish a framework for determining and reviewing quality objectives, to communicate them within the organization and ensure that they are understood, and to regularly review the policy to make sure that it is always appropriate.*

5.4 Quality planning – *This is the methodical organization of quality, acting at various levels of the organization such as in the:*

- *training plan;*
- *plan of programmed internal audits;*
- *production plan;*
- *maintenance plan, etc.*

Quality planning can be related to resource management, maintenance and renewal, but may also contain more strategic aims, particularly when it comes to a deliberate and progressive reduction of non-standard quality, etc. The standard indicates that quality objectives must be measurable and consistent with the quality policy defined by management.

5.5 Responsibility, authority and communication – *It is up to the general management to ensure that:*

a) *responsibilities for quality are defined, assigned and communicated within the organization;*
b) *a quality manager is appointed, whose tasks should be to:*
   - *ensure that processes relating to quality management are established, implemented and maintained,*
   - *report regularly to the general management on the performance of the QMS and any need for improvement,*

- *promote within the organization and among its employees a proactive attitude when it comes to customer needs;*

c) *ensure that internal ad-hoc communication is in place and that it particularly concerns the effectiveness of the QMS.*

5.6 Management review – *According to the standard, a regular, planned and documented review of the quality management system is required to ensure its relevance, effectiveness and appropriateness. It should include opportunities for change as well as modifications of the quality policy and objectives. This review can be seen as a process as it consists of:*

- ***Inputs:***
  - *results from internal and external audits;*
  - *feedback from customers (sales reports, questionnaires, claims, investigations, etc.).*
  - *data on process performance and product conformity;*
  - *a list giving the status of corrective and preventive actions;*
  - *actions identified in response to previous reviews;*
  - *changes performed (processes, suppliers, etc.) that may affect the product;*
  - *recommendations for improvement (suppliers, external expertise).*
- ***Outputs**, which are decisions and actions involving:*
  - *resource needs;*
  - *improvements to the effectiveness of the QMS and its processes;*
  - *product enhancements requested by the client.*

6 Resource management

Resources must be commensurate with the requirements of clauses 7 and 5.

6.1 Provision of resources – *It is specified that they must be able to satisfy customer needs and those of the QMS in a perspective of continual improvement.*

6.2 Human resources – *This point is discussed in more detail in Section 4.4 of this module.*

6.3 Infrastructure – *When it comes to infrastructure, the standard involves buildings and workspaces (and directly associated services), production equipment (hardware and software, in a more general sense than that of information technology), transportation, but also communication systems and information. The infrastructure should allow a product to be made that meets customer expectations.*

6.4 Work environment – *This should help to achieve consistent products. The environment covers the conditions under which production is implemented (e.g., physical environment, noise, humidity, temperature, lighting).*

7 Product realization

7.1 Planning of product realization – *Planning involves the need to develop appropriate processes for product realization and to have the right resources, but also suitable specifications, documentation, and the activities of verification, validation, monitoring, testing and recording. The planning document must be adapted to the organization and therefore be easily understandable.*

7.2 Customer-related processes – *The product must meet customer expectations, both implicit and explicit, as well as those of stakeholders (legislators). The specifications should be reviewed regularly and it must be verified that they are achievable before delivery. For example, a contract or contract modification is established with the client. Even in the absence of a specification document provided by the customer, the organization must give its formal consent (a thorough analysis of customer requirements and specifications can be found in Chapters 1 and 2). Communication must be clear (information, monitoring and possible modification of deliveries, claims handling and feedback).*

7.3 Design and development – *The organization must provide and implement a controlled design and development of the product (including the definition of design and development stages, validation and reviews, and also the responsibilities and authorities within the organization associated with each step). The interfaces between the groups involved in design and development should be clear in terms of communication and task responsibility. (See also Chapter 2.)*

7.4 Purchasing – *Mutually beneficial customers/suppliers relationships are based on transparent communications (in particular, full definitions and specifications when purchasing, an objective assessment of suppliers, audit integrity, inspection and control of supplied products/components/raw materials).*

7.5 Production and service provision – *The production process and its conditions of operation must be planned, controlled, documented and declared to comply with the specifications. The product must be identified and traceability established when necessary (for certain food products, for example).*

7.6 Control of monitoring and measuring equipment – *Monitoring, measurement and measuring instruments should be controlled, and their calibration, checks and possible deviations should be provided for and documented (also valid for integrated software).*

8 Measurement, analysis, and improvement

8.1 General – *All checks, measurements, analyses and process improvements should demonstrate that the product conforms to the requirements, and also to the quality system, and should enable the effectiveness of the QMS to be improved.*

8.2 Monitoring and measurement

8.2.1 Customer satisfaction. *It is necessary to have a continual perception of the degree of customer satisfaction (the standard does not require measuring the degree of customer satisfaction). This requirement is closely related to the first principle of management "Customer focus" (see ISO 9001, clause 5.2).*

8.2.2 Internal audit *(see also 4.12) for which the objectives are to:*

- *verify the adequacy of the management system with the objectives of short-, medium- and/or long-term achievements;*
- *verify the proper or efficient functioning of the system;*
- *understand any malfunction;*
- *verify that the system complies with the internal requirements of the company and with ISO 9001.*

8.2.3 Monitoring and measurement of the processes, *i.e., their efficiency and performance. One advantage of the process approach is the permanent control it provides over the relationships between processes within the process system, as well as their combinations and interactions. When used in a quality management system, this approach emphasizes the importance of:*

a) understanding and meeting the requirements;

b) considering processes in terms of added value;

c) measuring the performance and process efficiency;

d) continually improving processes based on objective measures.

8.2.4 Monitoring and measurement of product, *its features and how specifications are satisfied. This is important since it implies that the processes and business activities are not only analyzed and adjusted on the basis of noncompliance, but primarily on the basis of observed trends or deviations in the results of measurements performed on the products or services provided. This involves recording the measurement results correctly.*

8.3 Control of nonconforming product – *This must be identified, controlled and removed from production (process, decision-making authorities, responsibilities); its destruction, reuse or release is subject to an ad-hoc and recorded procedure. The causes of noncompliance should be sought and established, and a list of nonconformities must be kept. The nonconforming product should be retested after corrections (if it can be reworked).*

8.4 Analysis of data – *The data, outputs of clauses 5.2, 6.2, 7.2 to 7.5, 8.2 and 8.3, including information about suppliers and analyses of process trends, once formatted, make it possible to determine what kind of improvements can be implemented, and to demonstrate the relevance and effectiveness of the QMS.*

It is essential to develop approaches to performance analysis provided they are concerned with compliance. They should not be limited to an analysis of malfunctions. In this way, they are really a principle of prevention.

8.5 Improvement – *This is based on the quality policy and objectives, audit results, data analysis, corrective (clause 8.3) and preventive (clauses 7.3, 8.2.1 to 8.4) actions as well as management review. Any noncompliance must be investigated in order for its causes to be determined, which in turn must be eliminated by a corrective action, and its lasting effects must be reviewed and documented. The organization should also conduct risk analysis (although the standard does not specifically mention this) at regular intervals in order to implement any preventive action to overcome a potential and statistically plausible risk.*

4.7 ISO 9004:2009 – beyond quality management

ISO 9004:2009, *Managing the sustained success of an organization – A quality management approach*, is a standard that complements ISO 9001 by expanding quality management to include the expectations of stakeholders, taking into account the often unpredictable changes in the environment.

4.7.1 ISO 9004:2009, presentation

4.7.1.1 The objective of ISO 9004:2009: sustained success

ISO 9004 provides a formal approach enabling organizations to achieve sustained success[15] in a complex, constantly changing environment. It thus confronts management with constant challenges. Success is achieved through quality management, based on the eight principles defined in 4.2 of this book. The approach is applicable to any organization, regardless of its size, type or activity. The standard focuses on a company's ability to meet the needs and expectations of its customers and other stakeholders, in the long run and in a balanced manner.

4.7.1.2 A broader vision of an organization's performance

Compared with ISO 9001:2008, which deals with the quality management of products and services to improve customer satisfaction, ISO 9004:2009 provides a broader perspective of quality management, especially for performance improvement. It is useful to companies whose top management wish to go beyond the requirements of ISO 9001, with a view to continual improvement, measured through the satisfaction of customers and other stakeholders.

4.7.1.3 A cardinal tool: self-assessment

ISO 9004:2009 helps organizations improve product quality and delivery of services to their customers by emphasizing self-assessment as an important tool at different levels, especially to:

- compare their level of maturity with regard to leadership, strategy, management system, resources and processes;
- identify strengths and weaknesses;
- identify opportunities for improvement or innovation, or both;
- facilitate the strategic planning process of any business.

4.7.1.4 An essential complement to ISO 9001

ISO 9004:2009 complements ISO 9001:2008 (and vice versa), but they can also be used independently of one another. ISO 9004 is not intended for third-party certification, nor as a regulatory framework for contractual purposes, nor as a guide for the implementation of ISO 9001:2008. By enlarging the circle of clients to include stakeholders, ISO 9004, used in conjunction with ISO 10014 as well as with ISO 9000 and ISO 14000, helps promote sustainable development and social responsibility within an organization, by including many elements found in the *Triple Bottom Line* (see Chapter 13).

4.7.2 A model of maturity for lasting success

The standard uses an assessment of the performance level of the organization, i.e., its level of maturity. An organization is considered mature when it obtains results in an efficient and sustainable manner by:

- identifying and taking into account the needs and expectations of all stakeholders;
- establishing mutually beneficial relationships with suppliers, partners and stakeholders;

---

[15] Result of an organization's ability to achieve and maintain its performance in the long term.

- constantly involving stakeholders by informing them of the organization's activities and plans;
- using a wide range of approaches, including mediation, to balance the often competing or even conflicting needs and expectations of stakeholders;
- conducting careful monitoring of its environment;
- identifying the short- to medium-term risks associated with its operation by including the control of these risks in its strategy and policy;
- proposing a vision and long-term plan, and by determining and deploying strategy and policy;
- focusing on design objectives and goals related to strategy and policy;
- establishing appropriate processes, capable of adjusting quickly to environmental changes;
- optimally managing resources, and anticipating future needs from these resources, including human resources and the development of skills;
- demonstrating its confidence in its staff by offering numerous learning opportunities, which will result in increased motivation and involvement;
- developing and implementing the process of innovation and continual improvement.

4.7.3 Content of the standard

ISO 9004 defines six areas that contribute to the sustained success of an organization:

- management of sustained performance;
- strategy and policy;
- resource management;
- process management;
- monitoring, analysis and review;
- improvement, innovation and learning activities (see Figure 4.6).

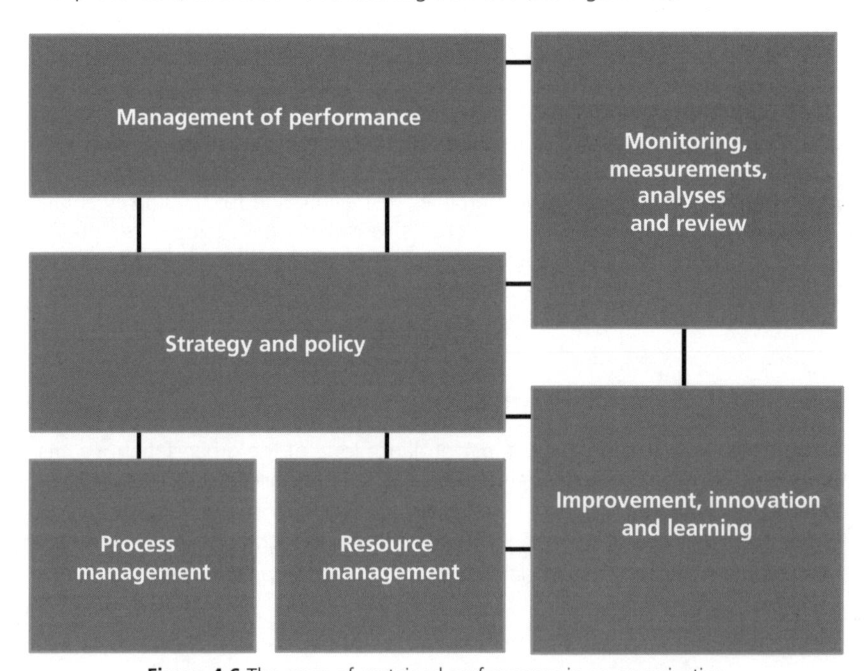

Figure 4.6 The areas of sustained performance in an organization.

4.7.3.1 Performance management (clause 4 of the standard)

The standard includes the first six points of 4.7.2 with emphasis on:

- the fact that each organization must identify its own specific stakeholders. The standard is broad in order to include society in general and its needs;
- the possible role of ISO 31000:2009, *Risk management – Principles and guidelines,* for handling risks from the environment and from stakeholders.

Information is, therefore, a cornerstone of ISO 9004:2009. This is one of the elements distinguishing it from ISO 9001.

4.7.3.2 Strategy and policy (clause 5)

Development

The organization must determine a mission (its reason for existence) and vision (the state it wants to achieve) and communicate them internally as well as externally.

In a second step, management should propose a strategy (a logical and structured plan to achieve the defined objectives). Vision and strategy should be reviewed periodically following the information obtained according to clause 8 of the standard.

Deployment

Strategy and policy are deployed by implementing processes that transform them into achievable and measurable objectives, with an established time plan. Relationships between processes, between organizational structures and processes, and problems arising from their interactions within the system must be given close attention.

Communication

Communication of the strategy and policy must be explicit, timely and continuous, adapted to specific business areas, and must benefit from feedback.

4.7.3.3 Resource management (clause 6)

Resource management (identifying needs, assessing potential shortages, supply planning, etc.) requires a process that provides, assigns, monitors, evaluates, optimizes, maintains and conserves resources.

Financial resources

The determination of financial needs in all their forms and their management is basic. Financial reporting can highlight non-effective or inefficient activities and trigger corrective actions.

Management of personnel

People are the lifeblood of an organization, and total quality emphasizes that they can make all the difference to competitiveness. Management should maintain a shared vision and an internal environment that allows staff to invest fully in life and to achieve the organization's objectives. It is essential to develop conditions to foster individual development, learning, teamwork, problem-solving, knowledge transfer, and the ability to transform the organization's strategies and goals according to individual work objectives. This will be achieved through personnel management that is planned, transparent, and responsible both ethically and socially.

Staff skills

Management must have a plan for staff development in order to identify existing skills, to determine which ones are missing or need to be developed, and to formulate an action plan associated with a *Deming* cycle. This plan should comprise a career plan, a continuous review of the level of satisfaction, needs and expectations of the staff, and opportunities for mentoring and coaching.

Partners and suppliers

Management must recognize that partners and suppliers on the one hand, and the organization on the other, are interdependent and that mutually beneficial relationships enhance their ability to create value. In any partnership that an organization develops with its partners and suppliers, it must consider different means, such as providing information, giving support by providing resources, sharing of profits and losses, etc.

The organization must also have a process for identifying, selecting and evaluating suppliers and partners. This process should be similar to that for evaluating suppliers according to ISO 9001.

Infrastructure and work environment

The *infrastructure* must be carefully planned and managed effectively and efficiently, taking into account the goals and objectives of the organization. A risk analysis should be performed and the resulting actions implemented. Much attention must be given to the dependability, health and safety of employees.

The work environment should allow people to deploy creativity, skill, and their potential for improvement and innovation. The work environment is a combination of physical and human factors, and should be comfortable, aim to boost productivity of the organization, and be consistent with legislation, regulations and agreements with stakeholders.

Knowledge, information and technology

The correct use of knowledge, of information and of technology is an essential factor in the competitiveness of an organization. The points cited below are vital:

- For *knowledge*, its foundation must be identified and protected. External knowledge sources, such as vocational training institutes or higher education, should be considered, and the implicit or explicit knowledge of staff, customers and suppliers must be collected.
- For *information*, robust processes to gather and use reliable data must be established; data should be formatted appropriately to inform interested parties about the organization.
- For *technology*, management must consider all existing possibilities to improve the performance of the organization as well as its operation.

Natural resources

These must be carefully evaluated, particularly in terms of availability and the impact they have on the sustainability of the organization. Without mentioning it explicitly, the standard suggests the implementation of an EMS (environmental management system).

4.7.3.4 Management of processes (clause 7)

The organization must establish effective and efficient processes that are specific to the business (product, market, market positioning, etc.) in a systems approach developed according to the eight principles.

Process planning and control

The process must be planned, controlled and verified, while taking into account the expectations of all stakeholders (including suppliers and customers) as well as environmental changes.

Responsibility and authority for the process

The standard specifies that each process is under the authority of a process manager, who is responsible for its functioning, possible changes and adaptations in a systems approach. The standard stresses that the role of these process managers must be recognized throughout the organization.

4.7.3.5 Monitoring, measurement, analysis and review (clause 8)

The organization should establish processes to *monitor* its environment (especially economic intelligence activities) and changes in stakeholders' expectations, thus widening the customer focus part of the eight principles.

Similarly, the organization must evaluate its performance in achieving its plans, formulated goals and objectives by means of a schedule. This requires a process for gathering data and for carrying out analysis through:

- definition of key performance indicators, which are quantifiable, to monitor all relevant activities of the organization;
- internal audits;
- self-assessment (see 4.7.4 of this book);
- benchmarking.

This process must be subject to review and systematic approaches to assess the available information and ensure that it is used for decision making.

4.7.3.6 Improvement, innovation and learning (clause 9)

In a fluctuating environment, the organization must be able to combine innovation and improvement, both of which might affect the products, processes, management systems, infrastructure and organizational activities.

Improvement activities may focus on elements of the organization or affect it as a whole, according to the *Deming* cycle. They must be integrated into the culture of the organization.

The organization must have processes that enable it to identify the need for innovation (and its urgency), to implement them effectively and efficiently, and provide the resources they require.

Training activities are one of the bases of a learning organization, integrating the capabilities of the employees into those of the organization.

4.7.4 The role of self-assessment

Self-assessment is a characteristic part of ISO 9004. It is a systematic and exhaustive review of the organization's activities and its performance in relation to maturity. This is different from an audit that considers the degree of conformity to a standard by identifying problems, risks and non-compliance.

The standard promotes self-assessment as an essential tool for assessing the level of maturity of the organization. This should apply to its leadership, strategy, management system, resources and processes in order to identify strengths and weaknesses, opportunities and threats, and determine opportunities for innovation and improvement. This type of self-assessment differs from that recommended by ISO 10014:2006, *Quality management – Guidelines for realizing financial and economic benefits*, in that the latter focuses specifically on the financial and economic benefits of a quality management system.

4.7.4.1 The self-assessment grid

ISO 9004 has established an assessment tool in the form of five maturity levels. The highest criteria provide information on best practices that the organization can decide to adopt. The assessment tool includes two types of tables:

- One for the self-assessment of key elements that should be performed periodically by management, and which gives the overall status of the situation. Below we give some questions with minimum and maximum maturity levels:
  - *Management*: What is the focus of management? *Products, some customers and shareholders only* (level 1), or *a balance of the needs of future stakeholders* (level 5)?
  - *Management*: What is the leadership approach? *Reactive and based on top-to-bottom instructions* (level 1), or *proactive, focused on learning with employee involvement* (level 5)?
  - *Strategy and policy*: How does one decide what is important? *By essentially considering elements of informal input from the market and other sources* (level 1) or *based on the need for flexibility, speed and lasting performance* (level 5)?
  - *Resources*: What do we need to get results? *An occasional management of resources* (level 1), or *management and use of resources that are carefully planned, deployed in an efficient manner and that satisfy stakeholders* (level 5)?
  - *Processes*: How are the activities organized? *Based on some fundamental instructions or procedures* (level 1), or *according to a quality management that supports innovation and benchmarking, taking into account stakeholder expectations* (level 5)?
  - *Monitoring and measurement*: How are the results achieved? *Randomly with occasional corrective actions* (level 1), or *are they above-average in the sector, with long-term sustainable improvement and innovation introduced at all levels of the organization* (level 5)?
  - *Monitoring and measurement*: How are outcomes monitored? *Exclusively by financial indicators and ones of productivity* (level 1), or *by key performance indicators integrated into the real-time monitoring of all processes, efficiently communicated to relevant stakeholders* (level 5)?
  - *Improvement, innovation and learning*: How are priorities for improvement established? *Exclusively based on claims, errors or financial criteria* (level 1), or *based on input from stakeholders* (level 5)?
  - *Improvement, innovation and learning*: How is learning applied? *Is it random and at an individual level* (level 1), or *is it planned in the organization to support creativity and innovation, to be shared with stakeholders* (level 5)?
- A deeper self-assessment of many elements, intended primarily for the management of operations and process managers, is based on 26 criteria each with levels of maturity from 1 to 5. It is of course possible for the organization to develop its own self-assessment grid.

4.7.4.2 Conducting a self-assessment: the method

ISO 9004 recommends following the sequence below:

1. determine the area of self-assessment;
2. determine the desired level of maturity for each criterion in the selected area;
3. appoint a manager and determine a date;
4. determine if the self-assessment is the result of teamwork or individuals. Is a coordinator or a leader required?

5. identify the level of maturity using the grid provided by ISO 9004 or a predefined grid;
6. gather information in the form of summary reports (including tables and illustrations);
7. evaluate the current performance and locate areas for improvement and innovation;
8. inform staff and management of the results;
9. determine an action plan in which a *Deming* cycle is included.

4.8 The ISO 9000 family

In addition to those already mentioned, the ISO 9001[16] standard is completed by the following documents (most of them contain good practices and advice, but do not lead to certification):

- **ISO 10014:2006, *Quality management – Guidelines for realizing financial and economic benefits.*** This fills the gaps between the financial world and that of quality, by measuring the results of the implementation of quality management from a financial perspective. The standard provides clear guidelines for making financial and economic benefits through a QMS based on ISO 9001. It is aimed primarily at the general management of an organization and is a complement to ISO 9004 for improved performance. It gives examples of benefits to be achieved by the implementation of ISO 9001 and identifies methods and tools that are available to help achieve such benefits.
- **ISO 10001:2007, *Quality management – Customer satisfaction – Guidelines for codes of conduct for organizations.*** This standard provides guidelines for planning, designing, developing, maintaining and improving codes of conduct related to customer satisfaction. ISO 10001 is applicable to product-related codes of conduct containing promises made to customers by an organization. It does not specify the actual content of the codes of conduct for customer satisfaction, nor does it govern other types of codes of conduct, such as those relating to interactions between an organization and its personnel, or between an organization and its suppliers.
- **ISO 10002:2004, *Quality management – Customer satisfaction – Guidelines for complaints handling in organizations***. This standard provides advice regarding the process for handling complaints relating to products within an organization, e.g., planning, design, operation, maintenance and improvements. The described process for treating claims is one of the processes in a global QMS.
- **ISO 10003:2007, *Quality management – Customer satisfaction – Guidelines for dispute resolution external to organizations.*** This standard provides guidelines for the planning, design, development, operation, maintenance and improvement of a process for efficient and effective conflict resolution for claims not resolved by the organization. It is applicable to claims concerning an organization's products that are intended for or necessary for customers. It specifies a complaints-handling process or a process for conflict resolution, and the resolution of disputes resulting from national and international activities, including those concerning electronic commerce.
- **ISO 10005:2005, *Quality management systems – Guidelines for quality plans***. ISO 10005 provides guidelines for the development, review, acceptance, implementation and revision of quality plans. It is suitable for organizations with or without a

[16] A **management system standard** provides a model to follow when establishing and carrying out a management system. This model incorporates features that have been agreed upon by experts as the international state of the art (in the sense of knowing the operational connotations).

management system complying with ISO 9001. It focuses primarily on product realization and is not a guide for planning an organization's QMS.

- **ISO 10007:2003, *Quality management systems – Guidelines for configuration management.*** This standard gives recommendations for the use of configuration[17] management within an organization. It is applicable to products from their design to their removal from production. It highlights the responsibilities and authorities needed and describes the process of configuration management, which includes planning, configuration identification, development control, recording of the configuration status, and configuration audits.
- **ISO 10012:2003, *Measurement management systems – Requirements for measurement processes and measuring equipment.*** ISO 10012 provides generic requirements and application guidelines for the management of measurement processes, and for verifying measuring equipment used to demonstrate compliance with metrological requirements. It specifies measurement requirements for a management system that an organization performing measurements can use and integrate within their overall management system. It aims to ensure that metrological requirements are met.
- **ISO 10015:1999, *Quality management – Guidelines for training.*** The ISO 10015 management system was implemented to help organizations, managers and human resources to identify "best practices" for the acquisition of skills. With such a procedure, supervisors must be involved to identify gaps and then, once the training is completed, learn to use the knowledge gained. In turn, employees who have undertaken training according to this standard should benefit as it leads to improved performances that are recognized by their superiors.
- **ISO/TR 10017:2003, *Guidance on statistical techniques for ISO 9001:2000.*** This is a set of best practices concerning the selection of appropriate statistical methods that may be useful to an organization during development, implementation, maintenance and improvement of a QMS similar to that of ISO 9001. This is done by determining which demands of ISO 9001 require the use of quantitative data, followed by identifying and describing the statistical techniques that can be applied to such data.
- **ISO 19011:2002, *Guidelines for quality and/or environmental management systems auditing,*** provides information on the principles of auditing, and the management of auditing programs for conducting audits of systems of quality management and/or environmental management (see the module on ISO 14000), and on the competence of auditors of these systems.
- **ISO 9000:2005, *Quality management systems – Fundamentals and vocabulary.*** ISO 9000 describes the principles of a quality management system and defines the terminology to be used.
- **ISO/TR 10013:2001, *Guidelines for quality management system documentation,*** provides information on best practices for developing and maintaining the documentation necessary to ensure an effective QMS, tailored to the specific needs of the user organization.

---

[17] ***Configuration*** is the set of functional and physical characteristics of a product defined by all the demands made by external and internal stakeholders. It is the sum of their implicit and explicit expectations that will be incorporated in the step of installing a QMS and implementing the House of Quality tool (see Chapter 2). These requirements therefore relate to those of customers, to criticality in terms of security, technology, new designs, interfacing with other items (so that everyone can move coherently), and the conditions of supply and support logistics. Configuration is a detailed *blueprint* for production.

4.9 Standards derived from ISO 9001

4.9.1 Specialized standards

ISO 9001 is a generic[18] standard. It applies to a wide range of companies, institutions and organizations. Some sectors have sought to develop more specific standards, based on the approach of the ISO 9000 family, including:

General industries

- ISO/TS 16949:2009, *Quality management systems – Particular requirements for the application of ISO 9001:2008 for automotive production and relevant service part organizations.*
- ISO/IEC 27001:2005, *Information technology – Security techniques – Information security management systems – Requirements.*
- ISO 90003: 2004, *Software engineering – Guidelines for the application of ISO 9001:2000 to computer software.*
- ISO/IEC 20000-1:2011, *Information technology – Service management – Part 1: Service management system requirements.*
- ISO/IEC 20000-2:2012, *Information technology – Service management – Part 2: Guidance on the application of service management systems.*
- (ITSMS) EN 9100, *Certification of management systems in aerospace, defense and space* (this standard has its counterpart in the United States: AS 9100).
- IRIS, *Certification of manufacturers and contractors in the rail industry.*
- ISO 13485:2003, *Medical devices – Quality management systems – Requirements for regulatory purposes.*
- ISO/TR 14969:2004, *Medical devices – Quality management systems – Guidance on the application of ISO 13485:2003.*
- ISO/TS 29001:2010, *Petroleum, petrochemical and natural gas industries – Sector-specific quality management systems – Requirements for product and service supply organizations.*

Industries for food and animal fodder

- ISO 22000:2005, *Food safety management systems – Requirements for any organization in the food chain.*
- ISO/TS 22003:2007, *Food safety management systems – Requirements for bodies providing audit and certification of food safety management systems.*
- ISO/TS 22004:2005, *Food safety management systems – Guidance on the application of ISO 22000:2005.*
- BRC, *Global Standard Food – Food safety certification by the BRC Global Standard.*
- IFS *(International Food Standard) – Food safety certification by the IFS International Food Standard.*
- IFS, Logistic Standard Certification of trade and transportation companies as required *by the IFS Logistic Standard.*
- BRC/IoP, *Global Standard Packaging Certification of suppliers of food packaging and other.*

[18] The adjective ***generic*** signifies that the same standard is applicable to any organization, whether small or large, regardless of the product or service provided, and in all activity sectors, whether a commercial business, a public organization or a government department.

- EN 15593, *Packaging – Management of hygiene in the manufacture of packaging for food products – Requirements.*
- Coceral GTP, *Certification according to the criteria of the European code of good business practices for food products and fodder for animals*[19].
- FAMI-QS, *European quality management for additives in animal feed*[20].

Other standards

- ISO 15378:2011, *Primary packaging materials for medicinal products – Particular requirements for the application of ISO 9001:2008, with reference to Good Manufacturing Practice (GMP).*
- FSC FM, *Certification of forestry.*
- FSC COC, *Certification of companies processing wood and paper.*

As previously stated, most of these are standards that are used internationally.

Depending on the country, other national standards may apply. They include ISO 9001 requirements supplemented by additional requirements or derived from ISO 9001 (e.g., in Switzerland: FSIO 2000, QuaTheDA, etc.).

4.9.2 Higher education and ISO 9001

Higher education also has a standard of quality management: ENQA[21] of the European Union – adopted by member countries of the Bologna process. Some standards are more specific, such as EQUIS[22] used mainly by the "Hautes Ecoles" of Management, Cti[23] principles of the Commission on French engineering degrees (which has a combined experience of over 35 years), without forgetting the EUR-ACE[24] reference for engineering schools in the EU. The EFQM Total Quality is also used[25] (especially by the "Hautes Ecoles Spécialisées" in Switzerland).

ISO 9001 is implemented in higher education and has been quite successful, particularly in Asia. The specifics of the education sector are addressed in the International Workshop Agreement IWA2: 2007 (E), *Quality Management Systems – Guidelines for the application of ISO 9001:2000 in education*, published by ISO. These guidelines focus on specific success factors such as the ability to adapt (to rapidly changing curricula), autonomy (the institution must have its own policy and its own assessment system), with focus on ethics, on safety and on environmental policy in education (students who are good citizens in an ethical institution and society).

[19] Standard available at: http://www.coceral.com/cms/beitrag/10010247/228174 (access Aug 2009)

[20] Standard available at: http://www.fami-qs.org/ (access Aug 2009).

[21] See the standard at: http://www.enqa.eu/pubs_esg.lasso

[22] Reference site: http://www.efmd.org/index.php/accreditation-main/equis/equis-guides

[23] Document available at: http://www.cti-commission.fr/-Documents-de-reference-

[24] Standard available at: http://www.enaee.eu/eur-ace-system/eur-ace-framework-standards

[25] For more information, see: http://www.efqm.org

4.10 ISO 9001 and the role of audits

4.10.1 Types of audits

Implementing ISO 9001 consists in identifying the key processes of the organization that are directly related to product quality, in applying the eight principles and implementing them, especially in the four sectors specified, and in documenting everything in a consistent, structured and pyramidal manner. The next step is to consider the *Deming* cycle of continual improvement, in which audits play a central part. Several types of audits are discussed in ISO 9001:

- *Internal (or first-party) audits* are performed by one or more qualified employees from the organization (who do not have any responsibilities in the domains they audit, so are neutral). They follow a schedule in accordance with the annual management cycle of the company. Their objective is to advise management on the compliance, effectiveness and relevance of the QMS in place.

  A few years after implementing a QMS, the internal audits should focus on how to manage changes and on the introduction of new products and processes. To renew audits of processes already implemented and optimized, or those pending (waiting for investment in new equipment for instance), can often be counterproductive and exhausting. Internal audits should instead validate the quality system.
- *Audits of suppliers (or second-party audits)* are, in most cases, strongly oriented toward processes and products. They are normally carried out by auditors who know the supplier's trade well and who have good knowledge of the system. The auditors are not necessarily (to a decreasing extent) management system specialists. Their goal is to ensure a high level of confidence in the ability of the supplier to provide products and services in the long run, and which are consistent with contracts, specifications and other (e.g., regulatory) requirements. Moreover, where appropriate, they aim to completely delegate conformity assessment (more reception control, ship online, etc.). These audits are strongly oriented toward processes and include a technical component that is more pronounced than in system audits. According to the results achieved, they can completely replace the partially controlled reception of goods or services provided.

  This type of audit renders it possible to verify that the supplier's quality system is adequate and to qualify it, but it should not be used for morally reprehensible intentions (giving information to a competitor, hostile use of know-how, etc.) or to negotiate discounts by suggesting risk taking.

  Audits of suppliers are facilitated if there is a QMS of the ISO 9001 type in place. Audits of suppliers located on another continent have gained in importance since globalization, but some incidents also show their limitations. These are sometimes avoided by the constant presence at the supplier's factory of a quality delegate from the customer organization (space and aeronautical industries, etc.). This approach, common in the past, is decreasing due to:

  - the costs it engenders;
  - the risk of completely or partially depriving the supplier of responsibility;
  - the risk of industrial espionage.
- *Audits of customers* are more rare, and are oriented toward processes and products, but can properly ensure that the delivered component or product has adequate specifications. Some functional aspects are only occasionally demanded for specialist products (e.g., packaging that is specific to a product or production line) from a supplier

who comes directly to the client (heads of purchasing departments may not be able to decide the detailed specification of a component).
- *External audits* can be performed by a potential client (second party), a manufacturer's trade association, or by a private or government accreditation or certification body (third party). Chapter 9.2 is dedicated to this type of audit, and it should be mentioned that its specifications are generic and applicable in whole or in part to the other audit types.

4.10.2 An audit and its procedures: ISO 19011:2002

Although the description below of how an audit is performed refers to the evaluation of a QMS such as that given in ISO 9001 or ISO 14000, the procedure described is also easily transposed to other types of audits.

The procedure is documented in ISO 19011:2002, *Guidelines for quality and/or environmental management systems auditing*. This standard provides information on the qualifications of auditors, the organizing and performing of audits, the drafting of audit reports and recommendations, as well as follow-up.

The audit must satisfy certain principles so that it is reliable and effective, and provides useful information for an organization that wants to improve its performance; the results must be largely independent of the team of auditors so that equal treatment between organizations is obtained.

The principles relate to the auditors and the mode of performing the audit. They are also valid, for example, for internal audits, but the number of auditors is generally smaller.

4.10.2.1 Quality of the auditors and selection of a committee

Ethics and general behavior. Auditors implement the following principles:

- Ethical conduct, *basis of professionalism:* trust, discretion, confidentiality, integrity.
- Fair presentation of results: *required to report in a fair, truthful and accurate (non-biased) manner.* Obstacles that are encountered during the audit and irreconcilable differences of opinion between auditors must be reported.
- Dedication: must be diligent with critical judgment and neutrality during the audit.

The application of these principles requires, besides the ethical provisions mentioned above, the following virtues (intellectual and moral): *diplomacy (and courtesy, the first virtue), flexibility, versatility, open-mindedness, curiosity, observational skills, independence, perseverance, and firmness in decision-making.*

Generic knowledge and skills. Auditors should have the following skills and knowledge of:

- principles, procedures and audit techniques;
- management and documentation reference systems;
- organizational management experience, enabling the auditor to understand the organization's situation, structure, culture, and processes and terminology;
- economic mechanisms of organizations;
- legal, contractual, regulatory requirements for a specific sector, or international conventions that the organization must follow.

Specific knowledge and skills:

- methods, techniques, terminology, principles of quality and its management, including common tools (problem-solving, risk assessment, statistical analysis of processes and products, etc.);

- processes and their technologies, products and services, best practices specific to the audited sector (which assumes profound knowledge of the sector[26]). In this way, it is possible to complete the external audit with a diagnosis audit.

To have this knowledge and expertise, a minimum level is required in terms of:

- scientific and technical training;
- mastering the quality landscape;
- experience in business;
- audit experience, which means that the auditor's training includes a junior stage.

Additional skills are required for the chairman of the audit committee (leadership, conflict management, planning). The auditors' expertise and knowledge must be updated and maintained.

4.10.2.2 Characteristics of an audit

In addition to the audit committee being regulated, or even formatted, the manner in which the audit is performed should also be standardized. Key points include the preparation, handling and procedure of the audit.

Preparation

An audit is a systematic, independent and documented process to obtain meaningful results. These enable an objective assessment of the audited organization so as to determine the extent to which audit criteria are met (here, the conformity of the organization's QMS to the contents of ISO 9001).

The following steps are part of the preparation, a process interacting closely with selection of the committee and carrying out the certification (see 4.10.2.1):

1. Who is ordering the audit? The audited organization, the administrative center, a third party?
2. What are the stated objectives of the audit, for example:
   a. priorities for management;
   b. commercial intentions;
   c. establishment of a management system;
   d. regulatory and legal constraints;
   e. stakeholder requirements;
   f. customer requirements;
   g. emphasis on the evaluation of suppliers;
   h. verification of good practice;
   i. emphasis on risk assessment.

3. What are the physical and organizational boundaries included in the audit? Are there relocated units? What is their status: already certified? If so, what is the reputation of the certification body and its hold on the organization?
4. What is the standard to be used for the audit? ISO 9001, ISO 16949 (automotive products) ...
5. Does the audited organization want the present standard to be supplemented by another one? For example ISO 14001.

---

26 This condition is naturally not valid for an internal audit.

6. How many documents and interviews have to be performed[27]? Who are the key individuals who will need extensive dialogue?
7. What preliminary studies are needed in order to go from a generic standard to one specific to the field? Are there additional documents to view, e.g., specific ISO standards or explanatory documents?
8. After establishing a first version of the audit schedule, what resources will the certification body need?
9. What will it cost for the certification body to carry out the audit?
10. What is the minimum preparation time for the audited organization?

The end of the preparation process implies a contractual offer by the certification body to the organization being audited.

Once the organization has accepted the conditions and starts preparing for certification, either independently or with a consultant, the establishment of an audit committee can begin. The last step is to draw up an audit guide, which will transform the requirements of the standard into a series of checklists which may include open questions, particularly for the interviews (see reference [4.10]).

Handling and procedure (Figure 4.7)

The handling and procedure of the audit can resemble that of a project. The first step is collecting and combining data, followed by production of the audit report and a possible list of recommendations:

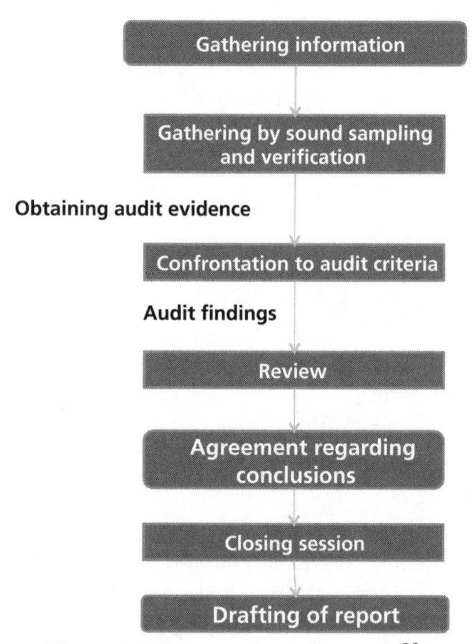

Figure 4.7 The procedure of an audit[28].

[27] A determining factor in order to evaluate the predicted duration of the audit.

[28] Inspired by Figure 3 of ISO 19011.

- Information gathering: a judicious sampling of information based on the objectives, scope and audit criteria. The data must be verified and recorded. Only verifiable information may constitute audit evidence. The methods of gathering information include:
  - interviews;
  - observation of activities, conditions and work environment;
  - reviews of documents: company policies, procedures, instructions, specifications, records, customer feedback and supplier assessments, etc.
- Development of audit findings: audit evidence is compared to the audit criteria to develop the audit findings. These can indicate either conformity or noncompliance with the audit criteria. They may also point out opportunities for improvement. Areas of compliance and noncompliance to audit requirements should be identified by summarizing the sites, functions, or processes that have been audited.
  Nonconformities can be classified. They will be reviewed with the auditee, in order that the audit evidence is recognized as accurate and the noncompliance understood. Efforts should be made to resolve any differences of opinion regarding the evidence and/or audit findings, and it is advisable to record unresolved issues.
- Preparation of audit conclusions: before the closing meeting, which outlines the key findings to the auditee, the members of the audit team should jointly conduct a review of the audit findings and other appropriate information obtained during the audit, reach an agreement on the audit findings, prepare recommendations (if this is specified in the audit objectives), and discuss procedures for audit follow-up.
- Preparation of an audit report that should include the following elements:
  - objectives and scope of the audit;
  - identification of the audit sponsor;
  - identification of the members of the audit team and the person responsible;
  - dates and locations where on-site audit activities were conducted;
  - audit criteria;
  - audit findings;
  - audit conclusions.

  The report may also contain the following items:
  - unresolved differing opinions between the audit team and the auditee;
  - recommendations for improvement, if specified in the audit objectives;
  - action plans for what happens after the audit, if agreed upon;
  - a statement on the confidentiality of the contents;
  - a mailing list for the audit report.

4.11 Certification and/or accreditation of the organization

4.11.1 Making the right choice

Implementing a QMS is good, but making it visible is better. Two scenarios are possible:

1. The performance of the organization must comply with an official government standard: this corresponds to *accreditation* by a body recognized by the state or any other official body (e.g., SWISSTESTING);
2. The QMS must comply with a recognized standard. In this case *certification* can be obtained (which does not comment on the level of product quality, but certifies that the organization has a system to achieve the goals set by management).

A variant involving simply the development of a QMS followed by internal communication is rarely used except for multinational groups that have their own inspection service. Certification, intermediate audits and renewal after 3 to 4 years gives rise to a residual pressure that facilitates the development of continual improvement, but it is not the key. The situation is better for accreditation, because in this case it is the performance, the company's primary mission, that is affected if the accreditation is not renewed or obtained.

Certification does not mean quality improvement; for this, it must be accompanied by a commitment to excellence in total quality. This momentum cannot be found in any document, form, profit, or alarmist vs. triumphant communication, balance sheet or annual report ...

4.11.2 Steps towards certification of an organization

Although the text below mainly concerns certification, it also applies to accreditation. Certification is a process[29] that has interactions with those of the (internal) implementation of a QMS. The steps, as seen from the point of view of the audited organization, are:

1. Choose a certification body: criteria such as international visibility of the certification, and its credibility, depend largely on the reputation of the chosen body (in Switzerland: SGS[30], SQS[31], Bureau Veritas[32], etc.).
2. Contact the certification body, to determine together with the auditee the scope of the certification, and appoint an audit team and leader.
3. Carry out a first session with the audit leader and the organization, with the objectives:
   a. to verify once again the scope of the audit;
   b. to estimate the time necessary to achieve the objectives set by the standard, which is essential planning information. The aim is to define the current status of the organization's QMS;
   c. to draw up a list of approved consultants, because the certification office cannot be both judge and jury, i.e., cannot both evaluate and advise. Establishing such a list is voluntary, but its use by the auditee facilitates subsequent work;
   d. to establish a planning outline.
4. Validate the planning, which ends the feasibility study for certification.
5. Prepare the audit by implementing ISO 9001 and verifying compliance with existing processes.
6. Determine the audit program (duration, schedule, assignments, etc.).
7. Hold an opening meeting of the audit, to discuss the outline of the project and present management policy.
8. Start the audit itself, in which the audit team divides between its members the tasks of evaluating processes and documents, including interviews with staff.
9. Prepare a summary of recommendations and nonconformities.
10. Hold a closing meeting in which the main conclusions and summary are presented to management and the employees (often in separate sessions).

---

[29] This list corresponds to the contents of ISO 19011:2002.

[30] SGS: Société Générale de Surveillance http://www.sgs.com.

[31] SQS: Association Suisse pour Systèmes de Qualité et de Management (SQS), http://www.sqs.ch founded in 1983, is one of the first organizations worldwide certifying and evaluating quality management systems. It carries out its tasks internationally and is leader in its branch in Switzerland.

[32] http://www.bureauveritas.com/wps/wcm/connect/bv_com/Group.

11. Distribute the audit report, which specifies precisely all remarks and nonconformities.
12. The organization's management takes a stance in response to the report, indicating what corrections will be performed and when.
13. The board of the certification body takes a certification decision, communicated to the organization's management, with the approximate dates of future audits (intermediate or for recertification).

4.12 Post-ISO 9001 certification

4.12.1 Certification is... only the beginning of a quality approach!

4.12.1.1 Post-certification napping (*or an organization's deep sleep*)

Certification of a company, which is external recognition of the effort of implementing a QMS, provides a moment of pride for the staff. This is also the case with its renewal. In some cases, unfortunately, it is also the beginning of a break that may lead to slackness (see reference [4.11]):

a) the organization's management, which asserted the necessary pressure for the establishment of the operation, has lost interest in it, and now searches for other sources of internal and external visibility;
b) management now wishes to use certification to reinforce other more profitable operations;
c) the consultant who had assumed the role of quality controller and brought visibility to the project, has distanced him- or herself, resulting in a gap in knowledge and expertise in the organization;
d) the new quality manager, freshly appointed, does not have sufficiently effective levers to assert him- or herself;
e) the organization is subjected to new challenges and is forced to redirect its energy into other activities;
f) staff turnover in key positions gradually erodes the quality culture of the organization if the HR department does not lend support to the quality manager;
g) the establishment of processes can lead to a heavy restructuring of the company, especially fiscal reforms that have been long desired, but never implemented. "The boat has been loaded" but the actual implementation does not follow; the finance department counts the cost of the operation and puts the brakes on any further involvement wishing to first obtain the early benefits of certification.

4.12.1.2 Breaking in the system

Once certification has been obtained, the QMS needs to be broken in, followed by maintenance, because:

a) a review of management and chosen quality indicators has not yet entered an active phase, and the same goes for internal audits; a sensitive part of the QMS has not yet been activated;
b) the new processes put in place are fragile, especially if they have shaken certain beliefs in the organization, whose followers sometimes show unexpected resilience;
c) certification is accompanied by recommendations, and it is easy to understand that these are significant and essential, since other difficulties, more easily addressed, were resolved or were in process of being resolved during the certification period;

d) the willingness for teamwork encounters resistance from middle management, which considers that there have been enough meetings about meetings...
e) the weight of habit, the skepticism of colleagues – against the fads and foibles of a management sometimes perceived as eager for change only to spark internal power struggles – mean the new procedures are not always respected, and informal approaches coexist with documented ones. This is especially true if the corporate hierarchy is consistent with a closed-door policy and the absence of a "teamplayer" management;
f) the organization is facing inevitable and often unpredictable changes; the QMS must be constantly revised, consolidated and adapted in real time;
g) internal process indicators, once tested, may not still be relevant, in which case new research is needed;
h) human resources, not formally included in the certification, prove to be of crucial importance and a process to extend the perimeters should be carried out.

The situation is not hopeless, because the quality manager can always remind his or her colleagues that the system must be certified at regular intervals, but there is an obvious risk of not benefitting from continual improvement. If a prospective client audits the quality system after certification has been used to seal the deal, and if many gaps then appear, the quality manager will be held responsible. This situation will be uncomfortable if the sales department can attribute the failure of one of its operations (in this case always the "most promising", although off to a bad start...) to production or R&D.

4.12.1.3 Collaborating with the certification body

The choice of certification body is essential, because the success of the post-certification depends on:

- the competence of the experts and whether they will maintain mutual loyalty;
- the ability to build relationships of trust;
- possible advice;
- the ability to listen and communicate with the body/expert between two certifications;
- the ability to collaborate.

4.12.1.4 Evolution of internal audits

After certification, internal audits aim to ensure the continuity, stability and robustness of the quality management system. Their essential content is specified.

In a subsequent step, an internal quality audit is to verify that the management system has successfully withstood the inevitable transformations of the company. The many previous closed questions (those for which the auditee responds with a yes or no) are replaced by open questions (worded to receive explanations of the objectives, of hows and whys). The current trend is towards constructive audits that transform negative reviews (finding differences that are not favorable to the auditee) into observations of areas for improvement (constructive and positive criticism) that help auditees improve their practices.

Given the experience gained since the implementation of an ISO standard, internal audits are often conducted in ways that are too formal, and are not sufficiently focused on the needs of the general management. This provides little added value, which is not or only poorly recognized by the company's managers. It is a good idea to develop a vibrant culture of internal audits, with emphasis on achieving the objectives that have been set, and for which content and form must be adapted. If necessary, more robust methods such as the Six Sigma approach can be proposed and initiated following an internal audit.

An internal quality audit must also take into account that the days are over when a company's products could reach the market containing defects: those companies are mostly dead or taken over. Quality is no longer a significant differentiator (in a chosen price range). The audit must have a global vision of the product and services associated with it to gauge the achievement of quality objectives. The conclusions of such an internal audit must always propose measures of progress and, in order for them to be accepted by all parties, the internal auditor must be able to assume the role of adviser and not be merely a controller. This role is also more rewarding, and makes it easier to find willing and motivated internal auditors.

It is useful to prepare the internal audit in order for it to gauge the effectiveness of each process for its six missions [4.12]:

- satisfy customers;
- achieve the objectives;
- implement the policy of the organization;
- optimize resources;
- continually improve;
- apply the rules.

Another critical factor is the management of skills by human resources. The auditor must find ways of managing staff skills, and develop training that will allow adjustment of these skills in a changing environment:

- Who detects the need for training and with what methods?
- On what criteria is the decision to train an employee based?
- Is the acquisition of skills formalized by a series of goals?
- Once training is completed, how can one be sure that the right skills have been acquired?

The implementation of ISO 10015:1999, *Quality management – Guidelines for training*, can help the organization to:

- identify and analyze training needs;
- design and plan training;
- implement training;
- evaluate training results;
- monitor and improve the training process.

4.12.2 The role of quality manager

Quality managers should not be surrounded by a large team since their mission is not strictly an operational one, but above all one of "getting things done" and giving advice. The chosen person should be a "person of many talents" with sufficient knowledge of the environment and product technologies involved in the process. This is in line with achieving, managing and securing quality. The person must also be responsible for carrying out other activities directly related to quality, giving him/her an interdisciplinary position (Figure 4.8).

Quality managers can make the mistake of forgetting to take the long view, thus becoming a fundamentalist when it comes to standards with a habit of applying normative (or legal) texts literally and not in the wider sense. They may be preoccupied with managing the documentation for compliance with customer requirements rather than carrying out continual improvement, especially if they are preparing for recertification. The standard is only a guide and not reality.

Figure 4.8 A quality manager's scope of action.

The usual role of a quality manager is built around five tasks:

- to communicate the concepts of quality management to the organization's senior management;
- to train employees in specific quality tasks (e.g., promoting simple methods of problem-solving, control charts, Pareto diagrams, internal audits, processing and monitoring of claims, processing of non-standard quality);
- to establish a quality organization as project manager and, together with the staff concerned, to plan, coordinate, support and verify implementation of the documentary structure and its relevance;
- to maintain the quality organization and its documentary structure, ensuring that any modifications are properly carried out;
- to raise awareness of staff and senior managers of the concept of continual improvement of processes, products and services, and to assist in its implementation.

Continual improvement, however, can require tougher skills. In this task, quality managers may have to act as experts to those in charge of processes by providing technical competence including, for instance, the design of experiments (i.e., Robust Design Methods or Quality Engineering), or statistical methods more sophisticated than control charts, histograms, analyses of mean values and variance, and Pareto charts (see Chapter 2). They may employ group methods of problem-solving, risk assessment, etc., and should also be on the lookout for changes in their own fields of competence (new methods and techniques, review of standards, etc.). These skills include increasing the efficiency and effectiveness of management, as well as the output of the organization, especially if they have the title of Master Black Belt or have gone through training according to Six Sigma Black Belt (see Chapter 11).

Quality managers need to develop communication skills, and must speak in simple words. In the best-case scenario, in addition to analytical and synthetic skills, and other features described in ISO 19011, they should have the following qualities and talents [4.13][13]: tenacity, open-mindedness, autonomy, common sense, a taste and gift for simplicity, a warmth and natural joy, generosity and openness to others (a form of empathy), because they are the person who "greases the wheels" of the organization. Their role of facilitator should not be underestimated.

4.13 Problem-solving after certification

Once certification is achieved, during the round of continual improvement with the first management review, non-standard qualities may persist. Further requirements are made by management, which cannot always be achieved through reflection based on the experience of employees, their rational approach and common sense. Some frequent cases are:

1. Corrective actions: *despite several attempts, all corrective actions of a process failed*[33]. An intense and collective mobilization of skills is necessary, by using a *method of problem-solving*. There are many methods for solving problems[34], but they all comprise the following steps:
   - *Identify the problem*; the group must agree on what the problem is and agree to address it. What is obvious to one person is not always so to a group (conflicts of interest, different views, priorities ...);
   - *Isolate the problem*, i.e., set its limits, define it precisely, especially with regard to form, frequency, and consequences;
   - *Investigate the causes* and rank them in order of importance and percentage effect on the problem – for instance: Has this problem always existed? If not, what has changed?
   - *Seek solutions*, using the least restricted thinking possible, preventing "idea killers" from running wild (*it has already been done, it's completely stupid, it's too expensive, etc.*); quantify the cost;
   - *Implement* the chosen measure(s);
   - *Measure the effectiveness and efficiency* of the implemented measure(s);
   - *Give the group recognition* and complete the action.

   Problem-solving usually requires the presence of a facilitator trained in group management and the specifics of the method. Many problems are solved quickly once all parties are gathered around a table, communicating in order to find a common solution, but this may require a step to resolve conflicts (long-term hostility and antagonisms, etc.).

   It may also be that the costs of improving the performance of equipment are prohibitive. The solution is then to strengthen quality control – which generally does not permit work according to the just-in-time principle – until the equipment can be replaced (or until the launch of a new product ...).

2. Non-standard quality may persist, *even though all involved parties agree on the element(s) of the process that are the cause*. Here, the methodology used is insufficient; a refinement is required. ISO 9001:2000 refers to statistical methods commonly used

[33] This is awkward if, in the excitement of certification, the quality manager was required to make investments to offset the lack of quality ...

[34] One widespread method of solving problems is Six Sigma, Motorola's registered trademark, largely used in the US. This method is based on the Deming cycle and the following steps: *Define, Measure, Analyze, Innovate/Improve and Control*. Six Sigma is a significant tool for statistical process control (see Chapter 11, dedicated to this approach).

in Quality Control and Assurance, grouped in ISO/TR 10017:2003, *Guidance on statistical techniques for ISO 9001:2000*, which cites 67 documents relating to the listed techniques.

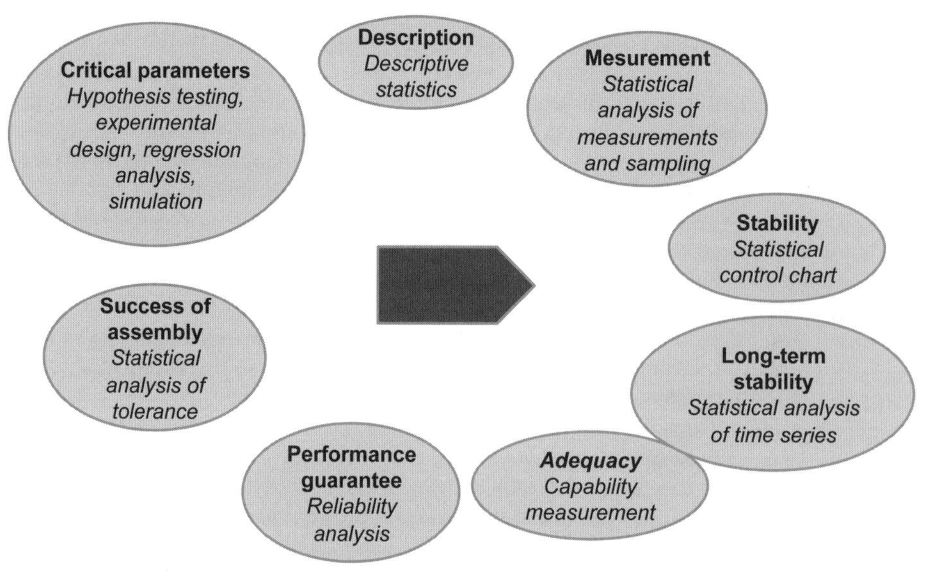

Figure 4.9 Process control and statistical methods.

These methods (Figure 4.9) are known by managers and are part of their training. The methods presented (not discussed in detail here since this book focuses more on quality management systems) are:

a. *Descriptive statistics*: Histograms, x/y "scatter" diagrams, trend graphs (abscissa time, ordered variable fluctuation), mean values, standard deviations, median values, etc. These basic methods have become widespread since the advent of spreadsheets such as EXCEL.

b. *Design of Experiments (DOE)*: This approach seeks to understand or verify the behavior of a system or a production tool when varying certain parameters. The results are put through descriptive statistical or mathematical analyses, such as variance analysis, permitting identification of the most important parameters, etc. DOE has been taken to a deeper level by the Japanese *Taguchi,* and its synonyms are Quality Engineering or Robust Engineering (see Chapter 8).

c. *Hypothesis testing*: This method involves developing a hypothesis based on a number of reasonable assumptions, then verifying it by comparing it to a group of empirical or experimental data. Descriptive statistics or mathematics is used for this comparison (*Chi-squared* test, etc.).

d. *Analysis (of uncertainties) of measurement*: Experimental data or control data is tainted with the uncertainty of the measuring instrument. What is this exactly? Is the difference between two measurements a significant deviation? If so, starting at what deviation and for what range? Is the instrument appropriate in this case? The uncertainties of a measuring instrument must be tested for each situation and measurement, and not from equipment manuals in a holier-than-thou fashion. The control of measurement processes is the subject of ISO 10012:2003, *Mea-*

surement management systems – Requirements for measurement processes and measuring equipment.

e. *Capability analysis of a process:* ***Capability*** is the field of *Statistical Process Control* (SPC). It aims to ensure that the process is adequate for its goals, that the process will perform over a range of determined operating parameters, and also that its performance will remain robust despite small variations in these parameters (typically the same order of magnitude as the uncertainties of measurement). A process is out of control when less than 99.73% of its output is in the range of specifications for data output.

f. *Regression analysis:* This is a quantitative statistical method for analyzing the possible relationships between a family of parameters and a chosen variable, and for quantitatively determining the influence of each. The statistical methods enable determination of the type of plausible mathematical relationships (linear, etc.).

g. *Reliability analysis*: This is the application of analytical methods and engineering outcomes to evaluation and prediction guaranteeing the performance of a process. It differs from capability analysis by a risk assessment. Preventive actions of ISO 9001 are a case of reliability analysis by the FMEA approach.

h. *Sampling*: Sampling can be used for process control or control of finished products (compliance) before a batch is released. Evaluation of a limited number of samples is a common method of sampling, although its statistical effectiveness is highly dubious. Sampling is based on statistical mathematics [4.14].

i. *Simulations*: Thanks to computers, simulation is certainly a new test method when it comes to strict sciences and engineering. However, it requires the development of a mathematical model and its validation, which can be time-consuming or expensive in highly complex situations (as is the case in production or during development when time is of the essence).

j. *Maps of statistical control procedures*. The best known is the control chart, resulting from BELL research laboratories. It is a graph of statistical trends with a tolerance range. Dividing the absolute tolerated deviation by 6 gives the standard deviation of the statistical distribution of the variable under control. The statistical control chart (Figure 4.10) is an application of the theory of errors, which here requires the distribution of measurement results to be normal[35].

k. *Statistical analysis of tolerance*: Any component or substance used to manufacture a product has a specification with a specific tolerance range. What is the effect of these fluctuations on the tolerance of the finished product? How can we quantitatively estimate it, or determine those of the components from the tolerances of the finished product? This is the purpose of this type of statistical analysis, based largely on the Gaussian distribution approach.

l. *Time series analysis*: This is the detailed study of fluctuations of a parameter over a prolonged time. The method is used when you suspect that a process is subject to variations over long time periods (cyclic, seasonal, daily). The idea is to compare the respective populations at an appropriate time to make projections (such as

[35] For an introduction to normal or Gaussian distribution: http://www.aiaccess.net/French/Glossaires/GlosMod/f_gm_normale_distri.htm in French and, in English, http://en.wikipedia.org/wiki/Normal_distribution.

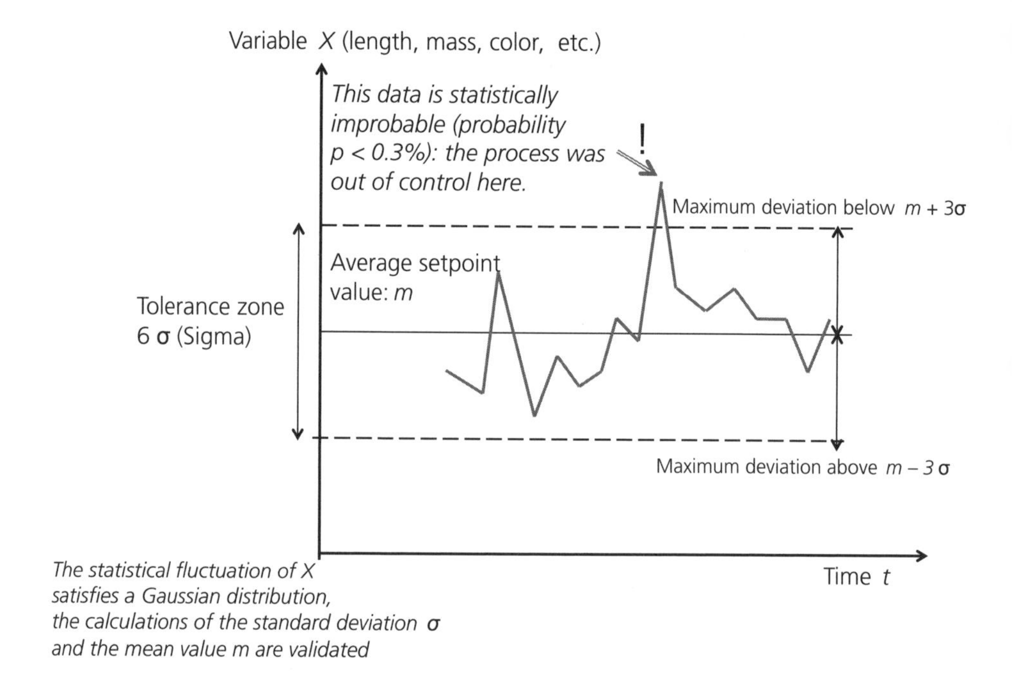

Figure 4.10 A statistical control chart according to the Deming-Shewart method.

fluctuating seasonal sales of ice cream, daily consumption of electricity, customer complaints, etc.).

3. Conflicts between process managers: The interdependence of processes and the internal customer/supplier relationships can reveal problems that were previously hidden, as well as power struggles that are not trivial. A mediator may resolve some of these difficulties. An internal customer/supplier relationship can sometimes resolve conflicts. Indeed, when considering a relationship between two processes (output data from one which becomes input data for the other), the process providing the data must listen to the client. It must identify the expectations before concluding a contract. In real life, the internal customer/supplier relationship gives priority to listening to the internal client.

4. Skills and talents of staff: the process/system approach can highlight deficiencies in the skills and talents of a person in a position that is essential for achieving quality. Sometimes the employee is not able to overcome these shortcomings, even with adequate training and coaching. This problem must be dealt with by HR and the supervisor, who may object to a transfer.

5. Preventive actions: management may insist that, following an unexpected incident, the program of preventive actions for a process be reviewed and reworked in more detail. This detailed analysis is based on specific methods and guidelines, and one of

the best known is *Analysis of Failure Modes, Effects and Criticality Analysis* (FMECA[36]), which is built around a systems approach. It is one of several risk management methods.

The FMECA method is applied to three targets during three stages of R&D:

a. FMECA Product: Ensure the reliability of the product by improving its design – *Carried out during product design by changing the specifications.*
b. FMECA Process: Ensure the quality of a product by improving its production operations – *Carried out during development*.
c. FMECA Means: Ensure the availability and safety of production equipment by improving their design, operation or maintenance – *Carried out during industrialization*.

For all targets, the FMECA method comprises the following steps:

- Identify and define the object to be analyzed.
- Search for plausible failures in a system approach by reviewing all components and production stages, and assigning a probability of failure (F), a severity index (G) of the consequences and a probability of non-detection (D).
- The criticality of each failure (C) is the product of F, G and D. The higher C is, the higher is the risk taken by the organization.
- It is then necessary to establish a program of actions, and decide which risk to eliminate:
  - by acting on the probability (modification of components or of the production tool, or maintenance improvement);
  - by lowering or eliminating the risk of non-detection (strengthening the quality control of either the process or product).

The FMECA method is time-consuming and resource-intensive, but many products make use of identical or similar components and production facilities. Its use is not necessarily expensive, because the implementation may simultaneously involve several product categories.
FMECA cannot judge the synergistic effect of two (or more) failures of low criticality. This is probably one of the reasons why the aviation industry inspects each component, lowering the detection risk to (almost) zero.

4.14 Conclusions

In 20 years, the QMS described by ISO 9001 and the portfolio of the ISO 9000 family has established itself as a European – but also worldwide – benchmark for quality tools.

The widespread use of the eight principles of the standard and the PDCA by *Deming* for other management systems (IT, security, environment, etc.) make it an essential step in the quality approach, although many people working with quality believe that it is primarily the gateway into the world of quality and that it is foremost a minimum safety net (which many CEOs are looking for).

[36] This method, developed by the American military in 1949, was formalized in 1993 by the AIAC (Automotive Industry Action Group).

However, the success of ISO 9001 is that of the success of implementing its instruments and tools. It is the management of an organization that, through commitment and talent, constitutes the engine and critical success factor for its implementation.

References

[4.1] The changes made to IS0 9001 in its 2008 version compared to 2000 are minimal. They are more clarifications than new requirements. Therefore, we do not show much interest in these changes as part of this introductory book. For more information, see for example the article by IAN CAMPBELL, Demnächst da – die ISO 9001:2008. ISO 9000er Normen, ein Erfolgweltweit, *Management und Qualität,* 10/08.

[4.2] Ibid, p. 34.

[4.3] The standard ISO 9001 is the subject of numerous books such as: HENRI MITONNEAU, *ISO 9000 version 2000*, ed. Dunod, 2007, or DANIEL BOERI, *Maîtriser la Qualité*, ed. Maxima, 2006.

[4.4] PMBOK Guide – *A Guide to Project Management (Fourth Edition)*, ed. Project Management Institute (2008).

[4.5] To understand the basics, see for instance, ROGER AÏM, *L'essentiel de la gestion de projet*, éditions lextenso, 2009.

[4.6] GILBERT J.B. PROBST, BETTINA S.T. BÜCHEL, *L'entreprise apprenante*, Les Éditions d'organisation, 1995.

[4.7] Taken from Figure 1 of ISO 9000: 2005, *Quality management systems – Fundamentals and vocabulary.*

[4.8] YVON MOUGIN et al., *Amour et management, le secret de la Qualité dans le service aux autres*, AFNOR.

[4.9] Detailed in the article by THOMAS TRÄGER, Human-FMEA. Fehlerquelle, *Management und Qualität* 10/08, p. 31.

[4.10] As an example, the audit guide developed in the book by DIDIER SEBILO and CHRISTOPHE VERTHIGEM, *De la Qualité à l'Assurance Qualité – Accompagner la démarche*, ed. AFNOR, pp. 123-176.

[4.11] For this reason AFNOR published a pamphlet, MICHEL BALLAÏCHE, *100 questions pour comprendre et agir – l'après certification ISO 9001*, ed. AFNOR, 2004.

[4.12] GENEVIEVE KREBS, YVES MOUGIN, *Les nouvelles pratiques de l'audit qualité interne,* ed. AFNOR, 2003, pp. 110 et seq.

[4.13] YVES MOUGIN, *Quel avenir pour les responsables qualité?,* ed. AFNOR, 2005, pp. 126 et seq.

[4.14] A groundbreaking book is W. EDWARDS DEMING, *Sample Design in Business Research*, by W.E. Deming, ed. John Wiley and Sons, 1960.

Chapter 5

Quality Management in Laboratories

Key concepts: GLP, GLP 07 (OECD), Ordinance on Good Laboratory Practice (OGLP), calibration, staff skills, Directive 2004/10/EC, equipment, sampling, test elements, reference elements, testing, calibration, reliability, training of personnel, testing facilities, ISO/IEC 17025: 2005, ISO 15189: 2007, handling, measurement, accreditation scope, accuracy, reagents, reception, outsourcing, storage, system testing, transportation and validation.

5.1 ISO/IEC 17025 is for who?

Many laboratories offer their services to the private and public sectors. They vouch for the accuracy, reliability and reproducibility of their results, while satisfying performance criteria. These organizations must demonstrate that their protocols comply with good laboratory practice and with official or recognized analytical methods. They must also withstand possible pressure from clients whose size and financial strength or policy might be able to sway the analysis results.

ISO/IEC 17025, *General requirements for the competence of testing and calibration laboratories* (previously EN 45001), helps in the implementation of a system of quality management, and makes it possible for a laboratory to be accredited by a private body or a public watchdog. The standard complements GLP (Good Laboratory Practice) of the OECD and the European Directive 2004/10/EC by placing laboratory activity more strictly in a market environment. Its implementation requires great care and attention to detail, qualities needed by the laboratory staff and management. ISO/IEC 17025 has certainly contributed to the development of numerous laboratories within the new EU members, which is proof of the effectiveness of the ISO approach when it comes to the globalization of trade. To avoid pressure on constituents, the standard specifies, for example, that statements with partial results must not be provided, but they must be sent all together in an indivisible report.

More recently, some academic laboratories carrying out analysis, calibration and testing have opted for a quality management system according to ISO/IEC 17025. The reason for this is simple: since these laboratories work with or for large groups in the private sector, their partners need to know that they provide the same proof of quality as the group's laboratories or their service providers. This is naturally the case if these laboratories are required to provide expert reports.

An effort to organize and streamline analyses in view of accreditation is particularly appropriate for a laboratory:

- if it uses accepted or official analytical methods;
- if it performs a large annual volume of measurements.

5.2 Measurements and tests: a process approach

A test or analysis includes the following sequential steps (main process, Figure 5.1):

- writing a protocol, an agreement (or contract) with the client or recipient of the service, specifying the types of analyses and testing, as well as planning (in research, this may be done by the head of the laboratory or the thesis advisor).
- having access to sampling tools;
- sampling according to the sampling plan and protocol;
- providing containers, packaging and transportation;
- cataloging and recording samples;
- storing samples temporarily;
- preparing the sample;
- testing or analysis;
- recording and processing results;
- archiving results;
- preparing the analysis report and archiving a copy;
- sending the assessment report;
- treating any additional requests or complaints.

All the above – steps must be controlled. The support processes below are incorporated into the quality management system:

- purchase of, and storage conditions for, equipment, reagents and reference materials;
- control of the premises and the storage conditions for the samples; cases of noncompliance with regard to storage should always be reported to the client;
- adequacy of external conditions, and calibration of measuring equipment;
- selection and implementation of appropriate analytical and experimental techniques;
- overview of the training and expertise of employees;
- document control (methods, protocols, test results, reports).

Three systems of quality management for laboratories are commonly used:

- a set of recommendations for good laboratory practice (GLP) from the OECD;
- the Directive 2004/10/EC;
- ISO/IEC 17025.

Also, for Switzerland, for example, there is the Ordinance on Good Laboratory Practice (OGLP; SR 813.112.1), which reproduces the OECD's GLP.

These three systems will be briefly presented; ISO/IEC 17025 accounts more specifically for potential conflicts between clients and testing facilities.

There exists a standard, similar to ISO/IEC 17025, but specifically adapted to biomedical laboratories: ISO 15189:2007, *Medical laboratories – Particular requirements for quality and competence*. This is a stand-alone standard and not an addition to ISO/IEC 17025. The

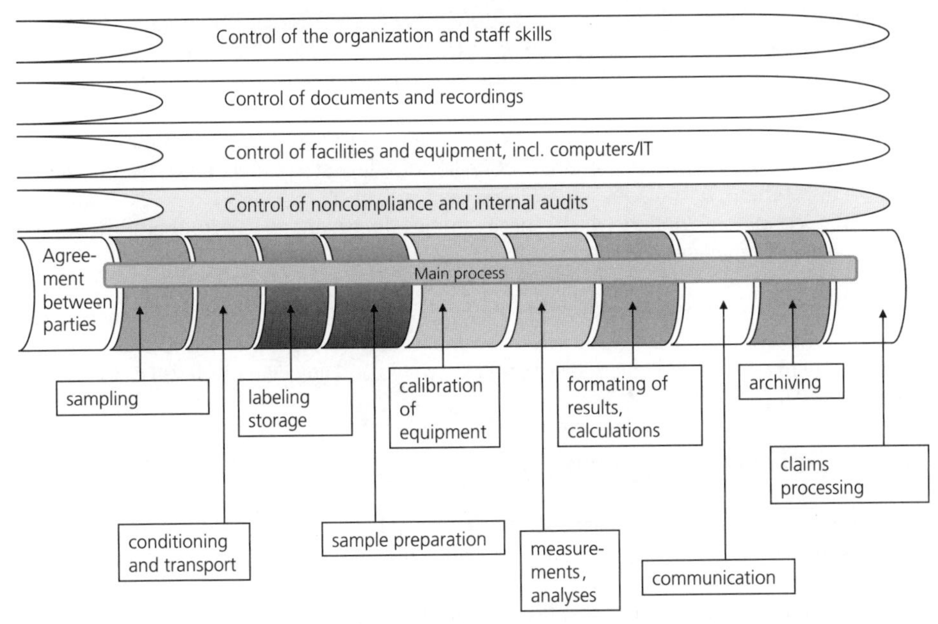

Figure 5.1 Quality management in laboratories: a process.

vocabulary is different, and focuses on the welfare of the patient. An appendix is devoted to ethics. Otherwise, it is very similar to ISO/IEC 17025.

5.3 GLPs of the OECD, the Directive 2004/10/CE and the Swiss OGLP

ISO/IEC 17025 is often applied using the principles of Good Laboratory Practice (GLP). These were published before ISO/IEC 17025, and can be used independently of the standard.

The first version of GLP was originally developed in 1978 by a group of experts from OECD, based on a GLP regulation published in 1976 by the Food and Drug Administration (FDA, United States). The application of these principles to data from tests on chemical products was recommended to Member States of the OECD in 1981. These principles have subsequently been reviewed, and an amended version was adopted by the OECD Board in 1997. The standard GLP 97 is supplemented by 14 additional standards[1] that address complementary aspects (such as sampling methods in the field, etc.).

For the European Community, the principles of GLP are defined by the Directive 2004/10/EC[2] regarding *the harmonization of laws, regulations and administrative provisions relating to the application of the principles of good laboratory practice and the verification of its applications for tests on chemical substances*. These guidelines are very close to those of the OECD that we will consider. GLP is also regulated by the Swiss Ordinance on Good

1 Available at: http://www.oecd.org

2 Available at: http://eur-lex.europa.eu

Laboratory Practice (OGLP, RS 813.112.1, 2005)[3], which is followed by organizations such as Swissmedics, the Federal Office of Public Health (FOPH) and the Federal Office of the Environment, who perform inspections in the biomedical and environmental fields.

What is the definition of GLP used by the OECD?

"Good Laboratory Practice (GLP) is a system guaranteeing quality when it comes to:

- the manner in which to organize non-clinical safety studies related to health and the environment, and
- the conditions under which these studies are planned, performed, monitored, recorded, archived and reported."

Good laboratory practice was introduced by the OECD to facilitate trade (primarily commercial) between countries and to avoid the replication of trials. The OECD's GLP involves physical, chemical and biological test systems, while the Directive 2004/10/EC focuses on chemical testing systems.

GLP 1997 of the OECD comprises 10 chapters:

Chapter 1 Organization and test facility personnel

(This chapter discusses the responsibilities of the management, study director, principal investigator and study personnel:

- *Management of any test facility must ensure compliance with the principles of good laboratory practice in the test facility.*
- *The study director is the sole person in charge of the study, and is responsible for the overall conduct of the study as well as for preparing the final report.*
- *The principal investigator ensures that the phases of the study that are delegated to him/her take place in accordance with the applicable principles of good laboratory practice.*
- *The study personnel are aware of the trial protocol, implement it, document the results, and report any deviation from the protocol.)*

Chapter 2 Quality assurance program *(in extenso)*

(A quality assurance program is a defined system, including personnel who are independent of how the study is conducted, and who aim to provide management of the test facility whilst ensuring that the principles of good laboratory practice are adhered to.)

General

a) The test facility must have a quality assurance program involving all relevant documentation needed to verify that the studies are conducted according to the principles of good laboratory practice.

b) The quality assurance program must be entrusted to a person or persons – with experience of the testing methods – designated by the management and directly answerable to them.

c) Such persons must not participate in the conduct of the study covered by the program.

[3] Available at: http://www.admin.ch

Responsibilities of personnel involved in quality assurance

The staff responsible for quality assurance has the following responsibilities. This list is not exhaustive:

a) *Maintain* copies of all study plans and approved standardized operating procedures that are used in the test facility, and have access to a current copy of the master plan.

b) *Verify* that the study plan contains all information required for compliance with the principles of good laboratory practice. This verification should be documented.

c) *Conduct* inspections to determine if all studies are performed in accordance with these principles of good laboratory practice. Inspections should also determine whether study plans and standardized operating procedures are made available to the study personnel and whether they are followed.

   These inspections may be of three types, as specified by the standardized operating procedures of the quality assurance program:

   – inspections related to the study,
   – inspections related to the facilities, and
   – inspections related to the process.

   Written summaries of these inspections must be archived.

d) *Inspect* the final reports to confirm that the methods, procedures and observations are accurately and completely described, and that the reported results accurately reflect and complete the raw study data.

e) *Report*, promptly and in writing, the inspection results to management and the study director, in addition to the principal investigator, as appropriate.

f) *Prepare and sign* a statement, to be included in the final report, which specifies the types of inspections and their dates. This should include the phase of the inspected study, and the dates the inspection results were communicated to management and to the study director, as well as to the principal investigator, if applicable. This statement will also serve to confirm that the final report reflects the raw data.

Chapter 3 Facilities

Two general requirements are described:

- Based on its size, construction and location, the test facility must meet the requirements of the study, and minimize any disturbances that would interfere with its validity.
- The spatial arrangements of the test facility should provide an adequate degree of separation of the various activities in order to assure that each study is properly performed.

Chapter 4 Equipment, materials and reagents

- The equipment, including validated computerized systems used for obtaining, storing and accessing data, as well as for controlling environmental factors involved in the study, must be positioned in a proper location, be suitably designed and have sufficient capacity.
- The equipment used should be periodically inspected, cleaned, serviced and calibrated according to standard operating procedures. Records should be kept of these activities. Calibration should, where appropriate, follow national or international standards of metrology.

- The equipment and materials used should not interfere adversely with the test systems.
- Chemicals, reagents and solutions should be labeled.

Chapter 5 Test systems *(excerpt)*

There are two types of test system (*a test system refers to any biological, chemical or physical system, or any combination thereof, which is used in a study*):

Physical and chemical

- Devices used to obtain chemical and physical data must be placed in a proper location, be of suitable design and adequate capacity.
- The integrity of the physical and chemical tests must be verified.

Biological

- Conditions suitable for the storage, housing, handling and maintenance of biological test systems must be created and maintained to ensure data quality.

Chapter 6 Test and reference elements

(A test element is an item that is the subject of a study, and a reference element or "control element" refers to any article used to provide a basis for comparison with the test element.)

Receiving, handling, sampling and storage

- It is essential to maintain records showing the characterization of test and reference elements, the date of receipt, expiration date, and quantities received and used in the studies.
- It is important to define methods of handling, sampling and storage, which will maintain homogeneity and stability as far as possible and avoid contamination or mixing.
- The storage containers must be labeled with identification information, the expiration date and specific storage instructions.

Characterization

- Each test and reference element should be appropriately identified (e.g., code, *Chemical Abstracts Service* [CAS] registration number, name, biological parameters).
- For each study, one must know the exact nature of the test or reference elements, including the batch number, purity, composition, concentration or other characteristics in order to appropriately define each batch.
- When the test element is supplied by the client, there must be a mechanism, developed in cooperation between the client and the test facility, to verify the identity of the test element.
- For all studies, one must know the stability of the test and reference elements under the conditions of storage and testing.
- If the test element is administered or applied in a carrier, one must determine the homogeneity, concentration and stability of the test item in that liquid.
- For test items used in field studies (e.g., blends in tanks), this information can be obtained through separate laboratory experiments.
- A sample of each batch of test items should be kept until the end of the analysis for all studies, except for short-term investigations.

Chapter 7 Standard operating procedures

- A test facility must have written standard operating procedures, approved by the management of the facility, which must ensure the quality and integrity of data generated by that facility[4]. Revisions to standard operating procedures must be approved by the management of the test facility.
- Each unit or area that is separate from the test facility should have immediate access to standard operating procedures concerning the activities being performed there. Books, analytical methods, articles and manuals may be used as supplements to these standard operating procedures.
- Deviations from standard operating procedures related to the study must be documented and recognized as applicable by the study director and/or the principal investigator, as appropriate.
- There must be standard operating procedures for the following types of activities at the test facility; the list is not exhaustive. Details given under each heading are only examples:
  - *Test and reference elements:* receipt, identification, labeling, handling, sampling and storage;
  - *Equipment, materials and reagents:*
    - *devices*: operation, maintenance, cleaning and calibration,
    - *computer systems*: validation, operation, maintenance, security, control of changes and backup,
    - *materials, reagents and solutions*: preparation and labeling;
  - *Record-keeping:* reporting, archiving, and retrieval; coding of studies, collecting data, reporting, indexing systems, handling of data, including the use of computerized systems;
  - *Test system (if any):*
    - preparation of the location and environmental conditions for the test system, procedures for receipt, transfer, proper placement and characterization, identification and maintenance of the test system, preparation of the test system, observations and examinations, before, during and at the end of the study, handling of specimens belonging to the test system found moribund or dead during the study (*biological*), gathering, identification and handling of specimens, including autopsy and histopathology (*biological*).
- Installation and placement of test systems on experimental sites.
  - *Quality assurance mechanisms:* assignment of personnel responsible for quality assurance in planning, scheduling operations, documenting and reporting inspections.

Chapter 8 Study plan

- Study plan
  - This must be established for each study before work is started, after approval by the study director for GLP compliance and verification by the QA staff.
  - Any amendments must also be validated.

---

[4] The number and the management of operating procedures must be carefully optimized; increasing these numbers at will does not necessarily improve the level of good practice (problems regarding redundancy, tracking updates, difficulty of ensuring consistency).

  - Any deviations from the study plan should be described, explained, acknowledged, dated and documented.
- Content of the study plan
  - Identification of the study, test item and reference element.
  - Information concerning the client and the test facilities.
  - Dates (approval of the study plan, start and end of the experiments).
  - Test methods.
  - Records (to be archived).
- Performing the study
  - The study and all related items must be given a unique identification.
  - Execution must comply with the predetermined plan.
  - Particular attention should be given to data records; statements must be signed or initialed and dated.
  - Any changes to data must not conceal prior records and must be motivated and documented.
  - If data is entered by computer, the persons inputting the data must be identified and the data must be traceable.

Chapter 9 Preparation of a study report
This should contain:
- The study plan for each study, signed by the persons in charge.
- An identification of the study (title) as well as the test and reference elements.
- Information relating to the client, the test facility and the people responsible for the testing and the phases of the study, as well as the name and address of scientists having contributed to the transcripts for the final report.
- Dates.
- Statements about the quality assurance program, including the types of inspections, dates, audited phases, and dates when inspection results were reported.
- Descriptions of materials and test methods.
- Results:
  - Summary.
  - All information and data requested in the study plan.
  - Presentation of results, including calculations and determinations of statistical interest.
  - Evaluation and review of results and, where appropriate, conclusions.

Chapter 10 Storage and conservation of records and materials
- The following should be preserved in archives for a period specified by the competent authorities:
  - Study plan, raw data, samples of test and reference elements and the final report of each study.
  - Report of inspections carried out in accordance with the quality assurance program.
  - Statement of the qualifications, training, experience and job descriptions of staff.
  - Documents relating to IT system validation.

  - Chronological file of standardized procedures.
  - Environmental monitoring reports.
- Archived material must be preserved.
- Personnel authorized to access the archives must be identified, and any consultations carried out must be traceable.
- Upon termination of the test facility work or of archiving, all data is transmitted to the client.

5.4 The key content of ISO/IEC 17025

ISO/IEC 17025 [5.1] aims to improve the process of sample treatment, as well as the quality of measurements and test results. As with ISO 9001 or ISO 14000, ISO/IEC 17025 does not in itself specify the level of accuracy and the reliability of analyses, measurements and tests, but it provides a management system that facilitates the achievement of the level described in the quality manual.

The content of ISO/IEC17025 differs from GLP in that the former comprises two main chapters: Conditions related to management (Chapter 4) and Technical requirements (Chapter 5). A comparison of the Good Laboratory Practice (GLP-1997 of the OECD) and the content of ISO/IEC 17025 shows many similarities. The requirements of ISO/IEC 17025 are more stringent, and consider the responsibilities of the company/laboratory in a social, legal, economic and political environment that is not to be underestimated. Among the requirements are careful documentation, archiving, traceability of samples, and staff training. Under these conditions, a measurement that is "troubling" to the customer should not be repeated, so that the credibility of the transnational results is guaranteed, thus reducing costs in a global market. However, a QA in accordance with ISO/IEC 17025 may be too expensive for small laboratories. Its heavy administrative load may discourage research laboratories. In academic work, a thesis advisor may prefer to have his PhD students search for more suitable methods in scientific literature rather than comply with official methods that might be more expensive and time-consuming. Nevertheless, for a laboratory that provides such expertise, flawless documentation is a sign of confidence and increases efficiency.

The most important information in the standard is contained in Chapters 4 and 5. The contributions of GLP-1997 of the OECD can be integrated. The full importance of ISO/IEC 17025 can be appreciated only by direct usage. However, a first approach highlights the essential elements, which will make study and analysis easier, which is the aim here.

5.4.1 Management requirements (Plan)

Organization and management:

- *Legal aspect*: The laboratory or the organization of which it is part must be an entity that can be held legally responsible (clause 4.1.1 of the standard). *This is an application of the adage "no authority without responsibility." It places the laboratory's activities in civil society and gives it impact on the functioning of the economy.*
- *Specific requirements of stakeholders*: It is the responsibility of the laboratory to carry out its activities ... so as to ... satisfy the needs of the customer, regulatory authorities or organizations providing recognition (clause 4.1.2). *This clause introduces specific requirements of the customer or regulatory authorities, which can complement the standard (waste management or storage of hazardous substances, for example).*

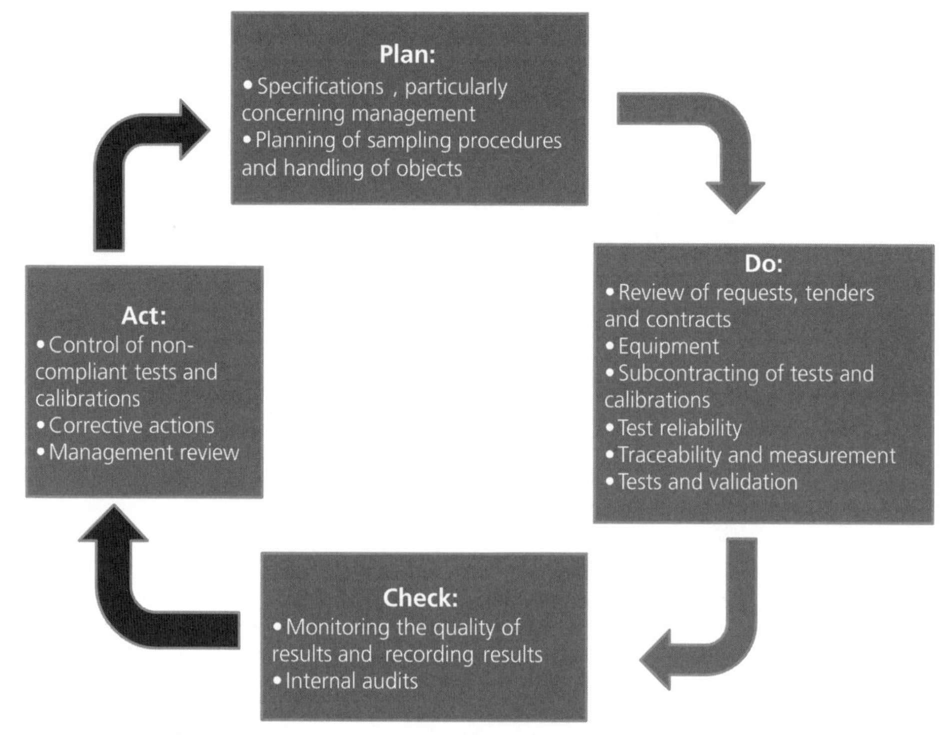

Figure 5.2 The Deming cycle and the main headings of ISO/IEC 17025.

- *Defining the scope of accreditation*: The laboratory's management system must cover work conducted in the laboratory's permanent facilities, at sites away from its permanent facilities, or in temporary or mobile facilities (clause 4.1.3). *In other words, the areas covered by the accreditation must be defined.*
- *Ethics and definition of tasks*: clauses 4.1.4 and 4.1.5 emphasize impartiality, integrity and confidentiality of personnel involved in testing or calibration. Conflicts of interest must be avoided, work sheets must be established, the functions of the quality manager must be defined, and procedures must be implemented to prevent management and staff from being pressurized externally or internally, which could lead to the quality of work being questioned, or even to work being falsified.
- *Quality and documentation system*: The laboratory must establish, implement and maintain a quality system appropriate to its field. The laboratory must document its policies (especially those issued by management), systems, programs, procedures and instructions to ensure the quality of the test results. Appropriate communications must be set up, so that the staff receives information and appropriate training (major quality risk). The quality manual must include strategies and objectives of the quality system (based on clauses 4.2.1 and 4.2.2).

5.4.2 Document control (Do):

- *Scope*: Control of the documentation should cover all documents (*printed or computerized*) that are part of the quality system (produced in-house or from external sources – *such as standard or official methods of analysis*. These include regulations, standards, other normative documents, test and calibration methods, as well as drawings, software, specifications, instructions and manuals (clause 4.3). *This clause makes it possible for the quality system of part of the laboratory work to be accredited, if appropriate*.

- *Approval and dissemination of documents*: The documents must be reviewed periodically, and their status must be indicated (for review, approved, released by, etc.). Dissemination of documents must be formalized and the list of people having copies should be recorded (clauses 4.3.2.1 and 4.3.2.2).
- *Modification of documents*: The review of documents must be completed with information that is not only tangible, but is of proven value to the staff. Moreover, they must be written primarily by the staff and updated; previous versions should be archived. Manual changes have special treatment (clauses 4.3.3.1 through 4.3.3.4).

5.4.3 Review of requests, tenders and contracts (Do):

- Strategies and procedures in place to reach an agreement (i.e. request for analysis) must ensure that the requirements, including methods used, are adequately defined, documented and understood. It must also be ensured that the laboratory has the resources to meet the requirements, and that the chosen method of testing or calibration is appropriate and capable of responding to customer requirements ... Each contract must be acceptable to both the laboratory and the customer (clauses 4.4.1 and 4.4.2). *This is to avoid the laboratory performing tests without knowing its precise commitments especially in terms of planning schedules and delivery times. Clauses 4.4.1 and 4.4.2 suggest the creation of a climate of dialogue between the laboratory and customer.*
- *Archiving of review records*: The different steps leading to the conclusion of a contract must be documented and archived, including records of work subcontracted during the contract period (clauses 4.4.3 and 4.4.4).

5.4.4 Subcontracting of tests and calibrations (Do)

- *Subcontractors*: the subcontracting of tests and calibrations can be either occasional (work with highly specialized requirements, temporary work overload) or regular (specific method that is part of a portfolio of offerings). The competence of the subcontractor must be assessed, and compliance with the present standard is an asset. The laboratory must demonstrate transparency by informing the client about the subcontracting and obtaining the client's authorization. The work carried out by the subcontractor must be registered and its conformity to the requirements of the standard must be documented (clauses 4.5.1 to 4.5.4).
- *Suppliers*: this also applies to those who purchase services, products and reagents that affect the quality of tests and/or calibrations. It is necessary to ensure that the delivered product, reagent or service meets the requirements of the contract, and that storage conditions for the product or reagent are adequate. These steps must be documented (clauses 4.6.1 and 4.6.2). Procurement documents must specify the level of quality of purchased products or reagents (clause 4.6.3). Suppliers must be evaluated (clause 4.6.4).
- *Customer service*: the laboratory should have discussions with customers to fully clarify a request. Visits to the laboratory, good technical communication, and attempts

to obtain information about any additional customer requests are encouraged (clause 4.7). *Here, the standard makes it clear that a relationship of trust must exist between the customer and laboratory: this confidence can be undermined when non-standard results are obtained. Transparency and cordiality help maintain a good level of trust in the relationship.*

- *Claims processing*: the laboratory shall have a procedure for handling complaints from customers and other parties. It shall maintain records of all complaints and the corrective actions that have been taken (clause 4.8). *This clause refers to the management of noncompliance. A first step is to verify whether the claim is justified. If not, one should carefully evaluate whether the complaint is because the result does not meet the customer's expectations. In such a case, it is necessary to treat the claim as indicated in clause 4.1.5. Often, the current confidence level enables the customer to immediately communicate the results that are considered not to correspond to what was reasonably foreseeable. Consistent management of these communications will benefit the client and enable the laboratory to function correctly.*

5.4.5 Control of nonconforming testing and/or calibration work (Act)

5.4.5.1 Curative, corrective and preventive actions

Addressing noncompliance requires an action. In the vocabulary of quality, there are three types of actions that need to be defined:

1. *Curative action* involves addressing the noncompliance of a product or service, or rendering it compliant. It covers the treatment of a nonconforming product, which may go as far as destruction ...
2. *Corrective action* is an action taken to prevent the recurrence of noncompliance, and involves cause analysis and the change of conditions prior to product realization.
3. *Preventive action* is an action taken to avoid the appearance of a potential nonconformity. This is a proactive process.

There is also a last action, rarely cited, *remedial action*, which is a corrective action of medical connotation or applicable in a context where it has no immediate marked effect. A health policy for prevention of smoking is remedial action, as the term corrective action is too strong.

5.4.5.2 Addressing noncompliance

Let us now consider how to address noncompliance and the content of the standard:

- *The existence of a policy, procedures, instructions and an effective framework*: The laboratory must have policies and procedures that are to be implemented when any aspect of its test or calibration work or the result of the work does not conform to its own procedures or the requirements agreed upon with the customer. These policies and procedures must ensure that the required actions in case of noncompliance (stop work, withhold test reports, recall the work results) have an adequate framework. In all cases, corrective actions must be taken immediately if the evaluation indicates that the nonconforming work is likely to recur, or if there is any doubt (clauses 4.9.1, 4.9.2, and 4.11.1).
- *An appropriate choice of actions, monitoring, additional audits and preventive actions*: The laboratory should identify potential corrective actions, and implement the actions most likely to eliminate the problem and prevent its repetition (clause 4.11.3). It should

monitor the results to ensure the effectiveness of the corrective actions implemented (clause 4.11.4). If the identification of noncompliance raises doubts about the laboratory's compliance with its own policy and procedures or with ISO/IEC 17025, an audit (usually internal) of the concerned sectors must be performed (clause 4.11.5). Finally, procedures related to preventive actions should include the initiation of such actions and the implementation of controls to ensure their effectiveness (clause 4.12.2). *This is a step that includes risk identification and risk control.*

5.4.6 Control of records (Do)

Proven methods and proper documentation are not sufficient to safely meet the expectations of stakeholders. Records must also be kept in a sustainable manner, with a specified shelf life for each document. This has been greatly improved and streamlined with the development of information management, but ISO/IEC 17025 is demanding on this point. This also relates to results of nonconforming testing and calibration:

- *Traceability of records*: The laboratory must establish and maintain procedures for the identification, access, filing, storage, retention and disposal of technical records relating to quality. *Quality records include external audit reports and management reviews, as well as records of corrective and preventive actions* (clause 4.13.1.1).
- *Storage conditions*: All records must be legible, stored, and maintained so as to be readily retrievable in secure and confidential facilities that provide a suitable environment to prevent damage and losses. The laboratory shall have procedures to protect, electronically save, and prevent unauthorized access to, or modification of, these recordings. The storage time for records must be established (clauses 4.13.1.2 to 4.13.1.4).
- *Records*: The laboratory must retain records of original observations and resulting data, and sufficient information to establish an audit chain, personnel records, and copies of each test report or calibration certificate issued during a specified period … If the laboratory has announced a limited storage time for records, they should be destroyed after that date. Otherwise, the shelf life should be indicated as at least x years. The records must identify the personnel responsible for sampling or the performance of each test and/or results of calibration and checking. Observations (data and calculations) must be recorded at the time they are made and must be related to the specific operation in question.

5.4.7 Internal audits (Check)

- *Managing internal audits*: Internal audits should be planned periodically and follow an adequate time schedule. It is part of the quality manager's job to perform these tasks. Audits must include all activities covered by ISO/IEC 17025. They must be conducted by an auditor who is uninvolved with the activities to be audited. The cycle for an internal audit should generally be completed over a period of one year (clause 4.14.1).
- *Corrective actions in case of deviations*: When the audit findings cast doubt on the effectiveness of operations or the accuracy of the laboratory's results, testing or calibration, timely corrective actions must be taken. The customer must be informed in writing if the investigation reveals that the laboratory results may have been affected (clause 4.14.2). Follow-up audit activities must verify and record the implementation and effectiveness of the corrective action (clause 4.14.4).

- *Content of the management review and the credibility of its actions*: To ensure that the quality system remains appropriate and effective, management must carry out reviews in order to introduce any modifications and improvements found necessary. The review should consider the following:
  - suitability of strategies and procedures;
  - reports from the people in charge, results of recent internal audits, and assessments by external bodies;
  - corrective and preventive actions;
  - results of comparative tests between laboratories or proficiency tests;
  - feedback from customers, claims;
  - any other information deemed significant.

  The results of the management review must be recorded. Management must ensure that these actions are implemented within an appropriate and agreed timescale (clauses 4.15.1 and 4.15.2).

5.4.8 Reliability of tests (Do)

- The standard (clause 5.1.1) recalls the conditions providing the reliability of a test or calibration in a laboratory according to the 5 M principle[5] of the Ishikawa diagram:
  - human factors (= **M**anpower; clause 5.2),
  - systems and environmental conditions (= **M**ilieu, clause 5.3),
  - methods of testing and calibration, and validation methods (= **M**ethods, clause 5.4),
  - equipment, measurement traceability (= **M**achines),
  - to which is added the treatment of the material to analyze or calibrate (= **M**aterial)
    - sampling (clause 5.7);
    - handling of test and calibration objects (clause 5.8).
- The standard recommends gauging the influence of these factors for optimum reliability – a process that can go as far as risk analysis (clause 5.1.2).
- Human factors (clause 5.2):
  - Staff skills: Employees must have the qualifications to carry out the tests (achievement tests, performance assessment and validation reports). These skills should be assured by management (clause 5.2.1), who should introduce staff tutoring or training by a colleague.
  - It is up to management to formulate objectives with regard to education, training and staff skills. Management must also have a process for identifying training needs, and provide training relevant to both the present and future (clause 5.2.3);
  - The laboratory must establish job descriptions for all employees, determining individual abilities (clause 5.2.4);
  - Skills must be documented and regularly updated, including personnel on contract jobs (clause 5.2.5).
- Facilities and environmental conditions (clause 5.3):
  - Test facilities[6] must enable correct realization of tests and calibrations; environmental conditions should not invalidate or impair the results; the relevant technical requirements must be documented (clause 5.3.1).

---

5 To which is sometimes added a 6th M, Management.

6 Definition *of a test facility according to the Ordinance on Good Laboratory Practices of May 18 2005 (RS 813.112.1):* premises, staff and equipment necessary to conduct a study; for studies carried out in phases over several sites (multi-site studies), the test facility corresponds to the site where the

- Environmental conditions must be monitored, controlled and documented. If non-standard conditions are recorded, tests and calibrations must be stopped (clause 5.3.2);
- If neighboring areas are not suitable for the activities, or may contaminate the testing or calibration, measures must be taken (clause 5.3.3).
- Access to areas affecting the quality of laboratory testing or calibration must be regulated (clause 5.3.4).
- Maintenance of laboratories must be carried out, including special procedures when necessary (clause 5.3.5).

5.4.9 Tests and validation (Do)

The following points are important:

- The laboratory must use appropriate methods and procedures for all tests or calibrations within its field of activity. It must also have instructions for the use and operation of all relevant equipment, as well as the handling and preparation of test objects (clause 5.4.1).
- The laboratory must use test and/or calibration methods, including sampling methods, which satisfy customer needs, and must ensure that the performed tests and calibrations are appropriate, preferably according to published international, regional or national standards (clause 5.4.2).
- Test methods developed by the laboratory must be performed by qualified personnel and development of the method should be subject to updates (clause 5.4.3).
- If non-standard methods are necessary, prior approval must be obtained from the customer and this should include a clear specification of the client and of the test object (clause 5.4.4).
- All laboratory methods must be validated[7]. The range and accuracy of the values obtained from the validated methods should match the customers' needs (clauses 5.4.5.1 to 5.4.5.3).
- A testing laboratory performing its own calibrations must have a single procedure that should be applied to estimate the uncertainty of all calibrations (clauses 5.4.6.1 and 5.4.6.2).
- Calculations and data transfers must be subject to appropriate and systematically performed verifications. Special requirements are implemented for automatic machines for which computers are used – software validation, data protection, maintenance, etc. (clauses 5.4.7.1 and 5.4.7.2).

5.4.10 Equipment (Do)

- The laboratory must have all items of equipment for sampling, measurements and tests required for the proper execution of testing and/or calibrations (clause 5.5.1).
- Equipment and associated software used for testing, calibration and sampling must be capable of achieving the required accuracy, and must comply with relevant

study director is, as well as all other test sites that can be considered, individually or collectively, as test facilities.

[7] Validation is the confirmation by examination and provision of effective evidence that the specific requirements for a pre-determined usage are met.

specifications for the type of testing and/or calibration in question. All devices and equipment must be calibrated and verified before use (clause 5.5.2).

- Equipment must be used by authorized personnel; ad hoc instructions must be available (clause 5.5.3).
- Each item of equipment and its software must as far as possible be uniquely identified (clause 5.5.4).
- Records must be established (identification of the element, identification of type, serial number, etc.) for each piece of equipment having an impact on the test (clause 5.5.5).
- The laboratory shall have procedures for handling, transportation, storage, use and planned maintenance of the measuring instruments to ensure proper operation and to prevent contamination or deterioration (clause 5.5.6).
- Any equipment that is not within specified limits due to overload, mishandling, or which gives suspect results, must be removed from service and clearly labeled (clause 5.5.7).
- Wherever practicable, all equipment under the control of the laboratory and requiring calibration shall be labeled, coded or otherwise identified to indicate its calibration status (clause 5.5.8).
- If, for any reason, some equipment is not subject to its direct control, the laboratory must ensure that the operating conditions and calibration status are consistent before putting it back into service (clause 5.5.9).
- If it is necessary to conduct intermediate verifications to maintain confidence with regard to the calibration status of some equipment, an ad hoc procedure must be established (clause 5.5.10).
- When calibration gives rise to corrections, these must be classified, operated, and used to provide values and not just raw data. The calibration is only archived when a new calibration document replaces it (clause 5.5.11).Equipment must be protected against possible adjustments that would invalidate the test and/or calibration results (clause 5.5.12).

5.4.11 Measurement traceability (Do)

- All equipment used for testing and/or calibration, including instruments for auxiliary measurements (e.g., environmental conditions) affecting the validity or accuracy of the test results must be calibrated before being put into service (clause 5.6.1).
- The calibration program of the equipment shall be designed and managed so as to link the calibrations and measurements performed by the laboratory to the International System of Units (SI) (clause 5.6.2).
- The laboratory must have a program and procedure for the calibration of its reference standards. These must be calibrated by a body that can provide traceability as defined under clause 5.6.2 (clause 5.6.3.1).

5.4.12 Planning of sampling procedures and handling of objects (Plan)

- The laboratory must have a sampling plan and sampling procedures for conducting sampling of substances, matrices[8], materials or products for subsequent testing or calibration (clause 5.7.1).

---

[8] A *matrix* is an agent that is used as a carrier to mix, disperse or solubilize the test item or reference item to facilitate its application to the test system.

- If the customer requires modifications to the sampling method, these must be documented and communicated (clause 5.7.2).
- The laboratory must use ad hoc procedures to record relevant data and operations related to sampling (clause 5.7.3).
- The same applies to the transportation, reception, handling, protection, conservation and elimination of test objects (clause 5.8.1).
- A system should be established to identify test objects, and the identification must be kept throughout the lifetime of the object in the laboratory (clause 5.8.2).
- Upon receipt of the test or calibration object, any anomaly or deviation from normal or specified conditions, as described in the test or calibration method, must be recorded (noncompliant object, clause 5.8.3).Procedures, facilities and suitable storage conditions must be put in place to ensure integrity and to prevent loss of test objects (clause 5.8.4).

5.4.13 Monitoring the quality of test results and documentation of results

- The laboratory must have procedures for quality control to monitor the validity of tests and calibrations undertaken (e.g., statistical techniques, clause 5.9).
- The results of each test or calibration should be documented in a clear, unambiguous, objective manner, and in accordance with specific instructions in the test or calibration methods. They should take into account the test conditions (clause 5.10), regardless of how the reporting is done (clause 5.10.9).

5.4.14 Summary

The contents of the standard are often summarized as an Ishikawa diagram.

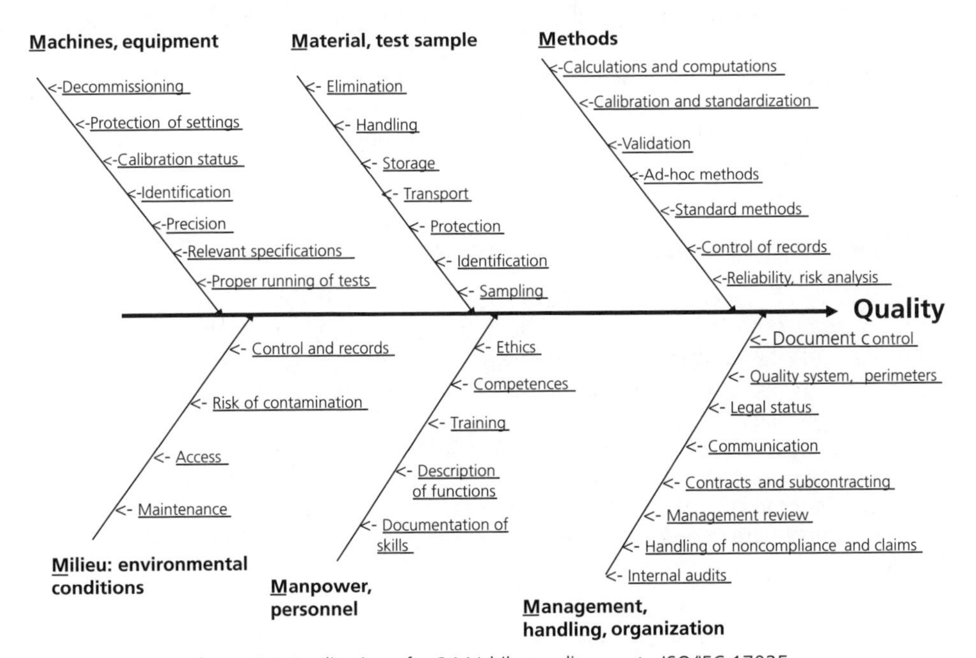

Figure 5.3 Application of a 6-M Ishikawa diagram to ISO/IEC 17025.

5.5 Quality and its management in leading research laboratories

Leading research laboratories rarely perform repetitive measurements; experimental conditions and analytical methods may change from publication to publication. This is why authors of articles describe their experimental procedures as faithfully as possible in the "Materials and Methods" section.

5.5.1 QA within research, a disputed tripartite method

How are quality assurance and quality control of these publications structured? There are three types:

- An article is generally co-signed by one or more senior scientists acknowledged by their peers (quality assurance and quality control during production of knowledge);
- Prior to publication, the article is reviewed by a peer committee that gives its go-ahead, with or without modifications being required (quality control before placing the knowledge on the market).
- Protocols for testing and analysis published in the "Materials and Methods" section are an implicit invitation, addressed to all peers, to replicate the results of the published experiments (call for a subsequent quality control by the recipient of the information or service).

This system was implemented very early in scientific research, but is it effective and, even more importantly, is it efficient?

This issue is regularly discussed in the media when it comes to scientific frauds that are now frequent. The problem is serious; the cases reported represent only the tip of the iceberg. In USA and Europe, moral and ethical appeals, as well as preventive actions against fraud and plagiarism, are organized.

These measures generally respond to a genuine concern rather than constitute a protective reaction from the direction of a university. However, they are often implemented retrospectively, and unfortunately appear to be primarily public relation stunts: no study has shown them to be effective.

5.5.2 Resistance to quality management

The resistance of academia to quality management systems is tenacious. As evidence: the documented reaction of peers in a faculty audit committee (1997) during the accreditation of a laboratory of a Swiss university of technology[9]:

"A research laboratory should not strive for accreditation under ISO/EN quality assurance. While this is justified for the private sector or possibly for a technical college, it is irrelevant at the university level (sic!)*. Service activities should be based only on the reputation of the laboratory, especially since these activities should cover only services that you cannot find in standardized forms in private laboratories. If the objective of accreditation is to force the staff to 'put its house in order' and there is no other way to do this, then it is time to close up shop."*

---

9 *Bulletin of the soils and rocks mechanics laboratories*, No. 22, 1999, available at http://lmswww.epfl.ch/en/pub/bult/fichier/no22.pdf (accessed January 2009).

This reaction of "academic bureaucracy" could surprise us, unless we keep in mind that management tasks in research laboratories, often small, are handled by secretaries, and therefore considered as second – or third-choice activities.

5.5.3 An idealized vision of the researcher

But experts also refer to a paradigm common in academic environments:

True quality cannot be produced by following standardized and bureaucratic patterns; it cannot be achieved by any accreditation. Rather, it comes from dedicated individuals who are willing to do whatever they can to the best of their abilities. It is not by regimenting them or closely controlling them that we nourish devotion or stimulate desire.

This paragraph refers to a generous, unselfish researcher, devoting his or her energies to the Greater Good. This is the ideal researcher, who puts all passions under wraps to devote him – or herself to research and the fight against prejudice and traditionalism.

But are we sure that we can show this attention and dedication every day, even if we have the sincere intention of doing so? And is this dedication really any different from being dedicated to acting with propriety, efficiency and effectiveness, and even going off the beaten track if necessary? Perhaps not. The authors forget that the dedication to quality required by accreditation goes hand-in-hand with a commitment to excellence when it comes to measurement accuracy. This goes beyond having just test results accredited in a laboratory: accreditation is also the continual improvement of analytical and experimental performance. Moreover, it is rarely mentioned that accreditations are of three types, and are increasing in scope:

- Type A: accreditation based on defined and fixed test methods.
- Type B: accreditation based on defined test methods that may be modified.
- Type C: accreditation based on defined technologies and measurement principles. Test laboratories of type C have procedures setting out the working methods and defining the responsibilities for the introduction and characterization of new test and validation methods for specific cases.

Type C accreditation corresponds to research on new methods, which can fit the scope of a technological university, but for each new method, additional accreditation is required. Few academic laboratories are keen to engage in such a lengthy and recurrent procedure, which is contrary to the desire to promptly publish the results of research and study the problem in depth. Many obtain accreditation for certain methods, relying on a halo effect on other protocols to improve the overall quality. It would therefore be preferable to have system-oriented, as opposed to method-oriented, accreditation.

5.5.4 A growing trend

Quality assurance of academic research, however, could expand in coming years. The French National Center for Scientific Research (CNRS) hopes to acquire quality managers.

Many research laboratories are already implementing Good Laboratory Practice (GLP of OECD-1997 or 2004/10/EC) and have a quality management system inspired by ISO/IEC 17025, covering the requirements of health and safety at the workplace (OHSAS 18001 and BS 8800).

This is true for laboratories that use reagents or samples that are toxic, hazardous or at risk of spreading infections, or analysis equipment presenting hazards to health and safety (X-rays or gamma rays, radioactivity, etc.).

This is also true for laboratories that perform research studies for companies. Quality assurance and quality management are often required, which naturally leads the laboratories in question to implement GLP of OECD or ISO/IEC 17025.

Quality assurance in small laboratories

Many *Hautes Ecoles*, technological universities, prefer small research laboratories (typically one professor, one to two senior scientists, a secretary, six to eight PhD students, and a technician) to large ones. How does one ensure quality in a small team? Referring to the Consensus Document of the OECD Quality Assurance (October 1999), one of the subsections deals with small test facilities (p. 11).

... In small testing facilities, it may be difficult for management to devote staff only to quality assurance. However, management must appoint at least one person, even part time, who is permanently responsible for coordinating the quality assurance function. Some permanence of quality assurance personnel is desirable to attain the required skills.

People participating in studies conducted under GLP can perform QA functions for studies subject to GLP conducted in other departments in the test facility.

It is also acceptable to use personnel from outside the test facility to perform functions related to quality assurance, provided that the efficiency needed to meet the principles of GLP is ensured.

Which system should be chosen by a research laboratory that wishes to get external recognition: ISO 9001 or ISO/IEC 17025? The choice depends on the needs and strategic goals of the team. For example:

- ***certification*** (ISO 9001) provides customers and stakeholders with a guarantee of robustness and sustainability when it comes to the quality of services. It also allows the laboratory to develop its own analytical methods (focus on products);
- ***accreditation*** (ISO/IEC 17025) provides stakeholders (other researchers, business clients) with a guarantee of the reliability and accuracy of the delivered results.

In conclusion, we can mention the case of laboratories that, although they do not carry out fundamental research, develop new methods of analysis or calibration. They can obtain special accreditation with regard to procedures on the basis of clauses 5.4.3 and 5.4.4 of ISO/IEC 17025 (Type C accreditation).

5.5.5 AFNOR's first attempt

But can research be subjected to a quality approach? The French national standards boy (AFNOR) affirms this in three manuals, published in the early 2000s, at the request of the French Ministry of Research:

- Documentation manual FD X50-550 (October 2001): *Quality management in research – General principles and recommendations.*
- Documentation manual FD X50-551 (November 2003): *Quality in research – Recommendations for organizing and carrying out a research activity in project mode, particularly in the context of a network.*
- Application Guide GA X50-552 (November 2004): *Quality Management Systems – Guide to the application of ISO 9001 in research organizations.*

These AFNOR documents led a French working group "Quality in research" (involving a large number of stakeholders, including members of the private sector and researchers) to publish in 1997 an "Experimental Guide for Quality in Research".[10]

[10] Available at: http://www.utc.fr (accessed in July 2010).

5.5.5.1 Aspects of research according to AFNOR

AFNOR's approach is primarily focused on organizational aspects of research, in order to allow all participants to clarify issues and possible approaches to quality in research. It does not focus on the intellectual activity of design, or consideration of advanced research. Moreover, the quality management of research laboratories can take many concepts from ISO/IEC 17025 or GLP of OECD.

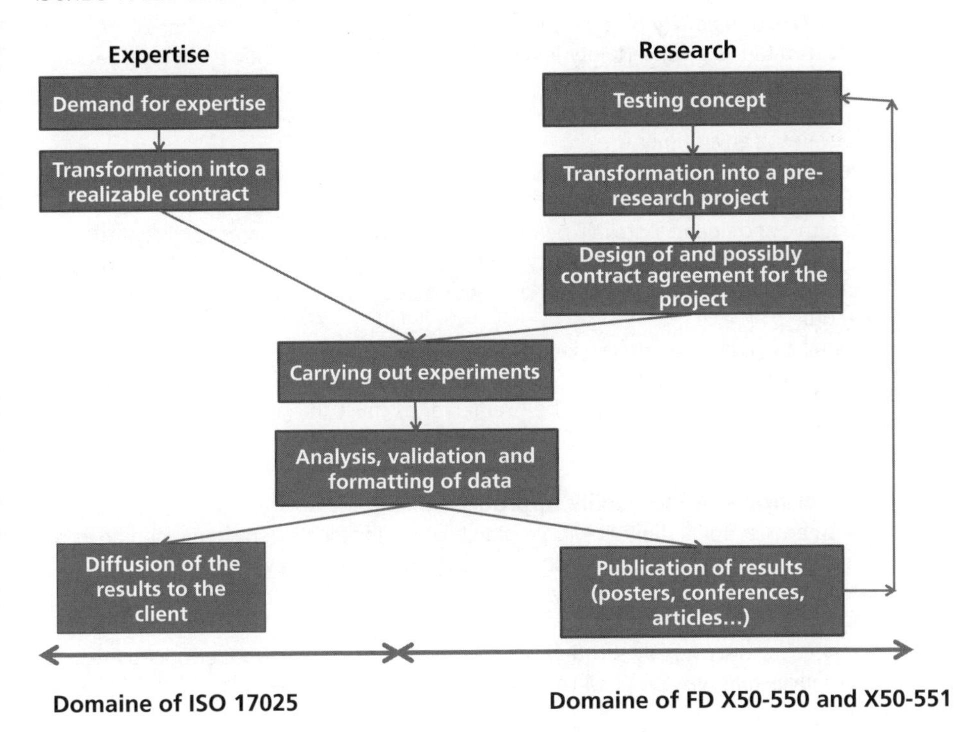

Figure 5.4 The scope of AFNOR's documents.

The AFNOR approach (Figure 5.4) is relevant especially for large projects requiring the collaboration of researchers and stakeholders when it comes to experiments. The checklists that can be found in FD X50-551 are particularly valuable for the leader of a major research project, composed of several teams located in different places, and for advanced facilities requiring the expertise of suppliers in order to work optimally.

This approach is generic, as it can be applied to all research activities (fundamental, applied, etc.). The goal is to provoke thought by staff who are closely (researchers, engineers) or indirectly (administration, suppliers, sponsors, institutional or private donors) involved in the research activity as part of their professional activities, ranging from scientific monitoring to data backup, dissemination and exploitation of results.

Published documents take into account the multiple challenges of research:

- Issues for the organization and researchers: the research stakeholders primarily place their confidence in their laboratories and researchers; increasing globalization of research, as well as the increasing number of researchers over the past decades implies that trust is no longer granted exclusively on the reputation of the "heads" of the laboratories or research institutes; more transparency of the processes in place is appropriate. AFNOR has identified two aspects:

- The ***eligibility*** of a research organization to conduct a program funded by third parties (research agencies of the state, the EU, private foundations, corporations): given the increasing competition in applying for funding, sponsors want assurances of the quality of the results provided. It is not certain that recognition of research units by research funding organizations will in the future continue to be based only on peer review.
- The ***responsibility*** of the research organization in relation to the knowledge produced (areas of uncertainty and validity): the diffusion of any erroneous data has a disastrous impact on the reputation of an organization, but can also lead to a dangerous and deleterious impact on individuals, society and the economy or even the environment.

- Scientific stakes: the organization of research to increase creativity and effectiveness in the face of limited resources.
- Economic and financial stakes: before embarking on a development activity (risk control), besides the desired efficiency, some sponsors want reliable results with maximum guarantees of accuracy, and knowledge of the validity of the investigations carried out.
- Societal and environmental stakes: the contribution of science and technology to society is now seen as ambiguous and controversial. The risk associated with the dissemination of research results must be examined, as must the risk of the uncertainty of published data.

5.5.5.2 The principle of the quality approach

AFNOR documents endorse the following research characteristics, often used when advocating the impossibility of the implementation of a QMS in research activities:

- the constant obligation to manage the unexpected, the uncertain, the risk of not achieving;
- the need to discover, by scrutinizing apparently disappointing or erroneous results, alternatives that make it possible to start up again and suggest new fields of promising research;
- a lack of determination and linearity in the researcher's approach, which involves constantly using his or her talent and creativity to complete investigations;
- a lack of a general research methodology that would provide a comprehensive and generic framework for quality assurance.

Given this situation, AFNOR uses three founding principles:

- *Pragmatism*: AFNOR finds that a research activity always goes through three phases:
  - The first step involves asking the "right questions" in order to identify the quality ***criteria*** to be considered when evaluating the level of success of the research process, in all its stages (e.g., originality of the research).
  - Then, favorable ***conditions*** are sought and defined to meet the quality criteria (e.g., evaluation of the objective of the research in relation to the state of existing knowledge.
  - Finally, practical ***solutions*** that meet these conditions are implemented (e.g., science and technology watch).
- *Teaching methods*, seeking to gradually make the quality approach a natural element in all research activities, particularly by incorporating it in graduate schools of higher education. Three elements are essential:

- the desire for self-assessment, by setting own standards, to correctly perform the research activity;
- continual improvement according to the Deming cycle;
- an information system, so that the experience is stored and accessible to all research players.

For this quality approach to be effective, suitable skills and resources are also necessary.

- *Integration*: researchers do not work in a closed environment, but are in contact with many internal (administrators, technicians, lab workers, other researchers) and external stakeholders (suppliers of equipment, etc.). A process must be undertaken to help them identify how their activities can affect the final performance of a research activity. In other words, they have to be integrated into a learning organization.

5.5.5.3 Implementing the quality approach

A research activity has three phases:

- *Definition of the purpose of the research*, which involves stakeholders' agreement, particularly in the context of research requiring precise management of teams, and of equipment (sometimes far from the laboratory, etc.). We must therefore have a suitable device (process, procedures) to ensure that consensus is reached. To determine this mechanism, researchers should ask themselves three questions:
  - Who are the stakeholders of the research objective?
  - What are their roles and responsibilities, but also their expectations, if any (participation in the publication of results, financing, etc.)?
  - How can their expectations and concerns be taken into account (time management, equipment availability, etc.)?

  As such, AFNOR considers that three basic criteria must be taken into account to construct the approach:
  - *Relevance, novelty, originality*, of the investigation objective in relation to the current state of research results (What is actually the question behind the research? Is its formulation adequate, does it make sense in the context? What new and innovative result can be provided by the research? What will be its likely impact on existing knowledge: marginal, significant, important?).
  - *Appropriateness of the objective*, i.e., how it can fit into the mission, strategy and programs of the organization or laboratory, as well as into the prevailing market conditions (e.g., is it a "hot topic", a subject that will capture the interest of the scientific community, a topic for which financial resources are available?).
  - *Feasibility*, a query that takes into account the skills and human resources that are at hand, the financial and technical means available, and of which the corresponding layout and solution include risk analysis to mobilize resources and planning of the preliminary design.

  By considering these three criteria, which are not restrictive since the researcher is invited to form his/her own thoughts, AFNOR recommends setting up a *means of validating an objective*, allowing an organization to control, and also to improve its scientific production.

- *Carrying out research*: one of the success elements of research is planned process control, in addition to the daily actions of research staff, whether in the laboratory or the entity defined by the perimeter of the stakeholders. However, one must always take into account the unexpected, the uncertainty and the risk of not succeeding, but ultimately, the process can only be perfectly defined in retrospect (managing the unexpected, the appropriateness, according to intermediary research results). AFNOR considers the issues below to be crucial at this stage of the project:

- Can the process used to conduct research be described in a realistic and useful manner?
- How can we control the stages and how should monitoring be implemented to ensure that the research is performed in a compliant fashion, provided that it can be described beforehand?
- How can we control the resources used and what devices are to be implemented to ensure that we maximize feedback, especially since the process can only be accurately described in retrospect?

This leads AFNOR to determine the following criteria for this phase:

- *Transparency in acquisition, recording and traceability of results*
- *Keeping to the schedule and time periods for the various stages*
- *Creativity in the use of deployed resources*
- *Minimization of the risk of failure, especially by the development of alternatives*

• *Development of results*: This plays a crucial role, especially because it is based on the recognition that research stakeholders can expect from their actions. This desire for recognition has already been scrutinized in the first phase, in the *definition* of the research, but only once the results are known is it possible to make concrete proposals to meet the expectations of researchers from the organization (researchers and staff of the institution, external stakeholders). The development of research results includes both their publication and their use for the transfer of knowledge and technology to the economy (aspects of intellectual property, patents, licenses, start-ups, spin-offs). AFNOR's ideas have led to the definition of two key steps:

- *Validating the results:* when the results have been gathered and edited, their consolidated content, with margins of uncertainty, must be submitted (unless there is a confidentiality clause specified in the first stage) to the researchers and stakeholders of the project. The thoughts of the different players and their synthesis make it possible to:
  - verify whether the results meet their expectations,
  - identify the lessons learned from abandoned research paths and unexpected results, and
  - create the necessary consensus on the evaluation criteria for the success of the research activity.
- *Transfer of results:* The transfer of results typically includes the following steps, which may be sequential:
  - The results are highlighted and developed formally:
    - in articles, talks, conference communications, forums, posters, for dissemination in the scientific community;
    - in any form determined by best practice when it comes to intellectual property (patent filing, etc.) for transfer to the economy.
  - The means of communication are determined so as to maximize the impact (which journal, which conference will be most suitable? In which country is it wise to file a patent, etc.?). The transfer activities to the scientific community vs. the economy should also be coordinated in order not to distribute information to the public sector that should be kept confidential for some time.

The retained criteria, mainly the validation of results and maximization of the impact, should be translated into ideas to consider practical solutions. AFNOR recommends putting in place a validation system, which can remain very light (e.g., an ad hoc committee). A summary of the AFNOR approach is reported in the FD X50-551 and illustrated in Figure 5.5.

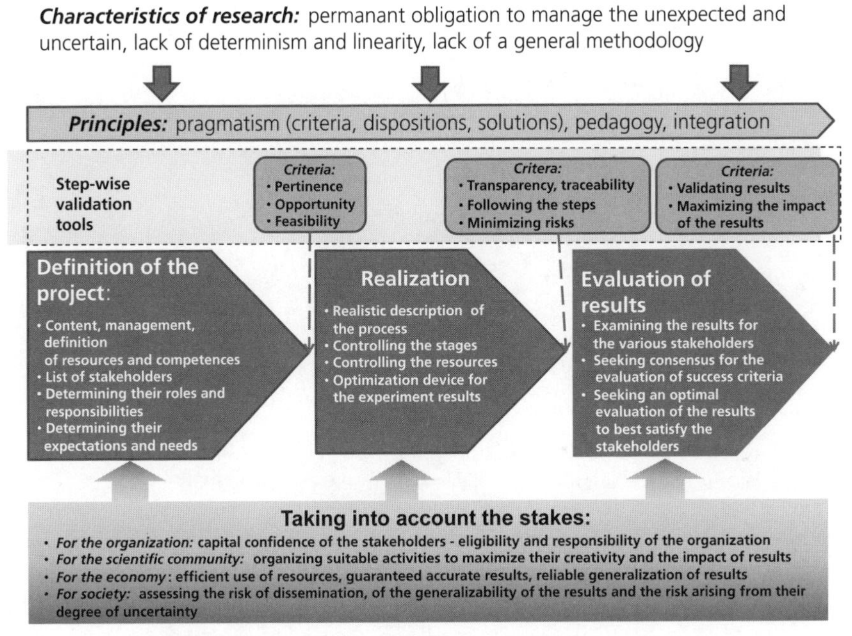

Figure 5.5 Schematic vision of the documentation FDX 50-551.

5.5.5.4 Conclusions

Although it is commendable, the AFNOR process remains largely at a pioneer stage. There is, however, room for hope: the complexity of current experiments (especially particle physics, which requires large facilities, or disciplinary investigations of life sciences), in addition to the pressure put on researchers to publish relevant and accurate observations as quickly as possible, in an ethic of integrity and appropriate objectivity, recommend a slow and progressive awareness of the requirements of quality, traceability, order and method that can provide quality first in applied research, and then in fundamental research.

Many researchers believe that the way research is carried out now is no longer in line with its objectives: to build sound knowledge about important areas, and not to publish documents that in the first place are intended to "excite" the scientific community and satisfy shareholders of the few private companies that own (the term racketeering would be incongruous, of course) the whole market of scientific publishing. The issue is complex, particularly in the form and content that international regulation could bring [5.2][2].

However, the quality approach is one of the few solutions to be proposed to the heads of institutions who are responsible for the scientific approach of their employees but incapable, in an inexorably growing amount of cases, to effectively control it, either in terms of ethics or simply with regard to the conformity of the process of producing results with good research practices. However, the quality approach must (given the instinctive reluctance of academia when it comes to administrative artfulness) be introduced in a flexible and light manner so as not to ostensibly add to the costs of research.

References

[5.1] GILLES REVOIL, *Qualité dans les laboratoires d'étalonnage et d'essais – une amélioration des processus*, ed. AFNOR, 2001.

[5.2] Cf. the article by LAURENT SÉGALAT, System crash – Science and finance, same symptoms, same dangers?, *EMBO reports*, vol. 11, N° 2, pp. 86-89 (2010).

Chapter 6

Safety of the Food Supply Chain and ISO 22000

6.1 Food safety and hygiene, a little history

6.1.1 Food safety, from antiquity to the Industrial Revolution

Ensuring that food is without risk of harm (contaminants, natural or industrial toxins, bacterial or viral infection) to the health of the population and without misleading consumers (reduced volumes, misleading scales and balances) seems to have been a concern for all societies, even in antiquity:

- Assyrian tablets specify weights and measures for wheat.
- Egyptian papyri specify the labeling of certain foodstuffs.
- The Republic of Athens had controllers to evaluate the quality and purity of beer and wine.
- The Roman Empire set up an inspectorate to ensure the compliance of food.
- In the Middle Ages, intergovernmental trade agreements, some of which are still current, specified the quality of foods such as eggs, sausages, cheese, beer, bread, and wine.

It is likely that food safety problems were one cause of the high mortality rate of the human race until the twentieth century. Some diseases have been known for a long time, such as lead poisoning from the pipes of urban water supplies during the Roman Empire. Contamination of water (and milk) by pathogens was certainly a regular source of disease until the late nineteenth century.

The agricultural revolution started in the UK at the end of the first half of the eighteenth century (and in the nineteenth century in France) and gradually decreased famine in Europe, with the first significant effect on mortality.

But it was in the nineteenth century, when the urban population started growing significantly, and when cities expanded and moved away from primary production centers, that food safety became more acute. It was also from this point that food safety became more closely associated with general hygiene, including clean water supplies in urban communities. One of the greatest cities of the time, London, with its misery and its high concentration of workers, would play a leading role.

6.1.2 Reduced diseases in the West: a primary cause, hygiene

The improved health and life expectancy since the nineteenth century are often attributed to modern medicine, especially vaccines and antibiotics (it is estimated that their use has extended human longevity by 10 years). But we often forget that although these weapons were undeniably decisive in the fight against infectious diseases in the twentieth century, this fight started long before. Some scientists argue that the increase in human longevity of the population of Western Europe can just as well be attributed to better diet and more comfortable living conditions.

In the 1800s, smallpox, scarlet fever, measles and diphtheria were ailments so common that they were considered to be associated with childhood. Epidemics of cholera and malaria were legion. Typhus and typhoid threatened the poor, tuberculosis both rich and poor. In Western countries in the early nineteenth century, the mortality rate during the "good years" without major epidemics was often four times that of today. In what is often described as the golden age of the British Empire, in *Rudyard Kipling's* Victorian England, the average age of death in the poor urban population was 15 to 16 years, given the high infant mortality rate.

To understand the magnitude of the revolution that then took place, we must remember the living conditions of the time. Rotting garbage, human and animal waste, stacks of dead carcasses, open sewers: the streets were veritable cesspits. Individual toilets were an exception: chamber pots were emptied into the streets or courtyards. The stench was almost unbearable (even for the olfactory senses of the time, not overly sensitive as ours), especially in summer. With regard to personal hygiene, the situation was not brilliant; perfumes were often used by the wealthy to mask repulsive odors. The frequency of washing clothes left much to be desired.Even though the source of infectious diseases was unknown before Pasteur's discovery in 1878, empirical observations associated living conditions with morbidity. Foul odors, miasma[1] were considered the cause of most diseases that decimated the population. "Dirt, poverty and disease" formed a merciless troika. That is why, paradoxically, the first battles against infectious diseases took place before scientists could explain their origin.

It was in the mid-nineteenth century that the health revolution began in the newly industrializing countries, particularly in the UK. It was also called the third medical revolution [6.1], with a marked increase in life expectancy, followed a few decades later by a substantial drop in infant mortality rates. Three key changes occurred:

- improved living conditions and, consequently, a reduction in overcrowding[2], not only human but also with regard to infectious agents and pests (rats, lice, fleas, bedbugs, poultry, livestock, etc.).
- improved nutrition through advances in agriculture and technology[3], and especially.
- improved personal and environmental hygiene.

In England, an informal coalition of social activists, reformers, doctors and scientists began to push for health reform. Under their leadership, legislation was changed in the 1850s and 1860s: public health authorities responsible for waste collection, distribution of water, the treatment of various types of environmental pollution, etc. were created (a decade later in America). A major breakthrough was the demonstration that the polluted water of

1 Putrid discharges from decaying bodies and substances.

2 It was not until the end of World War I that central heating and hot running water were introduced in Swiss households ...

3 This helped to shorten periods of famine, still very common in the early nineteenth century, which weakened the population.

the Thames was a source of disease, while supporters of the spread of diseases by miasma claimed that only air from the atmosphere was to blame.

A major sociocultural transformation of the nineteenth century was the behavioral change in personal hygiene, including cleanliness and laundry. The pioneers of this movement advocated personal hygiene. Their early efforts led to the reduction of the tax on soap (1833), the construction of public baths and laundries for workers in London (1844), and their widespread implementation throughout the country.

It is estimated that the increase in longevity is mainly due to the reduction in child mortality up to 1950, and then to a series of cultural, social, medical and food changes.

6.1.3 The introduction of food standards and regulations

Even though the first food laws and the creation of monitoring bodies date back to the early nineteenth century, it was not until the early twentieth century and the development of microbiology that food safety standards emerged, notably in 1903 when the International Dairy Federation established standards for milk and dairy products. These products were very susceptible to bacterial contamination and carried pathogenic germs under certain conditions (milking, transport, storage), especially the dreaded tuberculosis. Besides chemistry, bacteriology contributed decisively to the characterization of food hygiene and the minimization of risks of infection.

Different governments implemented more demanding food legislation, but national standards were often disparate and not always based on sound science (standards, analytical methods).

The most complete initiative is probably that of the Austro-Hungarian Empire, which, between 1897 and 1911, published a collection of food standards and descriptions, the *Codex Alimentarius Austriacus*. Although this document never had any legal value, it was widely used by legal authorities to identify standards or describe food.

6.1.4 FAO, WHO and Codex Alimentarius

6.1.4.1 Creation of FAO and WHO

Things changed significantly at the end of World War II with the creation of two international institutions:

- FAO (Food and Agriculture Organization) was established in 1945; one of its missions is to provide technical information on food based on solid data and conclusions obtained in an irreproachable manner.
- WHO (World Health Organization) was established in 1948; one of its activities is to define norms and standards relating to health, and ensure their implementation.

The establishment of these bodies was accompanied by growing public interest in objective data about food, now widely diversified and marketed by manufacturers – sometimes accompanied by boring and questionable information.

As a first step, food additives attracted the attention of WHO. In 1955, the organization set up an ad hoc committee that regulated their use, followed by, at a later stage, the permitted levels of contaminants, etc. In 1963, after a collaborative effort by the two agencies, the 16th World Health Assembly[4] adopted the joint WHO/FAO project to establish a global food standard, *Codex Alimentarius*.

4 Supreme governing body of WHO (World Health Organization).

Table 6.1 Safety of the food chain.

	Innovations, changes	Regulations, standards, organizational changes
Beginning of the XIX century	• Progressive urban densification, constitution of a working class • First food sterilization	
Middle of the XIX century	• Improvements in agriculture and technology, health revolution in the UK	• Creation of food laws and supervisory bodies • Use of analytical and scientific methods
End of the XIX century	• First frozen meat imports from New Zealand	• Use of analytical and scientific methods
Beginning of the XX century	• Inexorable progressive fall of the peasantry in Europe • Economic growth in international food trade	• Creation of food standards to facilitate international trade • The International Dairy Federation develops standards for milk and dairy products
Middle of the XX century	• standards for milk and dairy products Organization of the "food supply chain", generalization of conditioning • standards for milk and dairy products Rise of the distributors	• Creation of FAO (1945) and WHO (1948) • First HACCP design for NASA (1959) • Codex Alimentarius (CA; 1963)
End of the XX century	• Generalization of global food trade	• HACCP introduced in CE 93/43 CA adopted to resolve international conflicts (1995)
Beginning ofthe XXI century	• China and India disrupt economic exchange	• CE 178/2002, CE 852/2004, CE 853/2004 • Distributor standards, ISO 22000

6.1.4.2 *Codex Alimentarius* today

Codex Alimentarius is intended to protect consumers' health, while ensuring proper business practices in the global food trade, and coordinating national food standards and those of international organizations and NGOs.

Codex is a collection of standards, codes of practice, guidelines and recommendations. Some *Codex* texts are very specific, others more general. Some deal with detailed criteria for food or food groups, others with management and operating conditions of production processes, or the operation of government control systems working with consumer protection and food safety.

Codex publishes the following families of documents:

- *Standards for products*, which specify their composition, additives and their permitted concentrations, but also the maximum levels of contaminants, pesticides, potentially pathogenic microorganisms, etc.
- *Standards for food labeling* of prepackaged food products or staples.

- *Sampling and analytical methods*, which vary depending on the food under consideration.
- *Codes of Practice*, which define the process conditions and characteristics of manufacturing, transportation and storage of products or product families considered essential to ensure the safety of the product and its suitability for consumption. For food hygiene, the "canonical" text is the Codex General Principles of Food Hygiene, which introduces the use of HACCP (Hazard Analysis and Critical Control Points), widely used in food safety management systems.

Although *Codex Alimentarius* has no legal value, an international agreement was signed in 1995 stipulating that international standards and recommendations, including *Codex Alimentarius*, are formally recognized as reference materials to facilitate international trade and to resolve trade disputes under international law.

6.1.4.3 *Codex* in the hands of the powerful?

Codex Alimentarius is a respected and accepted standard but, in recent years, voices have criticized the almost exclusive influence of large pharmaceutical or chemical companies within its committees and its structures to the detriment of citizen rights and consumer protection.

Thus, according to a draft discussed at the 29th Session of the Codex Committee on Nutrition and Foods for Special Dietary Uses, held in Germany in 2007, practitioners of alternative medicine would no longer have permission to prescribe dietary supplements, or herbal and homeopathic preparations. Some see this as major industrial groups removing any alternatives to their way of defining medicine, medical care and healing, and thereby grabbing a market share that they didn't have before.

Alternative medicine therapists have little financial means to oppose this lobbying, which is as organized as it is dangerous, and guided mainly by commercial interests and unbridled financial greed that is not consistent with public health goals. But the intentions of lobbyists go further. They also plan to ban medical care by Chinese medicine, which has been practiced successfully for centuries. Let us bet that their schemes will be harder to carry out when it comes to this issue.

6.1.5 HACCP enters the scene

6.1.5.1 The birth of HACCP

HACCP is a system for inspecting manufacturing processes, which focuses on critical points in the process for food safety, and provides monitoring and continuous recording. The introduction of HACCP makes it possible to circumvent systematic control of the finished product.

HACCP is an addition to the risk management tool HAZOP[5] (HAZard and OPerability studies) when it comes to food safety. This tool was developed in the chemical industry for personnel safety and efficient operations. The two approaches were initiated in the same decade.

HACCP is a synthesis of two concepts [6.2]:

- The contributions by W.E. Deming on quality assurance, continual improvement and total quality.

---

5 Cf. British Standard BS/IEC 61882:2002, *Hazard and operability studies (HAZOP studies) – Application Guide* British Standards Institution.

- A concept developed for NASA by the Pillsbury[6] company. In the early 1960s, NASA was searching for a program to ensure zero defects in the production of food intended for astronauts. Drawing on critical points of manufacturing control for rockets developed by NASA, the research director of Pillsbury, the microbiologist Howard Baumann, set up the first version of HACCP. He then tested it within Pillsbury, in the scope of an accident: critical contamination of one of the company's products (glass fragments in baby food ...!). Given the success of the HACCP approach, it was then extended to all company production lines.

Pillsbury presented the HACCP approach to the American Conference on Food Safety in 1971. In 1974, the FDA (Food and Drug Administration, USA) introduced HACCP in sanitary regulations for low-acid products (i.e., sensitive to the proliferation of pathogens). From the 1980s, many food companies started using this approach.

In 1985, The National Academy of Sciences of the United States recommended that the HACCP approach be the basis of the assurance of food safety in the food industry. In 1988 and in 1991, the International Commission on Microbiological Specifications for Foods (ICMSF) and the International Association of Milk, Food and Environmental Sanitarians (IAMFES), also recommended the generalization of the HACCP system to ensure food safety.

6.1.5.2 *Integration of HACCP in Codex Alimentarius*

Recognizing the importance of HACCP for food safety, the twentieth session of the *Codex Alimentarius* Commission, in Geneva in 1993, adopted the Guidelines for the Application of the Hazard Analysis Critical Control Point (HACCP). In a second step, the internationally recommended Code of Practice – General Principles of Food Hygiene [CAC/RCP 1-1969, Rev.3 (1997), revised] – was accepted by the *Codex Alimentarius* Commission at its twenty-second session in 1997. The HACCP plan and guidelines for its application are annexed to this Code of Practice.

From the 1990s, in their ISO 9001 certification process, a majority of food companies have introduced HACCP in their management systems.

Finally, HACCP went beyond its original area of application and was introduced in the US for the design and manufacture of medical instruments and equipment[7].

6.1.6 Newly emerging risks

From the early 2000s, the European Union, international authorities and stakeholders in charge of global food safety identified new hazards. They found that although a massive attack on the health of European and North American populations and, to a (much) lesser extent the rest of the world, seemed to be ruled out, new factors changed the situation:

- the large-scale distribution of food can facilitate the spread of foodborne diseases;
- globalization of trade increases the spread of pathogens and calls for greater vigilance than before, especially when it comes to the storage and handling of food, which are ideal vectors of pathogens, both at home and where people eat collectively (canteens, restaurants, etc.); the larger number of meals eaten outside the home and the

---

6 An American food industry giant for bakery and confectionery products, cereal products, and precooked meals based on pasta (pizza, etc.). cf. www. pillsbury.com (accessed September 2010).

7 In 1997, the American FDA published the GMP for medical devices known as *Quality System Regulation (QSR)* in Chapter 21, section 820 of the *Code of Federal Regulations*.

generalization of cafeterias and restaurants increases the risk of pathogens spreading on a large scale;
- the depletion of the world's drinking water, particularly caused by pollution of water sources, may in the future worsen the resurgence of infectious diseases; the need for water, which increases with demographic evolution, may lead to a situation where the water needed for personal and domestic hygiene will be less available.

The profile of infections will also change by the spread of "exotic" pathogens, brought about by the globalization of trade, which can cause epidemics that are distant from each other in the world.

Even in a country like the US, monitored by the powerful FDA[8], according to figures released in 2011 [6.3] (annual statistics) there is still a significant deficiency when it comes to food safety:
- 48 million consumers became unwell or suffered from food poisoning in 1 year; for information, the US population is close to 310 million inhabitants – *one in six consumers is therefore affected annually…*
- 128 000 were hospitalized;
- there were 3 000 deaths (0.01 ‰ of the population); this is comparable to the homicide rate (number of murders and assassinations[9] per number of inhabitants), but less than the prison population (1 % of the US male population is in prison …)

Aware of existing and emerging risks, stakeholders have implemented new initiatives[10].

6.1.7 Reform of food legislation within the EU

In the early twenty-first century, the EU reorganized its food legislation (previously no less than 14 points) and established the "hygiene package" consisting of five Directives laying down requirements for the hygiene of foodstuffs and foods from animal sources:
- Regulation 178/2002 EC on the general principles and requirements of food legislation, called *Food Law*;
- Regulation 852/2004 EC on the hygiene of foodstuffs, which states food business operators (other than those engaged in primary production) must apply the principles of HACCP;
- Regulation 853/2004 EC on foods of animal origin;
- Regulation 882/2004 EC on official inspections;
- Regulation 183/2005 EC on hygienic requirements for animal feed.

The hygiene package requires economic players:
- to develop voluntary instruments such as guides to good hygienic practices (GMP: *Good Manufacturing Practices*, still called Prerequisite Programs, PRPs, in ISO 22000: 2005), and
- to reference the HACCP method.

---

8 US Food and Drug Administration: cf. http://www.fda.gov/: current programs of the FDA are no longer limited to food safety but also include dietary imbalance of much of the population – mainly the poorest – which causes severe obesity and chronic diseases (diabetes, etc.).

9 These deaths go unnoticed in the media – what Hollywood producer would dedicate a "bloody" series to this problem …?

10 Public health and the risk of pandemics are identified as two of the Grand Challenges that Europe will face. They are shown and listed in the Lund Declaration of July 2009, published during the Swedish EU Presidency (url: http://www.se2009.eu/polopoly_fs/1.8460!menu/standard/file/lund_declaration_final_version_9_july.pdf).

A management system for food safety, recognized in the European Union, can be established in a way similar to ISO 9001.

6.1.8 ISO 22000 and private frames of reference

Following the awareness of governments, media and consumers about food safety (goal: zero accidents), standards bodies of countries such as Denmark, the Netherlands, Ireland, Australia and Brazil developed national standards or audit standards on food safety management. Also, groups of economic actors took the initiative to develop (too) many private frames of reference (BRC[11], IFS[12], EurepGap[13] ...), which are more like auditing systems, or PRPs, than management systems.

This multiplication creates confusion among companies and organizations in the food business which, under pressure from external stakeholders, are forced to simultaneously apply multiple frames of reference that are not necessarily compatible. With the aim of harmonizing these, the Danish Standards Association (DS) submitted a proposal in 2001 to develop an international standard on a food safety management system. Work on ISO 22000 began in 2002 within ISO/TC 34, *Food products*. ISO 22000, *Food safety management systems – Requirements for any organization in the food chain*, was published in 2005.

It specifies the requirements for a food safety management system (FSMS) in the food chain, where an organization needs to demonstrate its ability to control food safety hazards in order to ensure that food is safe at the time of human consumption.

ISO 22000 specifies the requirements for an FSMS as a coherent set of processes to enable a company's management to ensure the efficient and effective application of its policy and its goals for improvement. The structure of ISO 22000 incorporates provisions contained in ISO 9001 to allow perfect compatibility and to complement the different management standards currently used by companies. It relies on four main blocks that are closely linked:

- the responsibility of management,
- resource management,

---

[11] *British Retail Consortium* system includes four technical frames of reference for the food industry, specifying the conditions to be met for an organization to be able to produce, package, store and distribute food products to consumers with maximum food safety. Standard available at: http://www.brcglobalstandards.com/

[12] IFS = *International Food Standard* is a frame of reference for auditing branded food product suppliers. IFS is imposed on producers by many distributors in France and Germany. It was established in 2003 by representatives of retail food products in Europe. IFS is used to review and certify systems to ensure food safety and the quality of food production. At an international level, IFS was developed specifically for the food industries that supply products of their own brand to distributors (business companies: "supermarkets", "hypermarkets", "food store chains", e.g., Carrefour). The requirements were defined by HDE (*Deutscher Einzelhandelsverband*, Federation of German retailers) and the French FCD (Fédération des Entreprises du Commerce et de la Distribution). IFS is based on the quality management standard ISO 9001:2000, to which are added the principles of good manufacturing (cleaning and disinfection, fight against pests, maintenance and training) and HACCP principles. Version 4 of IFS integrates the EU legislation regarding the use of allergens and genetically modified organisms (GMOs) http://www.ifs-certification.com/

[13] *EurepGAP* – frame of reference for good agricultural practices on farms – was created in 1997 at the initiative of supermarket distributors, members of the Euro-Retailer Produce Working Group (EUREP). The standard establishes requirements and procedures for the development of good agricultural practices. EurepGAP is based on the HACCP method (cf. www.eurepgap.org).

- planning and manufacture of safe products, and
- validation, verification and improvement of an FSMS.

The implementation of the standard requires the establishment of a team and an HACCP plan.

Major producers are now mobilizing to stop the proliferation of food safety standards, which complicate their tasks when it comes to management and production. They argue for the application and recognition by all stakeholders of ISO 22000 and *Codex* as privileged frames of reference.

6.2 HACCP: content and implementation

6.2.1 Prerequisite programs, good manufacturing practices

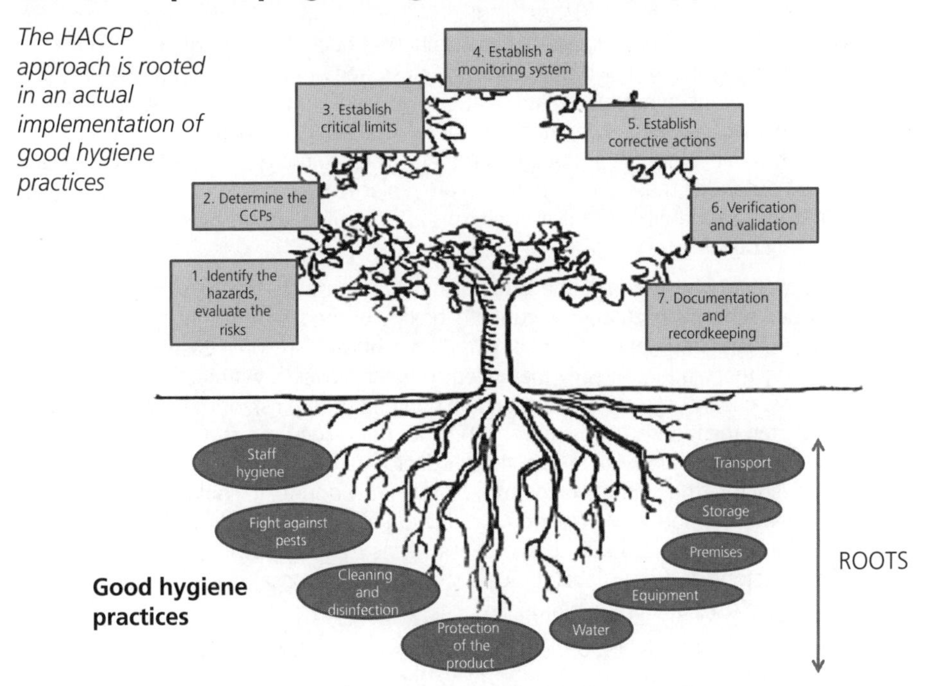

Figure 6.1 The tree describing the HACCP approach.

The HACCP system is developed under specific conditions: the prior implementation of a range of prerequisites (also called PRPs, prerequisite programs), grouped under a convenient term (e.g., Good Manufacturing Practices, GMP), closely related to good hygienic practices. The prerequisites support and root the HACCP method in the safety approach of the organization.

The number of prerequisites [6.4] may vary from one standard to another (but also from one food industry to another), but it is generally agreed upon that the following items must be mastered before formally proceeding with the HACCP approach:

a) *Employee hygiene*: Personal hygiene programs are designed to prevent contamination at the interface between the product and staff. They include immaculate clothing, cleanliness and disinfection of hands, wearing a cap, regular inspections to ensure that staff are not carrying pathogens (e.g., *Salmonella*, humans can carry and transmit the bacterium without being affected themselves), and prohibition of food consumption in production areas. The degree of severity given to staff hygiene varies with the food industry concerned: contact with the product is very sensitive in butchers' shops, and more moderate in dairies, where milk and its derivatives are separated from the staff by tanks, pipes, stainless steel, and the finished product is packaged by filling machines with sterile air and chemical disinfection/sterilization prior to packaging.

b) *Fight against pests*: birds, rodents, insects (especially flies and cockroaches) may alter, contaminate, soil (droppings) and chemically and/or bacteriologically contaminate the product. Programs to fight against their presence and leading to their (almost) complete elimination must be implemented and regularly repeated (often by a specialist company), as must local intervention of control and inspection (by a responsible person within the company).

c) *Cleaning and disinfection*: production facilities, utensils and equipment must be visually clean (dust, dirt and debris must be disposed of) and, to a large extent, free from bacteria, molds and yeasts. This is achieved by ad hoc programs at regular and planned frequencies. For installations (stainless steel) exempt from ambient air, such as tanks, pumps, pipes, and plate heat exchangers for industrial dairies or breweries, a sequential treatment with strong acids and bases at temperatures around 130°C enables sterility prior to production. When it comes to more sensitive facilities or ones placed outdoors, these drastic methods are replaced by a judicious choice of products annihilating all families of microorganisms after application (or by spraying hydrogen peroxide on surfaces).

d) *Protection of the product*: the product must be protected from all foreign matter, whether simply junk (paper, cardboard boxes, burned product, hair) or dangerous matter (sharp fragments of metal, glass, hard plastic). In some cases, the personnel handling the product directly may need to wear gloves[14]. As metal compounds are the most widely used in food production equipment, the risk of metallic foreign bodies is often the highest. Some production lines (baby food in jars, for example) employ a metal detector after filling. But the product must also be protected from chemical contaminants (the sulfites, mycotoxins, pesticides, non-toxic coolants and lubricants that are *food grade*; as well as from production equipment, acids, bases and products used for disinfection). A list should be made, as short as possible, of chemicals used in the factory, and their labeling, use, storage and disposal placed under control to reduce the risks of contamination.

e) *Water*: comparatively, food industries consume only little energy, but are voracious water users (cleaning, rinsing, etc.). The water used must be clean, clear and transparent and contain only few bacteria, or even be sterile. Control of the water supply and quality is essential. Finally, wastewater control is an important factor for ISO 14001 certification. The factors to master when it comes to water can also be applied by analogy to controls on ambient air and raw materials.

f) *Equipment*: Equipment must be designed to protect the product from undesirable contamination. It must also be easily cleaned or disinfected, while maintaining its integrity and not becoming altered. Currently, suppliers of equipment use industrial building standards that enable the use of ad hoc machinery and production tools.

---

14 Nevertheless, wearing gloves is not the best protection against contamination, as the staff lose much tactile sensation.

g) *Premises*: Guidelines for premises involve:

   a. *Their situation:* far from landfills that are burned, far from areas of intensive livestock feeding before slaughter, and other sites that may be unhealthy or infested with pests.

   b. *Adequate ground*: drained and paved (or tarred), with at least gravel on the ground around the production premises.

   c. *Outdoor lighting*: located as far as possible from the production facility to avoid attracting pests.

   d. *Indoor lighting*: designed to prevent broken glass from contaminating the product if the lights should break.

   e. *Insulation*: ventilation systems maintaining a slight positive pressure in the premises, with air flow leaving the finished product in the direction of raw materials, waste pipes, and windows and doors to prevent pests from entering.

   f. *Construction*: ease of cleaning and disinfection, walls and floors with ad hoc coatings, free of cracks and holes, water lines, gas and steam circuits readily available, without risk of harboring pests.

   g. *Provision of toilets* and hot water supply, detergents and disinfectants for cleaning hands, opening and closing of taps with pedals (to avoid recontamination). Disposable paper towels to dry hands (and not warm air hand dryers, unfortunately very common, that re-contaminate the hands …).

   h. *Storage*: raw materials, food components, but also the finished product, must be packaged, protected and stored so as to avoid deterioration and contamination. For perishable goods, cold chain storage (4°C or below) should be considered. The control of the storage temperature is very important here.

   i. *Transport*: raw materials, food components, but also the finished product, must be packaged, protected and stored so as to avoid deterioration and contamination.

These prerequisites are often supplemented by:

a) General programs of *maintenance* of the premises, equipment and utensils, as well as *inspections* and *audits* to ensure compliance with the prerequisites.

b) The monitoring of *suppliers* and how they are chosen: these are very close to the framework conditions set by ISO 9001, but with standards related to the food industry (absence or low amounts of pesticides and mycotoxins, no foreign bodies, maximum content of microorganisms per gram of product, etc.).

c) Programs for *calibration of measuring instruments* for the production and inspection of their accuracy (scales, thermometers, pH-meters, micrometers, pressure probes, etc.).

d) Ensuring the *traceability of a product,* to enable an account of the flow of raw material to the finished product in case of claims by customers (computer recordings, labeling including the date and exact time of production on the packaging or finished product, etc.).

e) *Training* of staff in good manufacturing practices and hygiene.

f) A system for processing claims inspired by the ISO 9000 family.

It is important that the prerequisites (PRPs) are implemented within the organization before the HACCP plan is introduced, otherwise there is a risk that some prerequisites will be erroneously included in the HACCP approach.

In practice, international food technology organizations, manufacturing groups or professional associations describe the prerequisite programs in their production area, characterized especially by the degree of importance attached to each category of PRP.

6.2.2 HACCP: seven principles

Codex Alimentarius describes HACCP as seven principles carried out in 12 steps. The seven principles listed below are illustrated by a simple practical example:

1. *Identify any hazards* that must be prevented, eliminated or reduced to acceptable levels (e.g., the bacterial flora of raw milk may contain pathogens that multiply rapidly).
2. *Identify critical points* at which control is essential to prevent or eliminate a hazard or make it acceptable (e.g., milk undergoing pasteurization to eliminate germs; the pasteurizer represents the control point, and the temperature measuring point of the milk undergoing pasteurization is especially important).
3. *Establish*, at the (critical) control points, the *limits* that differentiate acceptability from unacceptability for the prevention, elimination or reduction of identified hazards (e.g., a temperature range for which a lower temperature would render pasteurization inefficient, and for which a higher temperature would cause the milk to undergo organoleptic changes).
4. *Establish and implement* effective monitoring procedures at critical control points (registration of the pasteurization temperature; an alarm if the temperature falls outside the control zone).
5. *Establish corrective actions* to be taken when the monitoring system indicates that a critical control point is not under control (instruct the staff and provide an ad-hoc procedure stating what actions to take when the alarm goes off).
6. *Establish procedures performed periodically to verify the efficiency* of points 1 through 6 (e.g., review and calibrate thermometers, control the speed of pump flow, test the functioning of the alarm by pasteurization tests on water, with a simulated alarm going off to verify that the staff applies the procedure correctly, etc.).
7. *Establish documents and records* based on the nature and size of the company to *demonstrate the effective application of measures 1 through 7* (technical documentation of the facilities, record-keeping of the thermometer calibrations, records of the temperatures observed in production during pasteurization, test reports and simulated alarms, maintenance logs for equipment showing that everything is done to avoid cracks in the heat exchanger plates which may re-contaminate the product, etc.).

6.2.3 Identification of hazards

Hazard identification is performed by an interdisciplinary team involving several business functions (purchasing, production, R and D, logistics, sales, marketing, servicing and maintenance). The presence of a microbiologist and a food technologist is mandatory in practice.

In a first step, the team gathers in detail all the information regarding the manufacture of the foodstuff, the raw materials, composition, ingredients and the technology used. They then provide a flow diagram of all transactions, and identify all the risks for each step (microorganisms, contaminants, toxins, foreign bodies, etc.).

Table 6.2 Factors/steps to consider in the identification of risks that affect food safety.

Identification of good practices in the primary sector (agriculture)
Basic raw materials, storage conditions
Ingredients and composition of the foodstuff.
Intrinsic factors to the foodstuff.
Procedures and technologies used in the process.
Content of microorganisms in the foodstuff.
Environment, arrangement and production facilities (layout, etc.)
Design and use of equipment.
Awareness and training of production personnel
Packaging and conditioning.
Storage, logistics and transportation
Storage conditions at the customer / distributor/ seller.
Intended use.
Relevant consumer segment.

The table 6.2 lists the steps to be considered for hazard identification, and which must be analyzed according to the degree of severity and likelihood of occurrence, taking into account that, for many foods, a risk close to zero is often targeted.

6.2.4 Characteristics of CCP

A CCP (critical control point) is a step in the process at which an inspection must be performed in order to prevent or eliminate a risk to food safety or reduce it to an acceptable level. A CCP is distinguished from a CP (control point) by the fact that it refers to a stage in which biological, chemical or physiological factors must be brought under control. CPs do not affect food safety, but are critical to product quality (e.g., transit time/temperature of an oven to be checked to avoid degradation of food color or the occurrence of an unacceptable taste).

One often distinguishes between CCPs and CPs by asking two questions:

- If I lose control of this stage, is there a later one that can offset this defect (e.g., a freezing step, etc.)?
  - If so, then the next step is very probably the CCP.
  - If not, we formulate a second question.
- If I lose control of this stage, does this put the consumer food safety at risk? If so, then this point is a CCP.

Asking oneself these questions does not dispense with further analysis by using the HACCP decision tree (see Section 6.2.5).

CCPs are often identified as follows:

- ***For raw materials***: these may be contaminated with pesticides, toxins, herbicides, heavy metals, antibiotics, bacteria or pathogens. Quality assurance prior to production is frequently required to put good agricultural practices under control, and these are usually specific to the raw material.

Some food risks related to raw materials may be particularly complex. An example is the "mad cow" epidemic that originated from feeding cattle on meat and bone meal derived from uneaten parts of cattle carcasses and dead animals. The epidemic took a peculiar turn when scientists found in 1996 that there was a possibility of transmitting the disease to humans through consumption of beef products. The disease has to date 204 human victims, affected by symptoms similar to those of Creutzfeldt-Jakob disease, a disease of the same kind as BSE. Various measures have been taken to curb the epidemic and protect human health, such as the ban on using meat and bone meal in cattle feed, removal of consumer products considered at risk, even some animals (animals older than 30 months in the UK), screening for the disease in slaughterhouses and culling of herds in which a sick animal was observed. The legacy of the mad cow crisis has improved practices in the beef industry, through the ban on using bone meal in animal feed[15] and removal of parts of the carcass at slaughter during cutting, and has also led to enhanced traceability of the animals. With regard to public health, this crisis has also led to recognition of the following precautionary principle.

Faced with the microbiological contamination of raw materials, one frequently relies on the presence of heat treatment after production to eliminate bacterial or viral risks. However, this must be considered with caution: indeed, a high level of bacteria may change the properties of the food (toxin production) and affect its stability or its organoleptic properties after conditioning (acidification of milk by bacteria leading to a partial coagulation of the milk protein during filling or storage, for example). We must, therefore, consider the principle "*garbage in, garbage out*" to ensure compliance of the finished product.

- ***For storage***: some storage conditions (dry, concentrated, sterilized, strongly acidic, frozen, refrigerated raw material) reduce contamination (usually bacterial) of raw materials. For dry, concentrated, sterilized or acidic products, the permeability of the packaging to air and water constitutes the CCP. For frozen products, the CCP is primarily the storage temperature whereas, for refrigerated raw materials, CCPs are the impermeability of the packaging, and a maximum temperature before use. This brings about storage on a "first in, first out" basis, and the use of an expiry date that is to be strictly respected.

- ***For the process itself***: the CCPs in the production process are very often variables such as:
  - time,
  - temperature,
  - pressure (for example in a pressure cooker),
  - filling rate of tanks,
  - speed of rotation of agitators in tanks,
  - size of the solid ingredients used,
  - physical and chemical properties of liquids and solids in the process (e.g., viscosity) and, very importantly,
  - control of the formulation, the composition of which can affect the preservation of the product. This is most often done with organic, chemical or bacteriostatic agents, such as a specific rate of salt, of sugar, or a minimum pH, for instance below 6.4.5 for dairy products.
- ***For conditioning*** of the product and its packaging: use of an atmosphere low in germs or sterile during filling, or packaging under modified atmospheres, detection

[15] But not poultry, which, unlike cattle, are not exclusively vegetarian.

equipment for metals and foreign bodies, impermeability, sealing, inviolability and integrity of the packaging.

- ***For labeling***: control of the conformity of the label, especially the list of ingredients, some of which may be allergens. If they are not mentioned, ingestion of the product may jeopardize the health of sensitive people.
- ***For transportation and storage***. Here we find more or less the same factors as for the storage of raw materials, but with the risk that the consumer might not respect the storage conditions despite indications on the package (how many home refrigerators have a temperature of 4°C, for example?). Some areas are more delicate and the risks difficult to reduce. It is, for instance, virtually impossible to produce chickens without a certain amount of Salmonella adhering to their skin. Food contamination can occur if the consumer uses knives and utensils to prepare the chicken, and then switches to another cooking activity without cleaning them first (such as the preparation of mayonnaise).

6.2.5 HACCP decision tree

Once the production process has been described in detail (raw materials, ingredients, production equipment, operating parameters, conditioning and packaging), and risk identification carried out, the interdisciplinary team determines the CCPs.

In a first step, two questions should be answered, but one part of the answer has already been obtained in the evaluation and identification of risks:

- Does this process step contain threats (consider both the probability of occurrence and the severity of the effect) that are significant enough to establish an inspection? If so, the team moves to the next question. If not, the CP, if there is one, is not a CCP and they move on to the next risk.
- Is the risk controlled by a prerequisite program? This may be the case during automatic sterilization of production equipment prior to their use, for example. If so, the CP, if there is one, is not a CCP (but, as discussed in the context of ISO 22000, it can be an OPRP, Operational Prerequisite Program) and they move on to the next risk; if not, they proceed with the process below.

They then use a CCP decision tree for the risk(s) identified at each stage of the process. Figure 6.2 reproduces the four key issues and the HACCP decision tree recommended by *Codex Alimentarius* (1997), adopted by ISO 22000.

There are several HACCP decision trees, varying slightly depending on the countries that have defined and adopted them. However, they all have in common the following principles:

- Each step of the production process (the term "unit operation" is often used) must be processed sequentially.
- At each step, the decision tree must be applied to all identified risks.
- A decision tree should be implemented only when the risk assessment has been conducted by the interdisciplinary team and when a level of significance has been determined for each of them.
- A CCP can have several parameters (e.g., for UHT, temperature and time).
- More than one risk can be controlled by a specific CCP.
- There is no limit to the number of CCPs to be identified in a process.

In practice, CCP analysis is facilitated, in terms of equipment and technologies, by the fact that the supplier of production equipment designs its machines before proceeding with an HACCP approach. Some CCPs already exist in the production lines themselves so investigations planned according to principles 2, 3, 4 of the HACCP system are often facilitated.

Conversely, some contamination may occur within the equipment itself, including accidental cracks in heat exchangers (heat treatments are very common in the food industry),

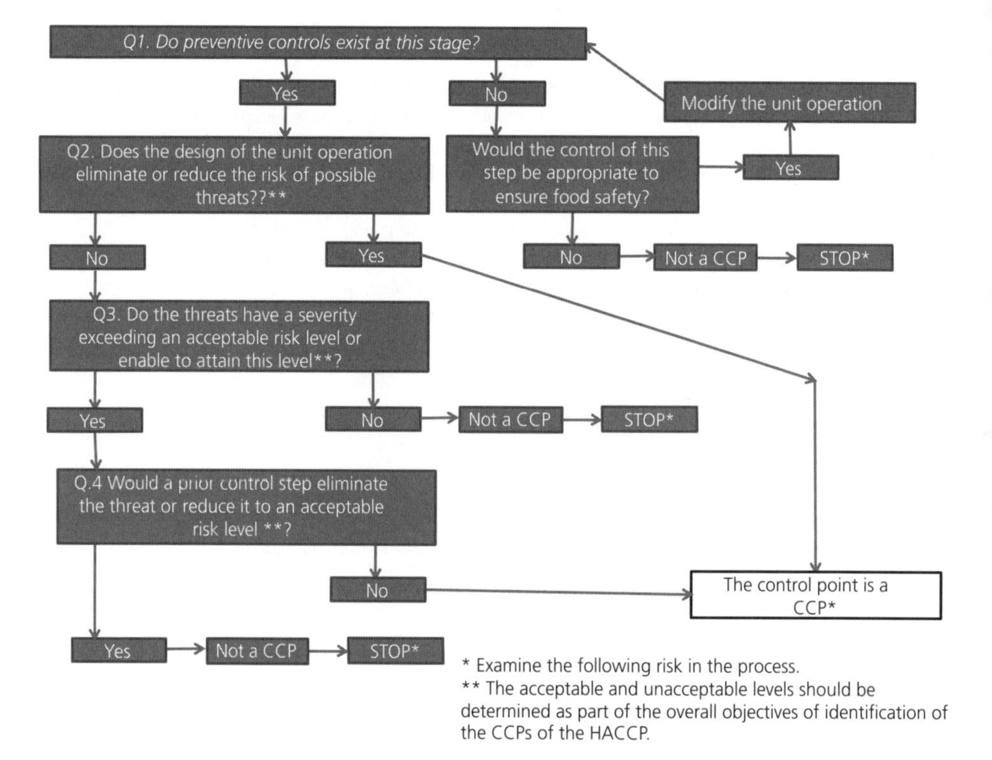

Figure 6.2 The HACCP decision tree.

which may then put the product in contact with the heat transfer fluid or coolant, thus contaminating it with chemicals or microorganisms. Only tight maintenance schedules are likely to reduce this risk, but unfortunately it cannot be completely reduced to zero.

6.2.6 HACCP: 12 implementation steps

Codex Alimentarius states: "The application of HACCP principles consists of the following tasks, as in the logical sequence of implementing an HACCP."

The logical sequence is that of 12 steps, as shown in Figure 6.3.

Steps 7 through 12 are identical to the principles 2 to 7. Steps 1 to 5, called descriptive, have been introduced to provide a solid foundation and enable the best application together with principle 1, which becomes step 6 of the HACCP system.

The constitution of the HACCP team, describing the product, its intended uses and the establishment of a flowchart according to the unit steps of the manufacturing process are the first five logical and necessary steps, as well as on-site verification of the flowchart. It is never advisable to use food technologists behind a desk or, worse, to consider hard cash production charts designed in an office. The flowchart should consider:

- the sequence and interaction of all stages of the operation, verifying the actual operating conditions (time, temperature, pressure, etc.);outsourced processes and subcontracted work;
- the point of introduction of raw materials, semi-finished products and ingredients in the production flow;
- points of effective retrieval and recycling;

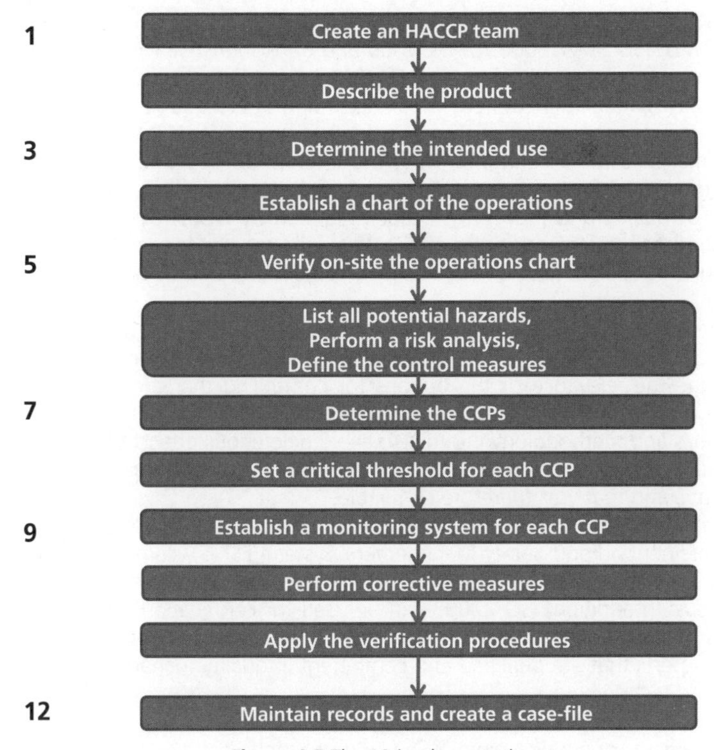

Figure 6.3 The 12 implementation steps.

- output points or disposal of finished products, intermediate products, waste and derivatives.

6.3 ISO 22000:2005

6.3.1 An overview of the standard

6.3.1.1 From ISO 15161 to ISO 22000

The aim of ISO during the development of ISO 22000 was for it to become the one and only text able to combine, in a structured, concise, yet complete, manner all aspects of the control and management of food safety, including:

- monitoring of and compliance with legislation;
- PRPs or Prerequisite Programs;
- OPRPs or Operational Prerequisite Programs;
- internal and external communications;
- the HACCP plan with control of OPRP and PRP;
- traceability;
- withdrawal or recall of products;
- knowledge management;
- emergency management and crisis management.

Previously, ISO added to the ISO 9001 family the standard ISO 15161:2002, *Guidelines on the application of ISO 9001:2000 for the food and drink industry*. This standard can be used in any company that has implemented an HACCP system. The application of the HACCP plan within a quality management system (QMS) according to ISO 9001 resulted in a system of food safety that was more effective than the application of only ISO 9001 or HACCP in increasing customer satisfaction and improving the internal efficiency of the organization. The application of HACCP for hazard identification and risk control is indeed directly related to the planning of quality and preventive actions required by ISO 9001. Once the critical points have been identified, the principles of ISO 9001 can be used for control and monitoring. The procedures for performing an HACCP plan could also be easily documented within the QMS.

However, the addition of this standard did not suffice: on the one hand, it was not intended for certification purposes and, on the other hand, auditors did not have the business skills to judge the relevance of the options for controlling the safety of food. It has indeed been proven that auditor and auditee must hear and speak the same language, have the same semantics and methodology in order to avoid deleterious effects on food safety. A flawed FSMS could exist within a quality management system that fully complied with ISO 9001 requirements, and that was therefore certifiable.

These difficulties led ISO to review the situation and prepare a specific standard to describe and implement a Food Safety Management System (FSMS), completely taking into account the HACCP approach while supplementing shortcomings detected after the use of *Codex Alimentarius*. In addition, ISO 22000 can be used for certification, which is required by organizations that have already been certified for ISO 9001 and ISO 14001(subject to the commitment of experts that are qualified and competent in quality management, as well as in food science and technology).

ISO 22000 specifies requirements to enable an organization to:

a) plan, implement, operate, maintain and update a food safety management system for products, and to provide products that, in accordance with their intended use, are safe for the consumer;
b) demonstrate compliance with legal and regulatory requirements applicable to food safety;
c) evaluate and assess customer requirements, demonstrate compliance with the requirements established in agreement with the customer that are related to the safety of food in order to improve customer satisfaction;
d) communicate effectively on issues of food safety with suppliers, customers and stakeholders in the food chain;
e) ensure compliance with its stated policy on food safety;
f) demonstrate such compliance to interested parties; and
g) certify/record the food safety management system by an exterior agency, or perform a self-assessment/self-declaration of conformity to this international standard.

The requirements of ISO 22000 are generic and designed to apply to all organizations in the food chain, regardless of size and complexity. To take into account the specificities of the various branches of the food industry, an annex is included listing Codes of Practice (often PRPs) of *Codex Alimentarius* that can be downloaded from the Codex website.

6.3.1.2 ISO 22000:2005, some concepts and definitions

ISO 22000 uses some of the following specific terms and definitions:

- *HACCP – Hazard Analysis and Critical Control Point:* system which identifies, evaluates and controls hazards which are significant for food safety [from ISO/TS 22003:2007]. This system must be established and followed by an HACCP team consisting of multidisciplinary specialists.
- *Food chain*: sequence of the stages and operations involved in the production,

processing, distribution, storage and handling of a food and its ingredients, from primary production to consumption [from ISO 22000].

- *CCP – Critical Control Point*: step at which control can be applied and is essential to prevent or eliminate a food safety hazard or reduce it to an acceptable level [from ISO 22000].
(The release of the product depends on the control of CCPs, thus enabling the avoidance of costly analyses of finished products.)
- *Critical limit*: criterion which separates acceptability from unacceptability [from ISO 22000].
- *Food safety policy*: overall intentions and direction of an organization related to food safety as formally expressed by top management [from ISO 22000].
- *HACCP plan*: a document based on HACCP principles, outlining specific measures, procedures and provisions to ensure permanent food safety.
- *HACCP system*: organizational structure, procedures, processes and means/resources necessary for the implementation of an HACCP plan.
- *Risk*: the probability of an adverse health effect (e.g. becoming diseased) and the severity of that effect (e.g. death, hospitalization, absence from work, etc.) when exposed to a specified hazard [from ISO 22000]. Risks can come from microbiological, chemical or physical agents.
- *Food safety hazard*: biological, chemical or physical agent in food, or condition of food, with the potential to cause an adverse health effect [from ISO 22000].
(e.g., bacteria, toxins)
- *Severity*: importance of a hazard.
- *Criterion*: parameter or requirement for one or more physical, chemical or microbiological characteristics of the operation or the product.
- *Risk management*: the process of balancing the different policies based on the results of the risk assessment and of implementing the appropriate control measures, including regulatory measures.
- *PRP – Prerequisite program*: basic conditions and activities that are necessary to maintain a hygienic environment throughout the food chain suitable for the production, handling and provision of safe end products and safe food for human consumption [from ISO 22000].
- *OPRP – Operational Prerequisite Program*: PRP identified by the hazard analysis as essential in order to control the likelihood of introducing food safety hazards and/or the contamination or proliferation of food safety hazards in the product(s) or in the processing environment [from ISO 22000].
- *Validation*: obtaining evidence that the control measures managed by the HACCP plan and the operational PRPs are capable of being effective [from ISO 22000].

6.3.1.3 ISO 22000 versus ISO 9001

The structure of ISO 22000 draws heavily on ISO 9001, allowing food companies that are already ISO 9001-certified to migrate without (much) difficulty to ISO 22000 certification. The difference is the emphasis on food safety instead of quality. In the food industry, quality can be highly subjective (chemical or physical-chemical analysis does not replace the palates of experienced tasters).

In terms of content, this difference is specifically expressed in clause 7 of ISO 22000, which focuses on the planning and realization of safe products, the importance of good manufacturing practices, PRPs and the implementation of HACCP; in ISO 9001, clause 7 deals specifically with design and development.

The correspondence table[16] on the next page shows the degree of similarity between the two standards. Although the chapter headings are often identical, their content may differ.

---

[16] Taken from Table A.2 of ISO 22000:2005.

ISO 9001:2000		ISO 22000:2005	
Introduction			***Introduction***
General	**0.1**		
Process approach	**0.2**		
Relationship with ISO 9004	**0.3**		
Compatibility with other management systems	**0.4**		
Scope	1	1	***Scope***
General	1.1		
Application	1.2		
Normative references	2	2	***Normative references***
Terms and definitions	3	3	***Terms and definitions***
Quality Management System (QMS)	4	4	***Food Safety Management System (FSMS)***
General requirements	4.1	4.1	General requirements
Documentation requirements	4.2	4.2	Documentation requirements
• General	4.2.1	4.2.1	General
• Quality manual	4.2.2		
• Control of documents	4.2.3	4.2.2	Control of documents
		7.7	Updating of preliminary information and documents specifying the PRPs and the HACCP plan
• Control of records	4.2.4	4.2.3	Control of records
Management responsibility	5	5	***Management responsibility***
Management commitment	5.1	5.1	Management commitment
Customer focus	5.2	5.7	Emergency preparedness and response
Quality policy	5.3	5.2	Food safety policy
Planning	5.4		
• Quality objectives	5.4.1		
• QMS planning	5.4.2	5.3	FSMS planning
		8.5.2	Updating the FSMS
Responsibility, authority and communication	5.5	5.6	Communication
• Responsibility and authority	5.5.1	5.4	Responsibility and authority
• Management representative	5.5.2	5.5	Food safety team leader
• Internal communication	5.5.3	5.6.2	Internal communication
Management review	5.6	5.8	Management review
• General	5.6.1	5.8.1	General
• Review Input	5.6.2	5.8.2	Review input
• Review output	5.6.3	5.8.3	Review output

ISO 9001:2000		ISO 22000:2005	
Resource management	6	6	***Resource management***
Provision of resources	6.1	6.1	Provision of resources
Human resources	6.2	6.2	Human resources
• General	6.2.1	6.2.1	General
• Competences, awareness, training	6.2.2	6.2.2	Competences, awareness, training
Infrastructure	6.3	6.3	Infrastructure
Work environment	6.4	6.4	Work environment
		7.2	Prerequisite Programs
Product realization	7	7	***Planning and realization of safe products***
Planning of product realization	7.1	7.1	General
Customer-related processes	7.2		
• Determination of requirements related to the product	7.2.1	7.3.4	Intended use
		7.3.5	Flow diagrams, process steps and control measures
		5.6.1	External communication
• Review of requirements related to the product	7.2.2		
• Customer communication	7.2.3	5.6.2	Internal communication
• Design and development	7.3	7.3	Preliminary steps to enable hazard analysis
• Design and development planning	7.3.1	7.4	Hazard analysis
• Design and development inputs	7.3.2	7.5	Establishing OPRPs
• Design and development outputs	7.3.3	7.6.	Establishing the HACCP plan
• Design and development review	7.3.4	8.4.2	Evaluation of individual verification results
• Design and development verification	7.3.5	8.5.2	Updating of the FSMS
• Design and development validation	7.3.6	8.2	Validation of control measure combinations
• Control of design and development	7.3.7	5.6.2	Internal communication
Purchasing	7.4		
• Purchasing process	7.4.1		
• Purchasing information	7.4.2		
• Verification of purchased product	7.4.3	7.3.3	Product characteristics
Production and service provision	7.5		
• Control of production and service provision	7.5.1	7.2	Prerequisite programs (PRP)
		7.6.1	HACCP plan
• Validation of production and service provision	7.5.2	8.2	Validation of control measure combinations
• Identification and traceability	7.5.3	7.9	Traceability system
• Customer property	7.5.4		
• Preservation of product	7.5.5	7.2	Prerequisite programs
Control of monitoring and measuring devices	7.6	8.3	Control of monitoring and measuring

SO 9001:2000		ISO 22000:2005	
Measurement, analysis and improvements	8	8	***Validation, verification and improvement of the food safety management system***
General	8.1	8.1	General
Monitoring and measurement	8.2	8.4	FSMS verification
• Customer satisfaction	8.2.1		
• Internal audit	8.2.2	8.4.1	Internal audit
• Monitoring and measurement of processes	8.2.3	7.6.4	System for monitoring critical control points
• Monitoring and measurement of product	8.2.4		
Control of a nonconforming product	8.3	7.6.5	Actions undertaken when monitoring results exceed critical points
		7.10	Control of nonconformity
Data analysis	8.4	8.2	Validation of control measure combinations
		8.4.3	Analysis of results of verification activities
Improvement	8.5	8.5	Improvements
• Continual improvement	8.5.1	8.5.1	Continual improvement
• Corrective action	8.5.2	8.5.2	Corrective action
• Preventive action	8.5.3	5.7	Emergency preparedness and response
		7.2	Prerequisite programs

6.3.1.4 ISO 9001 or ISO 22000 certification?

All food industries face this question: Should a QMS be established if an FSMS is in place? Or: Is ISO 9001 certification relevant if an FSMS according to ISO 22000 has already been approved by an external ad-hoc body? Although ISO 22000 certification establishes a management system that reduces to an acceptable level the risk to food safety, it has little control over what are unacceptable quality deviations for customers, despite possible negative effects on their health, relating to:

- unacceptable fluctuations in raw material composition;
- high amounts of dry matter or fat that do not comply with the requirements of food laws (e.g., milk, cream, cheese, etc.), where sanctions can go so far as withdrawal from the market;
- organoleptic properties, color of the finished product;
- deviation of the specific gravity (for powders), risk of caking of powders during storage;
- defective packaging;
- labeling presenting defects, etc.
- nonconforming logistics, leading to deterioration of the product.

All these elements require the presence of an operational QMS, which can be implemented in conjunction with ISO 9001, given the similarities between the two standards. This is why food companies generally have both systems, although many of them focus primarily on ISO 22000.

6.3.1.5 ISO 22000, a reference to the "fundamentals" of ISO 9001

ISO 22000 relies on the four main areas described in ISO 9001.

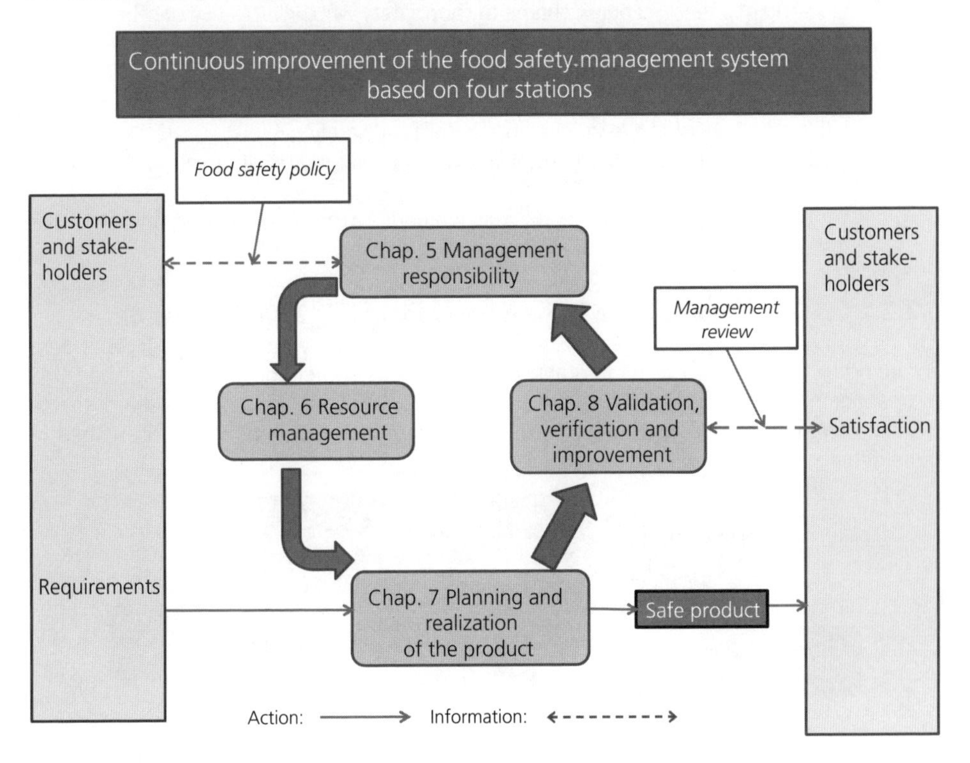

Figure 6.4 Continuous improvement of food safety.

Although the ISO 22000 family does not make explicit reference to the eight principles of ISO 9001, it is closely associated with them and relies on the process approach.

6.3.2 ISO 22000: some salient points

6.3.2.1 Driving force of management and the HACCP team leader

ISO 22000 puts management in the center of the food safety management system. The standard states that it:

- must provide evidence of its commitment to the development and implementation of the food safety management system and continually improve its effectiveness;
- shall define, document and communicate its policy on food safety;

- must ensure that:
  - planning of the FSMS is undertaken to satisfy the requirements and the objectives of the organization with regard to food safety, and
  - the integrity of the FSMS is maintained when changes to the system are planned and implemented;
- must ensure that responsibilities and authorities are defined and communicated within the organization to ensure effective operation and maintenance of the FSMS;
- must appoint a team leader in charge of food safety who, irrespective of other responsibilities, shall have the responsibility and authority for:
  - leading a team responsible for food safety and organizing its work,
  - ensuring the appropriate training, and both initial and continuing education, for the members of the team responsible for food safety,
  - ensuring that the FSMS is established, implemented, maintained and updated, and
  - reporting to the organization's management on the effectiveness and appropriateness of the food safety management system.

6.3.2.2 Continual improvement in the framework of ISO 22000 (clause 8)

Just as with a quality management system, the FSMS is in line with continual improvement according to the Deming cycle, integrating internal audits and periodic reviews of the FSMS by management. This approach is illustrated below, with great importance given to the HACCP Prerequisite Programs and Operational Prerequisite Programs (OPRP) described in more detail in Section 6.3.3.

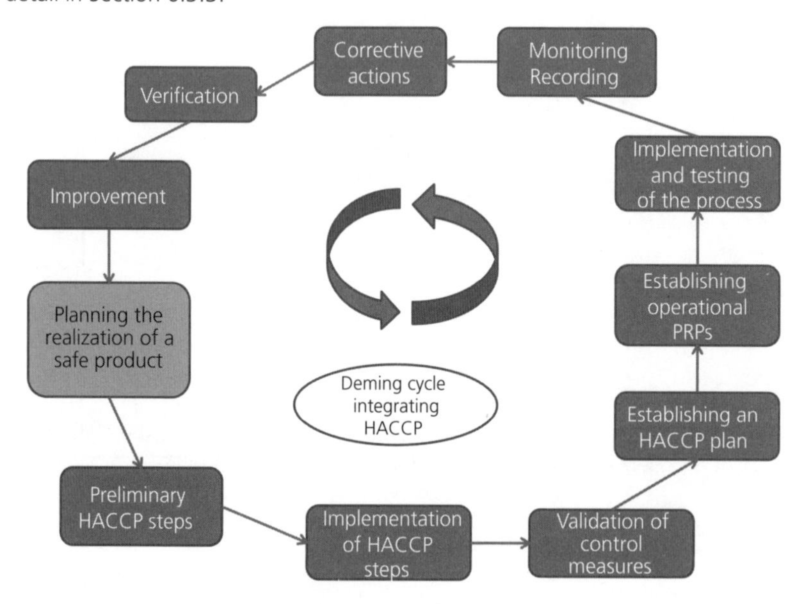

Figure 6.5 The Deming wheel.

6.3.2.3 Communication with stakeholders

Food safety requires ongoing communication, both internal and external, between the producer and stakeholders, which is why this is discussed in the body of the standard (clause 5.6, Communication).

Internal communication

Internal communication concerns, in particular, information that the food safety team must receive about any changes regarding:

a) products or new products;
b) raw materials, ingredients and services;
c) production systems and equipment;
d) production premises, equipment locations, and the ambient environment;
e) programs for cleaning and disinfecting;
f) systems for conditioning, storage and distribution;
g) level of staff qualifications and/or assignment of responsibilities and authorizations;
h) legal and regulatory requirements;
i) knowledge of food safety hazards and control measures;
j) requirements of customers, industry and others, observed by the organization;
k) relevant surveys of external stakeholders;
l) claims reporting food safety hazards associated with the product;
m) other conditions affecting the safety of food.

This is so that the team can adjust the FSMS to new conditions. Management should ensure that relevant information is included as part of the input elements to the management review.

External communication

The chart[17] below lists the main actors and their interactions in terms of external communication for a complex food supply chain; here the example is for meat products.

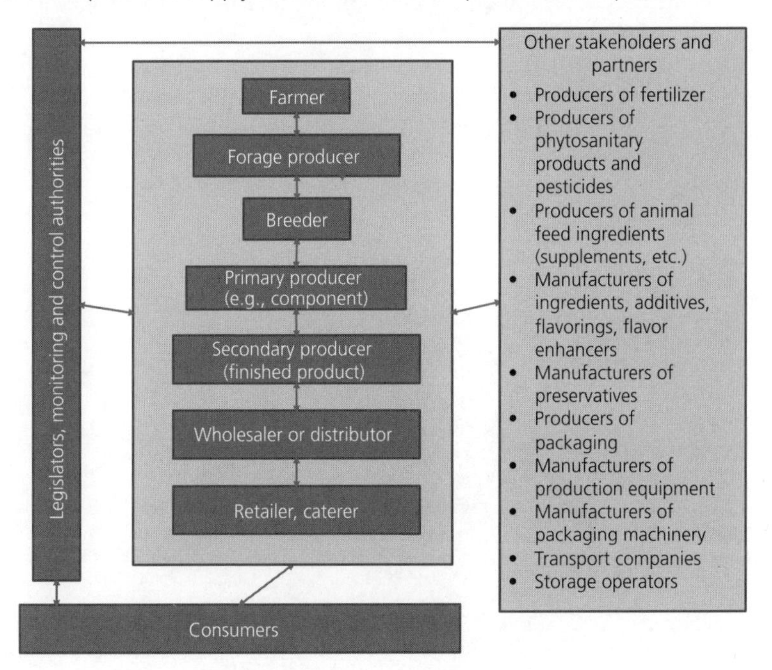

Figure 6.6 External communication chart.

17 Based on Figure 1 of ISO 22000: 2005.

Communications should provide adequate information concerning all aspects of food safety for this organization's products which may be relevant to other organizations in the food chain. This applies particularly to the known food safety hazards for which control is provided by other agencies involved in the food chain. Communication records must be retained. Information obtained through external communication must be integrated as part of the input elements for updating the system and management reviews.

Emergency situations

Finally, the establishment of a communication network must take into account emergency situations (blocking, withdrawal, information to the authorities, etc.). Top management of the organization shall establish, implement and maintain procedures to deal with possible emergencies and accidents that may affect food safety and which are relevant to the organization's role in the food chain. The management of emergencies is closely related to the adequacy of procedures set up for product traceability.

6.3.2.4 ISO 22000: Documentation

A pyramidal documentary structure similar to ISO 9001

Documentation for the implementation of ISO 22000 (clause 4.2) is similar to that of ISO 9001. The standard specifies that it must include:

- documented statements of the food safety policy (see clause 5.2) and related objectives;
- documented procedures and records required by this standard;
- documents needed by the organization for the effective development, implementation and updating of the FSMS.

Instead of a quality manual, there is usually a food safety manual, even though the standard does not explicitly mention one.

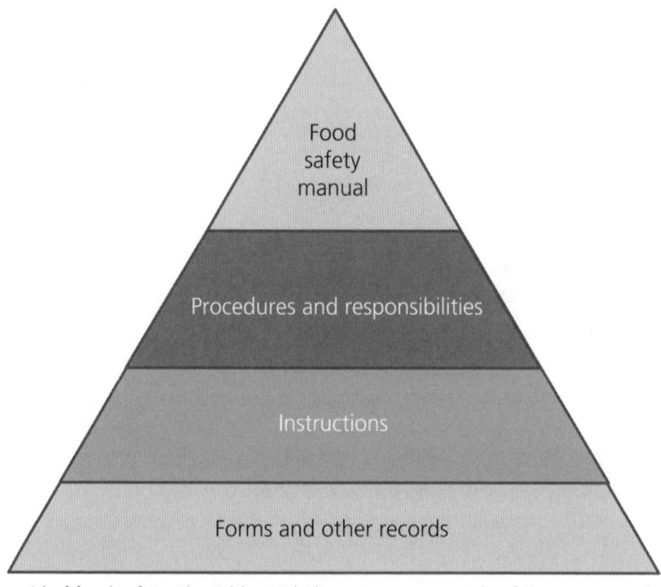

Figure 6.7 Pyramid of food safety. The Table 6.3 below gives an example of the contents of such a manual.

Table 6.3 Contents of the safety manual.

1. Introduction 1.1. Company Presentation 1.2. Application	**6. Resources** 6.1. Resource planning 6.2. Competences and training
2. Standards and definitions **3. Process approach** 3.1. Processes in our company 3.2. Process mapping **4. SDA management system** 4.1. General requirements 4.2. Documentation	**7. Realization of safe products** 7.1. Planning 7.2. Prerequisite programs 7.3. Initial steps 7.4. Risk assesment 7.5. ***Operational prerequisite programs*** 7.6. ***HACCP plan*** 7.7. Updating information 7.8. Planning of the verification 7.9. Traceability 7.10. Controlling noncompliance
5. Management responsibility 5.1. Management commitment 5.2. Policy 5.3. Planning 5.4. Responsibility 5.5. HACCP team leader 5.6. Communication 5.7. Emergencies 5.8. Management review	**8. Validation, verification and improvement of the FSMS** 8.1. Generalities 8.2. Validation 8.3. Monitoring and measurements 8.4. Verification 8.5. Improvement

Procedures, instructions and forms rely on the same general rules of implementation as in ISO 9001. In the EU, they are strongly influenced by the *acquis communautaire* in the field of food safety.

The HACCP plan

One of the key documents of ISO 22000 is the HACCP plan, which may be in the form, for each product or product category, of a table with the horizontal sections (for each column) associated with each critical control point (CCP) as follows:

- CCP No.
- Manufacturing step and the related unit operation (e.g., cooking).
- Description of the significant hazard (e.g., survival of *Listeria monocytogenes* bacteria).
- Critical control measure, CCP (e.g., heating process leading to the elimination of any *Listeria* that may be present).
- Critical limits for each CCP (2 min ± 15 sec heating at 100°C will heat the core of the product to 80 ± 2°C for 5 min ± 7 sec).
- Monitoring.
  - ➢ Parameter:
    - speed of the conveyor belt of the product,
    - temperature of the oven,
    - device recording belt velocity and temperature?
  - ➢ How:
    - measuring the belt speed with a stopwatch,
    - using a thermometer to measure the temperature of the oven,
    - visually inspecting the recording device.

- ➢ Frequency of monitoring:
  - frequency: after each production step,
  - temperature: continuous measurements,
  - visual observations: every hour.
- ➢ Who:
  - speed: operator,
  - temperature: automatic,
  - visual observation: head of the production line.

- Records:
  - ➢ speed: form C.1.1,
  - ➢ temperature: chart from the recording device, to be archived with forms C.1.1 and C.1.2, after the end of production,
  - ➢ visual observation: form C.1.2.
- Corrections, corrective actions and records, for example:
  - ➢ Isolate and label the nonconforming product for further analysis and possible release.
  - ➢ Describe and document the nonconformity in the ad-hoc register.
  - ➢ List and record the corrective action[18] applied (date and signature).
  - ➢ Determine the source of the problem, and take appropriate measures to prevent recurrence.
  - ➢ Retrain staff if necessary.
- Verification[19] (e.g.):
  - ➢ Daily verification of corrective activities by the quality manager.
  - ➢ Daily verification and analysis of data recorded for the oven by the quality manager.
  - ➢ The quality manager checks the measurements made by the operator by carrying out a control himself (once every 5 days).
  - ➢ The quality manager ensures daily the conformity of heating with the specifications.
  - ➢ Calibration of the measuring thermometer by the quality manager once a week.

The HACCP plan requires the existence of a detailed flowchart that describes all stages of the production process, and makes reference to the specification sheets for ingredients and components, and also to the packaging and finished product. The CCPs are included in the HACCP plan according to their position in the flowchart, from beginning to end of the production line. Each operation can be associated with one or more CCPs. The control applied to a point on a line may correspond to several different CCPs. Similarly, a CCP can neutralize several distinct hazards.

6.3.3 HACCP, PRP, ISO 22000 and OPRP

One weakness of the HACCP approach as advocated by *Codex Alimentarius* relates to the implementation of Good Manufacturing Practices (GMP) and Good Hygiene Practices (GHP),

---

[18] A corrective action includes:
– the process,
– identification of a nonconforming product,
– blocking times for the product,
– reprocessing the product,
– destruction of the product.

[19] Verification: confirmation, through the provision of objective evidence, that specified requirements have been fulfilled [from ISO 22000:2005].

called prerequisite programs in the terminology of ISO 22000. Indeed, a number of hazards can be identified but, based on PRPs, not be included in the HACCP plan. However, the Codex system does not strictly make provisions for the monitoring of control measures associated with these hazards. Rather, it is limited to audits. With the introduction of operational prerequisite programs (OPRPs), for which monitoring and recording are introduced, ISO 22000 has a three-legged system:

- HACCP
- OPRP
- PRP

By juxtaposing the key elements of clause 7 of the standard with the 12 steps of HACCP, the importance of creating OPRPs becomes obvious, since they introduce a monitoring step missing in the Codex system, while maintaining the control of the PRPs that are considered less critical.

The table below helps to distinguish between the key attributes of a CCP, a PRP and an OPRP.

Characteristic	***PRP***	***OPRP***	***CCP***
Mandatory nature	Yes	No	No
Consequence of hazard analysis	No	Yes	Yes
Key measurement that is specific to a hazard	No	Yes	Yes
Measurement validation	No	Yes	Yes
Monitoring of measurement	No	Yes	Yes
Records from the monitoring	No	Yes	Yes
Existence of critical limits	No	No	Yes
Monitoring for releasing or securing a batch at a process step	No	No	Yes
Product correction	No	Yes	Yes
Corrective action on the process or the environment	Yes	Yes	Yes
Verification of the implementation	Yes	Yes	Yes
Verification of the effectiveness	No	Yes	Yes

The OPRPs must be documented and should include the following information for each program:

a) the food safety hazards to be controlled by the program;
b) control measures (according to clause 7.4.4);
c) monitoring procedures that demonstrate that the operational PRPs have been implemented;
d) corrections and corrective actions to be taken if monitoring shows that the operational PRPs are not under control (according to clauses 7.10.1 and 7.10.2);
e) responsibilities and authorities;
f) monitoring records.

Finally, one must not confuse monitoring and verification of the OPRPs. Monitoring of OPRPs is always related to documents and records of evidence of control of the hazard to which the OPRP is associated, either periodically or continuously.
Verification of an OPRP is a periodic review of the application and effectiveness of the monitoring devices associated with the operational PRP.

6.3.4 Clause 7, the core of ISO 22000:2005

The specificity of the standard, compared to ISO 9001, is the HACCP approach, the choice of PRPs, and the control of OPRPs and CCPs, largely present in clause 7, the backbone of the standard. The diagram below illustrates the sequential steps for determining CCPs, PRPs and OPRPs, by juxtaposing the subclauses of clause 7 and the 12 steps of HACCP.

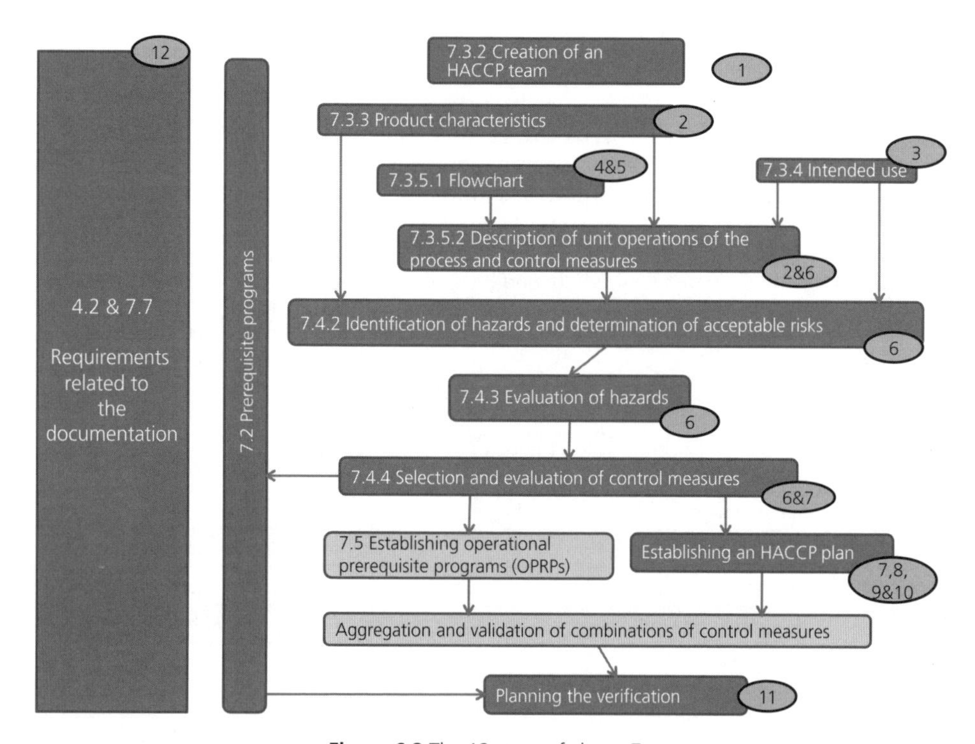

Figure 6.8 The 12 steps of clause 7.

Clause 7 describes four other key elements of food safety:

- The introduction of prerequisite programs (clause 7.2), widely discussed in 2.1 of this book.
- The establishment of operational prerequisite programs (clause 7.5), already presented in 3.3 of this book.
- The traceability system (clause 7.9): The organization shall establish and apply a traceability system that identifies the product batches and their relation to batches of raw materials, and records relating to processing and delivery. The traceability system must identify the direct suppliers of input elements and direct customers of the finished products. Records relating to traceability must be kept for a defined period for evaluation of the system to allow processing of potentially hazardous products and in the event of a withdrawal. This section is completed by ISO 22005:2007, *Traceability in the feed and food chain – General principles and basic requirements for system design and implementation*.
- Control of nonconformities (clause 7.10) and withdrawal from the market: The organization must ensure that in the case where the critical limits for the CCPs are

exceeded (see clause 7.6.5) or there is lack of control of the operational PRPs, the products concerned are identified and controlled with respect to their use and release ... Corrective actions must be taken if critical limits are exceeded (see clause 7.6.5) or when there is a nonconformity with the operational PRPs ... All batches of products that may have been involved in a situation of nonconformity must be kept under the control of the organization until they have been evaluated ... If a product that is no longer under the control of the organization is subsequently determined to be hazardous, the agency must notify interested parties and initiate a withdrawal (see clause 7.10.4) ... The withdrawn products must be secured or maintained under control until they are destroyed, or until they can be used for purposes other than those originally planned, or until they are determined safe for their intended use, whether identical or different, or until further processing ensures their safety.

6.4 Additional documents of the ISO 22000 family

ISO 22000:2005 is completed by three additional documents:

- The technical specification ISO/TS 22003:2007, Food safety management systems – Requirements for bodies providing audit and certification of food safety management systems. This technical specification:
  - defines the rules applicable to the audit and certification of an FSMS in accordance with the requirements of ISO 22000 (or other sets of requirements specified in the field). These are external quality assurance requirements complementing ISO 22000, which specify the conditions and context of the audit, the skills of the experts (especially for food engineering) and the chief auditor, etc. ...
  - provides customers with the necessary information on how to proceed with the certification of their suppliers and thus gives them confidence in this certification.

  Certification of an FSMS (designated "certification" in this technical specification) is considered an activity of conformity assessment by a third party. ISO/TS 22003 makes numerous references to ISO/IEC 17000:2004, *Conformity assessment – Vocabulary and general principles*, ISO/IEC 17021:2006, *Conformity assessment – Requirements for bodies providing audit and certification of management systems,* and ISO 19011:2002, *Guidelines for quality and/or environmental management systems auditing*, which are key documents specifying external quality assurance in the audit and certification process.
- ISO/TS 22004:2005, Food safety management systems – Guidance on the application of ISO 22000:2005. This provides guidelines for the application of ISO 22000 based on HACCP principles as outlined by the Codex Alimentarius Commission. It has been developed for application in conjunction with the standards set by Codex.
- ISO 22005:2007, Traceability in the feed and food chain – General principles and basic requirements for system design and implementation. This standard describes the existing specifications of ISO 22000, describes the principles and specifies the basic requirements for the design and implementation of a traceability system for the food chain. This system can be applied by an organization operating at any level of the food chain. It is designed to be flexible enough to allow organizations belonging to the chain to achieve their identified objectives. The traceability system is a technical tool to help an organization comply with its objectives, and it can be used, if necessary, to determine the history or location of a product or its components. The standard specifies that a traceability system must be:

- verifiable,
- applied consistently and equitably,
- results oriented,
- cost-effective,
- practical to apply,
- compliant with applicable regulations or policy, and
- compliant with determined accuracy requirements.

6.5 First experiences

An empirical study by AFNOR on the first initial certification audits [6.5] has identified the clauses and subclauses of ISO 22000 that are the most critical to the implementation. The study found that there were few discrepancies (nonconformities) to the requirements of clauses 4 (Food safety management system), 6 (Resource management) and, to a lesser extent, clauses 5 (Management responsibility) and 8 (Validation, verification and improvement of the FSMS). The majority of the discrepancies were identified in clause 7, which is not surprising because many of the companies audited were already aware of the contents of ISO 9001.

Difficulties in implementing the FSMS were focused on the following subclauses and non-conformities:

- Establishment of OPRPs (7.5):
  - no monitoring records for the OPRPs;
  - incomplete control measures managed by the OPRPs.
- Control of nonconformities (7.10):
  - incomplete control procedure, withdrawals forgotten or not tested;
  - corrections not recorded after a deviation.
- Initial steps for hazard analysis (7.3):
  - lack of information prior to the hazard analysis.
- Planning the verification (7.8):
  - verification results of the OPRPs not recorded;
  - shortfalls in the verification of the OPRPs;
  - frequencies, methods of analysis of tests undefined or unclear.
- Emergency preparedness (5.7):
  - absence of provisions for emergencies or insufficient corresponding procedures.
- Human resources (6.2.2):
  - required qualifications not identified for staff occupying tasks that affect product safety, and the effectiveness of training was under-evaluated.
- Establishment of the HACCP plan (7.6):
  - inconsistency between critical limits and the control steps;
  - HACCP plan inadequately documented;
  - CCPs not monitored;
  - lack of precision in identifying the CCP steps.
- Prerequisite programs (7.2):
  - regulatory requirements relating to unidentified PRPs;
  - drift in the application of good hygiene and manufacturing practices.
- Verification of the FSMS (8.4):
  - no internal audit program, and some requirements unaudited;

- no internal audit of part of the HACCP system (e.g., logistics);
- lack of action plan following an external audit.

Another study [6.6] provides similar findings to which can be added nonconformities with regard to the establishment and operation of the HACCP team (composition, roles and specifications, etc.).

These initial results show that, for pioneering companies (motivated and eager to learn), without prior knowledge of the implementation of a QMS, ISO 22000 certification requires a thorough study by the company of the core of the standard: food safety.

6.6 Conclusions

The publication of ISO 22000:2005 is undoubtedly a worldwide success. It was quickly adopted by the major manufacturers, and recognized by their customers and external stakeholders. There is no doubt that it helps to reduce hazards to the health of consumers. Again, the work initially done by the Technical Committee responsible for ISO 9001 has certainly had an unexpected impact by demonstrating the versatility of their work and the generic nature of the results. This confirms that management system standards are crucial innovations in management over the past 30 years, although their impact is ignored in the best business schools[20] and there are curiously no publications reporting on academic research on this subject...

One may wonder whether, for people in developed countries, breaches of food safety have become the first hazard to public health. The answer is probably negative, compared to the increased risk of unbalanced and energy-plethoric diets (substantial excess of salt, sugar and saturated fats directly related to physical inactivity), especially for households, families and single-parent households with low incomes, causing serious chronic ailments such as obesity, diabetes, cardiovascular disease, the incidence of mortality for which is considerable and pernicious.

The "mad cow" crisis officially caused 250 deaths, and this is certainly too many and totally unacceptable, but it is an insignificant number compared to the effects of an unbalanced diet in the medium and long term, and yet there was so much talk in the media and so many pugnacious actions taken by the US and the EU after the first case, compared with their torpor in the prevention of risks caused by totally inadequate diet.

Even though manufacturers have a clear responsibility (notably in terms of the formulation of food on the market), there is a wide choice of balanced foods on retailers' shelves, which should make it easy to have a healthy diet. The buyer has the last word, because fatty foods, sweet or savory, with a high palatability, are preferred. This is where education programs for consumers, which are hardly present in the actions of the US, should take over.

Two issues remain that must still get citizens and (some) authorities thinking:

- the large amount of additives from all sources that consumers eat daily, after buying processed food, the synergistic effects of which are unknown and may be damaging (but non-existent according to the manufacturers of these food products and additives...);

[20] It is true that management members, their degree in hand and having climbed the corporate ladder, prefer a controlling, top-down management (which spares them a little) to taking into account, in addition to their responsibilities and agenda, constraints formulated by a standard and (in addition) verified by an external body. As for the Boards of Directors ...

- the analogous situation for pesticide contents of cereals, fruits and vegetables, which may include several different substances, and this even if the rate of each pesticide, taken individually, is less than the statutory standard.

It seems clear that the authorities of Codex still have their work cut out for them.

References

[6.1] WILLIAM BYNUM, *The History of Medicine,* Oxford University Press (2008), Chap. 4, *Medicine in the community*, pp. 68-90.

[6.2] Food Quality and Safety Systems – A Training Manual on Food Hygiene and the Hazard Analysis and Critical Control Point (HACCP) System, publication FAO (1998), available at: http://www.fao.org/docrep/W8088E/W8088E00.htm. (access septembre 2010).

[6.3] E. SCALLAN et al. *Foodborne Illness Acquired in the United States – Major Pathogens Emerging,* Infect. Dis. 17, Jan. 2011.

[6.4] JOHN G. SURAK and STEVEN WILSON, *The certified HACCP Auditor Handbook*, ASQ Quality Press (2007), Chap. 15, *Prerequisite Areas for Food Safety*, pp. 155-173.

[6.5] P. BOUTOU, L. Lévèque, *Certification ISO 22000, les 8 clés de la réussite*, AFNOR (2008), Chap. 12. *Analyse des premiers résultats d'audit ISO 22000*, pp. 259-269.

[6.6] D. BLANC, ISO 22000, *HACCP et sécurité des aliments – Recommandations, outils, FAQ et retours de terrain*, ed. AFNOR (2009), Chap. 12. *Non-conformités les plus fréquentes*, pp. 397-409.

Chapter 7

The Off-shoots of ISO 9001

7.1 Introduction

The great success of the ISO 9000 family has led to the emergence of similar standards for management systems. Thus, after introducing a quality management system for laboratories (ISO 17025), ISO has developed a generic standard dealing with food safety management (ISO 22000) that is aligned with and can be used with ISO 9001.

The drafting of other management systems standards based on ISO 9001, but not directly related to quality management, was the next development. We briefly discuss an example of three well-known standards:

- ISO 31000:2009, Risk management – Principles and guidelines
- ISO/IEC 27005:2011, Information technology – Security techniques – Information security risk management systems
- ISO 50001:2011, Energy management systems – Requirements with guidance for use.

The implementation of an ISO management system can give rise to certification, but many companies introduce an MS without necessarily having this as a goal. These standards complement the management systems portfolio that is available to the Board of Directors of organizations in order to define expectations and challenges, facilitate decisions concerning these points, and implement and monitor their implementation, with a view to continual improvement.

In order to lead the key managerial activities of each organization, it is thus possible to have specific "Christmas tree"-type standards for system management (Figure 7.1), in which the "Christmas balls" are the standards that the organization deems appropriate to implement. The organization selects the appropriate standards for its business activities and for the products or services it provides to society. Thus, a company exclusively providing services of high added value will not consider it useful to implement the ISO standard for environmental management, for example, and a company in the food industry that only consumes little energy will use a simpler model for energy management, etc. The advantage of proceeding in this manner is that once the exercise has been carried out for ISO 9001, for instance, the similarities to other ISO management system standards will facilitate their implementation.

Figure 7.1 "Christmas tree" of management system standards.

The ISO management system standards also include the ISO 14000 environmental management family, and the ISO 26000 standards on social responsibility and sustainable development. The latter tops off, in a sense, the family of ISO management systems. These two standards are addressed further in two *ad-hoc* chapters.

When several management systems are implemented simultaneously in an organization, there is the question of their integration. Although this problem, to a large extent, initially touched companies that were ISO 9001-, ISO 14001- or ISO 22000- certified, generic approaches have been developed, notably in France with the publication of the standard AFNOR FD X 50-189:2004, *Management Systems: Guidelines for their integration* [7.1].

7.2 Risk management and ISO 31000:2009

7.2.1 Introduction to risk management

7.2.1.1 A brief history

The concept of risk is old, because our ancestors integrated it into their everyday life. They were not, however, in a position to manage it. The risk of food shortage, floods, epidemics and fires were serious threats, often perceived as divine punishment; it was nonetheless possible to control some risks, for instance by stockpiling food supplies to cope with possible shortages[1], or by carefully choosing habitat locations far from sites exposed to natural hazards, such as avalanches, landslides, overflowing rivers, etc. These actions were performed in the light of experience gained over decades. Moreover, in socioeconomic circles, whether or not to lend money is, for example, decided by assessing the risk that the debtor will be unable to repay his debt – a risk that one has evaluated for as long as debt has existed, thus since antiquity.

In the 17th century, philosophers and moralists still connected risk to the cardinal virtue of "Prudence". The notion of risk was introduced when probability was developed in the 18th century. It was also at this time that European society went from a concept of risk as a divine fate, against which human protection has only little impact, to that of a risk under control for which the corollary quickly became the right to safety.

Scientific progress and technological developments increased in parallel the risks relating to industrial and human development, and it was at this point that the notion of risk management appeared. Organizations have developed ways and means to address the irreparable or to minimize risk by *ad-hoc* measures, i.e., by taking out insurance to cover the damage that would result.

As a next step aiming at zero risks, modern society triggered an increasing demand for protection and insurance. The 1990s marked the beginning of a reconciliation of aspects of quality, safety, health and environmental protection. Finally, the implementation of globalization, coupled with the emergence of an information society, resulted in the large-scale communication of accidents, environmental damage, diseases, natural disasters and harmful consequences. This diffusion caused a feeling of insecurity among the sensitive population and induced an even greater demand for safety.

The interdependence of national economies also makes the global economy more unstable, sensitive to crises that are more difficult to predict and from which companies and organizations seek to protect themselves as far as possible. Risk management has thus become more professional, to the point of representing a function in its own right in a company or organization in the late 20th century.

7.2.1.2 Definition of risk

But what is risk? ISO/IEC Guide 73:2002 (E/F) gives the following definition: *a combination of the probability of an event and its consequence*, followed by two remarks:

a) The term is generally used only when there is at least the possibility of negative consequences.

b) In some situations, the risk arises from the possibility of a deviation from the expected results or event.

Specifically, the Guide indicates that probability is *the degree of likelihood that an event will occur* and that consequence is *the result of an event*, while the latter is the *occurrence of a particular set of circumstances*[2]. The scientific definition of risk includes a dual dimension:

1 An example dating back to ancient times (Gen. 41: 1-37) is that of the dreams of a Pharaoh concerning lean and fat cows, a dream interpreted by Joseph as a fertile period followed by barrenness of the land. He thus advised the Pharaoh to build up reserves during plentiful harvests to avoid starvation of the people in unfavorable years.

2 The Guide specifies that an event can be certain or uncertain, and could represent one or more occurrences.

hazards and losses, both with probabilities. As a result, risk is characterized by two components:

- the level of risk (probability of occurrence of a given event and intensity of the hazard), and
- the seriousness of the effects or consequences of the event assumed to affect the issues.

However, this concise definition can hide the complexity of the concept of risk [7.2].

a) Thus, we commonly define two types of risk, although in reality many of them are somewhere in between these two terms:

   i) **Normal risk**, which arises from a decision of an authority or a person associated with an act of management, whose goal is to make a profit, all the while aware that this action, if it misses its target (due to unforeseen events or an error in judgment, for example), can cause a loss. The field of normal risk, also called speculative risk or business risk, is that of management. A typical example is the business risk associated with the launching of a new product.

   ii) **Pure risk**, also called static or accidental risk, occurs suddenly and unexpectedly, resulting in damage and losses. While normal risks are inherent in the purpose and the functioning of an organization, pure risk is determined by its existence and activities.

b) Another risk segmentation is shown below:

   i) Risk as the **unexpected occurrence** of a hazard[3] (breaking a leg during downhill skiing, for example).

   ii) **Historical risk**, which is characterized by the appearance of a hazard, whose origins go far back in time, and were unknown, buried in oblivion or simply neglected. The long-term unknown adverse effects of drugs that have been on the market dozens of years are such risks. The potential deleterious effects of uranium of low radioactivity, unfit for extraction but used as backfill in the construction of roads in France, is another example.

   iii) **Emerging risk,** a threat whose potential deleterious effects have only been partially identified or evaluated. Thus, for many years, the release of CO_2 and its effect on global warming have been hotly debated. CO_2 production was, therefore, an emerging risk.

c) Risk is also divided into many categories; a classification of the most used follows, but does not cover the risks of the financial world for example:

   i. **Natural hazards:** damaging event with a given probability, or consequence of a natural threat affecting a vulnerable environment. Tsunamis and volcanic eruptions are natural hazards.

   ii. **Environmental risks:** these involve natural ecosystems, especially their integrity and sustainability. When human activities are the direct causes of this danger, they are classified as technological risks.

   iii. **Technological risks:** the possibility of occurrence of an accident caused by a technical system and which can lead to serious consequences for the staff, population, property, the working environment or the natural environment. The concept of

[3] Refers to any event, phenomenon or human activity that may cause unpredictable loss of life, injury, property damage, social or economic disruption, or environmental degradation.

industrial risk is used in situations where an industrial plant (chemical plant, power generation plant, etc.) is the source of the threat.

Finally, even if other management system standards have been implemented in an organization, the introduction of ISO 31000 can be confusing, particularly in the integration process. In fact, doesn't ISO 9001 aim to minimize the risk of poor quality; ISO 14001, damage to the environment; ISO 22000, the risk of food safety hazards for consumers, etc.? The concept of risk, its scope, the existence of additional and less generic ISO standards (e.g., ISO 27005), must, therefore, first be further developed by an organization that wishes to structure its management.

7.2.2 ISO 31000:2009, Objectives, goals and expected benefits

ISO 31000:2009 is an international standard that provides a generic approach to risk management. ISO 31000:

- provides principles and generic guidelines on risk management;
- can be implemented by anyone, any public or private company, any community, association, group or individual; it is not specific to any industry or sector;
- can be used throughout the life of an organization and for a wide range of activities, including strategies and decisions, operations, processes, functions, projects, products, services and assets;
- can be applied to any type of risk, regardless of its nature, and whether its consequences are positive or negative.

Although ISO 31000 provides general guidelines, it is not intended to impose uniformity when it comes to risk management within organizations. The design and implementation of organizational plans and the structures of risk management should take into account the varying needs of a specific organization, its mission, strategies, objectives, context, structure, activity, processes, functions, projects, products, services or particular assets, as well as its specific practices.

ISO 31000 should be used to harmonize risk management processes in existing and future standards. It provides a common approach to the establishment of standards addressing risks and/or specific sectors, without replacing those standards. Note that ISO 31000 is not intended to serve as a basis for certification.

According to the creators of this standard, its implementation can bring the following benefits to an organization:

- promotion of a proactive rather than reactive management;
- awareness of the risks identified in the whole organization;
- easier identification of opportunities and threats;
- improvement of the management of the organization, especially by establishing a stable base for decision-making and planning, while providing clearer financial reporting;
- increased confidence of stakeholders;
- improvement of the performance and operational efficiency, as well as inspections;
- improvement in the prevention of incidents, but also of health and safety;
- reduced losses;
- improvement in the organizational resilience and learning.

It should be noted that other management system standards exist that address risk. These are more sectorial, like the ISO/IEC 27000 family that will be briefly discussed later in this chapter. Others deal with medical devices (ISO 14971:2007), machinery (ISO 14121-1:2007 and ISO 14121-2:2008), ergonomics (ISO 15265:2006 and ISO 15743:2008), industrial risks when it comes to oil or natural gas (ISO 17776:2002, ISO 15544:2000), space systems (ISO 17666:2003), or project management (FD X 117:2003).

7.2.3 ISO 31000:2009, principles of risk management

In clause 4 of the standard, risk management is defined as follows from ISO Guide 73: 2009: *coordinated activities to lead an organization with regard to risk*; this is followed by the note: Risk management includes risk assessment, risk treatment, risk acceptance and risk communication. Within the ISO framework, a risk management system is defined as a *set of management systems of the organization concerned with risk management.*

For effective implementation, ISO 31000 recommends that management adhere to the following principles of risk management:

- Risk management creates value, by the expected benefits outlined above.
- It is integrated in organizational processes; risk management depends on the responsibility of management, and is thus not an independent activity that is detached from the key activities and processes of the organization.
- It is integrated into the decision-making process, as it is able to provide key data for decision-making, or facilitate the choice of priorities, either by indicating whether a risk is acceptable or if the risk treatment is appropriate.
- It explicitly addresses uncertainty.
- It is systematic, structured and useful in due course, and produces results that are consistent, comparable and reliable.
- It is based on the best information available, whether this comes from expert opinions, statistical or historical data, observations or forecasts; the decision makers of the organization must also learn to appreciate the degree of validity of the available information.
- It must be tailored, in light of the diversity of missions, strategies, objectives, goals, and the situation of the organization.
- It integrates internal or external human and cultural factors, as these are likely to facilitate or hinder the achievement of organizational objectives.
- It is transparent and participatory; for this, it is appropriate that all stakeholders are identified and consulted and their views taken into account in the determination of risk criteria.
- It is dynamic, iterative and responsive to change; it must indeed take into account changing objectives, but also that the environment in which the organization changes, which may happen quite suddenly and require a real-time adjustment of the organization.
- It facilitates the continual improvement and development of the organization, all the more so if the organization has already integrated management systems of the ISO type with the dynamic energy of the Deming cycle.

7.2.4 ISO 31000:2009, the organizational framework of risk management

Risk management should be part of an appropriate organizational structure, which will give it the desired efficiency and enable implementation of the risk management process (clause 6 of the standard). This organizational framework is presented in clause 5 and includes a full PDCA, falling within continual improvement (see Figure 7.2).

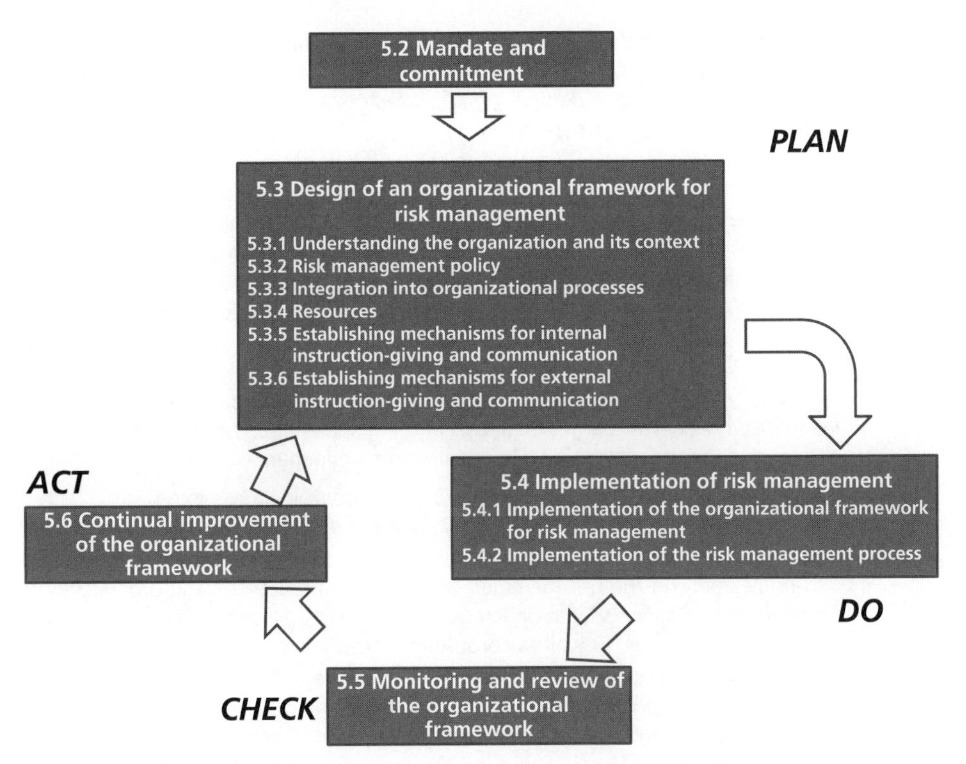

Figure 7.2 Organizational framework and the PDCA cycle.

The organizational framework of risk management is addressed in six sub-sections:

- The implementation of risk management involves, as in all ISO management system standards, a commitment from management, discussed in clause 5.2, Mandate and commitment. Management must articulate such a policy and approve the risk management policy, define performance indicators, ensure an alignment of the objectives of risk management, ensure that the allocated resources and the defined organizational framework are suitable, communicate the benefits of risk management, and ensure that the legal and regulatory requirements are met. Note that, unlike ISO 9001 does not recommend the appointment of a manager for for this task.
- Regarding the "Design of an organizational framework for risk management" (clause 5.3), several steps are considered.
  - *Understanding the organization and its context* (5.3.1): First, it is important to understand the organization and explicitly take into account internal and external contexts. The internal context includes the flow of information, internal stakeholders, pursued policies and strategies, and the existing values and culture. When it comes to the external context, this brings together external stakeholders, the various environmental aspects, and factors influencing the objectives of the organization…
  - A second step in the design of the framework is to have a *risk management policy* (5.3.2) clarifying the objectives and commitment of the organization by defining the different responsibilities, motives, how to deal with conflicts, methods and tools used, and how to measure the performance of risk management.

  - The third stage of the design involves *integration into organizational processes,* which consists in generalizing the risk management over all parts of the organization. In this way, integration will affect all sectors and will be addressed in a comprehensive risk management plan.
  - The fourth step is *financial responsibility,* which involves designating the persons competent in the implementation of the organizational framework, defining performance measures and establishing the levels of approval and sanctions.
  - The fifth step involves *defining means for allocating resources*, such as information management systems.
  - The last step is the establishment of *mechanisms for internal* (5.3.5) and *external* (5.3.6) *instruction-giving and communication*, ensuring the relationship between the organizational framework and results, and providing stakeholders with a consultation process where useful information is always available; also the exchange of data with external parties on a legal and transparent basis enabling the building of trust in the organization while maintaining a minimum of privacy.
- The implementation of risk management (5.4) can be divided into two parts:
  - *Implementation of the organizational framework for risk management* (5.4.1), which is managed as a project and includes steps such as defining a schedule and strategy, applying the risk management policy and process to the organizational procedures, observing legal and regulatory constraints, organizing training and information sessions, as well as consulting stakeholders.
  - *The implementation of the risk management process* (5.4.2) consists in ensuring that the process is integrated with the organization's business activities and practices at all levels and all functions directly affected by the risk management.
- The section Monitoring and review of the organizational framework (clause 5.5) emphasizes the importance of a regular and continuous agreement between the policy and the risk management plan on the one hand, and the internal and external context of the organization on the other hand, through measures of performance and progress with regard to the work plan.
- Finally, the last element is the Continual improvement of the organizational framework (clause 5.6) by making decisions to improve the framework itself, the risk management policy and plan, and this by using the results of the examination carried out during the monitoring and review of the organizational framework.

7.2.5 ISO 31000:2009, Risk management process

Risk management is ultimately performed by means of comprehensive and systematic logic. The different stages defined in clause 6 of the standard are shown in Figure 7.3.

Figure 7.3 shows that the activities discussed in clause 6.2, *Communication and consultation*, involves all stages from 6.3 through 6.5. What can be said about this? Internal and external communication with stakeholders must be performed at all stages of risk management. This involves the preparation and implementation of a communication and consultation plan that addresses the areas of the risk itself when it comes to consequences and the measures taken to manage it. In addition, a consultative approach based on teams has the following advantages:

- enabling the most accurate definition of the context;
- ensuring that the interests of stakeholders are identified and taken into ac count;
- ensuring, to a great extent, through the pooling of several different areas of expertise for risk analysis, that the risks are properly identified;

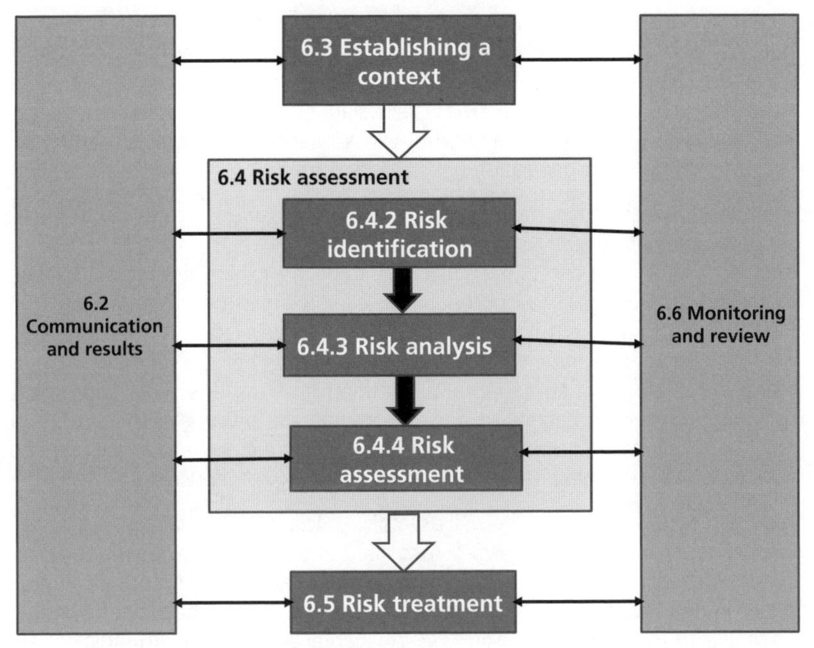

Figure 7.3 Risk management process.

- ensuring that the risk assessment takes into account the most appropriate points of view expressed;
- strengthening the approval and support in favor of a procedure for addressing and developing a communication plan, as well as appropriate internal and external consultation.

Communication and consultation with stakeholders also makes it possible to identify, document and take into account each stakeholder's perception of the risk[4], which can vary widely depending on their professional, economic, political, and sociocultural environment. As their opinions may have a significant impact, the fact of knowing them and taking them into account is a critical success factor for the establishment of an appropriate risk management system.

The first step in risk management is the establishment of context (clause 6.3), which more deeply addresses analysis of the same topic conducted to determine the organizational framework (6.3.1). Here, the internal (6.3.2) and external (6.3.4) parameters must be detailed, in particular with regard to their impact on the particular application domain of the risk management process.

Establishing context also includes the following additional elements:

- Context of the risk management process (6.3.4): this mainly consists in defining the strategies, objectives, areas of application and activity parameters of the organization or parts of the organization to which the risk management process is applied. This

[4] The perception of risk is defined by ISO as *how a stakeholder considers risk from a set of values or concerns*. Risk perception may differ from objective data, but it also depends on the stakeholder's needs, issues and knowledge.

context can include the definition of responsibilities for the risk management process, but may also identify methods of risk assessment, as well as the definition of the method of performance evaluation of the risk management system.
- Developing risk criteria (6.3.5)[5]. These values are often directly related to the resources and objectives of the organization, but can also arise from legal, regulatory or other constraints that the organization has freely accepted.

The next step, which is the most complex, is the *assessment of the risk* (clause 6.4)[6], which can be divided into the successive activities below:
- Identification of the risk (6.4.2)[7]: At this stage an exhaustive list of risk sources[8], impact areas, as well as their potential causes and consequences[9] is produced.
- Risk analysis (6.4.3)[10]: In this phase, the aim is to develop an understanding of the risks in order to determine whether these risks should be considered in the risk treatment and, if so, what strategies and treatment methods are most appropriate. The risk sources, their probability of occurrence, and the severity of their consequences must be taken into account here, while also considering the reliability of the information, as well as possible divergent opinions of the experts consulted.

The last part of this process is the *risk assessment* (clause 6.5) that, on the basis of the risk analysis, consists in determining which risks need treatment and establishing an order of priority for the implementation of the treatment. In many cases, this order of priority is established by multiplying the severity of the event (on a scale) by the probability of occurrence, in a manner similar to the approach for FMECA in Chapter 4. The standard indicates that risk assessment:
- should take into account the tolerance of a wider context of the risk and the risk tolerance of stakeholders other than those of the organization affected by the risk;
- can result in a status quo by maintaining the status quo of existing means of control, or alternatively can lead to a more detailed analysis. This decision depends on the risk appetite of the organization, its risk attitude, or the selected risk criteria.

Then comes the most visible part, which can also be the most expensive for the organization: *risk treatment* (clause 6.6)[11]. This process is iterated until the level of residual risk meets the expectations. The treatment options may include:
- Avoiding the risk, either by stopping the activity under scrutiny or by never initiating it (abandoning a production process that is too polluting and replacing it by a "greener" one).
- Removing the source of the risk, for example by destroying a dangerous staircase and replacing it by a safe means of access. An alternative is to reduce the occurrence of risk to an acceptable probability of occurrence. Since zero risk does not exist for sure, this scenario is common.

---

5 Reference terms for assessing the level of risk.

6 Overall process of risk analysis and risk assessment.

7 Process enabling to find, list and characterize risk elements. These items can include sources, dangerous phenomena, events, consequences and probabilities.

8 Element or activity that has potential consequences.

9 Result of an event, which in itself is caused by the occurrence of a particular set of circumstances.

10 Systematic use of information to identify sources and to estimate the risk. Risk analysis provides a basis for risk assessment, risk treatment and risk acceptance. The information used in this analysis may be empirical data, theoretical analysis, well-founded opinions and concerns of stakeholders.

11 Process of selection and implementation of measures aiming to modify the risk.

- Changing the nature and importance of the likelihood of the risk.
- Changing the consequences by, for instance, treating a polluting effluent before discharging it into the environment.
- Sharing the risk with several parties; the pooling of risk-taking in the creation of a corporation of the type "société anonyme par action" is one example.
- Voluntarily taking the risk, such as organizing a large festive event, while accepting the small risk of fire, for example. In this case, even if preventive measures are taken, the risk is never zero and the voluntary risk is usually the chosen path.

The two operations of risk treatment are *selecting options* and *preparing* (6.5.2), *and implementing risk treatment plans* (6.5.3). For the former, a detailed assessment of the cost/benefit ratio, as well as the appropriateness of the measure in relation to the expectations of the stakeholders should be considered. For the latter, documented risk treatment plans must be presented to and discussed with the stakeholders.

Subsequently, the standard is also intended to cover all types of risks: natural, technical, financial, regulatory or human. However, we can easily imagine that they require different approaches, assessments and treatments. The fact that this standard covers a wide range of companies and activities and a variety of risks makes it difficult to apply in its current state. We believe that working to adapt the standard for each individual case is necessary for it to be as efficient as possible.

As with any element of a management system of the ISO family, the process of risk management includes continual improvement, and monitoring and review (clause 6.6) applied at all of its stages. The last item of the standard includes *recording the risk management process* (clause 6.7): the text focuses on the importance of records for the traceability of processes, but also the sensitive nature of the information contained therein, which also implies a regulation of access. Recording the process is of course a source of information for its monitoring, review and continual improvement.

7.2.6 Discussion

This standard covers a very wide field of risk: risk associated with any activity, process, function, project, product, service or asset. The standard can be broken down further to be adapted to the types of risks and objectives: production, services, public institutions, etc. This is the case for risk in the field of informatics but also in the fields of quality, environment, laboratory work, and food safety, and their appropriate standards, even if the term risk is not present in their content. Numerous areas remain to be specified, perhaps by additional documents or ISO standards. The introduction of risk management in a reasonably large organization is a specialty in its own right, and should not be performed without the help and careful supervision of a risk specialist for the field in question.

The following remarks apply to risk management in general, for which the standard is the formalization, a distillate of best practices and the current state of the art:

- The continuous involvement of stakeholders at all levels is a strong advantage of the standard, but in reality, the participation of external stakeholders may be limited:
  - because of the confidentiality of data transmitted during the consultation;
  - by the fact that sensitive data, if known, could discourage potential investors to buy equities;
  - by the profound differences in the perception of risk between the organization's management and certain external stakeholders, such as NGO activists, whose impact on the media and potential large public demonstrations can be seen as a threat. Any dialogue is difficult here. On the other hand, the disclosure of a risk

analysis over too wide a circle can also affect the development of the risk management system if it falls into the hands of malicious people, who may look for vulnerabilities. It is, therefore, understandable that many organizations are tempted to apply the saying: when in doubt, do without …

- Conflicts are always possible between the bodies responsible for achieving production targets or such like, and getting results, and the structure of risk management. Theoretically, this should be integrated in the organizational process. But the assigned requirements are not all of the same nature and can sometimes be contradictory. Indeed, the major risk for an organization is often not achieving its objectives. However, an organization that does not achieve its objectives endangers its very existence. An organization can have excellent risk management but fail to reach its strategic objectives and thus jeopardize its future, and vice versa. Thus, tobacco manufacturers have taken dozens of years to accept the risk they exposed their consumers to, but also to accept their responsibility in the huge sums that public health has had to spend given the deleterious effects of their products. It is easy to see that their business would have been threatened if they had accepted with candor and good faith the results of numerous clinical studies (which they did not, as history shows …).
- The concepts of acceptability and risk tolerance are delicate. The standard does not in fact give a very clear definition. Aside from the legal and regulatory requirements, the organization can specify its own requirements and submit to them voluntarily. However, the business objectives often outweigh considerations relating to potential risks. For example, some organizations, although well aware of certain risks, intentionally choose not to manage them because such management would be more expensive than the indemnification if the risk occurred. In addition, some of the risks they impose on society are such that, if a major incident occurred, they would be unable to comply with the "polluter pays" principle, because:
  - Certain damage cannot be compensated by money.
  - These organizations, usually corporations with limited liability, have no equity or opportunity to borrow money or pay risk premiums to compensate injured parties. So after a serious accident, once bankruptcy is declared, in most cases (remember the saying "too big to fail"), it will be up to the community to bear the consequences of the accident or disaster. This was largely the case, for example, with the Fukushima nuclear plant incident.
- Finally, risk assessment is, in many cases, an approach that combines objective data with assumptions, projections, and speculations. These are marred by subjective factors: the economic and sociocultural environment, political beliefs, initial education, and the professional experience of experts. However, risk assessment in a complex or high-technology field requires long experience in the domain. For instance, experienced experts, recruited by a government agency to monitor economic activity, could reflect all or part of the beliefs of their previous employers, who they are now supposed to keep watch on. We see that their "neutral" view as an external stakeholder, when requested for a risk assessment or consultation, may not fully comply with the principle of neutrality.

It can, therefore, be concluded that risk management and ISO 31000 unfold in a sensitive and delicate environment, and on changing grounds, but this approach has become essential given the complexity and unpredictability of the social, cultural and economic landscape of the 21st century.

7.3 ISO/IEC 27005:2009 and the management of risks relating to information security

7.3.1 Threats to information systems

Before the 1930s, the threat to information systems (IS) was generally only about malicious access to protected data by either espionage or interception, the production of false documents, the issuance of false news and the possible alteration of official documents.

Since then, information and communication systems have grown exponentially thanks to the computer, and face a growing number of threats:

- Destruction of hardware and software, especially by sabotage;
- Illicit use (hardware and software);
- Theft of hardware, software, strategic information or material carrying this information;
- Interference by electromagnetic radiation;
- Passive listening, which allows third parties to obtain sensitive information;
- Alteration of the software by snares, particularly Trojans, spyware, viruses, logic bombs;
- Extortion of data by phishing, which means to obtain sensitive information (such as access codes or passwords) by expressing a false demand (such as an inspection by an IT manager), or by tempting one or more users with an advantage that will prove to be false;
- Saturation of the computer system, for example by a bombardment of e-mails (spam) or saturation of hard drives;
- Deliberate emission of false information to destabilize the organization and/or its stakeholders;
- Alteration of data, for example by changing the amount in a bank account;
- Abuse and usurping of rights: each user has a limited number of rights allowing access to the IS or permitting modification of it. Abuse consists in excessively using this right, while an act of usurping means using unallocated rights;
- Denial of action, which violates the traceability of the actions of IS users and, for instance, deletes them from the register.

The sources of these threats are mainly malicious individuals, users of the organization, illegal programs, but also natural phenomena, accidents or unforeseen technical malfunctions.

7.3.2 Elements involved in an information security system

System security simultaneously affects several areas of the information system. On the one hand, it needs to be addressed from a global perspective and, on the other hand, be applied in-depth to each element of the information system. The most important are:

- data security;
- network security including the transmission of information between different computer systems;
- security of operating systems;
- security of telecommunications;
- application security especially during programming operations;
- physical security or security of the hardware, facilities and their access rights, etc.

7.3.3 Sensitive information and security criteria

To develop a security system, we must first determine which sensitive data the organization needs to protect. This data contains the intellectual capital of the organization so the threats and vulnerabilities for each of them must be assessed, as well as the degree of severity in cases of damage or loss.

The security criteria to be applied in this analysis are mainly:

- Integrity: the elements should be considered as accurate and complete at all times and only be changed by authorized persons.
- Confidentiality: only people with ad-hoc access rights can view or edit the data (according to the rights granted to them).
- Accessibility of information in a timely manner by the people with access rights.
- Traceability of consultation or modification (records).

7.3.4 ISO/IEC 27001:2005 and information security management systems

The ISO/IEC 27000 family of standards

ISO/IEC 27001 is part of the ISO/IEC 27000 family [Information Security Management System (ISMS) standards]. This suite includes all the standards on information security. It contains all the information needed to initiate the implementation and maintenance of an ISMS. The family includes 11 key standards listed below (but the ISO/IEC 27000 family comprises over 30 standards):

- ISO/IEC 27000:2009, *Information technology – Security techniques – Information security management systems – Overview and vocabulary*
- ISO/IEC 27001:2005, *Information technology – Security techniques – Information security Management Systems – Requirements*
- ISO/IEC 27002:2005, *Information technology – Security techniques – Code of practice for information security management*
- ISO/IEC 27003: 2010, *Security techniques – Information security management system implementation techniques*
- ISO/IEC 27004:2009, *Information technology – Security techniques – Information security management – Measurement*
- ISO/IEC 27005:2011, *Information technology – Security techniques – Information security risk management (we shall briefly present this)*
- ISO/IEC 27006:2011, *Information technology – Security techniques – Requirements for bodies providing audit and certification of information security management systems*
- ISO/IEC 27007:2011, *Information technology – Security techniques – Guidelines for information security management systems auditing*
- ISO/IEC TR 27008:2011, *Information technology – Security techniques – Guidelines for auditors on information security controls*
- ISO 27799:2008, *Health informatics – Information security management in health using ISO/IEC 27002*
- ISO/IEC 27011:2008, *Information technology – Security techniques – Information security management guidelines for telecommunications organizations based on ISO/IEC 27002*

This very complete family can be compared with those of ISO 9000 and ISO 14000. This shows the care that ISO took when defining and disseminating the state of the art and best practices in this field.

7.3.4.1 The goals and objectives of ISO/IEC 27001:2005

According to the official statement, the generic standard ISO/IEC 27001 specifies the requirements for establishing, implementing, operating, monitoring and reviewing, updating and improving an information security management system (ISMS). This must be documented in the context of the overall risk associated with the activity of the organization. The standard also specifies requirements for the implementation of security measures tailored to each organization or parts thereof.

ISO/IEC 27001 is designed to ensure the selection of adequate and proportional security measures to protect assets and give confidence to interested parties. It includes the PDCA of the Deming cycle.

7.3.4.2 Overall structure and content of ISO/IEC 27001:2005

The table of contents provides information on the structure of the standard, of which some items are familiar to those who know the content of the ISO 9000 family. Annex C also contains a table comparing its contents with those of ISO 9001 and ISO 14001:

0. Introduction
1. Scope
2. Normative references
3. Terms and definitions
4. Information Security Management System
4.1. General requirements

This subsection refers to Figure 1 of the standard, which has many similarities with Figure 1 of ISO 9001 which shows the dynamics of the PDCA (Figure 7.4).

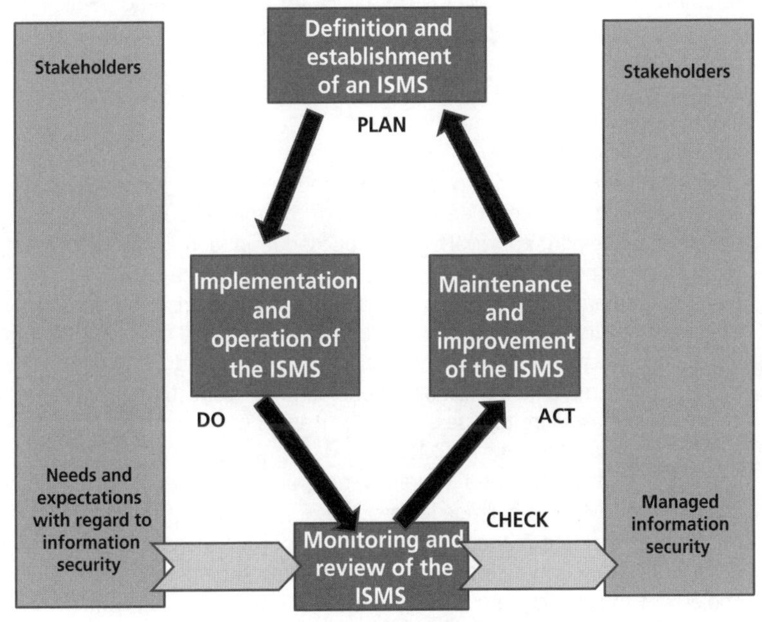

Figure 7.4 The Deming cycle applied to an ISMS.

7.3.4.3 Establishment and management of the ISMS

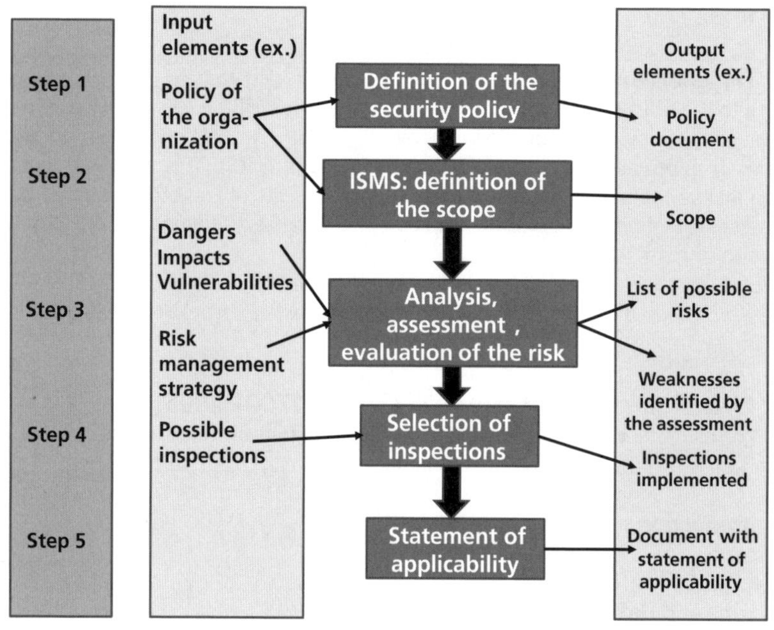

Figure 7.5 Definition of an ISMS.

The goal of Figure 7.5 is to:

- Define the scope and boundaries of the ISMS;
- Establish an information security risk policy for the organization which is congruent with its general risk policy;
- Define the risk criteria;
- Announce the method of approach for risk assessment;
- Identify, analyze and evaluate the risks;
- Identify and evaluate options for risk treatment;
- Select the inspection objectives but also the inspections. Annex A of the standard contains a long and almost exhaustive list of these two items;
- Obtain management approval of the proposed residual risks, as well as authorization to implement the ISMS;
- Prepare a statement of applicability that should summarize the decisions about risk treatment and contain a list of inspection objectives, and selected and implemented inspections, with justification for those listed in Annex A but have not been selected;
- Introduce and implement a risk treatment plan with inspection objectives and selected inspections; make sure to assess the effectiveness of these inspections (this item is in fact the establishment of the ISMS);
- Implement staff training and awareness programs;
- Operate the ISMS and manage its resources;
- Establish an incident management system (procedures and inspections) designed to quickly detect security-related events and react accordingly;
- Implement monitoring and review of the ISMS at regular intervals by reviewing the risk assessment process and establishing a program of internal audits;

- Maintain and improve the ISMS, taking into account the results obtained during operations of monitoring and review.

7.3.4.4 Requirements with regard to the documentation

Without recommending the writing and updating of an ISMS manual, the standard suggests:

- documenting the management decisions and establishing their traceability with regard to actions undertaken;
- including all key documents, such as the scope of the ISMS, the risk policy, the statement of applicability, a description of the methodology used for the risk assessment, procedures and inspections, and the risk treatment plan;
- inspecting the documents and making any changes traceable, etc. These requirements are very similar to those of ISO 9001 for documentation. The same goes for the records.

7.3.4.5 Management commitment

- Management must be able to prove its commitment when it comes to the establishment, implementation, operation, monitoring, review, maintenance and continual improvement of the ISMS.
- Resources should be identified and released to allow the proper functioning of the ISMS at all of its stages.
- Staff in direct contact with the ISMS must have appropriate skills (the initial status of these must be assessed before any training). Necessary training should be identified and carried out, with traceability of its progress. Its effect must be assessed.

7.3.4.6 Audit program

The recommendations for audit programs are similar to those in ISO 9001 and ISO 14001. Moreover, the standard specifies that ISO 19011:2011, *Guidelines for quality and/or environmental management systems auditing*, is a useful source when planning and implementing internal audits.

7.3.4.7 Management review

Here too, the recommendations regarding the review of management are similar to those in ISO 9001.

7.3.4.8 Continual improvement

The previous remark also applies to this clause: the organization must be able to prove that the information at its disposal, including that presented at the management review, is used for continual improvement. The standard also provides for corrective and preventive actions.

Remarks

Risk assessment is a sensitive part of ISO/IEC 27001. It is, therefore, appropriate to devote a complementary standard specifically for information security risk management. In contrast to ISO 31000, ISO/IEC 27001 can lead to certification and many companies have obtained this.

7.3.5 ISO/IEC 27005:2011

The goals and objectives of ISO/IEC 27005

ISO/IEC 27005 is intended to assist users in implementing ISO/IEC 27001, a standard for information security management systems that is based on a risk management approach. It is important to understand the concepts, models, processes and terms set out in ISO/IEC 27001 and ISO/IEC 27002 to comprehend this standard.

The standard is closely aligned with ISO 31000:2009 in order to assist organizations wishing to manage their risks related to information security in a manner analogous to the management of their "other" risks. However, the standard specifies that its contents do not contain a specific method of management for information risks and that it is up to the organization to determine what methodology to use based on its business activities and general context.

Overall structure and content of the ISO/IEC 27005:2011

The table of contents of ISO/IEC 27005:2011, *Information technology – Security techniques – Information security management systems – Requirements*, comprises:

0. **Introduction**
1. **Scope**
2. **Normative references**
3. **Terms and definitions**: these are taken from ISO Guide 73:2002 (E/F).
4. **Structure of the international standard**
5. **Background**
6. **Overview of the risk management process**; this item presents the generic process described in ISO 31000 or ISO 27001
7. **Establishing the context** (this is quite close to that of ISO 31000, but provides additional information, particularly in terms of impact criteria, criteria for risk acceptability, the scope and organization of the risk management. Annex A provides additional information on security risk assessment)

   7.1 General description of the security risk assessment

   7.2 Risk identification

   This section includes the following steps:

   - Identification of assets; the standard distinguishes between primary assets (business processes and activities) and secondary assets (hardware, software, networks, personal, etc.). It states that assets should be identified and known in order for one to conduct a risk analysis with this information. This item refers to Annex B that contains information on assets and also deals with the estimation of their value.
   - Identification of hazards (Annex C provides many examples).
   - Identification of existing inspection measures.
   - Identification of vulnerabilities (Annex D provides a list of common items).
   - Identification of the consequences.

   7.3 Risk analysis

   This section, which also refers to Annex E, addresses:

   - Methodologies for risk analysis, distinguishing between qualitative and quantitative approaches.
   - Evaluation of the consequences, which includes an assessment of the assets in question, not only in terms of their intrinsic value, but also the losses to business activities which may be caused by their possible dysfunction.
   - Evaluation of the probability of an incident.
   - Assessing the level of risk, especially by considering the likelihood and severity of the consequences for each scenario.
   - Risk assessment, resulting in a list of priority incident scenarios.

8. **Treatment of information security risks**
   8.1 General description of risk treatment (cf. Figure 7.6)

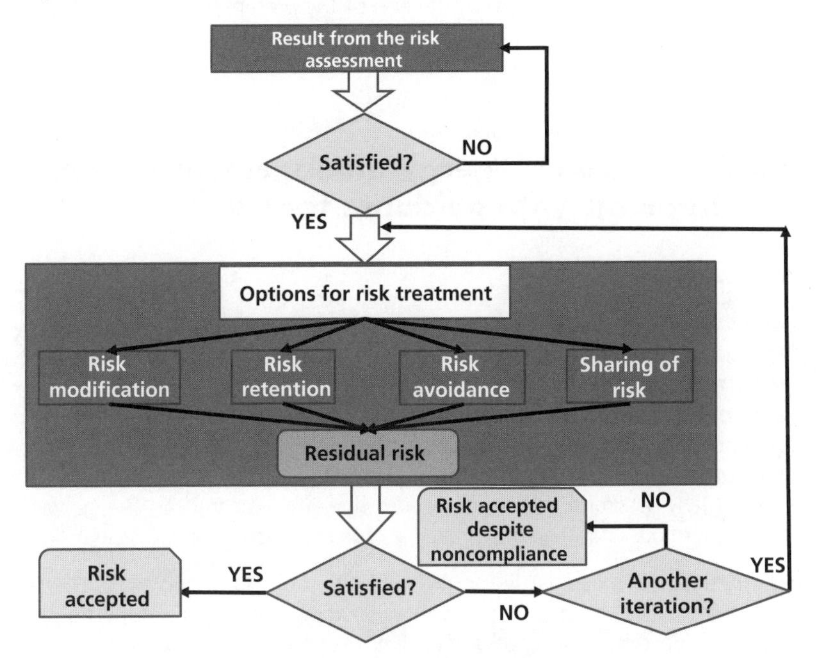

Figure 7.6 Key steps of risk treatment.

   8.2 Risk modification (Annex F gives very useful information)
   8.3 Risk retention
   8.4 Risk avoidance
   8.5 Sharing or pooling of risk (for example, by purchasing insurance)
9. **Information on the acceptability of a security risk**: it is at this stage that the treatment plan and assessment of residual risks must be submitted to and accepted by the management of the organization. The resultant document is a list of accepted residual risks, which can be added to those that the organization accepts despite a risk level that does not meet its requirements (it may be that the desired level of residual risk is not yet achieved but that the risk must be taken to continue the business).
10. **Communication and consultation with regard to the information security risk**; this important item is discussed in a similar manner in ISO 31000.
11. **Monitoring and review of information about the security risk**: this clause is divided into two parts:
    - Review and monitoring of risk factors
    - Monitoring, review and improvement of risk management.

Remarks:

- The contents of ISO/IEC 27005 follow the outlines of ISO/IEC 27001 and ISO 31000. Much of its added value is in the Annexes, which specify the generic content of the standard in the field of information technology and communication. These constitute more than half of the standard.

- The standard specifies that risks that do not have an acceptable residual level might be accepted by management, especially if the business could be in danger. This clause refers to the general comments at the end of the presentation of ISO 31000 on the limit of risk management without having a neutral external view, able to overcome the "Betriebsblindheit" (subjective view) of the organization.

7.4 ISO 50001:2011, Energy management systems – Requirements with guidance for use

7.4.1 Introduction

The increasing access of the world's population to the well-being and utilities previously reserved to the West, together with global warming, depletion of non-renewable resources, and the limited capacity of renewable energy to meet our needs, has revived the interest of the political, academic and economic sectors in research and development of new energy means. Wind, solar and tidal power, the use of biofuels of the first and soon second generation, the increasing use of heat pumps for heating, the market entry of LED lighting, are initial responses to this request. Many pessimistic observers wonder, however, whether this change will enable us to meet the enormous challenge of sustainable development.

Within ISO, this new focus has led to the publication of ISO 26000:2010, *Guidance on social responsibility*. Since the exploitation of new sources of energy is only part of the problem of sustainable development, another factor has yet to be explored, namely, the careful management of the energy consumed and its optimization. Within ISO, a consensus has emerged: significant gains and savings in energy consumption could be achieved if we had at our disposal a system of energy management that was efficient, responsive, progressive, and with the possibility of continual improvement. This element was missing in the ISO portfolio, which already included standards for management systems.

ISO 50001 was published in 2011, shortly after ISO 26000. Less than a year after its introduction in late January 2012, about 100 organizations in 26 countries had already obtained external certification, including Bouygues Telecom (France), Hyundai Motors (Korea), Lindt & Sprüngli (Germany) [7.3]. The production company Coca-Cola Enterprises Ltd in Wakefield, England, which is the largest manufacturer of beverages in Europe, is among the front runners: the company claims to have already decreased its energy requirement by 16.5%. In addition, after applying the concept of the standard to its water consumption, this was reduced by 10%.

Implementation of this standard appears to be promising for the future.

7.4.2 Goals and objectives of ISO 50001:2011

ISO 50001 specifies requirements for the design, implementation, maintenance and improvement of a system of energy management that enables organizations to achieve, through a systematic approach, continual improvement of its energy performance, which includes efficiency, use and consumption.

The approach specifies requirements for the use and consumption of energy, including measurement, documentation and reporting, design and procurement of equipment, as well as systems, processes and personnel that contribute to energy efficiency.

It is applicable to all factors affecting energy performance which the organization can monitor and influence. It does not prescribe specific performance criteria for energy. It is designed to be used on its own, and can be aligned or integrated with other management systems.

It can be used for any organization that wants to ensure compliance to a formally fixed energy policy and provide proof thereof. A self-assessment or self-declaration of conformity or the certification of the energy management system by an external organization can attest to this.

It also provides, for information purposes, recommendations for its implementation. The standard covers two aspects:

- Energy efficiency, which establishes for each item of energy consumed how to improve its effectiveness and obtain an equivalent performance with less energy.
- The overall energy performance of the organization, which includes efficiency, use and consumption. This analysis helps the organization to determine how it can optimize its operations to reduce energy consumption. Should we rethink the logistics, the heating, or a certain technology? Is the current choice of energy sources related to the use made of it? Is it optimal? How can we improve it, and at what cost?

Thus, the standard enables an organization to have an overview of its energy performance and consumption, and to identify, describe, manage and optimize them. For many organizations, this will be the first time they have considered this exercise.

In conclusion, the implementation of the standard, which can lead to certification, enables an organization to:

- develop a policy for a more efficient use of energy;
- set goals and targets to implement this policy;
- rely on data to better understand the use and consumption of energy and make decisions relating to this;
- measure the results;
- examine the effectiveness of the policy;
- continually improve the management of energy.

7.4.3 Structure and content

The table of contents, which consists of four sections of very unequal length, has subsections with headings resembling those of ISO 9001 and ISO 140001 (Annex C has a comparison table). It has the following structure:

Foreword
0 Introduction

This provides a summary of the standard and recalls the PDCA cycle (see Figure 7.7).

1 Scope
2 Normative references
3 Terms and definitions

Some definitions of concepts specific to the standard are:

- baseline consumption: quantitative reference as a basis for the comparison of energy performance (e.g., the annual energy consumption of an organization expressed in joules);
- energy consumption: amount of energy used;
- energy efficiency: ratio or other quantitative relationship between a performance, a service, goods or energy products and energy intake (note that both the input and the output must be clearly defined in terms of quantity and quality, and be measurable);

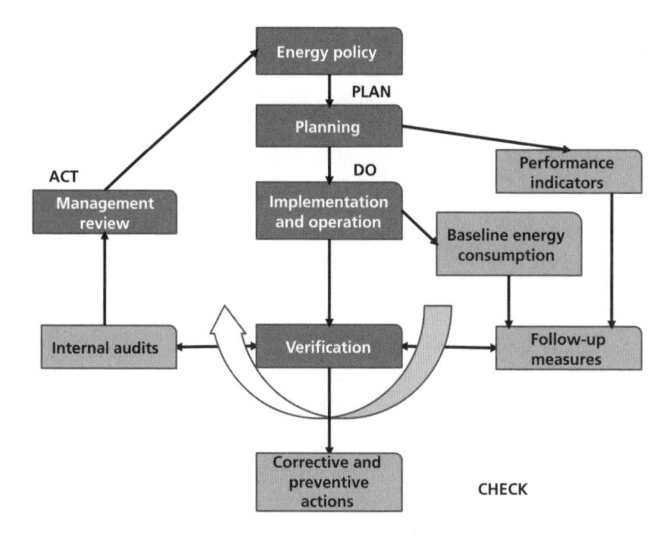

Figure 7.7 The Deming cycle and ISO 50001.

- energy management system (EMS): set of correlated or interactive elements for developing an energy policy and objectives, as well as processes and procedures to achieve these goals;
- energy goal: specific outcome or result set to achieve the energy policy of the organization and improve the energy performance;
- energy performance: measurable results related to energy consumption and its use (see Figure 7.8).
- energy performance indicator (EPI): quantitative value or energy performance measure defined by the organization;
- energy policy: general intentions and guidelines of an organization on its energy performance as formally expressed by management (the energy policy constitutes a framework and definition of objectives and energy targets);
- energy target: detailed energy performance requirement that is quantifiable, applicable to all or part of the organization, based on an energy target that must be set and met in order for this objective to be achieved;
- energy use: type or mode of energy application, such as ventilation, lighting, heating, cooling, transportation, production or business equipment (computers, etc.);
- significant energy use: energy use representing a significant portion of the energy consumption and/or with the potential to considerably improve the energy performance.

7.5 Requirements of the energy management system (EMS)

7.5.1 General requirements

7.5.1.2 Management responsibility

This states that management must:

- delineate the perimeter of the EMS;
- appoint a person responsible for the EMS whose tasks have many similarities with those of the ISO 9001 quality manager and which are specified in 4.2.2, Roles, responsibility and authority;

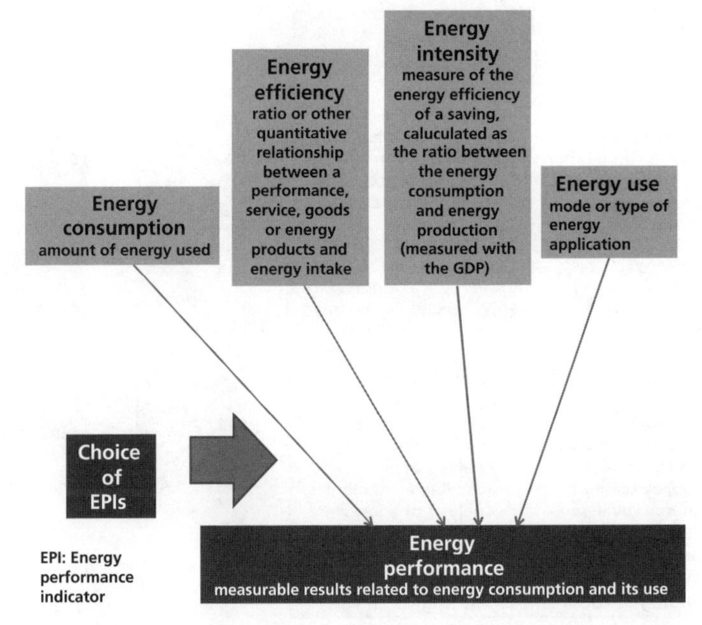

Figure 7.8 Relation between EPIs and energy performance.

- communicate to the organization the importance of energy management;
- ensure that the objectives and targets of the energy performance are defined and that the energy performance indicators (EPIs) are appropriate for the organization;
- include energy considerations in the long-term planning, as appropriate.

7.5.1.3 Energy policy

The energy policy must be an expression of the commitment of the organization to improve its energy performance and should especially:

- comply with the nature and size of the organization's energy use and the impact of its policy;
- include a commitment to continual improvement of energy performance
- set the framework within which energy objectives and targets are set and reviewed;
- encourage the purchase of energy-efficient products and services.

7.5.1.4 Energy planning

The graph in Figure 7.9 shows the dynamics of the energy planning process described by the standard.

A documented energy plan must be developed (it should also be revised at regular intervals, especially during the implementation of new installations). It includes:

- legal and other requirements to which the organization is subject, in order for them to be taken into account in the development, implementation and operation of the energy management system;
- an energy review that should include the methodology and criteria for the development of this review; the review comprises the following steps:
  - analysis of energy usage and consumption – in the past, present and future (projected) – from measurements and other data;

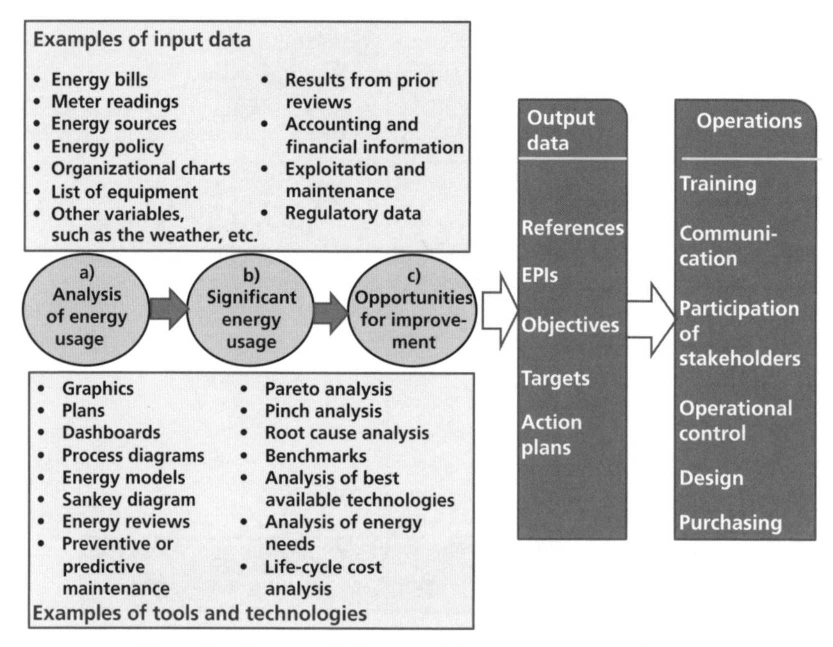

Figure 7.9 Conceptual diagram of the energy planning process.

- identification – based on the analysis of energy use – of the areas of significant energy use and consumption, and a determination of the actual performance of systems and equipment related to the significant energy uses identified;
- identification, prioritization, and recording of opportunities to improve energy performance, including – where appropriate – potential energy sources, use of alternative or renewable energy sources (it goes without saying that the person responsible for the EMS will start by optimizing the most significant energy usage);

- a baseline consumption established from information from the initial energy review, over a relevant period for the organization's energy use; this must be documented;
- energy performance indicators, suitable for monitoring and measuring energy performance, which must be reviewed and adjusted periodically.
- measurable objectives and targets, documented for each relevant function, level, or installation process in the organization; they must be met within a specified time and be consistent with the energy policy;
- action plans, documented and periodically updated, which contain:
  - an assignment of responsibilities;
  - the means and timeframe for achieving each target;
  - a statement of the method by which the improvement of the energy performance must be verified ; and
  - a statement of the method for verifying the results of the action plan.

7.5.1.5 Implementation and operation

For this, the organization must employ the action plans for energy management obtained from the planning. The organization should be particularly concerned with:

- training of the staff directly involved in the operation of the EMS;
- informing the employees about the role and the positive effects of the EMS;
- deciding whether or not to communicate externally about its energy management system and its energy efficiency; if the organization decides to communicate externally, it should develop and implement a plan for this;
- drafting and updating documentation including the scope of the EMS, the energy policy, objectives, targets and action plans when it comes to energy, as well as the plans to achieve the energy objectives and targets; this documentation must, as for ISO 9001, be archived and traceable;
- operational control, i.e., identifying and planning the operations associated with significant energy use that are in line with its energy policy, objectives, targets and action plans to ensure that the organization is allocated the necessary means and that operations will take place according to the specified conditions;
- opportunities for improving energy performance during design of facilities, equipment, new, modified and renovated systems and processes that can have a significant impact on energy performance;
- similar considerations made for the purchase of energy, energy services, products and energy-consuming equipment.

7.5.1.6 Performance verification

7.5.1.6.1 Monitoring, measurement and analysis

- The organization must consider the evolution of legal obligations such as to adjust periodically the EMS and its requirements accordingly.
- Periodically, the organization must conduct internal audits at planned intervals to ensure that the EMS complies with provisions for energy management, including the requirements of ISO 50001, and that it is properly implemented and maintained.
- An audit program must be developed, taking into consideration the status and importance of the processes and areas to be audited, as well as the results of previous audits.
- The organization must develop, implement and maintain procedures for handling actual and potential nonconformities and make corrections as well as conduct corrective and preventive actions.

7.5.1.6.2 Nonconformities, corrections, corrective and preventive actions

- The organization must establish and maintain records necessary to demonstrate compliance with the requirements of the energy management system and of ISO 50001, as well as the achieved energy efficiency. It should also define and implement arrangements for the identification, removal and preservation of these records.

7.5.1.7 Management review

Management reviews, analogous to those under ISO 9001 must, among other things, include the following points:

a) follow-up of actions from previous management reviews;
b) a review of the energy policy;
c) a review of the energy performance and the corresponding EPIs;
d) evaluation of legal compliance as well as changes in legal and other requirements with which the organization abides;
e) level of achievement of energy objectives and targets;

f) results of the audit of the energy management system;
g) status of corrective and preventive actions;
h) energy performance scheduled for the coming period, if applicable; and
i) recommendations for improvement.

Among the output elements of the management review, one should find any decision or action related to:

a) changes in the energy performance of the organization;
b) changes in the energy policy;
c) changes in the EPIs;
d) changes to the objectives, targets and other elements of the energy management system consistent with the commitment to a continual improvement of the organization. Discussion and conclusions

7.6 Discussion and conclusions

ISO 50001 is promising, as it concerns a very topical area, where the search for savings is a priority, and also since it has a well thought-out structure, and is logical and simple in its implementation. Companies that have already implemented ISO 9001, ISO 22000 or ISO 14001 have become accustomed to the content and implementation of ISO management systems, and should be able to implement it without much difficulty.

To finish, some elements of critical thinking are:

- The industrial world has not awaited the emergence of this new standard to strive to reduce its energy bill. This effort was started in the 1970s... with the first oil shock! For companies that have already made an effort in this direction, the added value of the standard is therefore to provide a global vision.
- ISO 50001 certification is not the equivalence of an energy performance certificate that provides a score in order to classify organizations according to their energy consumption (such a score exists for houses, for instance). The reason for this is simple. A company that is energy-intensive due to its trade (e.g., a steel mill), even one that is solicitous about its consumption, can never compare with service companies, even those that bear little attention to their energy consumption.
- As a consequence, ISO 50001 certification does not mean that the certified organization is a pioneer in energy saving, but that it will gradually get there, if it plays by the rules, through the continual improvement included in the standard.
- Even though it is true that large organizations, both private and public, are sometimes voracious consumers of energy, it is the logistics, the travel of users and the heating/cooling of homes that cripples the energy performance of countries. Therefore, it may be overly optimistic to imagine that the contribution of ISO 50001 will be cardinal for sustainable development, even though it would be very unwise to challenge its usefulness.

References

[7.1] *Management des risques*, AFNOR ed. (2008), p. 211.

[7.2] J. CHARBONNIER, *Le risk management, méthodologie et pratique*, ed. l'argus de l'assurance (2007), pp. 20-37.

[7.3] Article by IAN JOHNSON, PHILIP FULLERTON, REFHAN RAZEEN & MAURICE AHERN, ISO 50001 s'embrase – La norme sur le management de l'énergie se globalise, March 14, 2012, http://www.iso.org/iso/fr/home/news_index/news_archive/news.htm?refid=Ref1537 (accessed in October 2012).

Chapter 8

Total quality, personalities who stand out

Key concepts: Quality improvement, understanding of variations in quality, designs in tune with the market, psychological knowledge of human nature, the Deming cycle of total quality, product design, parameter design, process design, quality design, system design, tolerance design, *kaizen*, noise factors, control factors, signal factors, quadratic function of loss of quality, annual *hoshin*, quality engineering, breakthrough management (BT), WV model, thought and system approach, loss of quality, quality planning, Deming's fourteen points, four levels of management by breakthrough, four revolutions of management by breakthrough, seven mortal wounds, total quality control, theory of knowledge.

8.1 The triumvirate of the 1980s

8.1.1 Management according to Deming

8.1.1.1 The four areas of managers' in-depth knowledge

The field of total quality began with the global thinking that *Deming* developed from the 1950s to the 1990s. *Deming* recommended that all managers transform their approach to the environment and master, without being experts, a system of in-depth knowledge, based on four areas in constant interaction [8.1]:

1. *Thought and system approach*: understanding all processes of a company and their interactions as a system. *Deming's* definition of a system is a network of interdependent components that work together towards a common goal under conditions of constant interdependence. Certain corollaries result from this definition:
   - individualism and cronyism are counterproductive;
   - the system will deviate from its optimum if one of its processes is defective;
   - implementation of the system leads to a plateau performance of its outputs, in which employees, who are small cogs, do not play a very large role; focusing on employee productivity is not a cure-all;
   - interactions between processes cause a lack of performance of the system under consideration, and as these interactions are orchestrated by managers, most poor quality results from management. Might not this observation apply to the blindness of senior bank managers who allowed inclusion of toxic assets in their

portfolios without understanding their content, and then urged their employees to sell them to even less informed clients (global financial crisis of 2008)?;
   - improved quality and consequently enhanced outputs of the system are obtained through better cooperation (positive interactions) between the units in the process and not by exhorting employees to work harder and make fewer mistakes (since, according to Deming's observations, they are responsible for only 15% of them).

2. *Understanding variations* in quality (degree and causes), and the usefulness of statistical sampling. *Deming* believed that the variation of a process (he considered life a process) consisted of two parts:
   - a variation of normal and incompressible distribution, which is generated by the process itself and which can only be overcome by replacing all or part of the process by another more effective one (often more expensive); attempts to reduce this scatter by other means is a waste of time and energy;
   - a special variation due to poor quality that can be eliminated and which is beyond the six sigma level of the Gaussian distribution of observed values.
3. *Theory of knowledge* (epistemology) explaining the genesis of knowledge and the limits of what can be known: the complexity of situations makes intuitive thinking difficult. Therefore, the logical application of the Deming cycle, without hasty decisions, is the key to correct knowledge. Some seemingly simple things are not easily defined, leading to the relativity of knowledge, which may be highly influenced by the manager's intentions.
4. *Psychological knowledge of human nature:* managers can only understand human nature through the reactions of human beings in a social environment that is often alienating and oppressive. This is often the case in a company. Thus, it is not true that job insecurity and internal competition enhance employee performance. Managers have this impression because employees in this case seek security, recognition, visibility and worthiness from them (cronyism), rather than efficiency of teamwork, risk-taking in innovative projects and learning new skills, which are much less visible but more productive for the company. Competition and the lack of recognition of the quality of a person's work block their ability to spontaneously enjoy performing well and doing so in groups. Material rewards, if not accompanied by recognition, are counterproductive.

One of the interactions between pillars 2 and 4 is that every human being is unique and learns through different methods and at various speeds. Variations in the performance of direct associates also follow a natural variability that it is futile to try to decrease. Moreover, their performance is highly dependent on the quality of the social relationships they have with others, and even the quality of interactions in the process where they work with others (interdependency of pillars 1 and 4).

The system formed by these four interacting pillars is the basis for understanding and application of the 14 points of management described below[1].

8.1.1.2 The 14 points

In 1982, the activity of *W.E. Deming* intensified. He suggested 14 points [8.2] to restore dynamism to American companies, and these points catalyzed a chain reaction [8.3], the steps of which are shown in Figure 8.1.

[1] The concept of interdependence of phenomena is rooted in Buddhism, the doctrine of which is immersed in Japanese society. Is this where Deming found his inspiration? Moreover, Deming acknowledges that human beings have a strong urge to do good. Is he here referring to the radiant nature of the mind (again a Buddhist concept), or to an approach consistent with the psychology of Rogers?

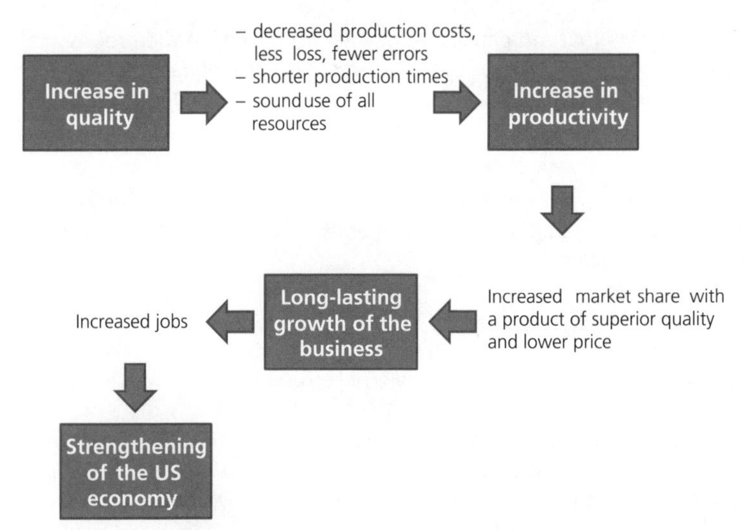

Figure 8.1 Deming's chain reaction of quality.

The 14 points are sometimes reminders of common sense (as well as good practices tested in Japan), which was neglected by US companies during their attempts to maximize return on investments in the short term, and whose vision was clouded by Fordism (this seems, in 2012, to still be the case):

1. *Create an environment favoring improvement of the quality of products and services by constant efforts.*
2. *Adopt a new "philosophy", which does not accept mistakes and negativity in the workplace and which targets excellence.*
3. *Reject quality control inspections on a massive scale and promote rather process improvement, as well as the introduction of quality at the design stage.*
4. *Reject the purchase of components and materials with price as the sole criterion; promote mutually beneficial supplier/customer relations and look for savings throughout the process.*
5. *Constantly and continually improve the quality of the system of production and services, to improve quality and productivity, and thus constantly decrease costs.*
6. *Train employees on the job; too many people learn their work from badly trained colleagues.*
7. *Enhance executive leadership instead of focusing exclusively on control, so that employees receive assistance in case of difficulties, and have dynamic management.*
8. *Drive out fear and build confidence to work more effectively; too many people are afraid to ask questions for fear of being labeled unqualified, so errors persist.*
9. *Remove the ivory towers, and break down the barriers between business units to facilitate teamwork and a learning organization.*
10. *Eliminate slogans, exhortations and short-term goals for employees, all of which have no or a negative impact on the quality of work.*
11. *Remove quantitative quotas and create a better quality of management; this method does not take into account the impact on quality since an increase in productivity can be achieved by other means.*

12. *Remove barriers that rob employees of their right to pride themselves in their workmanship; there are too many barriers preventing this (unqualified supervisors, unsuitable equipment, non-compliant basic materials and components, etc.).*
13. *Instruct the staff and management with regard to quality, statistical techniques and teamwork.*
14. *Carrying out this transformation is everybody's job; the staff alone is not able to make this transformation, as it requires management commitment and a strong signal from the board of directors.*

8.1.1.3 The seven deadly plagues

Deming believes, however, that the implementation of these 14 points will be unsuccessful if the organization does not eliminate seven deadly plagues (referring to the plagues of Egypt sent by Yahweh):

- Lack of consistent effort: *actions of the type "fix it" or "fire and forget" have no effect – they demonstrate to the staff that management is not reliable.*
- Considering exclusively short-term profits: *this is directly related to the lack of consistent effort.*
- Annual performance assessment of the staff: *this reduces to zero teamwork, staff initiatives and trust within the company; for Deming, the greatest successes of human beings are obtained in the absence of competition …*
- High turnover rates of management and managers: *this prevents them from seeing the results of their actions and taking responsibility.*
- Managing an organization by applying only visible facts and indicators: *a lot of data is intangible, such as the multiplier effect of a satisfied customer.*
- Excessive medical costs: *here Deming refers to the expense some companies devote to health.*
- Prohibitive costs of legal disputes and precautions to prevent them, *fueled by lawyers paid a percentage of the compensation obtained.*

Deming points out that the last two plagues only concern the United States.

8.1.1.4 The 13 obstacles

Certain behavior slows down the transformation of a company:

- forgetting strategic planning and long-term plans *in order to devote oneself exclusively to the emergency (business as usual) – believing that these documents are only "paperwork";*
- being convinced that only technology, gadgets, automation and new equipment can solve the problems (*i.e., forgetting the human factor*);
- seeking solutions from examples of other companies; *these solutions are perhaps not applicable and are just lazy thinking (argument against benchmarking); moreover such actions do not provide understanding about how this process will succeed or fail in the company;*
- spontaneously thinking, when faced with a possible solution, that it is not applicable in the context (*the "not invented here" syndrome*);
- believing that the education offered by business schools, which claim to deliver knowledge and generic skills in management, replaces knowledge gained in the real world of the company and production *(emergence of a caste of theorist managers full of their own knowledge, but who dare not confront the real world of production);*
- trusting only the quality department and relying solely on its responsibility;
- blaming and holding the staff responsible for errors, *while 85% of errors, on average, cannot be assigned to them in any way and are attributable to the management;*

- reducing the quality function to merely inspection;
- false start; *e.g., beginning quality cycles unrelated to the commitment (and support) of management;*
- using and entering data into a computer without thinking in advance how to treat it - *cheaper, less sophisticated, less glamorous methods may give equally good results...;*
- contenting oneself with meeting specifications and satisfying compliance, *which has never given rise to an increase in quality or productivity;*
- settling for partial tests with prototypes, *and testing them only in the laboratory...;*
- thinking that we should know everything about a process in order to improve it; *antidote: to trust the neutrality and novelty of a critical eye.*

Based on this list, one could imagine that *Deming* contented himself with giving useful advice. This would be to forget his role and his principles in the Japanese prize of excellence called the *Deming Prize*, which was to become the foundation of Japanese quality and a strong influence on the American *Baldridge* Award. In view of the constant changes[2] in management methods, the recommendations of *Deming* are still relevant.

8.1.2 Ishikawa and TQC (Total Quality Control)

8.1.2.1 People, not machines

Difference in quality comes from people and not from machines!

This response from a Japanese manager of *Sony* in the 1980s, summarizes the book *Total Quality Control* by *Kaoru Ishikawa*, published in Japanese in 1981, and in English in 1985 [8.4]. The book quickly became a bestseller, since it exposes more clearly than *Deming* (who communicates almost exclusively in terms of production of the secondary sector) the principles of total quality, applied in Japan since the late 1950s.

8.1.2.2 Deming's cycle of (total) quality

Ishikawa cites the contributions of *Deming* and *Juran* from the end of World War II (p. 17 to 20) and recalls *Deming's* cycle of quality, on p. 56 of his book, which includes activities affected by total quality called Total Quality Control in Japan. The cycle shown is not the commonly depicted PDCA, but a cycle incorporating the key quality functions of the company and stakeholders. This point is sometimes forgotten in the West (Figure 8.2).

This Deming cycle of (total) quality was presented to the Japanese in the 1950s and was one of the bases of the *Deming Prize for Total Quality Control*. It sets the scope of total quality: R&D (quality and design), production, sales and marketing, form a cycle permitting constant change in the standard of quality expected for a product.

8.1.2.3 A critical, skeptical and sarcastic mind

To the surprise of American managers at the time, who thought of Japanese culture in terms of conformity and collectivism, *Ishikawa* shows a sarcastic, critical mind, especially with regard to the technical norms and standards. Below are some excerpts from his warnings for the implementation of a system of Total Quality Management [8.5]:

---

2 *The new trans-national elite require that all individuals "move" to follow the globalizing movement, accelerate their own movement, and from now on live in an era of "globalization". The standards are simple, even perfunctory: consume ever more, communicate more rapidly, trade in an optimally profitable manner.* From the frontispiece of the book Résister au bougisme, de P.-A. Taguief, 1001 nuits, 2001.

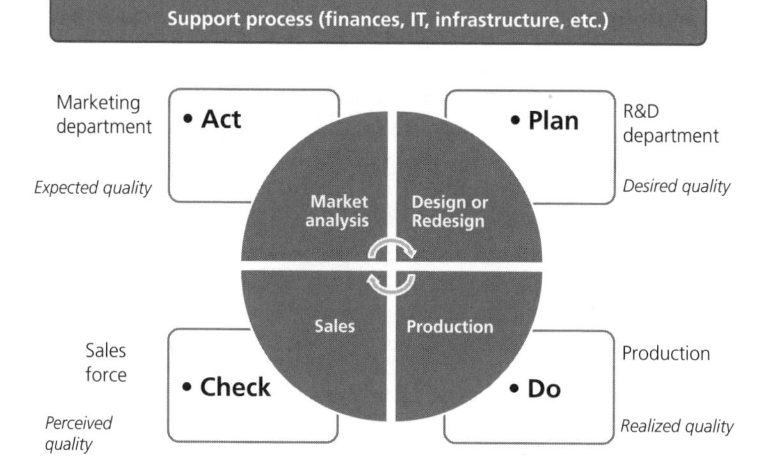

Figure 8.2 Deming's cycle of total quality as cited by Ishikawa.

- the passivity and unwillingness of top management to take real responsibility is one of the most significant causes of poor quality;
- if someone presents you with the quality standard of a product, be skeptical:
- if someone presents you with the quality standard of a component or raw material, be skeptical;
- if someone shows tolerance gaps on a diagram, be skeptical;
- if someone presents the results of chemical analysis or measuring instruments, approach them with suspicion.

But also:

- total quality starts and ends with training [8.6]·

8.1.2.4 An iconoclastic vision of quality in the western world

As a corollary to the Deming cycle of total quality, *Ishikawa* reminds us that Japanese industry applies *Total Quality Control* in:

- purchasing, including supplier qualification by assessment of their quality system and good practices;
- marketing activities; "intuitive and creative" American managers with a prima donna temperament certainly responded unenthusiastically to this perspective;
- sales and after-sales services;
- the activities of middle and senior managers of the organization; also here, the prospect of equal treatment of the elite and ordinary employees probably did not arouse much enthusiasm.

Ishikawa cites quality circles, the loyalty of managers to their company (and vice versa), and the emphasis on the education of all Japanese citizens, especially stimulated by the difficulty of learning how to write without phonetic assistance, which contributes to striving towards excellence.

Being an astute observer, he openly criticizes labor practices in the US and even stigmatizes their core values. He points out that the West, by religion, believes that man is fallen, which is why he cannot be spontaneously trusted. According to *Ishikawa* this explains the division between quality control and production in the US and why the focus is placed on the power of the former [8.7]. In Japan, influenced by Buddhism and the Neo-Confucianism of *Mencius*, it is natural to think that man can be good. *Ishikawa* therefore spontaneously connects the Puritanism of Ford to his basic design of quality control.

8.1.2.5 Total Quality Control; an expression dating back to ...1961

Ishikawa used the term *Total Quality Control*, which is the English translation of the Japanese expression. However, he did not create this expression. It was used by another quality pioneer, who apparently has left little traces for posterity: *A.V. Feigenbaum* of the company *General Electric.* This is the definition he gave as early as 1961:

> "*Total control of quality is an effective and efficient system integrating the quality of design and development, the quality of maintenance, and efforts to improve the quality implemented by stakeholders of an organization. Its objective is to fully satisfy the customer while allowing the production and services to operate at minimum cost* [8.8]."

8.1.3 Big Q and Little Q, the contribution of J.M. Juran

In 1983, *J.M. Juran* published the book *Juran on Quality by Design* [8.9], which introduced total quality, the Big Q, by differentiating it from the quality assurance traditionally confined to production (and sometimes R&D), the Little Q. The topic is more systematic, more concrete than those of *Deming* and *Ishikawa*, with the motto:

> "*Quality must be planned and up until now, when it has been, it has been done by amateurs* [...]" [8.10]

The subject of the book is how to get out of this situation. For this, *Juran* uses a trilogy of interrelated actions – presenting similarities with the Deming cycle of continual improvement, which applies to every stage of planning:

- quality design: design and adjust the product features according to the implicit and explicit expectations of customers;
- quality control: maintain the status quo of the defined quality standard, especially by feedback actions;
- quality improvement: in the sense of a breakthrough, by a decisive improvement, by the achievement of an exceptional level of performance, not by small increases in improvement. This concept will be developed by Shiba (Sect. 8.2).

Quality planning involves the following steps, and the output of each step is generally the input of the following one:

- establishing objectives for the quality policy of the company;
- introducing means of measurement for all stages of planning;
- identifying customers;
- identifying their needs, desires and wishes, including cultural aspects that may interact;
- developing the characteristics/criteria of product quality;
- developing the characteristics and quality criteria of the process;
- determining controls of the process, starting production and testing it.

In parallel, *Juran* determined the transverse support processes aimed at improving the quality performance of the entire system, from the macrocosm of the company to the microcosm of the employee:

- strategic quality planning, which mobilizes the management responsible for running the company system;

- strategic quality management, which aims to optimize and implement the objectives of the company's quality policy by senior managers;
- quality planning of macro-processes, which involve several business functions, controlled by the heads of the process;
- quality planning of processes and micro-processes in the company's units, groups and at the staff level;
- organization of the memory of the company (data);
- motivation and training of staff.

The work of *Juran*, *Deming* and *Ishikawa* formed the basis for the Baldridge Award (Figure 8.3).

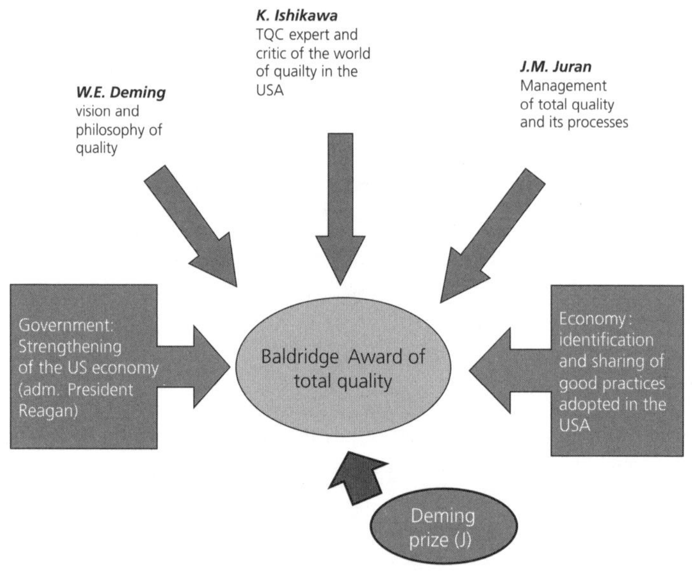

Figure 8.3 The motors behind the creation of the Baldridge Award.

8.2 Total quality according to Shoji Shiba

8.2.1 From kaizen to hoshin

Considered one of the gurus of Japanese total quality, *Prof. S. Shiba*[3] developed breakthrough management (or BT; *hoshin* [8.11] in Japanese) in the mid-1990s. *Shiba* found that quality management through continual product improvement (*kaizen* in Japanese), as described by Deming's PDCA cycle and included for example in ISO 9001, is not a sure-fire way to increase the competitiveness of an organization or company and is certainly not a guarantee of its survival.

Indeed, the abrupt changes that occur outside a company (new socio-economic conditions, emerging technologies, changes in raw material costs, etc.) mean that the future of a market cannot be scrutinized by examining only the wishes of current customers and key stakeholders. To survive, especially in companies with rapid cycles, we must move from a the presently developed activity A, to a new activity B. This is *Shiba's* idea for the new field

[3] *Shoji Shiba*, TQM Professor at the University of Tsukuba, Invited Professor at MIT, individual Deming Prize winner in 2002.

of application of total quality. He develops this concept by expanding the ideas of *Juran* on quality improvement.

8.2.2 The WV model

In addition to the PCDA cycle, *Shiba* advocates the use of the WV model of continual improvement, taken from the Japanese *Jiro Kawakita*. The WV model reaffirms the need for a rational approach to problem solving by highlighting the dialectic between the level of thinking (mental activity, reflection) and the level of experience (research data, facts). The WV model is used for three types of improvements:

- *process control*, which is achieving a defined standard (PDCA cycle becomes SDCA where STANDARD replaces PLAN);
- *reactive improvement*, following detection of sub-standard quality, which leads, for instance, to the standard being changed;
- *proactive improvement*, which can begin with an impression or intuition, without really knowing at first which improvement to support.

These three types of improvements include seven steps, in the shape of WV (Figure 8.4). Even though these steps seem trivial to the reader, the fact of their being formalized and respected is essential for teamwork under time pressure, the result of which is often a hasty decision based on firmly rooted opinions rather than on facts and careful reflection. It is normal to begin an analysis by an opinion, but it must be validated by facts.

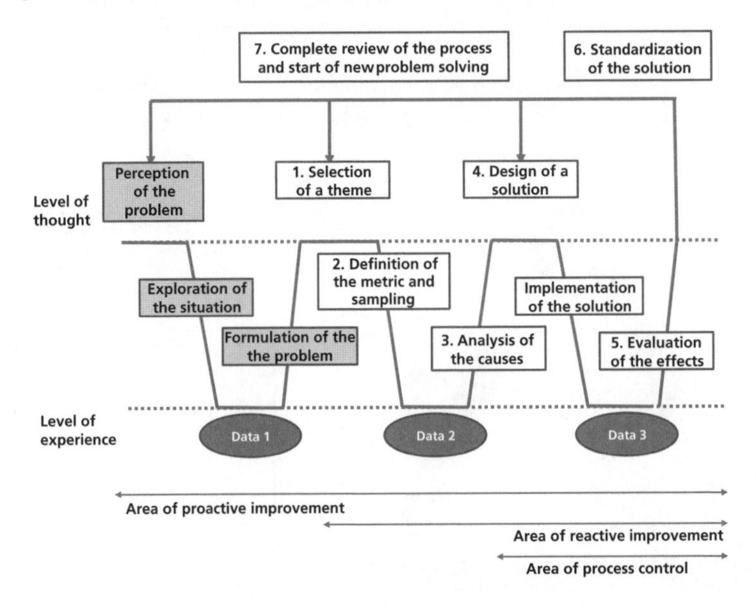

Figure 8.4 The WV model and the Deming cycle.

8.2.3 Total quality by four management revolutions

Although *Shiba* states that every company must find its own way of implementing total quality, he nonetheless observes that organizations that have done so successfully share four revolutions of management thought and four levels of practice (Figure 8.5):

- the revolution that causes the organization to focus its efforts and activities on the customer, making it possible to mobilize resources towards the key objective: customer satisfaction and loyalty;
- the revolution that causes all employees to seek continual improvement of services and products, using both an empirical (fact-finding) and scientific (analyzing and systematic) approach and proceeding in stages, with tools for planning and control (horizontal coordination);
- the social revolution in the company, which consists in mobilizing and motivating the full participation of staff to achieve the two above goals (vertical integration);
- the revolution of society-based learning which involves building a network of customers, suppliers and companies committed to the continual improvement of services and products, making it possible to benefit from their knowledge and expertise and to gain valuable time.

Shiba favors four levels of implementation:

- the level of the employee, who must be convinced that the performance of daily tasks is not sufficient to satisfy the customer if the continual improvement of services is not included;
- the level of teams and work groups, for which the organization of daily tasks should favor collaborative activities (team work) and the process approach (optimization of units);
- the organizational level, which defines and implements innovative improvements that are consistent with the organization's purposes, and are a driving force for the mobilization of personnel to carry out the objectives of the organization (organizational alignment);
- the society-based level, which aims to develop a network of partners who communicate about best practices, which may also be national standards.

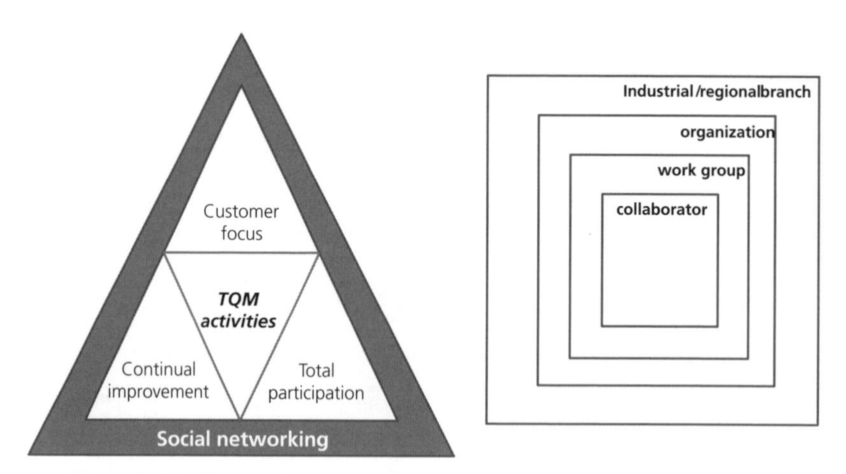

Figure 8.5 The four revolutions and the four levels of breakthrough management.

8.2.4 BT, integration of strategy and forecasting in TQM

BT (breakthrough management) is not the result of a linear progression, although it has a process. Rather, it proceeds initially by a breakthrough, followed by an implementation phase respecting *Deming's* PDCA cycle. Its main challenge is to perceive signs of change (a discipline known as early detection), which are often tenuous (or at the opposite end of the observational perimeter of the organization), with intensities only slightly above background noise, and to translate these signs into routes for a breakthrough.

BT is the stage of TQM (total quality management) which includes the strategic activity of the organization; this activity, run by management, is not necessarily on the agenda of management reviews provided by ISO 9001, for example, which might relate only to continual improvement. BT can move from a reactive improvement to a proactive one (Figure 8.6).

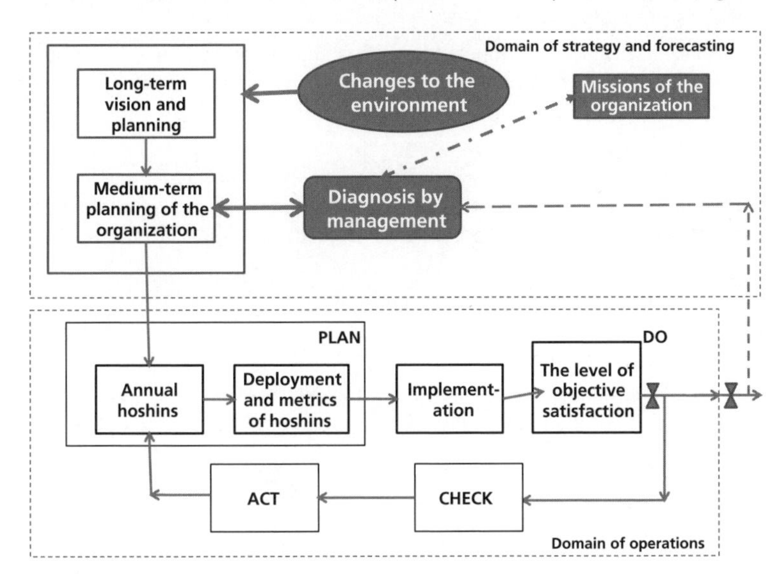

Figure 8.6 Breakthrough management, annual *hoshins* and the PDCA cycle.

Four phases of BT

Shiba described four phases of BT [8.12]:

PHASE1: identify areas of breakthrough: these areas, limited in number, are determined as a result of a collective thought process which includes:

- analysis of the past, with regard to two facets:
  - performance problems, dissatisfaction of customers, difficulties relating to products/services and internal management;
  - difficulties relating to the process(es) of improvement;
- analysis of the evolution of the environment;
- representation of what is desired for the future; a vision that should become a shared ambition and which requires coordination within the organization, and teamwork. The vision (defined as expectations of future customers in the form of products and/or services) is available as a medium-term plan and results in an annual plan with a maximum of three defined breakthroughs (the vital few, concentration of the activity on the key points necessary for success), as well as ambitious but achievable objectives.

Each area of breakthrough is defined by specifying:

- an orientation;
- the necessary means and resources;
- an indicator for measuring;
- a deadline for the implementation.

PHASE 2: the deployment and adjustment of the breakthrough areas:

- ***Deployment***: the achievement of consensus and the identification of measures to be implemented within the organization, consistent with analysis carried out in the field for every level and unit regarding orientation, measures, indicators of target values, and deadlines, which also improve the internal communication of the organization. The following are also part of the deployment:
  - the choice of metrics, from which indicators are established to measure achievement of the objectives;
  - mobilization and communication within the organization, accountability of the players and motivation of team work, often in the form of interdisciplinary work;
  - networking and external partnerships (e.g., determining new standards, searching for investors).
- ***Consistency***, which involves reconciling breakthrough management, continual improvement management and the management of daily tasks (business as usual) by examination of personal missions and the objectives assigned to them.

PHASE 3: the guiding of breakthrough management:

- monitoring by indicators that form the basis of the guidance; these indicators are adapted to each hierarchical unit;
- Shiba recommended at this stage to proceed with caution and avoid drastic corrections, even if this means adjusting ongoing actions or the annual (or even medium-term) plan.

PHASE 4: the president's diagnosis:

- this is conducted annually by top management, with site visits;
- the investigation focuses on results, but also on how to obtain them; the objective is to improve the process of progress itself, and the outcome of this initiative is visible in the medium to long term;
- this diagnosis enables one to adjust or even complement the annual plan for the following year.

8.2.5 Other concepts of breakthrough management

Breakthrough management may also be developed in an environment that does not initially lend itself to this type of management because the processes used and technologies are not consistent with the projected innovation, which is beyond the scope of the current activity of the organization. In this case, other concepts are used, sometimes in addition to those described above:

- The phases can be broken down into three steps [8.13]:
  - *the breakthrough of a visionary*, who introduces a new concept, generally an idea for a new product; the visionary develops his idea from a new reading of the existing situation, which may begin with a simple feeling, and which focuses not on a detailed description of the situation but on what is present and what is lacking in order to seize new opportunities;
  - *the breakthrough of a development group* – adventurous characters from within the organization, who will convince the entire institution of the merits of the theses of the visionary (remember that the first IBM PC was designed by a small team in a garage);

  - *the breakthrough of a joint-development partnership* – with suppliers and/or customers, which leads to commercialization.
- A cycle or a creative mechanism for BT. For each of the three above steps, there are a series of iterative similar cycles:
  - dedication – for passionate devotion to the proposed BT;
  - mental breakthrough, i.e., to develop an innovative vision that constitutes a break with the past:
    - unlearning, not getting caught up in past circumstances;
    - keeping some distance from the current field of activity;
    - making a mental breakthrough often consists in detecting a factor of change in the environment and being able to distinguish it from background noise; this ability can be stimulated by heuristic methods, including one used by *Shiba*, which was based on the education curriculum of the Museum of Modern Art (MOMA) in New York: strategic visual thinking consists, in a given situation, in asking oneself three questions:
      - What is happening in this situation?
      - What are the reasons that have led to the formulation of this statement/these statements?
      - Are there other significant aspects that have escaped the initial analysis? [8.14]
  - technological breakthrough – leading to a technical revolutionary resolution plan.
- The organizational infrastructure, which accelerates the process of BT and protects its implementation.

Shiba's strategic perspective complements the continual improvement process of *Deming*, which originally involves striving for operational excellence. Both activities are orthogonal and necessary for the sustainable development of an organization (Figure 8.7).

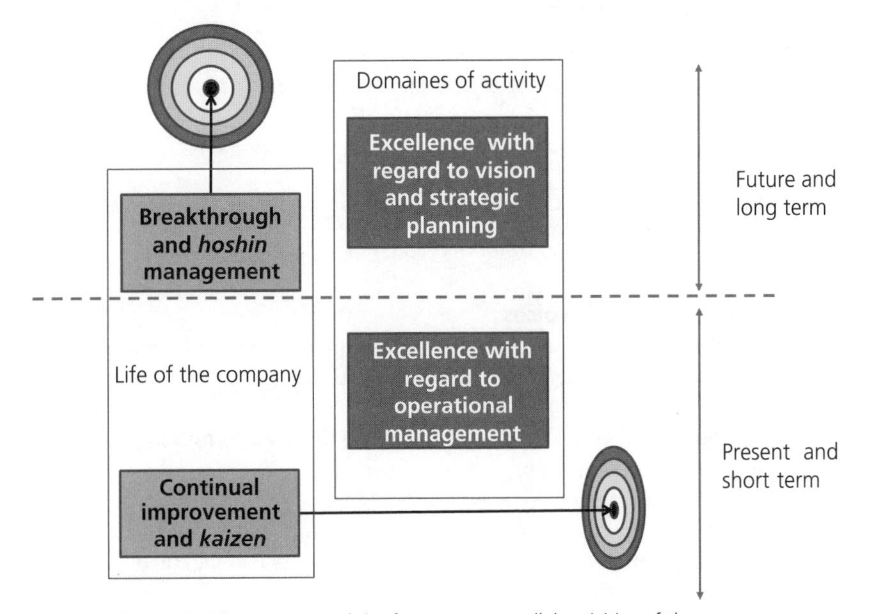

Figure 8.7 The present and the future, two parallel activities of the company.

8.2.6 Voice of the customer

All systems of quality management have customer orientation as a central point in the approach. This activity is entrusted primarily to the marketing and sales departments of a company, many of which are, however, not clear about the choice of methods for determining customer expectations. Within total quality, *Shiba* developed the voice-of-the-customer algorithm [8.15], a method developed to define a service or product with the best positioning. The process flow follows the WV model [8.16] (Figure 8.8), and the detection of faint signs that are indicators of change factors.

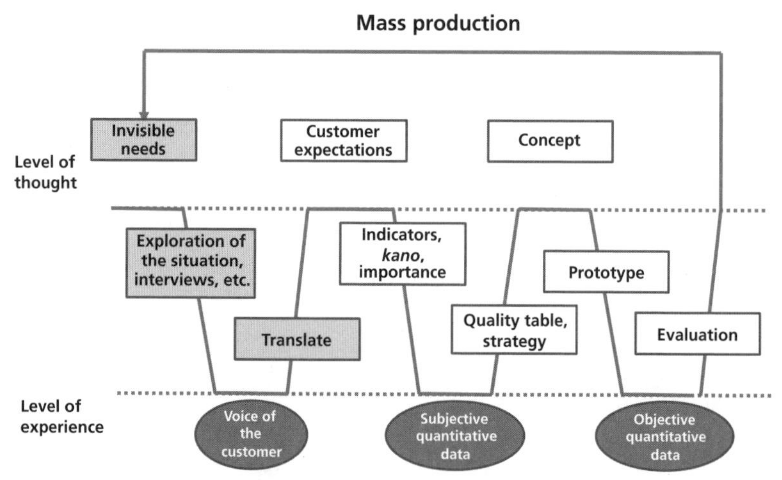

Figure 8.8 The WV model applied to determination of the concept of a product or service.

Shiba's algorithm comprises five steps [8.17] (Figure 8.9):

1. Capturing customers' voices.
2. Transcribing customers' voices into customer expectations.
3. Qualitatively analyzing customer expectations.
4. Defining a strategy for the product or service.
5. Defining a concept for the product and service.

The steps of implementation and distribution are not explicitly mentioned.

8.2.6.1 Capturing customers' voices

This step involves the organization of visits, observations and interviews with a limited number (between 12 and 20) of customers. Experience shows that beyond 20, the amount of new information obtained is not significant.

However, an appropriate sample of customers should be taken corresponding to:

- the different market segments;
- the manner in which they follow novelty trends: pioneers, followers (the main body), laggards (resistant to innovation); customers who are pioneers are particularly interesting since they are the ones who provide the largest contingent of remarks and ideas for future innovation;
- the overall perception they have of the product: satisfied, indifferent, dissatisfied;

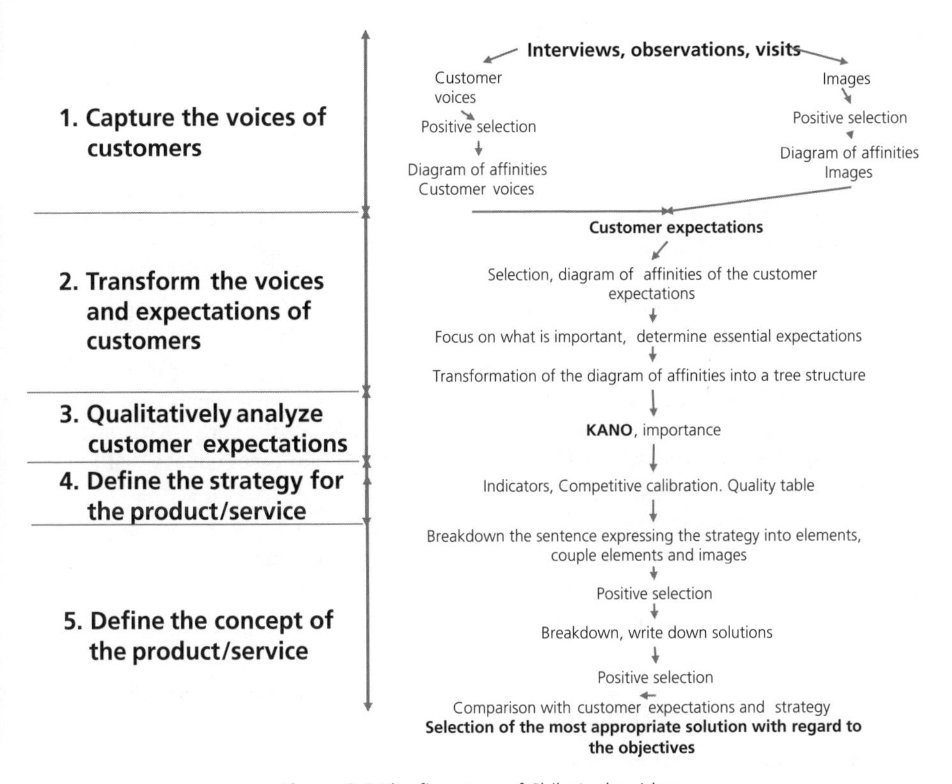

Figure 8.9 The five steps of Shiba's algorithm.

- in the case of consumer products, include the sociocultural environment if this factor is not reflected in the market segment.

After training the people in charge of the interviews, it is common to prepare a list of relevant questions to be put to the customer, and to determine the validity of this list in three pilot interviews.

It would, however, be wrong to assume that to capture customers' voice according to *Shiba's* approach merely involves a series of questions and answers. Indeed, the interviewer must be prepared to capture the wishes and concerns of customers and get them to clarify these. For this, the interviewer must have an empathic and neutral character, to encourage customers to freely express themselves, and get the message across that he is attentive. This sometimes includes restating what the customer communicates, especially if he feels that this is key information. The interviewer must also let himself go, use intuition, rely on luck and consider customers in their entirety, including place of work, body language, and hesitations. Certain customers express opinions that need to be refocused in order to obtain qualitative information and facts about the product/service, especially from the customer's experience – in short, get the customer to tell one or two stories.

After the meeting, the items of discussion are transcribed onto cards. Sometimes the client evokes images, or the interviewer may ask for them. In this case, these are drawn on index cards (there may for instance be a total of 60 to 100) after the interview. The duration of this documentation work is about the same as for the interview.

Initial work of sorting through the information involves selecting, from among the large number of sheets, the 20 that appear the most significant. One method used is the *positive selection* of the Japanese ethnologist *Jiro Kawakita*. This is a group effort for selecting, over several rounds without discussion, the cards considered the most important – according to data from the interview. Each person notes the relevance of each sheet/card, and those that have not been retained by at least one participant are excluded. This step is stopped when the number of selected records is about 30. In the final round, each participant has a limited number of choices, and cards that have not been selected by at least one participant are eliminated.

The last step consists in making a diagram of affinities: this is done by grouping the cards in columns with a header corresponding to a family or a common element. This last part requires a thorough discussion and confrontation of the opinions of team members.

Shiba's method ignores the fact that certain customers or stakeholders are weighed, either individually (the buyers for major retailers for instance) or as groups (NGOs and consumer associations) more heavily than isolated customers. Giving them equal weight in the subsequent analysis is inadvisable.

8.2.6.2 Transcribing customers' voices into customer expectations

Customers' voices reveal a content that is sometimes vague and elusive. This is why they must be transformed into more clearly formulated expectations. *Shiba* recommends the following procedure:

a) linking the customer's voice to the context of the service or product, notably using the images produced by the customer;

b) identifying one or two key points for each card;

c) translating these into customer expectations;

d) ensuring that drafted expectations correspond to the original voice and image.

Sometimes you have to iterate in order to draft a list that the team considers clear. In order to obtain a list of customer expectations that can be treated effectively in the following process steps, expectations should be formulated so as to:

- reflect a nuanced thought, because the reality is rarely "black" or "white";
- have a specific content that accurately reflects the situation because generalizations lead to a loss of valuable information;
- express needs and not solutions, because on one hand the latter is covered by other steps downstream in the overall process and, on the other hand, there can be multiple solutions to meet the need, and the described expectation is not necessarily the most suitable. However, sometimes the customer already wants a system that meets his expectations, for example a camera, a GPS, etc.

EXAMPLE

Suppose a public transport company surveys its users and finds that the customer's voice espresses: "I do not like taking the bus in bad weather because I invariably get wet when waiting for it". Here, the key point is "protection against bad weather" and the customer's expectation is "stay dry while waiting for the bus".

Once this analysis has been carried out, it is probably necessary to perform one or more round(s) of positive selection of the expectations of the clients interviewed to retain only the expectations considered essential.

8.2.6.3 Qualitatively analyze customer expectations

Three types of expectations

In this step, customer expectations are categorized. One can also achieve this by asking clients to rank their expectations at the end of the interview according to importance. This is not the method advocated by *Shiba*, who prefers that of *Prof. Noriaki Kano* of Japan.

Prof. Kano stated that, as part of the quality of a product or service, the satisfaction level should be proportional to the expected level of response. The better these expectations, proportionally, are satisfied, the higher is the level of satisfaction. However, *Kano* states that this is not by far the only factor to consider. According to him, there are also:

- expectations corresponding to mandatory functions, which if not satisfied lead to rejection of the product and service (e.g., the phone function of a cell phone is mandatory), regardless of the level of satisfaction achieved for other expectations;
- potential expectations, corresponding to appealing functions, not expressed by the client but waiting to be satisfied. For example, the SMS function of cell phones was integrated into the devices only as an ancillary service and manufacturers had no idea of the global success that this service would have. It has since become a mandatory feature.

By seizing the opportunity to satisfy new and emerging expectations, manufacturers can differentiate themselves from their competitors. Over time, many appealing features satisfying potential expectations progressively become models, first proportional, then mandatory (one wouldn't buy a car without airbags, whereas a GPS is considered, in 2009, as a function satisfying proportional expectations).

The Kano questionnaire

How then can one identify these families of expectations? Use a questionnaire developed by *Kano* where the idea is to ask two supplementary questions to a sample of customers with regard to each expectation:

a) How would you react if the function XX existed in the product?

b) How would you react if the function XX was absent from the product/service?

The customer is asked to give a response according to table below:

1. I find this feature very appealing.
2. Such a function is mandatory.
3. I do not care.
4. I would be satisfied.
5. I would be dissatisfied.

The answers will help determine if the proposed function is appealing (**A**), proportional (**P**: consumer satisfaction increases when performance of the function increases) or mandatory (**M**). This method can also help detect the functions for which the client feels no interest (indifference: **I**), or even hostility (opposition: **O**). Sometimes the juxtaposition of the results of two questions leads to a contradictory answer (equivocal: **E**), which requires careful analysis to rule out any inattention of the customer.

The expectations are then tabulated (horizontal lines) with the type of answer (A, P, M, I, O, E) and the score for each response in each case for all the customers. At the end of each line is put the type of response that was given most often (Figure 8.10).

A cell phone should have	P	M	A	I	O	E	Result
a videocamera	9	4	3	3		1	**P**
a camera	2	15	2	1			**M**
a flashlight	4		3	11	3		**I**
a cigarette lighter	2		2	4	12		**O**
a speed-camera detector	7		9	3	1		**A**
a GPS	8	2	7	2		1	**P/A**
an SMS encrypting system	1		2	5	2	10	**E**

Score for the 20 customers

The high proportion of equivocal responses suggests that the concept of encryption is not understood by the customer

Figure 8.10 Hypothetical *Kano* table for the functionalities of a cell phone.

The quality of the synthesis is highly dependent on:

a) the sample of customers;
b) communications made before and during the distribution of the questionnaire;
c) a pretest conducted on a few "guinea-pig" customers;
d) possible weighting of expectations by the customer at the end of the questionnaire.

Shiba's method is consistent with the marketing approach that involves uncovering customer expectations in a qualitative manner (not including assumptions), and then confirming these expectations by a questionnaire containing closed questions.

Once the expectations have been identified, a last step involves assigning one or more *quality indicators* to each expectation (in the second case, we get a tree). This step requires creativity, brainstorming and often experience. The indicator should be selected according to its relevance (Is it an adequate measure?) and its feasibility (Is this action easy to carry out? Is it reliable?).

8.2.6.4 Define a strategy for the product or service

In this step, we construct a quality table that compares the expectations and indicators, thus rendering it possible to verify:

- whether some indicators correspond to more than one expectation;
- whether each expectation has at least one relevant, reliable and easily measured indicator.

The next step is to focus on the expectations that have a strategic advantage in terms of a competitive position in the market, and implement them first. To determine these, one generally has to analyze and compare:

- results from the Kano questionnaire;
- changes in the market and its segments (are there any new ones, e.g., emerging markets?);
- the current performance of the competition (product analysis) and what may be accomplished (contacts with common suppliers, recent patent filings, presentations at fairs, or any appropriate economic information).

This research requires additional work. The result is that some expectations are prioritized and these are the ones that will primarily be integrated into the strategy of the product or service. The strategy is then made up of a few sentences comprising all the expectations that the customers want to see fulfilled and which the organization is prepared to achieve.

EXAMPLE
We want to provide students a university of technology that is prestigious, recognized internationally with cutting-edge research in the most promising and emerging areas, located on a lively and vibrant campus, and that has affordable semester fees that do not necessarily require obtaining a scholarship.

8.2.6.5 Define a concept for the product and service

The last step consists in dividing the strategy into product and/or service features. Some are derived directly from the quality table, once the strategic expectations have been selected.

It is nonetheless useful to define the features from the strategy statement, as some of them may not have been determined during the investigation or after its sorting. Ideas are generated by a carefully chosen group (the most critical success factor) that notes ideas and images on post-its. After a series of rounds of positive selection, and maybe putting together a final affinity diagram, it will be possible to determine the important features and to compare them with results from the Kano survey.

The product definition ends but its design, as well as its production, can continue. The ball is now completely in the court of R and D and production.

8.3 Taguchi's quality engineering

8.3.1 A talented and somewhat isolated researcher

Genichi Taguchi was born in 1924 in Japan. After university, he worked at the Institute of Mathematical Statistics for the Ministry of Education and then, in line with the concerns of the US occupation authorities with regard to the quality of Japanese communications, at the electrical communications laboratory of the Nippon Telegraph and Telephone corporation (1950 to 1962). Here, he developed and taught experimental design, thus building the foundation for what was to become Quality Engineering. In 1957, he published the first edition of a book that would be revamped and revised until 1976, *Design of Experiments* [8.18]. Just like *Shiba*, he received the individual Deming Prize in 1960.

In 1964, shortly after defending his thesis, he was appointed Professor at the College of Science and Engineering, University of *Aoyama Gakuin,* and he kept this position until his retirement in 1982. *Prof. Taguchi* has been a visiting professor at Princeton University and MIT, which indicates certain academic recognition. However, the prestigious American universities were only vaguely interested in quality management and statistical methods developed by Taguchi, preferring to focus on developing new emerging high-tech technologies, which were large potential sources of income. Is this indifference of the American academic community of the best technological universities a subtle form of arrogance with regard to the more mundane problems of the world of production? It is not forbidden to think so, as it is certain that this disinterest deprived the American industrial community of a competitiveness it would desperately need.

Taguchi developed his expertise in the 1960s without having regular contact with the rest of the community of statisticians in Western countries. Nevertheless, he was in contact with *Deming,* who spread his ideas, his approach and methods to US manufacturers. His methods have remained little known to the academic world, even if they have been applied in several US and Western industries: *Ford, ITT, Texas Instruments, Xerox,* etc. [8.19]. This isolation explains the fact that his methods were challenged at the beginning of their use by statisticians from scientific establishments, in particular by those who did not take into account the contributions of *Shewart* and *Deming*...

8.3.2 A new definition of quality – Loss of quality

Concerning quality, a director of *Honda once* said the following:

> *"If you reduce quality defects from 2 to 1% of each batch produced, you will be tempted to say that you have done a good job. However, this is not the opinion of the President and founder of Honda, who reminds us that the customer buys one product at a time: therefore, for the customer who purchases the defective vehicle, 100% of the product is unsatisfactory."*

In the same frame of mind, *Taguchi* stated another, more comprehensive definition of quality, and linked it to a loss, *one that the company that manufactured the product will impose on society once it is delivered*. This definition has the virtue of including the price paid for the product, its impact on the environment during use (maintenance, repair, energy, materials and labor required), the cost of breakdowns, but also, once the product is rejected by the consumer, it includes the cost of waste treatment and recycling (even if this addition was not made by Taguchi himself in his definition at the end of the 1950s). This definition of quality can therefore be included in the perspective of sustainable development (*Triple Bottom Line*, Chap. 15).

It also links quality to cost, thus putting an end to the harmful decoupling of financial aspects and the quality of manufacture of a product. But *Taguchi* developed his quality approach even further.

8.3.3 The quadratic function of the loss of quality

In the manufacturing world, specifications determine the features of a prototype for a new product; these specifications, which are themselves described by standards or reference values, are assigned to each component. The corollary of mass production is that the actual values of the specifications for the product component differ from the set-point standards. They should be within a limit corresponding to the accepted range or span of the tolerance (higher or lower than the set-point value). Keeping the manufacture of components within the tolerance range is one of the tasks of production, either through control processes, or by costly post-production inspections.

The span of the tolerance range is usually determined empirically, bearing in mind that a component with a set-point value outside the tolerance range will certainly lead to a malfunction and/or reliability problem in the finished product. Certain tolerance ranges are determined by adopting those of a previous component.

Taguchi rejects this "All or Nothing" model by introducing a quadratic function for the cost of quality. This indicates that any deviation of a real value x from the set-point value x_a of a specification gives rise to a cost (L: a loss for the client and/or the company), which grows as the square of the deviation (Figure 8.11):

$$L = b(x - x_a)^2, \text{ where } b \text{ is a constant}$$

And for a component that should meet several set-points:

$$L = \sum_{a=1}^{n} b_a(x - x_a)^2$$

Taguchi believes that any deviation introduces poor quality, because the fault does not appear ex nihilo, i.e., if and only if the value is beyond the tolerance range: there must be a form of continuity between full compliance and confirmed poor quality. Thus, for a component associated with mechanical movement, any deviation, while remaining within the norms of tolerance, can produce an annoying noise for the client and subsequently cause a decrease in the lifetime of the product.

If Δ is half of the total deviation of tolerance, then:

$$b = L_o/\Delta^2$$

where L_o is the cost of failure induced by a sub-standard value.

Adopting a quadratic rather than a linear function of cost is deduced from the fact that the square is the first symmetrical term of the Taylor series for any loss function in the industrial world, and thus is a reasonable first-order approximation, since the value of the tolerance range is lower than the set-point value.

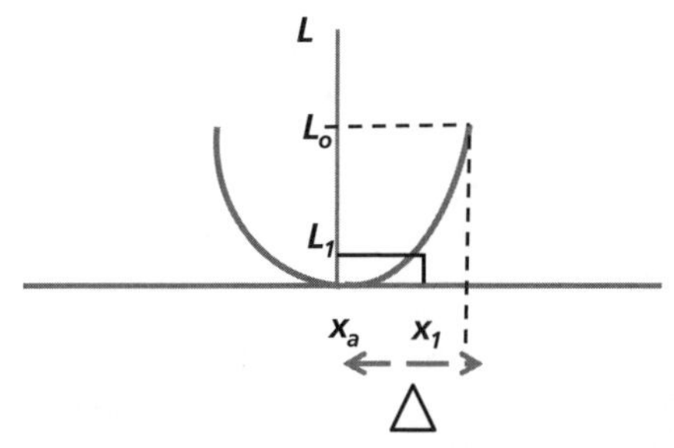

Figure 8.11 Quadratic function of quality according to *Taguchi*.

Determining the tolerance ranges of a component

Before *Taguchi*, there was no method for determining the optimal process tolerances. Using the quadratic function, one can determine the tolerance range of a set-point value in a rational manner. The idea is that society should not bear higher costs than that for fitting a product to its set-point values.

Let us suppose that a company produces yellow paint. To determine the loss function, we have to determine the functional tolerance Δ_0 (the deviation at which one of two consumers will complain or return the product) and the loss function. A_0 is here the recorded loss for the functional tolerance Δ_0. Let us suppose that the set point is 200 g of pigment per 50 kg of paint, that the consumer is not satisfied if the amount of pigment is not in the range 200 ± 10 g and that the cost is 10 USD for the value of A_0. The cost to society of a product with only 185 g of pigment is:

$$L = k(y - m)^2$$

$$A_0 = \$10$$

$$\Delta_0 = 10 \text{ g}$$

Because $(y - m) = \Delta_0$ and:

$$k = A_0 / \Delta_0^2 = \$\ 10/(10 \text{ g})^2 = \$\ 0.10/\text{g}^2$$

$$L = \$0.10\ (y - 200)^2 \text{ dollars / 50 kg of paint}$$

$$y = 185 \text{ g}$$

$$L = \$0.10\ (185 - 200)^2 = \$\ 22.5$$

To deliver a product with only 185 g of pigment costs the company 22.5 USD.

Let us thus calculate the tolerance deviation: it is determined by assuming that the cost of poor quality withstood by the company should be less than or equal to the cost of adjustment, by production, of the manufactured paint such that its pigment concentration again corresponds to the standard (rework). Suppose that the cost of rework is only 1 USD. Outside of what tolerance should the manufacturer then adjust the amount of pigment?

The tolerance deviation is determined by setting L = 1 USD and:

$$\$1 = \$0.10\ (y{-}200)^2 \text{ dollars / 50 kg of paint}$$

Consequently:

$$y = 200 \pm \sqrt{\frac{1.00}{0.10}} = 200 \pm 3.1$$

As long as the paint includes 200 ± 3.2 g of pigment, the manufacturer should not invest 1 USD to adjust its content, since the loss without rework would be less than 1 USD. The difference of ± 3.2 represents the tolerance at which the product can be delivered. If the manufacturer delivers paint comprising a pigment content of 185 g, it saves 1 USD but imposes a cost of 22.5 USD on the company, which is expresssed in product returns and consumer dissatisfaction, leading to a deterioration of the brand and market share.

8.3.4 Targeting the set-point value

But the quality loss function generates another consequence, especially if one realizes that the deviations from the desired values of the component specifications have consequences that are difficult to assess when they are cumulative (and when we want a default proportion of the order of a millionth): the production must thus seek to achieve the set-point value and not be content to remain within the tolerance range.

This assertion was tested by the company *Sony* that made color television screens in the late 1970s in Japan and US [8.20]. Consumers preferred the screens made in Japan, as they found them to be of higher quality. The production was however the same in the US and Japan, so how can we explain this preference? Curiously, the proportion of delivered screens outside the tolerance range was higher for Japanese screens than for their US counterpart, although this difference was minimal: 0.3%. The answer found by the quality department was the following:

- the screens manufactured in the US showed an almost uniform distribution of color within the tolerance range. The US manufacturing method was classic: remain within the tolerance range;
- the screens made in Japan showed a normal distribution, centered around the set-point value, for which the standard deviation was a sixth of the total tolerance range. This had the result that half of these TVs had a color density within ± this deviation. The majority of Japanese televisions therefore had a color density closer to the set-point value than the US televisions, explaining the consumer perception that the quality of Japanese screens was higher; the manufacturing process in Japan was above all focused on targeting the set-point value.

8.3.5 Quality by design

One of the innovations introduced by *Taguchi* comes from the observation that the best method to eliminate variations in the parameters of a product relative to a reference value is to do it during the development and associated production processes. According to *Taguchi*, a manufacturer does not sell a product, but first and foremost a *design*. The manufacturer has thus developed a methodology for engineering quality that can be used in both contexts. The originality of this approach is that the optimization of an industrial process aims to simultaneously reduce costs and increase quality. Quality and financial performance are therefore no longer on opposite sides, which makes the acceptance of quality control within an organization easier and connects it intimately with a routine function under the supervision of the head of finance: the cost of the product.

In fact, optimizing the quality and simultaneously reducing production costs is no small matter:

- relationships between the many parameters of the product and process and expected quality characteristics are only partially known, if at all;
- external factors influencing the process or product, the cost and quality of raw materials and components used in the assembly, and the tolerance ranges of production parameters are uncertain.

The process developed by *Taguchi* has three stages [8.21] (Figure 8.12), and covers both the *product and production process*. It is also called *Off-Line Quality Control*, as it is a quality control subsequent to production. The control of production parameters during manufacture is called *On-Line Quality Control*:

1. system *design*;
2. parameter *design*;
3. tolerance *design*.

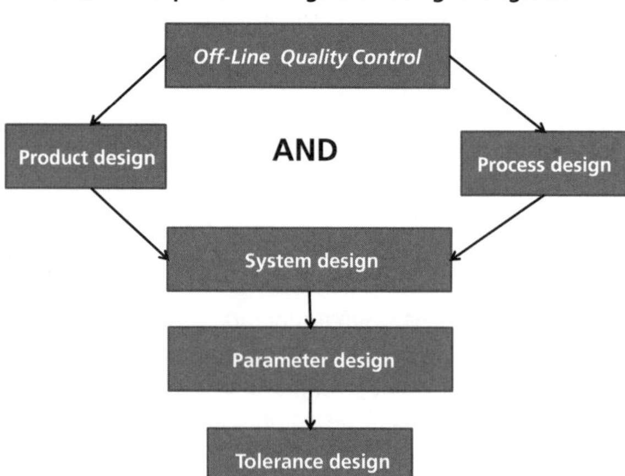

Figure 8.12 The three design stages according to Taguchi.

Let us briefly examine these three steps, implemented by a working group uniting all internal stakeholders (and sometimes external, such as certain key customers) of the organization:

1. System design

This first development step targets the design of the basic architecture of the process and/or product. In this phase, engineers use their knowledge and expertise to create the product that is to be delivered. Very often, an engineer creates the product design process and establishes precedents for it, while bringing added value in relation to the characteristics of the new product/process. It is at this stage that the scientific and technological knowledge in the field of engineering is required the most. It is now possible to carry out a first study on reducing manufacturing costs, but also to partly minimize the effects of external factors on the manufacturing process. During this stage, components and sub-systems of the product are selected with a minimal level of quality. This level will be adjusted at the third step: that of tolerance design. Two tools, the *Quality Function Deployment*[4] (QFD [8.22]) and *Pugh's*[5] matrix selection method, help in system design. US engineers spend approximately 70% of their time on system design – the Japanese spend 40%.

---

[4] ***Quality Function Deployment*** (QFD), developed in Japan by *Yoji Akao* in 1966, is a matrix approach to product (or service) design to satisfy customer expectations as best as possible. This method takes into account all the market needs and/or "desires" of future users in the design of a product (or service) and allows the planning of the best manufacturing process and development in accordance with the quality requirements set. This method is not a variant of the House of Quality, but the latter is a summary of its main tools (see Chap. 2).

[5] The *Pugh* matrix tool is used to facilitate process design and selection by a team. A project reference is initially selected as the current most powerful concept. Other concepts are then developed and evaluated (grade or qualitative assessment) according to their strengths and weaknesses relative to the reference design. A comparison of scores gives valuable information for finding the best alternative, which may be a hybrid of all the concepts.

2. Parameter design

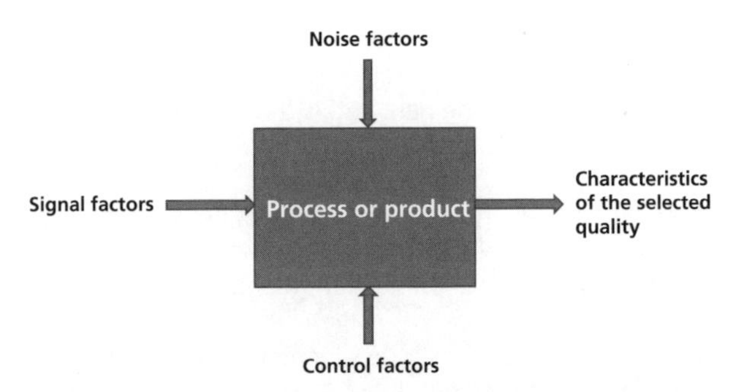

Figure 8.13 The four parameters of the process or product.

This step (Figure 8.13) consists in optimizing the system by performing a series of experiments (DOE: *Design of Experiments*), in order to be free, as far as possible, of environmental factors present during use, which are uncontrollable [8.23]:

- The first phase is to identify the quality characteristics that best reflect the performance of the system, especially those that are most relevant for client expectations. We would, for instance, expect a car to start regardless of the outside temperature in temperate regions. Taguchi also proposed to denote as signal factors (Figure 8.13) the parameters determined by the user to define the expected response value of the product (e.g., turning the ignition key when starting a car).
- The second phase is to identify the controllable factors and those that fluctuate statistically without the possibility of action by the manufacturer (in our previous example, these include external factors of the product, such as humidity and outdoor temperature, the altitude range at which the car will travel, the type of fuel, but also the unpredictable ways the client will start the vehicle, etc.). Taguchi calls these uncontrollable factors noise factors and these are the ones that lead to deviation from the nominal values of the defined quality characteristics. Taguchi also proposed calling two other factors of variation noise (Figure 8.14): internal noise (e.g., progressive deterioration of the power of the vehicle's battery) and product-to-product noise (e.g., manufacturing imperfections induced by statistical variation of key parameters of the battery). Some noise is uncontrollable when using the product, but can be determined during testing (external temperature and humidity, ambient air pressure, fuel type, etc.).
- The third phase is to identify factors that can be controlled to minimize the effects of all noise (in the case of a car, these controllable factors are the primary ignition voltage, adjusting the starter speed, the system parameters of the air/fuel mixture, etc.). These parameters are called control parameters or factors.
- The fourth phase is a testing stage, followed by high-performance statistical analyzes, for which Taguchi's contribution is significant:
  - The tables from the test are in the form of orthogonal matrices, which can limit the number of trials to a minimum; each column is assigned to a control parameter and each row corresponds to a state of the control parameter (e.g., minimum, average, maximum).
  - Outside the matrix are the noise parameters that can also be modulated.

- The test grid enables one to determine:
  - which control parameters have the greatest effect on the chosen quality characteristic;
  - which control parameters are very sensitive or insensitive to noise.
- Accordingly, the engineer is able to set the control parameters to a state that adjusts the characteristics of the selected quality level, while choosing those that are less sensitive to noise. The parameters that are insignificant can also be determined, enabling component costs to be controlled.

As for the tolerance design (see below), engineering science can be judiciously used to modulate the test matrices and parameter values. Far from being simply a statistical method, preparation and planning is an art, whose implementation is the result of a working group made up of all stakeholders concerned (production, R and D, quality, technical service, purchasing and sometimes sales).

Japanese engineers occupy 40% of their development time with design factors, whereas US engineers use only 20% for the same task.

3. Tolerance design

In this step, engineers determine by how much the performance levels of certain factors must be increased to be in line with the required quality level of the characteristic. They can discover, for example, that the battery is heavily used and that it is necessary to install a model delivering more power, which is thus more expensive. Tolerance design once again makes it possible to determine which factors to adjust (which ones increase production costs) and which ones are irrelevant (and where savings are possible). They finally determine the maximum tolerance deviations of the control parameters which will not affect the set-point value of the quality parameter to be optimized. Narrower tolerance ranges may be possible, but this may increase manufacturing costs.

The results of these three steps are then analyzed by using the quadratic quality loss function to determine the savings obtained by the optimization. Japanese engineers invest 28% of their development time in tolerance design – US engineers only 20%.

In conclusion, *Taguchi's* thinking is the key to Japanese industrial competitiveness. Although developed for the production of tangible products, the *Taguchi* approach can also stimulate the service economy. Another key point is revealed by the "philosophy" and corporate culture of the *Toyota* company.

References

[8.1] These four domains were only formalized and published at the end of his life, in W.EDWARDS DEMING, *The New Economics for industry, government education*, MIT publications, 1993, Chap. 11, *A system of profound knowledge*, pp. 94-118.

[8.2] W. EDWARDS DEMING, *Out of the Crisis*, MIT Press,1982, Chap.2, *Principles of transformation*, pp. 18-90, as well as: MARY WALTON and W. EDWARDS DEMING, *The Deming Management Method,* Berkeley Publishing Group, 1986, preface by Deming. pp. 34-39.

[8.3] W.EDWARDS DEMING, 1982, ibid., p.3.

[8.4] KAORU ISHIKAWA, *What is Total Quality Control - The Japanese Way*. Prentice Hall,1985, published in Japanese in 1981.

[8.5] Ibid., p.46.

[8.6] Ibid., p.13.

[8.7] Ibid., p.31. This point was contested in the text by the translator himself, proof that Ishikawa touched a sensitive topic... (but probably less than if he had written it today!).

[8.8] ARMAND V. FEIGENBAUM, *Total Quality Control - Engineering and Management*, McGraw-Hill Book Company, 1961, p.6.
[8.9] JOSEPH M. JURAN, *On Quality by Design - The New Steps for Planning Quality into Goods and Services*, Free Press, 1983.
[8.10] Ibid., p. 25.
[8.11] SHOJI SHIBA, *A New American TQM – Four Practical Revolutions in Management,* Productivity Press, 1993, Chap. 14, *Hoshin Management*, pp. 411-460.
[8.12] See the work of SHOJI SHIBA et al., *Le management par percée*, INSEP CONSULTING, 2007.
[8.13] SHOJI SHIBA , MARTINE MOREL, *Le management selon Shiba – capter les signaux du changement pour une performance durable,* Ed d'Organisation, 2007, Chap. 2, *Le processus du BT*, pp. 45 et seq.
[8.14] The method VTS (*Visual Thinking Strategy*) is described in the work of SHOJI SHIBA et al., *Breakthrough Management*, published by *Confederation of Indian Industry*, 2007, pp. 131 et seq.
[8.15] SHOJI SHIBA (with the cooperation of MFQ), *La Conception à l'Ecoute du Marché (CEM)*, INSEP Editions, 1995.
[8.16] Ibid., p. 105.
[8.17] Ibid., p. 101.
[8.18] GENICHI TAGUCHI, SUBIR CHOWDURY, YUIN WU, *Taguchi's Quality Engineering Handbook*, John Wiley & Sons Inc. 2004.
[8.19] See the work of LANCE.A EALE, *Quality by Design: Taguchi methods and US industry*, ASI Press, 1998, pp.167 et seq.
[8.20] See the book by MADHAV S. PADHKE, *Quality Engineering using Robust Design*, PTR Prentice Hall, 1989, pp. 15-16.
[8.21] Ibid., pp 33-40.
[8.22] YOJI AKAO, *QFD: Quality Function Deployment - Integrating Customer Requirements into Product Design*, Productivity Press, 2004.
[8.23] See especially WILLIAM Y. FOWLKE and CLYDE M. CREVELING, *Engineering Methods for Robust Product Design – Using Taguchi Methods in Technology and Product Development,* Addison Wesley, 1995.

Chapter 9

Total Quality Management: Awards

Key concepts: Pathfinder map, criteria for performance excellence, the EFQM excellence cycle, EFQM degree: Commitment to excellence, EFQM degree: Recognition of excellence, European Framework for Quality Management (EFQM), RADAR logic, Japanese medal for quality, management model prizes, weighting of criteria, principles of performance excellence, Deming Prize, EFQM Award of Excellence, Malcolm Baldridge Award, frame of reference for awards, sub-division of excellence criteria

9.1 The Deming[1] Prizes

9.1.1 The various types of prizes[2]

In 1950, *Deming* taught an eight-day course in Japan on statistical quality control (SPC). A written version of the course was compiled and sold to the participants. Deming decided to turn over his royalties to the Japanese Union of Scientists and Engineers (JUSE). In his turn, the Director of JUSE decided in 1951 to establish a Deming Prize, in honor of the contribution and friendship of Deming toward Japan and also to boost the growth of total quality control (TQC) in Japan.

In a second step, the copyright of the book *Some Theory of Sampling* [9.1] covered part of the costs of the prize. Since then, the Deming Prize has grown and all operating costs are covered by JUSE.

The Deming Prize, which can be given to non-profit-oriented organizations (public economy, NGOs, foundations, etc.), is divided into several categories:

a) *The Deming Application Prize* that rewards companies or their divisions (managed independently) which have obtained a significant increase in their annual performance by implementing TQC. This award has been open to non-Japanese companies since 1984. *Toyota* received the prize in 1965. Between 1964 and 1990, one-third of companies of

---

1 Information taken from the JUSE website: www.juse.org (accessed June 2009).

2 According to the pamphlet, *The Deming Prize Guide 2009 for Overseas*, downloaded from the JUSE website.

significant size which received the Deming Prize were controlled by *Toyota*. Since 2001, the companies which have received the Deming Prize are Asian, most often Japanese. In 2007 and 2008, three Indian companies received the Deming Prize, including *Tata*. Many large and prestigious Japanese companies have not received the Deming Prize (*Honda, Kawasaki, Mazda, NEC, Panasonic, Sony, Toshiba, Yamaha,* etc.)

b) *The Deming Prize for Individuals* (or small groups), rewards experts
   - who have made an outstanding contribution in the field of total quality management (TQM) studies, or statistics used for TQM, and those
   - who have contributed to the spread of TQM in an unusual way.

   This prize is open only to Japanese citizens.

c) *Quality Control Award for Operations* rewards operational units of companies that have demonstrated outstanding performance improvements over a particular year, thanks to the implementation of quality management/control in a total quality effort. Since 1997 this award is open to non-Japanese companies.

Finally, JUSE awards two other distinctions:

- The Japan Quality Medal (since 2012 called the Deming Grand Prize), established in 1969, which is aimed at organizations that have already received the Deming Prize. Three years after this distinction, they must show that they have again significantly increased their implementation of TQM. For companies that practice TQM over a long period, the Japanese medal is the ultimate prize. Toyota has been awarded this medal twice: in 1970 and 1980.
- In 1954, Nippon Keizai Shimbun (a Japanese economics journal) created the Nikkei QC Literature Prize. This prize recognizes outstanding publications in TQM studies or statistical methods used in TQM, which have been recognized as contributing to progress and the development of the management or control of quality.

9.1.2 The procedure for the Deming Application Prize

This prize is awarded if the *ad hoc* panel of JUSE is convinced that the company applying for the prize has established a most efficient TQM system, optimized for the type of commercial activity, both present and future, in which the company is involved. It is important that the company determines what challenges it faces, and that it is able to address them with appropriate processes and TQM.

JUSE defines TQM (Japan has since adopted the terminology of TQM instead of TQC which was the first designation) as follows (the footnotes give the clarifications provided by JUSE):

TQM is a set of systematic activities[3] *implemented by the entire organization*[4] *to effec-*

---

3 Organized activities that aim to achieve the corporate objectives, which are carried out by a management oriented towards leadership, and guided by clearly formulated medium – and long-term visions and strategies, as well as adequate quality strategies and policies.

4 Taking into account all employees in all sectors and at all levels of the company, so as to achieve the objectives quickly and efficiently with minimal resources. This requires an appropriate management system that includes a quality assurance system at its center, and its integration into other transverse systems such as cost, safety, environment, product delivery. Respect for employees motivates the company to develop human resources to support its key technologies, its vitality and agility. The company maintains and improves its processes and operations by using statistical methods and other

tively and efficiently satisfy the goals and objectives of the company[5]*, and to deliver*[6] *products and services with a level of quality*[7] *that satisfies the customers*[8]*, at an appropriate time and price.*

Companies determined to obtain the Deming Prize can see their hopes fulfilled if:

1. They have planned strategies and business objectives that are ambitious, challenging, even provocative, in close relation to the explicit and implicit client expectations, based on an explicit management philosophy of the company. These must be perfectly adapted to the business environment, and formalized with leadership by top management, and must fully take into account the type of business activity, its structure, environment, requirements and scale.
2. TQM has been properly implemented to achieve the objectives and deploy the strategies mentioned in 1.
3. Outstanding performance results have been obtained by the implementation of 2 and by the achievement of the objectives in 1.

As for the other prizes, the Deming Application Prize analyzes the performance excellence according to a system of points assigned to the various key functions of the company (Figure 9.1).

In order to apply for the prize, the organization must turn in an *ad hoc* document on February 20 of each year. The document, called Description of TQM Practices (DTQMP) describes the content and implementation of TQM since the time of its introduction until the application.

The application prize subcommittee examines the DTQMP, and evaluates the level of implementation and quality of the presented TQM. From this analysis, the committee determines the candidates for the prize. If the application is successful, an assessment is made by the subcommittee at the company (visit), with the DTMQP as a documentary base. The examination is conducted in Japanese (!), but documents can be presented in English.

The decision and analysis results of the subcommittee are forwarded to the award committee who determines the winners. The prize is awarded in mid-October.

Three years later, the winning company is requested to submit a brief report on the status of its TQM. A one-day visit is scheduled. This procedure enables the company to apply for the Japanese medal for quality.

---

tools. Based on facts and data, the company manages its business by a sound use of the Deming cycle. It resizes/redefines its management system by implementing appropriate scientific methods and by using information and communication systems.

5 Including the assurance of adequate long-term profits through continual and systematic customer satisfaction, but also that of all stakeholders that comprise the company as a whole: employees, suppliers and shareholders.

6 Not limited to product distribution to customers and consumers, but also includes surveys, research, planning, development, design, product preparation, purchasing, production, installation, inspection, collection of orders, sales and marketing, maintenance, customer service, product disposal and recycling after use.

7 Including the usefulness (by taking into account the functional and psychological dimensions), security and reliability, but also the expectations of stakeholders, environmental perspectives, as well as sustainable development.

8 Including stakeholders.

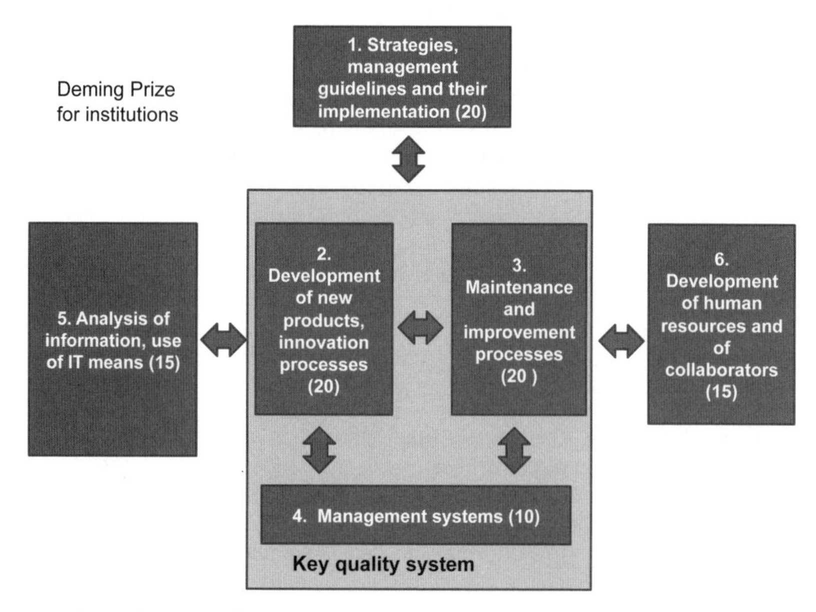

Figure 9.1 Relevant functions of the company and their weighting in the Deming Application Prize.

9.1.3 The expected benefits

According to information from the website of JUSE[9], obtaining the Deming Application Prize favorably influences the performance of the company (these statements are accompanied by mostly very positive testimonials from presidents and CEOs of companies that have received the prize), especially:

- stabilization and improvement of quality;
- reduced costs, increased productivity;
- increased sales;
- increased profits;
- improved efficiency of planning and business management;
- ability to fully realize the ambitions of the Board of Directors;
- strengthening the organization at the expense of setting up disciplinary silos within the company;
- improved motivation in management and acceleration of standardization;
- improved power of the organization, both with regard to effectiveness and efficiency, especially through the implementation of TQM, in product development and in the allocation of tasks to each employee.

9.1.4 Does the Deming Prize really increase performance?

But in 1999, a dissenting voice spoke [9.2]: after analyzing the subsequent performance of companies awarded the Deming Prize between 1964 and 1989, it turned out that companies not belonging to the *Toyota* group did not improve their performance after receiving the

9 See: http://www.juse.or.jp/e/ (accessed in Aug 2009).

Deming Prize (rather, the opposite was observed).

The author concluded that these companies did not have prior experience of TQC and that they were not able to take advantage of the effects of the prize:

- either because preparing for it required an almost obsessive mobilization taking up too much of the capacity of the company's top management, and this over long periods of time, so that management lost sight of other objectives of the company;
- or because the implementation of TQM was accompanied by the establishment of a bureaucracy generating heavy and significant documentation, rendering the company less flexible, and demotivating staff by making them lose creativity and flexibility.

The author recommended that managements stop trying to obtain the Deming Prize if they did not have a prior culture of TQM. Such reflections were also issued later about the Baldridge Award.

The following chapters of this book devoted to Toyotism and Six Sigma can shed some light on why *Toyota* and *Motorola* improved their performances with the Deming and Baldridge Prizes, while others saw their performances decline.

9.2 The Malcolm Baldridge Award

9.2.1 The aim of the award is performance, not distinction

Established in 1987, the *Malcolm Baldridge*[10] *National Quality Award* was established by the administration of President Reagan in order to restore the US economy with the support of American companies. The first company to win this distinction was *Motorola* in 1988: the tool that allowed the company to achieve excellence was later given the name Six Sigma. According to the rules of the award, the company communicated its good practices.

But the essential purpose of the Baldridge Award, reserved exclusively for US companies and organizations, is not the distinction of one or more companies. The objectives, as those of the Deming Prize, are to stimulate the emulation and the diffusion of good practices in quality and performance, leading to growth of the US economy and its businesses according to the model of the Deming cycle. For this reason, companies that have received this distinction are few compared to those using the model: there are many companies who conduct internal or external assessments using the tool without trying to win the prize.

The award is aimed at six distinct categories: education, health care, manufacturing, non-profit/government, services, and small business. The award has undergone considerable changes in its criteria, but also in its vocabulary, since the term "quality" was replaced by "performance" in 1995. This change was not anecdotal, but demonstrated a paradigm shift: the award was not given exclusively to those incorporating some of the most essential quality criteria, such as quality standards, quality planning, staff training, team work for quality, or obtaining a product with the goal of zero defects and customer satisfaction. Rather, the challenge was now to balance all aspects of organizational performance: profitability, growth, security, market shares, employee morale and commitment, innovation, etc. The reason for this change was that companies can have excellent product quality and satisfied customers, but at the same time, be underperforming in other activities of the organization, leading to serious financial difficulties.

The approach is complex, the criteria are numerous and not always easy to access. For these reasons applying to obtain the award cannot be the goal of solely a quality department

[10] The name of the US Trade Minister who had died the year of the creation of the award.

and does not always motivate small and medium-sized companies. The implementation of the approach, which may lead to organizational changes that are difficult to put into action, requires strong commitment from management as well as effective and comprehensive corporate communications.

ISO 9001 primarily seeks compliance of a quality management system to requirements comprising accumulating experience and good practices, whereas the Baldridge Award rather demands an increased overall performance by the applicant organization.

9.2.2 The frame of reference of the Malcolm Baldridge Award

The frame of reference of the Baldridge[11] Award (Figure 9.2) is a management system of performance based on the concepts of the pioneers of total quality. It relies on 12 core values and concepts, drivers of systematic processes leading to the following performance measures:

- leadership with vision;
- focus on the future;
- system perspective;
- process management;
- knowledge management;
- innovative management;

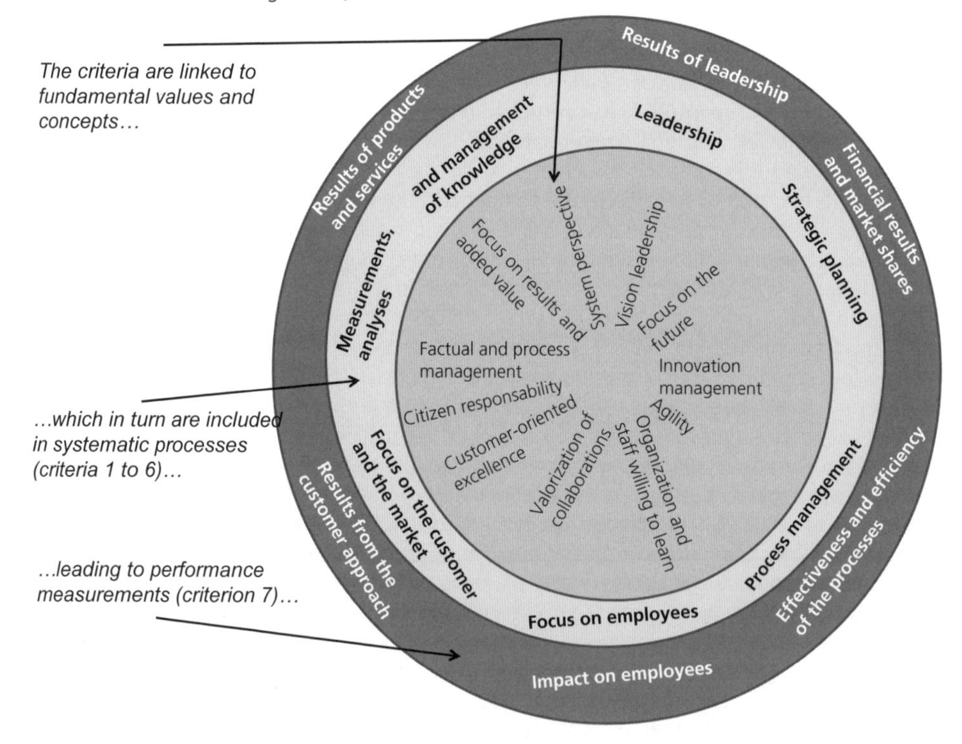

Figure 9.2 Baldridge Award: role of the 12 concepts and their impact on performance.

[11] The Baldridge Award includes six distinctive frames of reference, albeit very similar (url: http://www.quality.nist.gov/baldrige).

- organization and workforce willing to learn;
- recognizing the value of collaborations and partnerships;
- civic responsibility;
- customer-oriented excellence;
- focus on results and created value.

A majority of these principles have already been cited by *Deming, Juran* and *Ishikawa* (Chap. 8).

A model of the total quality management or performance of a company is represented as a system of seven criteria of excellence in interaction with given data, setting the framework for the company's activities, and the profile of the organization, environment, relationships and challenges (Figure 9.3). The choice of criteria, formulating open questions associated with them, and the expectations of experts, who judge the value of applications, represent the best state of the art in strategy and management.

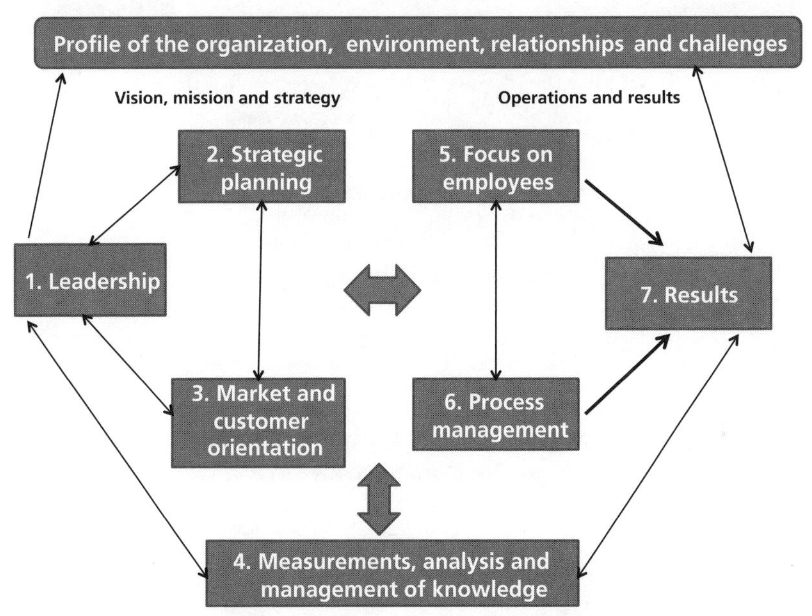

Figure 9.3 Framework of performance excellence criteria for the Baldridge Award.

The frame of reference for the award includes not only interactions, but dynamics ranging from leadership to results, with a control loop provided by the transverse criterion 4: Measurement, analysis and knowledge management, the memory of the organization (Figure 9.4).

The performance rating of the organization according to the frame of reference for the Baldridge Award is based on the degree of excellence expressed in points, broken down for each criterion, to reach a maximum of 1000 points (Figure 9.5). However, it is an accomplishment for an organization to exceed 750 points.

The weighting of the various criteria have varied. From 1997 to 2008, the vision, mission and strategy have become more important (the best course of action in case of bad weather cannot be deduced from a boat's performance, but really comes from the captain).

The weighting of one factor, that for the staff, may appear surprising. With a total of 85 points out of 1000, it seems low for a future company concerned with information and the knowledge economy, where personnel costs represent the majority. Perhaps we should interpret this as the legacy of *Deming*, who said that focus should be placed on system performance rather than on that of the employees.

Figure 9.4 Dynamics of the excellence criteria of the Baldridge Award.

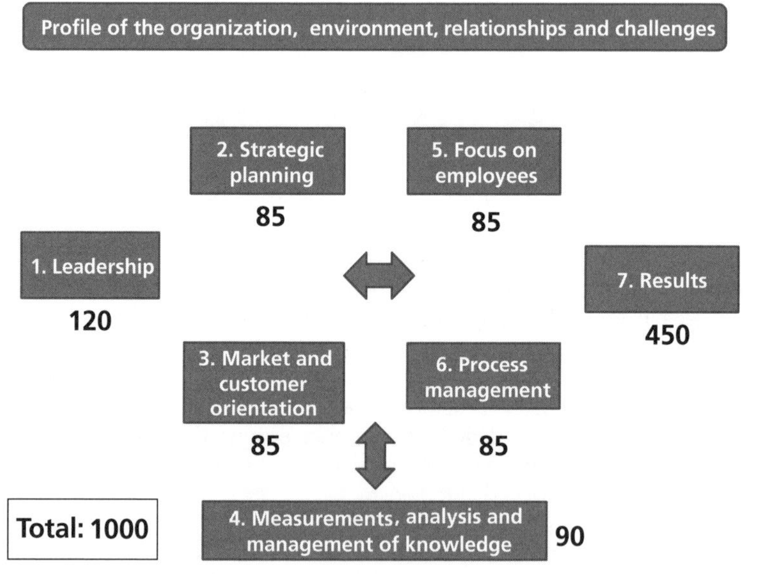

Figure 9.5 Weighting of the performance excellence criteria for the Baldridge Award.

9.2.3 Subdivision of criteria

To allow for a fine grading, each criterion includes subdivisions, which is appropriate considering that criterion 7, Results, represents 450 points all by itself.

1. Leadership (120 pts.):
   1.1 Organizational leadership (80 pts.)
   1.2 Governance and social responsibility (40 pts.)
2. Strategic planning (85 pts.):
   2.1 Strategy development (45 pts.)
   2.2 Transforming strategy into action plans, research data, preparation, content, formulation and justification of strategy (40 pts.).
3. Market and customer orientation (85 pts.):
   3.1 Knowledge of the market (and expectations) of the clients (40 pts.)
   3.2 Customer satisfaction and customer relations (45 pts.).
4. Measurement, analysis and knowledge management (90 pts.):
   4.1 Measurement, analysis and improvement of organizational performance (45 pts.)
   4.2 Management of information, information technology and knowledge (45 pts.).
5. Focus on employees (85 pts.):
   5.1. Employee commitment (45 pts.)
   5.2. Staff environment (40 pts.).
6. Process management (85 pts.):
   6.1 Design of the work system (35 pts.)
   6.2 Management and improvement of work processes (50 pts.).
7. Results (450 pts):
   7.1 Results of products and services (100 pts.)
   7.2 Results of customer orientation (70 pts.)
   7.3 Financial results, market outcomes (70pts.);
   7.4 Results of staff orientation (70 pts.)
   7.5 Results of efficiency and process efficiency (70 pts.)
   7.6 Outcome of leadership (70 pts.).

9.2.4 Grading the subdivisions and sections of the prize; expectations

Each subdivision is divided into sections that contain open-ended questions.

For example, for the subdivision 3.2, *Customer satisfaction and customer relations*, the full title is:

Describe how your organization uses its relationships with customers to attract, satisfy and retain them by increasing their loyalty (3.2.a). Describe how your organization determines customer satisfaction (3.2.b).

The frame of reference of the Baldridge Award includes:

- The main question (3.2): How does your organization implement customer relations and does this increase their satisfaction and loyalty?
- The section on implementation of customer relations (3.2.a), which is subdivided into open-ended questions (four in this example):
  - 3.2.a.1: How do you build relationships to acquire customers, meet their expectations and even go beyond these, to increase loyalty and retain it, and to get the customers to give positive recommendations?

  - 3.2.a.2: What is your primary access mechanism allowing your customers to seek information, to close a deal and file complaints? What are the main access mechanisms? How do you determine the conditions that generate customer contact for each mode of access in place? How do you ensure that these contact conditions are deployed to all personnel and processes involved in the feedback path to the client?
  - 3.2.a.3: How do you handle customer complaints? How do you ensure that complaints are handled promptly and efficiently? How do you minimize customer dissatisfaction and, if necessary, the loss of their loyalty? How are claims treated, aggregated and analyzed to improve the performance of your organization by including potential partners?
  - 3.2.a.4: Considering all of the above approaches 3.2.a.1 to 3.2.a.3, how do you assess and improve them, especially in relation to the needs and new directions of your company?
- Section 3.2.b is also divided into several open questions.

The company that fills out this detailed questionnaire, which is explicit and involves sometimes tedious teamwork, knows that behind every question lurk expectations that are found in *ad hoc* guides published each year [9.3]. For example, here is the (very) long list of expectations (17!) in connection with the series of open questions 3.2.a.3:

- the system for collecting feedback and complaints has good traceability, and is easy to understand;
- the approach for collecting and documenting data on customer complaints and comments is objective;
- all employees who have direct or phone contact with clients have a simple but specific documentation system for their secondary comments, or complaints about the products or services of the organization;
- data on customer complaints and comments from various sources are collected for an overall assessment and comparison;
- data from client assessments and observations are forwarded to relevant staff within an appropriate period of time;
- customers are confident that the comments and claims they make to any employee will be documented and included in reports;
- a formal and logical process exists to solve customer complaints;
- procedures for the mobilization of managers exist if customers feel that their claims are not resolved by the staff;
- improvements have been made to decrease the processing time of complaints in past years;
- objectivity and reliability of data on the level of customer satisfaction in relation to claims processing are established;
- diligence and in-depth analysis of claims processing is proved;
- a systematic and organized process is implemented for the handling of claims;
- the employees responsible for processing claims have suitable knowledge and skills;
- information on the causes of customer complaints are forwarded to relevant employees who carry out the appropriate corrective actions;
- it can be demonstrated that claims data are used to start improvement projects that prevent future claims and loss of customers;
- it can be shown that analyses of causes of complaints are used to make changes in processes, products and services;

- claims data are transmitted to partners of, and suppliers to, the organization if this is appropriate.

The sum of expectations and questions included in each section makes it possible for experts to obtain a finely detailed grading by noting the score of points for each criterion.

Faced with such a detailed approach, it makes sense to ask whether the sum of the expectations underlying the criteria, sub-criteria and sections of the Baldridge Award does not *de facto* correspond to a standard of quality? One can almost say yes, because the sum of expectations looks like a list of good practices, but it does not include the formalization of an ISO standard for example.

9.2.5 What to do to receive the Baldridge Award?

In a first step, the applicant organization prepares a self-assessment document of 50 pages that addresses the 32 sub-criteria in question, covering the seven criteria, which include 203 questions. This team effort should be carried out by those in the company who are involved and should be managed as a project. The person writing is thus the project manager. This activity must be reviewed by management and includes the following internal phases [9.4][4]:

- project planning;
- gathering data and information;
- preparing the initial draft of the document;
- reviewing the initial version before release;
- revisions and corrections;
- final formatting, layout preparation, finishing touches and final verifications of tables and graphs;
- sending the document to the experts of the Board of Examiners for the award.

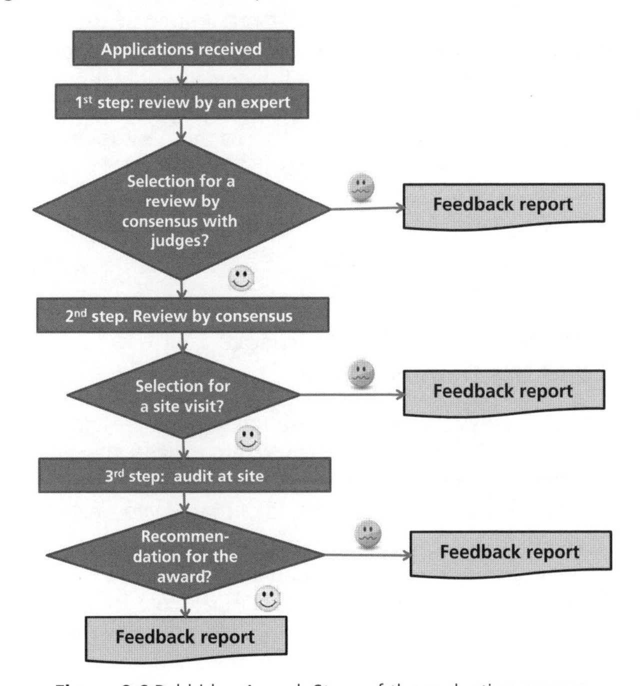

Figure 9.6 Baldridge Award: Steps of the evaluation process.

Once this is done, the process involves four steps (Figure 9.6).

- All applications are reviewed by at least four members of the Board of Baldridge Award Examiners (which includes 200 experts, certified and bound by a strict code of ethics), who are members whose profession is related to that of the applicant organization.
- The application files which have been subject to a first grading with a sum of points are looked over again by a senior expert who reviews the rating, and examines the differences in scores and potential conflicts. This done, the senior expert calls the four experts for a conciliation meeting where a final decision is made by consensus. It is then that the members of this session issue a recommendation on whether to conduct a site visit, and this recommendation is analyzed by an independent panel of judges, who are highly qualified experts of international repute, with extensive experience when it comes to the award. In general, the majority of candidates receive a score less than or equal to 500 points, and a visit is usually organized if the application receives a score of 600 points or more.
- A visit of 3 to 5 days is organized on the site, which is similar to an audit and includes inspections, interviews and presentations, especially by the CEO(s) of the company. The objective is to verify and clarify the information contained in the self-assessment document, but also to elucidate any uncertainties that may have arisen from the review of this document. The audit results are recorded in an audit report which is forwarded to the Board of the award;
- The judges on this Board determine whether the applicants are entitled to a prize and determine the appropriate category. They make a recommendation to the National Institute of Standards and Technology, which transmits its notice to the American Secretary of Commerce.

All applicant organizations receive a detailed report, written by the experts, which summarizes the strengths and weaknesses identified during analysis of the self-assessment document. This is probably the most useful outcome of the process. It is also particularly advantageous as the external cost to the applicant organization is 3000 to 6000 USD, whereas an assessment made by a consultant costs at least 10 times that. This is why organizations that know that their performance is well below 600 points still submit their application.

The award itself includes a trophy, national recognition based on the excellence of the organization receiving the prize, and an opportunity to disseminate its good practices during the annual conference *The Quest for Excellence*.

9.2.6 Differences between the Baldridge Award and the Deming Prize

Both models highlight the satisfaction rate of customers, employees and stakeholders. Both also insist on seven key areas for quality: leadership, strategy and planning, customers, suppliers, employees, processes and results. However, their perception about what these areas mean and cover can vary. In addition, the weighting of these factors also differs.

Thus, in the years 2000-2010, compared to the Baldridge Award, the Deming Prize accorded more importance to process control, and less weight was given to knowledge of the market and clients. Another difference is that the final score for the Baldridge Award is the arithmetic sum of the number of points for each criterion, while for the Deming Prize, three categories of criteria must give a relative value of 70% of the maximum value for the application to be taken into account.

9.2.7 Is the Baldridge Award in decline?

Since 2004, the nominations and winners of the Baldridge Award have primarily been companies from the public sector rather than the private economy. A certain weariness has overtaken some large US companies which claim to have invested substantial efforts without the benefit of any improvements.

Examination of several cases, however, shows that numerous initiatives have only been performed by isolated divisions of a company, often the quality function, usually without any real support and interest from the Board of Directors. In some cases, little or none of the recommendations suggested by the experts have been implemented by the management of the applicant organizations.

A study of the market value of companies having received the award, compared to those of companies not having applied for it, has generated much controversy[12]. This has been enough to discourage the top management of many organizations, although such a comparison mainly concerns the winners before the paradigm shift of 1995 (performance quality). A lack of communication by the decision-making body of the Baldridge Award has also been discussed.

But there may be another reason for the decline of the award: the way the Japanese have taken control of quality – especially by Toyotism – is associated with drastic reductions in costs. Such results are hailed by company stakeholders with opposite interests: marketing and sales (representing clients) on the one hand, and the board of directors (representing the shareholders) on the other.

The copying of Toyotism in the US (dare we mention it) initially led to the Six Sigma approach, and then, in the middle of the 1990s, to the Lean Six Sigma system, which is preferred by American companies.

9.3 EU, the EFQM model and the EFQM 2013 Excellence Award

9.3.1 A foundation created by European companies

EFQM stands for the *European Foundation for Quality Management*. EFQM is a Brussels-based organization, founded in 1988 by 14 European companies (*Bosch, BT, Bull, Ciba-Geigy, Dassault, Electrolux, Fiat, KLM, Nestlé, Olivetti, Philips, Renault, Sulzer, Volkswagen*), with support from the European Commission. The EFQM, a non-profit organization, has over 450 members from all over the world.[13]

The EFQM points out that, regardless of their size, sector, structure or maturity, organizations must establish an appropriate management framework to succeed. The proposed model, EFQM 2013, is a frame of reference that is easy to grasp, holistic in its definition, and compatible with the other tools, methods and approaches to management. Since it is not prescriptive, it enables organizations to:

- assess where they are on the path towards excellence by helping them identify their strengths and areas for improvement in relation to the announced vision and mission;
- provide a common language and common read mode, which should facilitate communication;

12 See these results at: http://www.quality.nist.gov

13 See: http://www.efqm.org/en/PdfResources/MemberList.pdf

- integrate existing and/or planned initiatives by eliminating any duplicates, but also by identifying any insufficiencies;
- provide a generic framework for the management system of the organization.

The EFQM 2013 Excellence Model assumes that an organization of excellence respects:

- the European Convention on Human Rights (1953);
- the European Social Charter (revised in 1996);
- the United Nations Global Compact (2010, see the chapter on social responsibility).

The EFQM 2013 approach involves:

- a definition of excellence that is based on eight fundamental concepts;
- the EFQM[14] 2013 Excellence Model for which the foundation stands as guarantor – a guideline allowing companies to qualify and reach their level of excellence;
- methodological and generic EFQM frameworks and assessment tools, such as the *RADAR* logic.
- the *EFQM Knowledge Base*[15], which is both a network, a subscription to *News* and access to a database of best practices and management tools – reserved for members.
- certificates reflecting a gradual shift towards excellence and an annual award for the best organizations: the EFQM Excellence Award.

9.3.2 A definition of excellence and a call to 8 concepts

The EFQM 2013 model is based on the following definition of excellence:

> *"Excellent organizations achieve and sustain superior levels of performance that meet or exceed the expectations of all their stakeholders."*

Truly excellent organizations meet their stakeholders' expectations:

- by the results they achieve;
- by the way they achieve them;
- by what they are likely to achieve thereafter;
- by reassuring their stakeholders of the sustainability of these results.

Achieving (and maintaining) excellence means to fully adhere to a list of basic concepts and transmit them with all the necessary leadership skills. These concepts are the principles based on which the organization will emulate its behavior, activities and initiatives. When the organization puts the principles into practice, the path to a sustainable status of excellence becomes accessible. These basic concepts are 8 in number (and have similarities with those of the ISO 9000 family):

- Principle 1: Create value for customers by anticipating and understanding needs, expectations and opportunities.
- Principle 2: Contribute to a sustainable future, in short, have a positive impact on the world that surrounds the organization thanks to improvements of its performance by fostering the economic, environmental and social communities it interacts with.
- Principle 3: Stimulate creativity and innovation: The organization creates value and increased performance levels through continuous improvement and innovation by

---

[14] The EFQM model has been applied to higher education by the Centre of Excellence of the *Sheffield Hallam University* (order at: http://www.shu.ac.uk/research/integralexcellence/outputs.html)

[15] http://www.efqm.org/en/PdfResources/EFQM%20Knowledge%20Base%20one%20pager.pdf

developing a generalized creativity in their stakeholders[16].
- Principle 4: Manage in a visionary and inspiring manner, with integrity; noticeably, the leaders shape the future and make it a reality, by acting in an exemplary manner for their values and ethics.
- Principle 5: Develop the capacity of the organization, by optimizing its organizational capacities both inside and outside the organization.
- Principle 6: Manage with agility, i.e., have the capacity to identify and respond effectively to opportunities and threats.
- Principle 7: Succeed thanks to the talent of the staff; by valuing the staff and creating a culture of independence, the organization enables the staff to achieve both their own goals and those of the organization.
- Principle 8: Support remarkable results by sustainably achieving exceptional results that meet needs both in the short and long term for all stakeholders in their operational environment.

These concepts are not immutable; their list and their numbers are not fixed. They simply summarize the current state of the art. Changes will appear progressively as the performance is improved and with the degree of innovation of companies.

Before attempting to adopt the EFQM Excellence Model, a management team must first ensure that its members are comfortable with these concepts. If they are not fully understood and/or accepted, the chances of success when implementing the model would be very small.

9.3.3 The EFQM Excellence Model: a system of 9 criteria

The organization wishing to adopt the EFQM approach must, in addition to the basic concepts, also study the EFQM 2013 Excellence Model.
This model is an operational instrument and has several functions.

- It is a self-assessment tool.
- It facilitates benchmarking with other organizations.
- It identifies areas/processes to optimize.
- It is a basis for vocabulary and a common way of thinking.
- It is a generic structure of the organization of the management system.
- It is a system approach, a comprehensive understanding of cause and effect in the organization.
- It is a basis of assessment of the levels for the *Excellence*, *ESPRIX* and *EFQM Excellence Award* (EEA).

The model assumes that high performance that respect both the customer and society are achieved through leadership able to effectively guide the strategy and policy of the organization. Here, the latter is implemented by the employees, by involving partnerships, a process approach and other resources (infrastructure, equipment, etc.).

The model recognizes that excellence can be achieved by several pathways, and for this reason it does not consist of firm requirements. The excellence model is based on nine criteria:

- five are **factors** that represent what the company does;
- four are based on **results**.

[16] Note that innovation and continuous improvement are grouped in this key concept, so that ISO 9001 for instance is based primarily on continuous improvement. Stimulating the creativity of stakeholders here generalizes the principle of ISO 9001 to develop mutually beneficial relationships with suppliers.

The results are the effect of implementation of factors, and these factors are optimized by feedback from the analysis of the results. There are some similarities between the criteria of EFQM and those of the Baldridge award. The EFQM model introduces a segmentation of excellence for each criterion as follows:

1. **Leadership**: Leaders of excellent organizations shape the future and make it become reality by acting as role models of values and ethics and inspiring full confidence. They are flexible, allowing the organization to anticipate and respond in a timely manner to ensure its continued success.
2. **Strategy**: Excellent organizations implement mission and vision by developing a strategy focused on the stakeholders. Policies, plans, objectives and processes are developed and deployed as contributions to the strategy.
3. **Staff**: Excellent organizations value their people and create a culture facilitating the achievement of their individual and collective goals, as well as those of the organization itself in their mutually shared interest. They develop the skills and talents of their staff by promoting fairness and equality of opportunity. They lend them their constant attention, communicate with them, reward and recognize them so as to motivate them, encourage their involvement and enable them to use their skills and knowledge in profit of the organization,
4. **Partnerships and resources**: Organizations plan and manage their partnerships, suppliers and internal resources to support their strategy, policy and the organizational performance of their processes. They provide an effective management of their environmental and societal impact.
5. **Processes, products and services**: Excellent organizations design, manage and improve their processes, products and services to generate increasing value for customers and other stakeholders.
6. **Customer results**: Excellent organizations achieve and maintain outstanding results that meet or exceed the needs and expectations of their customers.
7. **Staff results**: Excellent organizations achieve and maintain outstanding results that meet or exceed the needs and expectations of their staff.
8. **Societal results**: Excellent organizations achieve and maintain outstanding results that meet or exceed the needs and expectations of their stakeholders in society in the broadest sense.
9. **Economic results**: Excellent organizations achieve and maintain outstanding results that meet or exceed the needs and expectations of economic stakeholders.

The EFQM Excellence Model is represented as a schema of 9 interacting rectangles. It is thus a system with the processes of the organization at its center. As for the *Baldridge* Award, which obviously served as inspiration, each criterion is weighted by a key derived from the experience of managing institutions in the private or public sector (50% with regard to factors, 50% with regard to results). In the diagram below (Figure 7.7), the arrows emphasize the dynamic model and mutually beneficial influences:

- from the implementation of the strategy to results, on the one hand,
- and from feedback leading to improvements as well as to innovation when it comes to the factors, on the other hand.

Factors and results have the same weight, i.e., 10%, with the exception of key results and performances, as well as those related to customers, which each contribute to 15% of the final assessment.

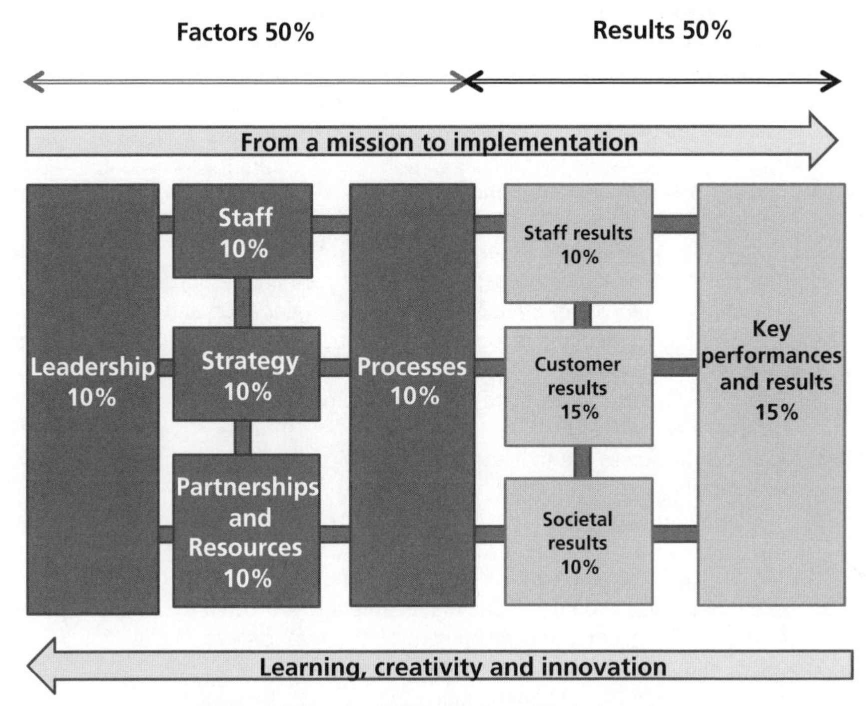

Figure 9.7 The nine EFQM criteria and their dynamics.

9.3.4 Itemization of the nine criteria

In order to obtain a detailed description and analysis of the organizational structure of the management system of the applicant organization, each criterion is broken down into 32 sub-criteria[17] (which corresponds to a sectoral application of the eight principles[5]) in a manner analogous to that of the Baldridge award. Each factor has five sub-criteria, whereas each result has only two.

1. *Leadership*:
   a. The leaders develop the mission, vision, values and ethics and should be exemplary;
   b. The leaders define, monitor, evaluate and guide the improvement of the management system and the performance of the organization;
   c. The leaders are involved with external stakeholders;
   d. The leaders reinforce a culture of excellence among their staff;
   e. The leaders ensure that the organization is flexible and that it manages the change effectively.
2. *Strategy*:
   a. The strategy is based on understanding the needs and expectations of each stakeholder as well as of the external environment.

[17] EFQM – Introduction à l'excellence, v. 2.1 Fr 1999-2000, downloadable at http://www.efqm.org/uploads/inex-fr.pdf.

   b. The strategy is based on understanding the internal performance and capacity.
   c. The strategy and the policies that it is broken down into are developed, reevaluated and updated.
   d. The strategy and policies that it is broken down into are communicated, implemented and driven.
3. *Staff:*
   a. The plans and policies of managing human resources support the organization's strategy.
   b. Knowledge and skills are developed.
   c. The staff is managed with a sense of commitment and responsibility.
   d. The staff communicates effectively across the organization and at all levels.
   e. The staff is rewarded and recognized. It is a focus of attention in the organization.
4. *Partnerships and resources:*
   a. Partnerships and suppliers are managed from the viewpoint of sustainable and profitable exchanges.
   b. Financial resources are managed from the viewpoint of sustainable security and profitability[18].
   c. The structures, equipment, materials and natural resources are managed responsibly[19].
   d. Technology is managed and developed in support of the strategy[20].
   e. The information and knowledge management is structured to effectively promote decision-making and the development of organizational capacities.
5. *Processes, products and services:*
   a. Processes are designed and managed to maximize the value for all stakeholders.
   b. Products and services are developed to create optimum value for the customers.
   c. Products and services are promoted and marketed effectively.
   d. Products and services are produced, delivered and managed.
   e. Relationships with the customers are managed and enhanced.
6. *Customer results:*
   a. Perceptions (of the customers in a broad sense), e.g., the perception of the reputation and image of the organization, the value of goods and services, etc.
   b. Performance indicators, for example, claims management, etc.
7. *Staff results:*
   a. Perceptions (of the staff towards the organization), such as results of a staff satisfaction survey, etc.
   b. Performance indicators such as the rate of absenteeism, training and career development, etc.
8. *Societal results:*
   a. Perceptions (towards the organization of society in the broadest sense), e.g., perception of the environmental impact, of the labor market, etc.
   b. Performance indicators, such as the mass of waste produced, but also the performance when it comes to health and safety, etc.

---

[18] The pillar "economy and finance" of sustainable development (see the chapter on this topic in this book).

[19] The pillar "environment and parsimonious use of natural resources" of sustainable development (see the chapter on this subject in this book).

[20] Which itself should be the subject of continuous monitoring.

9. *Economic results:*
   a. Strategic results (key results, whether financial or not, that demonstrate the success of the strategy and its implementation), for example, profits, fiscal performance, etc.
   b. Business performance indicators (measuring the organizational performance of the organization), such as performance indicators of key processes, etc.

Each sub-criterion has the same coefficient within each main criterion, e.g., sub-criterion 3b corresponds to 2 points out of the 10 points allocated to criterion 3 *Staff*. There are two exceptions:

- For criterion 6 *Customer results*, criterion 6a *Perceptions* represents 75% of the coefficient assigned to this criterion. Qualitative perception therefore outweighs the performance indicators.
- The same is true for sub-criterion 7a *Perceptions* in criterion 7 *Staff results*. In the world of management, which is highly oriented towards quantities, it is fortunate that assessments often deemed subjective receive the greatest significance.

The eight fundamental concepts have been related to the five factors of the nine criteria and EFQM has produced an ad-hoc table [10.6]. Thus, the fundamental concept of "Creating value for customers" is connected exclusively to sub-criteria 5b, 5c, 5d, 5e, while the criterion "Developing the capacities of the organization" is linked to sub-criteria 1c, 2b, 2d, 3d, 4a, 5d.

9.3.5 *RADAR* logic and performance evaluation

RADAR logic, very inspired by the Deming cycle, postulates that an organization works sequentially by:

- defining the Results[21] it wants to achieve with its strategy (PLAN);
- planning and developing an integrated set of relevant Approaches for results expected now and in the future (PLAN);
- systematically Deploying these approaches to ensure their implementation in full, in order to (DO);
- Assess them, evaluate them (CHECK) and;
- Revise them (correct them) (ACT).

The RADAR Logic aims to identify more fully the strengths and weaknesses of an organization and initiate a phase of continuous improvement or continue a process already underway. It can also be used as a method of problem solving throughout the company.
Factors and results are analyzed separately by a RADAR grid:

- the five EFQM factors are analyzed by the elements of the RADAR grid, i.e., **A**pproach, **D**eployment, **A**ssessment and **I**mprovement and
- the four results according to specific elements of the RADAR grid, i.e., Relevance and Usefulness on the one hand, and Performance on the other hand.

The grid elements have attributes associated with guidelines that specify the desired performance characteristics.

The elements and attributes of the RADAR grid are used to determine the level of excellence of an organization, by allocating the number of points provided by the criteria and

[21] Other systems prefer, before the results, that management defines the mission of the company, which is its reason for being, but also its vision (its institutional USP).

sub-criteria:

- shows no evidence of achievement;
- shows some evidence of achievement;
- shows strong evidence of achievement;
- shows evidence of extensive achievement,
- has such a level of achievement that the organization can be considered as a model.

Note that for the five factors, the rating of each of them should not be greater than that assigned to the element Approach. This therefore gives particular importance to the overall way in which the factor is deployed and realized when it comes to structuring processes and actions in a framework consisting of principles and policies.

9.3.5.3 Standards and tools

Using external standards

The EFQM system is an assessment tool that enables a company to carry out a system approach. Implementing excellence can be accompanied by the use of external standards:

- **ISO 9001** for factors pertaining to leadership, staff, policy and strategy, partnerships and resources, processes, staff results and customer results;
- **ISO 9004 and ISO 14000** for societal and process results;
- ***Investors in People***[22] for leadership, partnerships, staff, policy and strategy, staff results and key performances.
- ***Charter Mark***[23] for staff, policy and strategy, partnerships and resources, processes, customer results and key performances.

Internal frames of reference

As stated by EFQM, many roads lead to excellence and the EFQM model only describes the macrocosm of a management system of excellence. Can one then employ generic good practices while remaining within the EFQM frame of reference? The EFQM offers the following services:

- generic EFQM methodological frameworks to facilitate the achievement of excellence in the management of certain activities (currently 5: risk management, knowledge management, innovation management, social responsibility management, management of external resources);
- other "toolbox" documents that facilitate the diagnosis and implementation of the deployment or management of processes, and the implementation of the RADAR[24] logic;
- the EFQM Knowledge Base (see 9.3.1);

---

[22] This standard is a tool for improving the performance of an organization's human resources (its staff). The standard can be dowloaded at: http://www.investorsinpeople.co.uk/Documents/Branding2009/IIP_FRAMEWORK09.pdf

[23] *Charter Mark* was a standard of the UK government for excellence in customer service and enabled the organization to focus on improving customer service. It was replaced in 2008 by the Customer Service Excellence Standard, with the final Charter Marks expiring in 2011.

[24] See list at: http://www.efqm.org (accessed in Dec. 2012)

9.3.6 The stepwise path to EFQM excellence

9.3.6.1 Striving for high-quality performance

The EFQM recognition system has been designed to proceed in steps towards excellence in order to:

1. carry out a progress approach;
2. set new challenges;
3. boost a quality approach that is dozing off and manage the post-award period.

There are two complementary formats of recognition: certificates and awards. The objective for both is to provide guidance along the quite long path that leads to excellence

9.3.6.2 The certificates

There are two of them (general processes, Figure 9.8) [9.7].

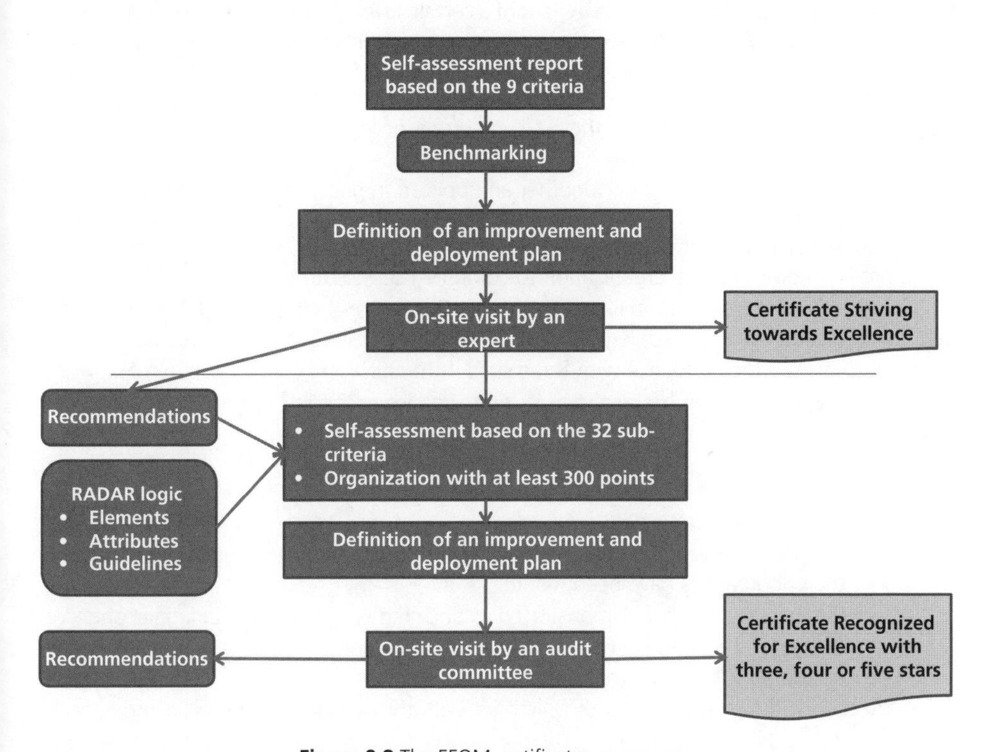

Figure 9.8 The EFQM certificates: processes.

The EFQM certificate: Striving towards Excellence

The certificate is awarded for a period of two years, and targets organizations just starting their journey towards excellence. Its objective is to facilitate the assessment of the current level of performance, identify strengths and weaknesses of the organization and establish priorities for improvement.

Organizations having started this process, that takes six to nine months, are usually in the following situation:

- they have made occasional initiatives for improvement and would now like to integrate them into a perspective that concerns the entire organization;
- they evaluate their performance primarily through internal measures without external data and comparisons that could serve as a basis when setting more ambitious or adapted objectives.

The process begins with a self-assessment according to the nine criteria of the model (with advice from an EFQM expert). An overview of the situation is created, often in comparison with model companies, and areas for improvement are found.

The second phase begins with the implementation of the improvement plan. An external EFQM evaluator visits the organization, verifies the actual deployment of the plan and proposes a set of recommendations. If the result is satisfactory, EFQM issues the certificate *Striving towards excellence*.

The EFQM certificate: Recognized for Excellence

This certificate confirms the achievement of a certain level of excellence and is intended for organizations that have met the requirements for the certificate *Striving towards Excellence* (for instance, Clinique de la Source in Lausanne in 2008). It provides organizations with a methodical approach for identifying strengths and areas for improvement in their organization. This level takes into account the 32 sub-criteria of excellence and primarily uses the RADAR logic for the 32 sub-criteria.

The procedure involves putting together an application pack, and then being paid a visit by two or three certified EFQM-evaluators. The team reviews the application pack, writes an inspection report with recommendations for further improvements, the corresponding score, and a profile in order for the organization to compare itself to others.

If the result of the evaluation is positive, the national representative of EFQM assigns the certificate "Recognized for Excellence" with the following distinctions:

- Recognized for Excellence with three stars if the following conditions are met:
  - the organization has analyzed its performance according to the EFQM Excellence Model 2013, and finds that the number of points awarded is more than three hundred;
  - key results show a significant improvement and are subject to comparisons to put them in context;
  - key processes are defined, inspected, and reviewed regularly and improvements in their approaches have been made;
  - the organization has demonstrated its ability to carry out and manage change;
  - many good management practices are in place and have had a significant impact on the results of the organization.
- Recognized for Excellence with four stars, with the following additional conditions:
  - the organization shows good performance in several areas and manages the changes resolutely;
  - the organization successfully carries out actions and demonstrates that it has invested in several practices that lead to success;
  - the results show significant improvements
- Recognized for Excellence with five stars, with the following additional requirements:
  - the organization is particularly efficient in the management of change;
  - it enhances, refines and simplifies the practices used to achieve its objectives;
  - it obtains results that are in line with its strategy.

9.3.6.3 The awards

In the form of a competition, the awards recognize European flagship organizations and have three progressive categories (finalist, award winner according to 1 criterion, all-category winner). The number of recipients is 10 to 20 annually. There are 5 categories of candidates: big business, business unit, public sector, small group subsidiary, independent SME.

The process is similar to that for the certificate *Recognized for excellence*:

- The company has previously been Recognized for Excellence with five stars;
- It puts together an application pack of 75 pages in English and submits its application;
- A panel of experts from the EFQM association assigns a theoretical grade to the application pack;
- If the score is greater than a specified minimum, 3-5 qualified EFQM evaluators visit the organization during one week to make an assessment;
- A detailed assessment report is sent to the jury;
- The jury decides whether to present the EFQM Award, which will be awarded at an ad-hoc ceremony.

The winners receive new avenues for improvement and a more in-depth vision, but also an increased visibility in what is referred to as gray media (professional and technical).

9.4 Quality system: EFQM or ISO 9001?

9.4.1 A non-exclusive approach

The two approaches are not mutually exclusive, because the ISO standard can structure the quality system of the company, enable its visualization onto which, where applicable, a path of EFQM excellence can be grafted.

The introduction of an EFQM approach, however, may be faced by the reluctance of management and the board of directors who may question the benefit of such an investment. Management also knows that its remuneration is above all based on the profit and share value of the company. It might therefore be reluctant to implement a model whose impact on profitability can be challenged by the shareholders, despite that a large part of the EFQM approach takes into account the results.

What does such an approach bring the shareholder? This is one of the questions that will be on the table, and management, as well as the board of directors, will expect measurable productivity performances.

The EFQM model can first be tested discreetly on certain criteria and processes that are considered capable of being optimized with a high chance of success. Finally, an excellence approach differs from a compliance control approach by the quality of the experts and the chairman of the audit committee. Indeed, the recommendations they will give will depend primarily on their experiences more than their adhering to a control protocol and to a formal quality culture. The success of EFQM also relies on the ability of the EFQM organization to attract reputable managers or directors as experts.

Just like the Baldridge Award, the EFQM approach emphasizes excellence taking into account all facets of business, as well as the complexity of all the activities and functions of the company. The performance of an organization is not the result of excellence of merely one of its parts. It is the entire system that must be considered, optimized and managed, including the management, the conduct of senior directors and the vision of the board. The

EFQM approach favors an overview of the organization's performance and identifies limiting factors.

EFQM awards are only valid for two years. This period is related to the expected growth of the organization that will regularly submit progress in excellence and go through all the steps warranted by the various awards. To compete for the award encourages emulation, but perhaps also disappointment within the organization if it does not win. In addition, the EFQM approach assumes that over the years of growth, a culture of sustainable excellence is developed within the organization and will continue beyond the EFQM activity. This requires continuity from management in the pursuit of excellence, and in light of its current turnover in some companies, the risk that this effort might be somewhat vilified by the next round of actors can unfortunately not be underestimated.

9.4.2 From ISO 9001 certification to EFQM

ISO 9001 is today the most common reference (> 900,000 certified companies in the world) in terms of system quality management, but also in terms of organization within a company.

Some companies may ask the question of choosing one or the other approach: certification with ISO 9001 or striving towards Total Quality with the EFQM Excellence Model? Even though these approaches are different, they are not contradictory. Table 9.1 below shows a brief comparison of the two models.

Table 9.1 Comparison of the ISO 9001 and EFQM models.

ISO 9001/CERTIFICATION	EFQM
PRINCIPLES	
Third-party certification (external audit)	System of self-evaluation and/or external recognition
Formal requirements	Non-prescriptive
Certification generally required	Voluntary participation for the various awards
Voluntary or imposed (e.g., by customers) approach	Voluntary approach
Focused primarily on process control (Organization/Procedures), continuous improvement, and staff skills	Process control and significant consideration of the criteria «results»/strive for efficiency
Periodic reassessment: – annual follow-up audit – a renewal audit every 3 years	Open periodic reassessment (not imposed)
OBJECTIVES	
Conformity of the product and customer satisfaction (except when considering ISO 9004)	Product conformity and satisfaction of all stakeholders (customers, partners, shareholders, collaborators, community)
RISKS	
Not implementing continuous improvement once certification achieved.	Complacency and overestimation of performance (except in the case of an Excellence Award: external evaluation)
Losing the certificate	Losing motivation

These two approaches can be complementary, as with the *Baldridge* Award and the Six Sigma approach in the US. Indeed, the ISO 9001 approach is more analytical and focused on the details of the process, while that of EFQM is more concise and comprehensive. Note, however, that none of these approaches is strictly focused on reducing costs and simultaneously improving quality.

There is one significant difference: two years of continuous improvement may be sufficient for a company to achieve a level of ISO 9001 certification, but it will take 2-3 additional years to reach a level that corresponds to EFQM award entitlement.

References

[9.1] W. EDWARD DEMING, *Some Theory of Sampling*, ed. Wiley and Sons, 1950.

[9.2] A. L. LAQUINTO, Can Winners become Losers? *Managerial Auditing Journal,* 14, 1/2 [1999] 28–35.

[9.3] The most famous is the book by MARK GRAHAM BROWN, *Baldridge Award Winning Quality – How to interpret the Baldridge Criteria for Performance Excellence*, Productivity Press, 2007 (copy consulted), published annually.

[9.4] A detailed review of procedures and project organization for the application for the Baldridge Award can be found in the book DAVID W. HUTTON, *From Baldridge to the bottom Line – a road map for organizational change and improvement,* ASQ Quality Press, 2000.

[9.5] MICHEL WEILL, *Le management de la Qualité*, La Découverte (2001) ISBN 2-7071-3467-8, pp. 110-112.

[9.6] *The EFQM Excellence Model*, EFQM Publications (2012), *THE* document on which the model is based and that one needs to have handy in order to implement it (sub-chapter of this book is only an introduction, a teaser ...), p. 24.

[9.7] *EFQM Recognition –Become a global role model*, EFQM editions (2012).

Chapter 10

The Toyota Way and the Toyota Production System (TPS)

Key concepts: learning, continual improvement, autonomy, A3 (report), visual inspection, challenge, genchi genbutsu, 3g, heijunka, jidoka, just-in-time, kaizen, kanban, total productive maintenance, 4P model, 7 muda, 3M (muda, mura, muri), poka yoke, 5 why, respect for individuals, 5S program, SMED, Toyota Way 2001, Takt time, RPM, TPS, teamwork, Toyotism, TWI, 5 zeros.

10.1 The Toyota phenomenon

10.1.1 A brief history of Toyota

The history of *Toyota* begins in the late 19th century when *Sakichi Toyoda* invented the first mechanized loom in Japan, which revolutionized the textile industry (as a comparison, the first mechanized Jacquard loom was European and dated back to 1801).

In January 1918, *Sakichi* founded the Toyoda Spinning and Weaving Company, with the help of his son *Kiichiro Toyoda*. In 1924, he realized his lifetime dream: building an automatic loom. Two years later, he created the *Toyoda* Automatic Loom Works.

Sakichi Toyoda established his business in a small town in Japan (Kosai), and not in the urban outskirts. This is why the founders of *Toyota* have always claimed to have a rural culture and background, accustomed to challenges and capable of bearing hard working conditions with perseverance.

Just as the inhabitants of the town of Kosai, *Saikichi Toyoda* was influenced by the Buddhist monk *Kichiren* (a polemicist and influential reformer of the 13th century), and a farm leader of the 19th century *Sontoku Ninomiya*, a symbol of perseverance and hard work. The values derived from Buddhism and a form of glorification of labor became the basis of the culture of Toyotism (toyota is the Japanese word for "abundant rice"). The values professed by *Saikichi* were summarized in five precepts:

- whatever your position, work together to fulfill your duties faithfully and contribute to the welfare of your nation;
- always position yourself upstream through research and creativity;
- avoid frivolity; be strong and sincere;

- be friendly and generous; strive to build a family atmosphere;
- be respectful and lead your life expressing appreciation and gratitude.

These values are not comparable with the catechism memorized in corporate courses or in business schools. They are not "good recipes", or good practices to be applied. They represent merely the written formalization of an attitude, behavior deeply rooted in the individual and which in the end is naturally expressed without conscious reference to these principles. Thus, memorizing them is not the same as practicing them.

A manager of *Toyota* is believed to have said: "*The principles underlying the operation of Toyota are simple to understand, but the training of our employees takes about 10 years.*" This component of Toyotism should never be overlooked when trying to analyze it. It also explains the relative success of the implementation of the Toyota culture in Western companies, whose core values are different (e.g., the focus on short-term financial results, the extreme simplification of worker tasks in Fordism, the high turnover rate of American managers who favor individualism, etc.). The juxtaposition of two such different cultures leads to failure, especially if the new values are introduced by external consultants and advocated by a Board of Directors that is obsessed with quarterly financial results.

Fig. 10.1 *Sakichi Toyoda.*

Fig. 10.2 *Kiichiro Toyoda.*

Like his father, *Kiichiro* was an inventor and, on the occasion of his visits to Europe and the US in the 1920s, he became deeply interested in the automotive industry (Ford's book *My Life and Work* was his favorite reading). Using the £100, 000 that *Sakichi Toyoda* received as payment of dues from his patent on the automatic loom, *Kiichiro* laid the foundations for the *Toyota Motor Corporation* (TMC) which was established as an independent company in 1937. The first motor car models, the A1 and G1, were produced in 1935. At that time, industrial productivity in Japan was nearly one-tenth of that of the United States. The challenge was therefore very serious.

From looms to cars, the *Toyota* experience now involved a constant preoccupation for pushing the boundaries of manufacturing. *Kiichiro* did not, however, get the opportunity to carry out field tests on the principles of Ford. Caught in the chaos of the Sino-Japanese and Sino-American war, his company manufactured mainly military trucks and only a few cars, all still largely hand crafted. During his lifetime, *Kiichiro* himself wrote manufacturing instructions and technical notes, instilling the sense of keeping written records of good practice, which would later be integrated into the culture of *Toyota*.

At the end of World War II, *Toyota* was facing a situation different from that of vehicle producers in the United States:

- Employees had an almost life-long commitment, but harsh social conflict and financial difficulties led Kiichiro to resign from the company. As an upside to this situation, the staff was committed to being active in their workplace, to taking initiatives and being flexible in the tasks assigned to them.

- The price of land was high, limiting areas for manufacturing, storage and components of cars produced.
- Vehicles had to be manufactured in small quantities, but over a wide range of products in a Japanese market that was protected from 1945 to 1960, but still limited.
- Investment capacity was low for the acquisition of production equipment (given the general economic situation in Japan after the war) limited the number of assembly lines.

How could one obtain the benefits of mass production while integrating all these constraints? The challenge was enormous for a small company like *Toyota*, without significant industrial experience. The disparity between *General Motors* and the Japanese company exceeded even that between David and Goliath.

Fig. 10.3 *Eiji Toyoda*, le pionnier du TPS

Fig. 10.4 *Taichii Ohno* (1912-1990) de la *Toyota Motor Company*.

But there were more ideas to come in the *Toyota* family. Results would show that the courage and perseverance of *Kichiren*'s disciples were not empty rhetoric (Who said that Buddhism, a religious form of nihilism, disconnected from the world?). The ideas that followed also validated the adage: *creativity and innovation are the daughters of necessity*. *Eiji Toyoda*, a nephew of *Kiiricho*, took a study trip to the US in 1950. He visited *Ford's* Rouge plant in Michigan, which produced 7000 vehicles per day (while the total number of cars manufactured by *Toyota* in 1950 was 2685...). He was impressed by mass production, but he also found things to improve in the world of Fordist production [10.1]:

- Absenteeism in the factories exceeded 10%: would it be possible to have a more motivated staff by changing the working environment?
- In an environment of questionable order and cleanliness, a significant stock (too much?) of components were stored in large quantities here and there in the factory waiting to be used. They were manufactured in batches and inspected after the fact: were these stocks necessary?
- The production staff and operators, who were able to verify product quality during manufacturing, did not have the freedom to react in cases of poor quality and stop the line. This decision was exclusively up to the engineer of the assembly line, whose office was rarely close by. Often, the assembly line was not stopped and the fault had to be corrected, after inspection, over the entire batch. This operation was expensive (it took up 20% of the surface of the factory and mobilized 25% of the employees' time). Moreover, its success was not guaranteed. However, in their statistics of the

quality produced, plant managers only indicated the poor quality that remained after this adjustment... Was there no way to respond more quickly?

- The production staff, whose pay was the lowest, was specialized in performing a single task. Despite their potential responsibility, managers reminded them that their presence was made necessary only because the automation of the factory was not yet complete. Was this really the best way; shouldn't the production staff have some responsibility?
- The production was not very flexible because of the time required to change tools (press molds for the body work, for example), which might take from one to several days. To overcome this problem, the number of production lines was large. Could this time be reduced?
- Technicians and engineers involved in the maintenance of production equipment had higher status and pay than the operators. Two consequences were evident for Eiji:
  - Maintenance operations were given priority over production, and the downtime of the lines was not considered a critical factor.
  - As operators and technicians rarely interacted, the operators had insufficient knowledge of their production tools. Moreover, they did not feel directly responsible, and thus called upon technicians and engineers at the slightest hitch, resulting in an overload of the service. Also, the production tool was not improved, possibly because there was no mechanism for the experience gained by the production staff to be channeled to the development of the equipment. Continual improvement was non-existent.
- The specialization brought on by Fordism multiplied the types of specialized activities, not related to each other: operators were not responsible for the cleanliness of their workplace. This task was delegated to cleaners. In short, the production staff was completely without a sense of responsibility.
- This specialization also applied to managers, who become specialists in a fraction of the making of a car, rendering it difficult to exchange experiences and information among engineers. Design and product development suffered from this fact, because engineers had problems collaborating when defining the functions of operations and associated research criteria.

From the United States, *Eiji* wrote a letter to the administrative center indicating that he thought that improvements in the production system were possible. Back in Japan, *Toyoda* and his production engineer, *Taïichi Ohno*, an organizational genius of the stature of *Ford* and *Sloan* (but also a tough man who inspired fear...), went to work. They realized that mass production was impossible in Japan and chose another path.

They began by giving more responsibility to the production staff by grouping the employees into teams and teaching them the principles of continual improvement. They devoted themselves to patiently solving all the malfunctions of Fordism, one problem after the other, and to carrying out the industrial production of small quantities. Toyotism took shape. In the early 1960s, most of TPS was defined and its concepts were gradually adopted by other Japanese manufacturers. The Americans only became aware of it more than a generation later.

In 1965, *Toyota* received the Deming Prize; the efforts required to win the prize had three effects:

- an increase in the quality of the vehicles,
- an increase in exports and Toyota's market share in Japan, and
- decreased costs.

In 1995, *Toyota* introduced TQC in its research and development units. It launched its first vehicle on the US market in 1957: the Crown model. In 1964, the global share of vehicles produced by Japanese companies was 8%; 10 years later it was 20%, and in 1988 it reached 28%. In 2005, ten of *Toyota's* vehicles were in the forefront of the best American rankings, while *GM* and *Ford* had only four and two, respectively.

10.1.2 Two key ideas of the Toyota Way 2001

Based on the ideas of its founders, *Toyota* defined a common set of business values, references and methods acting as the lifeblood of the company, generating its development in a continuous manner. Unlike other companies, whose transmission culture consists of oral transmission and mimicry, this set of values is well documented, which is why some authors have even talked of *Toyota's* "corporate DNA". The values are rooted in the thinking of the founders of *Toyota* and are carefully instilled in all employees. They are passed on to newcomers by officers, directors, managers and senior employees. Company training courses are also scheduled.

The Toyota Way, leading ideas of the *Toyota* culture, was described in 2001 by the Board of Directors in the form of a booklet [10.2] in order to share the *Toyota* methods that made the company a success, particularly in fulfilling its business objectives with all members and partners of *Toyota* worldwide and easing the dissemination of a common culture.

The Toyota Way 2001 is based on two concepts (Figure 10.5): *Continual Improvement* and *Respect for People*.

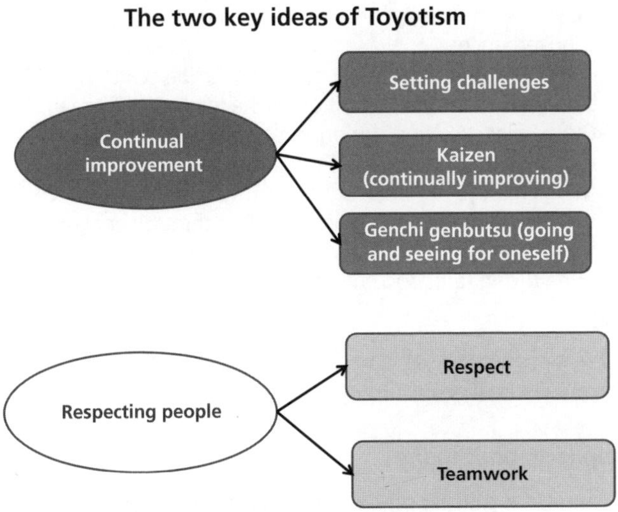

The two key ideas of Toyotism

Continual improvement
Setting challenges
Kaizen (continually improving)
Genchi genbutsu (going and seeing for oneself)
Respecting people
Respect
Teamwork

Figure 10.5 The two key ideas of Toyotism.

1. **Continual improvement:** *We are never satisfied with the current status and we are constantly striving to improve our company by continually advancing new ideas and working to the best of our abilities.*
   a) ***Challenge*:** *We forge a long-term vision, and meet challenges with courage, determination, and creativity, being preoccupied with added value to achieve our dreams.*
   b) ***Kaizen*:** *We continually improve our business operations, driven by innovation and the evolution of best practices, in a lean management perspective.*
   c) ***Genchi Genbutsu*:** *We get to the bottom when establishing the facts which leads to correct decisions, resulting in consensus and achievement of our goals.*

2. **Respect for people**[1]**:** *We respect all Toyota stakeholders (community, members and partners) and are convinced that the success of our company is generated by the commitment of individuals and good teamwork.*
   a) ***Respect*****:** *We respect others, take our responsibilities and do our best to create understanding and mutual trust.*
   b) ***Teamwork*****:** *We stimulate personal and professional development, share development opportunities and maximize individual as well as group performance.*

10.2 Operational excellence as a strategic weapon

10.2.1 The 4P model [10.3]

The foundations of Toyotism revolve around 14 principles, grouped into four families, formalized by the American *Jeffrey K. Liker*, and are often called the 4P model (even though there are five!) (Figure 10.6):

- Long term Philosophy (1st P) is the basis of the investment in continual improvement and learning (the term "learning" includes both the preoccupation for constant learning and the willingness to learn).
- Good Processes (2nd P) generate good results: processes with a constant flow are the key to achieving the best quality at lower costs with increased safety and higher morale. Many Western companies have borrowed this aspect of Toyotism to integrate it into their governance.
- Value is added to the organization by taking proper care of Personnel (3rd P) and Partners (additional 3rd P): the ambition of the leaders of Toyota is to "build" people, not just automobiles. We find the principle stated by Ishikawa on the importance of staff training in Japanese industrial production.
- Resolving Problems (4th P) at their root, striving for consistency and continuity of effort, is the center of the learning organization; through analysis, reflection and sharing of lessons learned, the Toyota Way takes continual improvement to the highest level by standardizing (and documenting) the most efficient best practices.

10.2.2 The 14 principles of the Toyota approach: milestones towards excellence [10.4]

Section I: A long-term philosophy

Principle 1. A long-term philosophy should inspire all management decisions, even at the expense of short-term financial goals.

- All decision-making in the short term is dependent on the philosophical sense of the objective.
  - The entire organization must be founded on and grow in alignment with a common objective that goes beyond making money.
  - It is important to understand the place one occupies in the history of the company and try to pass this on to the next level (higher).

The philosophical mission is the foundation of all other principles.

[1] One of *Toyota's* slogans is: *Customer first, then the supplier, and only then the manufacturer.*

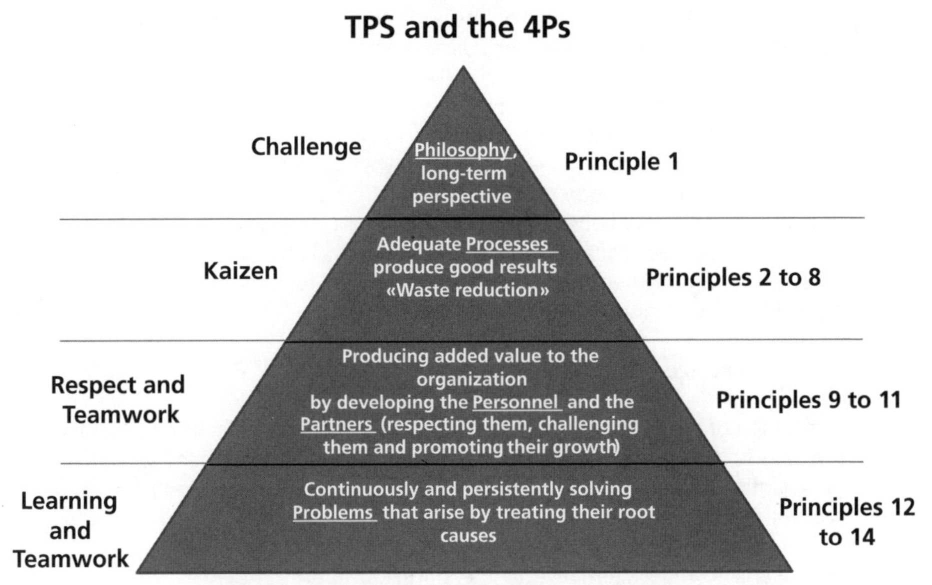

Figure 10.6 Pyramid of the 4Ps and the 14 principles.

- Starting point: create value for the customer, company and the economy.
  - Each function of the company must be evaluated in terms of its ability to accomplish this.
  - Being accountable: to ensure control over one's own destiny. *In Fordism, the worker is responsible only for the microscopic task assigned to him.*

Section II: Good processes lead to good results

Principle 2. The creation of a continuous flow of processes highlights the problems.

- Reconsider the work process to achieve a continuous flow of high added value.
- Strive to reduce lost time and waiting time of every task to zero. In Fordism, production lines were often stopped for organizational reasons. Sometimes however, under the pressure of a planned production, the production line operated even if poor quality had been detected. This problem was corrected by inspection and costly corrections because having a faulty component in a product made from hundreds of different parts is often risky.
- Create flows that are able to transmit information and materials as well as connect people and processes so that problems surface immediately.
- Make the workflows clear throughout the organization. This is the key to a truly continual process improvement and also to the development of staff.

Principle 3. Use "pull" systems to avoid overproduction

- Provide customers, after production, with what they want, whenever they want it, and as much as they want. This is the principle of "just-in-time" (JIT), which is based on replenishment of materials ordered only for customer needs. In the mass production of Fordism, components manufactured before customer needs were known to pile up, without the certainty that they would be used. This led to overproduction (to

give value to the produced parts) or losses (destruction or recycling of components). In the first case, sales are forced to sell the stock at a lower price, with a reduced margin, which decreases the profitability of the company.
- Adjust to a minimum the work in progress and limit the inventory to a small number of components or commodities, and replenish these frequently following the customer's actual purchases. This was vital for Toyota since the company was faced with extremely high square meter costs for land.
- Respond to daily variations in customer demand rather than relying on scheduled rates or computer tracking systems to the waste inventory processes. This was also suitable for Toyota, because JIT production limited the storage of vehicles parked around the plant waiting to be sold.

Principle 4. Regularize (smooth) the workload (heijunka). ("Work like the tortoise, not the hare")

- Elimination of waste (muda) represents one-third of the success of the TPS approach.
- It is vital to reduce the overburdening of personnel and equipment and to iron out irregularities in the production rate. This fact is often misunderstood by companies seeking to apply the principles of "Lean Management", whether in administration, production or in the company as a whole.
- Heijunka is Toyota's alternative to the start/stop approach of batch productions of plants whose mode of operation is derived from Fordism. These large batch productions often contain poor quality goods that have been detected only after the component is integrated in the automotive construction. Producing small quantities limits stocks, storage costs and inventory, and renders it possible to detect the root of the poor quality.

Principle 5. Install shutdown devices to fix problems, and get the quality right the first time

- Quality for the customer guides the value offer.
- Use all available modern methods of quality assurance.
- Provide equipment with features for problem detection and automatic shutdown (poka-yoke). (It would be even better if it were designed by the production staff.)
- Introduce a visual alert system indicating the breakdown of machinery or a process.
- Jidoka (intelligent machine) is the foundation of integrated quality (autonomation).
- Provide the organization with systems to help solve problems and implement corrective actions. Thus, Ohno decided that a fraction of the working time of the production staff should be dedicated to problem solving. He introduced quality cycles recognizing the importance of the production staff.
- Convince and adhere to the philosophy of stopping or slowing the production line in order to get quality right the first time, which in the long run will increase productivity. Early in the implementation of this rule, the production lines of Toyota stopped regularly. Very quickly, the correction of errors resulted in a productivity of the production line that was much higher than that of Western countries, with products of superior quality.

Principle 6. The standardization of tasks is the foundation of continual improvement and the autonomization of staff

- Always use methods that are stable and repeatable to maintain predictability, a regular rate and a constant result for the process. This is the basis of the just-in-time approach. The standardization of tasks is easily done by the just-in-time approach, since with the stabilizing effect of heijunka, it becomes possible to define a maximum time (Takt Time) to perform each operation. The progressive decrease of the Takt Time also puts pressure on employees to remove any non-productive task (elimination of waste).

- Preserve the accumulated knowledge of processes by standardizing best practices.
- Encourage individual creativity to improve standards that complement the existing one, ensuring transmission of this knowledge during times of staff turnover. Since the staff is able to complete several tasks, the production also becomes more flexible, while breaking the monotony of repetitive work.

Principle 7. To resolve problems, nothing beats visual inspection

- Use simple visual indicators to immediately determine whether a process deviates from the standard conditions. This is the basis for line control, which does not require the intervention of an engineer or a qualified technician.
- Avoid using a computer screen if it distracts attention from the worker's workstation. The observation of what really happens during production takes precedent over the sophistication of management tools.
- Develop a simple system of the workplace as support to the just-in-time approach.
- Reduce reports, whenever possible, to a single page (A3 reports). Even though Toyota values documentation, it should not be heavy.

Principle 8. Put proven and reliable technologies at the service of the personnel and processes

- Technology is of use to the employees but does not replace them. It is often better to work manually on a process and use technology as backup. This also makes it possible for the staff to keep pace with changing technology and to avoid as far possible social conflicts due to technology that staff no longer understand.
- New technologies are often unreliable and difficult to standardize; they may jeopardize the flow. A proven process running well takes precedence over new and untested technology. Toyota avoids choosing exotic emerging technologies (or sophisticated production gadgets).
- Conduct full-scale tests before adopting a new technology for business processes, manufacturing systems or products.
- Reject or modify technologies that conflict with the corporate culture or even disrupt the stability, reliability and predictability of the process.
- However, encourage staff to consider new technologies when developing their work. Do not delay implementing technology that has been carefully assessed and confirmed by tests, and is able to improve process capability and reliability.

Section III: Add value to the organization by taking care of personnel and partners

Principle 9. Strengthen supervision that includes in-depth work, live the "Toyota philosophy" and teach others

- Recruit managers in-house rather than from outside the organization. This approach is not Toyota's own, but one of multinational companies whose growth and profitability are steady.
- Do not reduce the role of managers to that of a mere human resource capable of accomplishing tasks competently. Managers should be role models embodying the

philosophy of the company and the manner in which to do business. Through them, the values and philosophy of Toyota are transmitted and guaranteed. They are also mentors.

- Good leaders must understand the daily work in every detail so as to be the best teachers of the company's philosophy. Engineers confined to their offices will not have a relationship of trust with the production staff, or be capable of solving technical problems satisfactorily.

Principle 10. Train exceptional staff and teams to follow the philosophy of the company

- Create a culture in which strong and stable values and belief in the company are widely shared and experienced over several years.
- Train exceptional individuals and teams able to work within the framework of the corporate philosophy and implement non-standard results.
- Work with all the perseverance necessary to continuously strengthen the culture.
- Interchange teams to improve quality and productivity and increase the flow by solving technically complex problems.
- The autonomy of teams is the result of staff having a free choice of the company tools for improvements.
- Constantly strive to teach individuals to work together toward common goals.
- Teamwork is something that is learned.

Principle 11. Respect an extensive network of partners and suppliers by challenging them to improve

- Have respect for partners and suppliers by treating them as an extension of the company. In the mass production of Fordism or the way it was practiced at General Motors, suppliers were not considered as partners but as interchangeable subcontractors in constant competition, to whom detailed plans of principal components would be delivered without any information on the environment in which the component would be used. At Toyota on the other hand, partnerships were entered into; sometimes Toyota acquired an equity share of the supplier, or some managers of Toyota were hired by the supplier, or Toyota would agree to loans. Information on policy and the car model were provided. Further down the line, customers and distributors were directly involved in defining and designing new vehicles.
- Encourage external business partners (suppliers, for example) to grow and develop. This shows esteem. Determine high enough goals and assist partners to achieve them. In lieu of detailed specifications, Toyota provided specifications of a component based primarily on its functionality (e.g., a brake system capable of stopping a vehicle of one ton in X meters when traveling at y km/h) and key dimensions corresponding to its assembly. Since Toyota did not exclusively choose suppliers based on price, the company was able to build a long-term relationship of trust, limiting the entry of new suppliers and poor quality components that an inexperienced provider regularly provides in early deliveries. As fierce competition between suppliers was rare, suppliers could also exchange experience and information. Result: their productivity and the quality of their shipments rose steadily, reducing the purchasing and production costs at Toyota.

Section IV: Continually solving basic problems is instructive to the organization

Principle 12. Going to see for oneself for understanding the situation as it really is (*genchi genbutsu*)

- Solve problems and improve processes by going to the source and personally verifying and analyzing (actual) data rather than theorizing from those that others or a computer provide.
- Think and communicate only about those items checked personally.
- Even managers and the most senior executives should see for themselves (on-site) to acquire a less superficial knowledge of the situation. Certainly, managers and leaders should not focus exclusively on operations, as this causes them to lose their ability for breakthrough management cherished by Shiba. However, too many leaders make strategic decisions without strong knowledge of the company's products and processes. This can be catastrophic.

Principle 13. Make decisions slowly by consensus, by carefully examining all options, and then apply (made) decisions without delay.

- Only commit to a specific direction after carefully weighing the alternatives.
- Once a direction has been adopted, commit quickly but carefully.
- Nemawashi is the process of discussing problems and potential solutions with all stakeholders, thus gathering their ideas to reach consensus about the next move. This consensus process, although time-consuming, helps broaden the search for solutions. When a decision has been made, all elements are in place for a quick implementation. Here we find the Japanese rule about devoting the time necessary for reflection and the design of an idea or product, but also the importance of gaining the consent of employees.

Principle 14. Become a learning organization through unrelenting reflection (hansei) and constant improvement (kaizen)

- Once a process has been stabilized, use continual improvement tools to determine the root causes of inefficiency, and apply (immediate) corrective actions.
- Define processes that require the least amount of stock possible, which allows everyone to identify wasted time and resources.
- Once the waste has been identified, employees must eliminate it by a process of continual improvement (kaizen).
- Protect the knowledge base of the organization by forming a stable workforce, by slow promotions and a very cautious succession system. This condition is difficult to establish in a culture with fast staff turnover, such as in the US.
- Give intense reflection to key stages of a project, but also to its conclusion, in order to identify any weaknesses. Apply appropriate measures to avoid making the same mistakes again.
- Learn by standardizing best practices rather than reinventing the wheel with each new project or with each new manager. The DNA of the Toyota culture involves documenting principles and practices and transmitting them.

10.3 The Toyota Production System (TPS)

10.3.1 The two pillars of the Toyota production system

TPS is the deployment of the 14 principles in production. It was formalized by *Taiichi Ohno* (Figure 10.7) [10.5] as late as in 1988, despite the fact that the system was developed in the 1950s, and then slowly adopted (not without resistance) by *Toyota* in 1962. According to *Ohno* himself, influenced by *Deming*, it took more than four decades for Toyotism to mature.

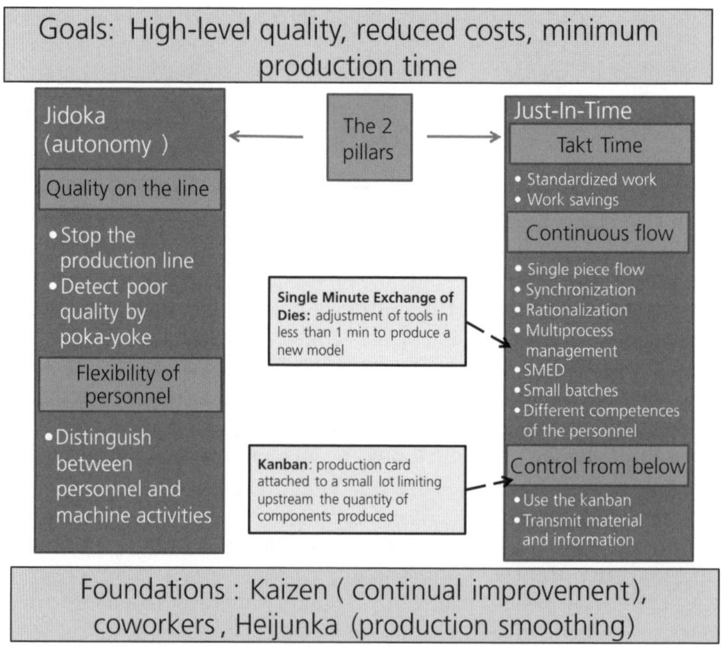

Figure 10.7 Objectives, foundation and pillars of TPS.

The basic idea of TPS is the complete elimination of waste (in Japanese, waste is *muda),* but also of the **3Ms**:

- *Muda* (no added value, waste),
- *Mura* (variability), and
- *Muri* (overburdening employees or machines).

These are the causes of poor performance (when it comes to productivity and quality). Figure 10.8 shows the foundation, the two pillars and the objectives of TPS:

- "just-in-time" production, and
- the ''auto-activation" of the production (or autonomation).

10.3.2 Heijunka

TPS avoids fluctuations in production, leading to an increase in the 3Ms, i.e., factors of poor quality, productivity loss, and waste of human resources and materials. It implements *hei-junka* or production smoothing, because the flexibility provided by the JIT-approach makes the system very sensitive to fluctuations in production. Furthermore, this smoothing facilitates the management of staff, eliminates on-call work and reduces overtime (which has higher pay). Smoothing is based on the volume and portfolio of the product (production of several different models in one day on a single line thanks to SMED).

Production smoothing considers an average order calculated on the basis of previous days. *Heijunka* may conflict with other aspects of TPS, including the fact that *Toyota* should produce only vehicles that have been ordered by the customer. However, the abandonment of batch production means that the interval between an order and delivery is reduced, since each product is available every day. The number of failed deliveries and products in stock are thus minimal.

10.3.3 Just-in-time (JIT) production

Just-in-time assembly of an automobile means that each component reaches the assembly line at the right moment, at the right place, is of an appropriate quality and in the right amount. If this can be done step by step throughout the company, it becomes possible to achieve at the same time the conditions of "zero inventory", thus reducing the production area, the inventory and the inventory costs by transferring components of suitable quality from one station to another.

This idea was issued by *Ford* in *My Life and Work*, and was partially implemented by the company. However, *Ohno* was inspired instead by the practices of a US retail chain: *Piggly Wiggly*.

According to the principle of "just in time", each workstation must take from the station preceding it in the production process only material that is strictly necessary. So why doesn't the preceding workstation produce merely the parts taken from it?

Kanban

In practice, this rule utilizes a system of cards called *kanban*. A card (*kanban*) is attached to a container that holds a limited, but fixed, number of parts coming from upstream. The *kanban* generally provides the following information: reference number, heading, name of customer or subsequent workstation, number of parts per container, and name of supplier or preceding workstation. A batch of products from the daily production of components used by the downstream workstation corresponds to a fixed number of containers. When the downstream station uses the parts from one container, the *kanban* is removed and put in a box. Periodically, an employee retrieves the cards to give them back to the preceding workstation. This corresponds to a sort of production order of the number of batches to be delivered. The upstream workstation does not produce anything without a *kanban*. With this simple method, *Toyota* was able to minimize the areas for production and storage of components (very expensive given the land prices in Japan).

The principle is the same for components delivered by suppliers. Nevertheless, transport, generating costs and pollution should be taken into account. Instead of a truck, a van may, for example, be sufficient to ensure delivery.

Total Productive Maintenance (TPM)

Toyotism reduces waste due to stops and failures, leads to higher production rates and increases the quality by redefining the maintenance of production equipment. TPM is based on three pillars: availability, performance and quality. It can be explained as follows:

- Total: consider all aspects of maintenance and involve everyone, including operators.
- Productive: ensure that the maintenance has minimum interference with production. This prevents long shutdowns of production to carry out maintenance operations.

- Maintenance: maintain the production tools in an optimum state of operation, changing the necessary parts, repairing, cleaning and agreeing to spend the time prescribed by the equipment manufacturer.

The TPM approach goes beyond intelligent maintenance. Its purpose is to modify and improve the production tool. The involvement of line staff, which is one of the original points of TPM, increases their level of expertise and allows them to be thoroughly familiar with the production, while making it possible for them to perform basic maintenance. This results in personnel savings (maintenance personnel often have higher qualifications than production staff).

To evaluate the performance of equipment, a metric, the overall equipment effectiveness (OEE), is used. This is the product of three rates (each expressing the ratio of the duration of unacceptable quality and not the total time of operation of the tool):

- the availability rate, whose unacceptable quality consists of breakdowns and tool changes;
- the performance rate, whose unacceptable quality comprises stops and changes in rates;
- the quality rate, whose unacceptable quality concerns defects and loss of components in the beginning and end of production, but also start-ups since a production tool does not work optimally during the first moments of operation after shutdowns.

Each of these rates varies between 0 and 100%, and the OEE is thus between 0 and 100%. The closer the value is to 100%, the better is the efficiency of the tool.

SMED

Production smoothing at *Toyota* not only consists in adjusting the daily quantities produced, but also in increasing the types of products manufactured in one day on a line. This requires modification of the production tool. In Fordism, the adjustment of tools was a tedious and delicate process that took at least several hours or even one to three days. If the time required for this change were close to zero, it would be possible to produce multiple vehicles on one line without decreasing productivity. Moreover, if the changeover times between series were zero, the manufacturing of units would be performed without increasing costs.

Adjustment of a production tool comprises the following steps:

- preparation of the tool, of the workstation dedicated to adjustments, and of the equipment;
- verification of the measuring instruments;
- dismantling/assembling the equipment;
- settings;
- realization and control of test parts, to ensure the conformity of the product component;
- cleaning;
- keeping tidy the workstation dedicated to adjustments.

By separating the operations that could be done with the production running from those that required shutdowns, *Toyota* engineers managed to significantly reduce the time needed for adjustments. In a second step, they devised solutions to transfer operations that were carried out with a running production tool to a group of operations that were carried out independently of this. Finally, they further decreased this time by a thorough study of the production tool. They named this new adjustment *Single Minute Exchange of Die* (although the time was generally less than 10 minutes), because it was the body parts manufactured with presses that presented the most problems.

Toyota was thus able to solve one of its basic problems: to produce a wide range of automobiles without significant additional costs with a limited number of production lines. Later, this advantage would allow *Toyota* to manage a worldwide range of products at costs that its Western competitors could not match. Taking advantage of their pricing policy with a wide range of vehicles and an unparalleled robustness, *Toyota* caused its competitors to significantly reduce their margins, while *Toyota* left the process as a winner and increased its market share.

10.3.4 Self-activation or autonomation

The other pillar of the Toyota Production System is ***autonomation***. This is different from automation, and is also called self-activation.

In "Fordist" production, most machines run by themselves from the moment they are plugged into a power source (or activated). The trouble is that they continue to produce in the case of poor quality. With automatic machines designed for mass production, we cannot therefore avoid mass production of defective components.

At *Toyota*, a "self-activated" machine is a machine equipped with an automatic shutoff in case of anomalies. These machines are equipped with various systems – simple ones, where possible – to prevent defective products, also known as *poka-yoke*, which gives them a limited form of artificial intelligence.

It is not necessary to maintain an operator at the machine while it is operating normally. It is when the machine stops after an anomaly that it needs to be attended to. Moreover, a single operator can attend to several machines, which significantly increases the production efficiency.

10.3.5 Eliminating waste and the 5S program

The 7 mudas

TPS is devoted to the elimination of waste (*mudas*) which are 7 in number:

1. *excessive or poorly planned production:* producing too much or too soon;
2. *wait:* waiting for parts, or for a machine to finish its cycle, etc.;
3. *unnecessary transportation and handling:* any transport in the production site is essentially a waste and should be minimized;
4. *unnecessary machining:* any action of added value that is not done right the first time or in a simple manner;
5. *stocks:* more material and components than the minimum required to do the job;
6. *unnecessary movement:* any movement of a component or product in production that does not directly contribute to adding value;
7. *corrections:* any repair or adjustment after inspection is a waste.

The 5S program

The elimination of the 7 mudas is facilitated by the 5S program [10.6] (for *Seiri, Seiton, Seiso, Seiketsu, Shitsuke*), which was very important to *Ohno*. The cleanliness of a production facility permits visual inspection, making it possible to detect an anomaly faster, and thus leading to better performance.

The 5S program, applied in Japan, is the basis of the exceptional order in production workstations noticed by visiting Americans from the same trade, who claimed that it was possible to "eat off the floor".

The 5S program involves the production workplace and consists of:

- Sorting, in order to keep at the workstation only what is strictly useful; the following algorithm is a typical procedure:
  - *everything* that is used less than once a year is available on request;
  - of what is left, *everything* that is used less than once per month is stored in another location (e.g., administrative office, or central storage of the plant);
  - of what is left, *everything* that is used less than once a week is stored in close proximity (typically in a cabinet in the workstation, or storage place in the workstation);
  - of what is left, *everything* that is used less than once a day is at the workplace;
  - of what is left, *everything* that is used less than once an hour is at the workstation, directly within reach of the operator;
  - and whatever is used once an hour, the operator keeps with him.

This differentiation of tools and work aids logically leads to the next step:

- Straightening of the workplace in an optimal manner using the principle: "a place for everything and everything in its place". A procedure often used to implement this step is:
  - organize the workstation rationally so as to facilitate the operators' activities (proximity, heavy objects easy to access or already mounted, etc.);
  - define rules for storing tools or work instruments;
  - make clear the placing of objects, with tips for easy retrieval;
  - objects used frequently should be close to the operator;
  - arrange tools and props according to their order of use;
  - standardize the workstations;
  - favor 'FIFO' (first in, first out), i.e., see to it that for a class of objects or components, those that arrive first at the workstation to undergo an operation providing added value should also be the ones to leave first (this is particularly suitable for perishable products such as in the food industry, etc.).
- Shining to discover anomalies by clearing and rehabilitating. When the workstation is tidy and free from unnecessary things, cleaning is possible. Dirt can cause irregularities, fluctuations or breakdown of machinery, and may also affect the health and safety of employees (dust, sharp objects, toxic substances evaporating or spilling on the floor, etc.).
- Standardizing and repeating on a regular basis the three above steps. Order and cleanliness must be visible at all times.
- Sustaining: maintaining order and cleanliness must be accompanied by staff self-discipline, and should be subject to continual improvement initiatives, but also management audits to maintain pressure.

The 5S program improves staff morale and the working environment (working in a clean and tidy workstation is more motivating). It also reduces the time and energy spent, and the risk of accidents and/or health issues (which can be very expensive in every sense), and often improves the quality of production. However, it has little effect on productivity, and it is wrong to liken it to a method of Lean Management.

The 5 zeros

This is the elimination of all forms of waste aims to achieve 5 zeros:

- zero paper: reduce the structures as far as possible by the maximum simplification of

procedures and paperwork, whether manual or automated;
- zero delays: prevent loss of time from when the order is registered to when the product is finished and shipped;
- zero failures: ensure that no failures or incidents occur;
- zero defects: quality is manufactured, it is not limited to inspections;
- zero inventory: only receive the right parts, and only when they are needed.

"Perfection is achieved not when there is nothing left to add, but when there is nothing left to remove." (Antoine de Saint-Exupéry)

10.4 Training of staff at Toyota

Ohno wrote [10.7]: *"At present, I am aware that members of a company tend to forget and to distance themselves from training needs. Of course, if the talents to be acquired are not creative and challenging and if they do not mobilize the best, training may seem unnecessary. But let's face it: no objective, regardless of size, can be achieved without training.*

Toyotism puts an emphasis on training of staff, whose skills and expertise are inextricably linked to the development of its production system. For *Toyota* – unlike many Western companies wishing to apply Toyotism – lean production is inconceivable without focusing on training. Too often in the West, training needs are not a subject of serious research to the company's HR. Nobody really cares about the actual expertise of the personnel. Indeed, HR professionals are hardly capable of determining what courses would be useful for employees – they are too alienated from them. At the management level, they are supposed to have learned everything in school or during their inevitable passage through an institute delivering MBAs. Continuous training in companies is often considered as merely *nice to have*, providing only basic tools.

But which method does *Toyota* offer this knowledge and determine what type of knowledge is needed? We have seen that managers at *Toyota* have not only a controlling function but also advice-giving and coaching tasks. There are also tools for problem solving, one of which is common to an entire company, from the top to the bottom of the hierarchy: the A3 report (see ref. [10.8]).

The basic idea of *Toyota* contradicts the system of stars and the search for outstanding individuals often advocated in the US: *Toyota* develops skills from a solid and talented base, but without looking for extraordinary individuals like a needle in a haystack, because egotistical stars do not make good teams. On the contrary, a set of participants with good skills, eager to learn and working in unison, make outstanding teams. And it is above all the teams that are the basis of corporate success.

The basis of training at *Toyota* was borrowed from the Americans at the end of World War II: many workers had been recruited and sent to the front and were replaced mainly by women who did not have much know-how. An American training program was set up by the human resources commission of the war, and called the TWI [10.8] (Training Within Industry), as of August 1940. According to its founders, the ultimate goal of this training program was not exclusively directed towards the production of weapons and defense material, but would also contribute to the growth of employees at the mental, moral, spiritual and, of course, technical level.

Shortly after the war, this method fell into disuse in the US, but was spread by American advisors in Japan from the occupation authorities. *Ohno* adopted and introduced certain changes. Again, one detects a feature of the *Toyota* culture, i.e., continuing to use and develop proven methods by highlighting their effectiveness and efficiency before modernity. Such an approach would be difficult in a Western company, which focuses on innovation and novelty to boost business. One rarely sees a succession of CEOs continue staff training with the same method for over 60 years. They would be the laughing stock of all magazines celebrating the exceptional performance of business executives.

What is the content of the training method of TWI? It comprises the following elements:

- Job Instruction: the tasks are segmented into individual elements and the important points are identified; the tasks are demonstrated until employees are able to carry them out themselves.
- Job Methods: these are techniques for determining the need, sequence and responsibility for each task; it is a form of elimination of waste that makes the optimization of the employees' work possible.
- Job Relations: this module provides the basic knowledge for managing and supervising a team, including:
  - how to give feedback to employees;
  - how to address the concerns of employees;
  - how to communicate about events and changes;
  - how to take advantage of their qualities.
- Program Development: training reserved for experts (sensei) who have developed one of the three programs above.

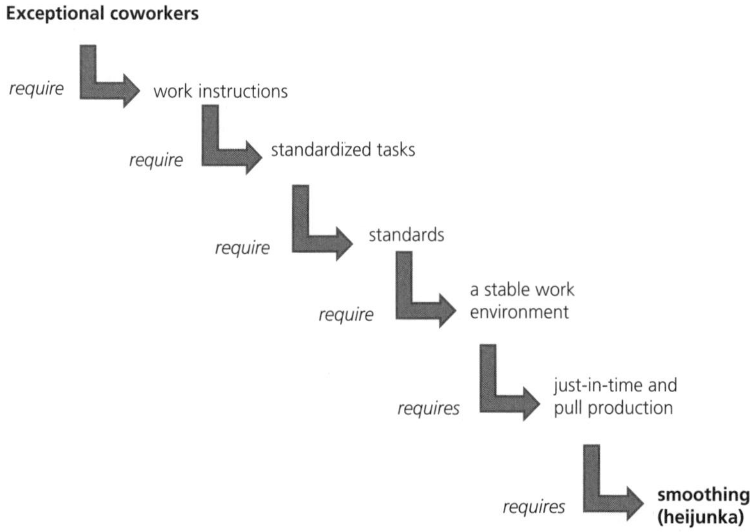

Exceptional coworkers

require → work instructions

require → standardized tasks

require → standards

require → a stable work environment

requires → just-in-time and pull production

requires → **smoothing (heijunka)**

Figure 10.8 Training, cascade of TPS elements.

Training at *Toyota* is the beginning of a cascade of elements included in TPS (Figure 10.8). Currently, TWI has gained renewed interest in the US. Many employees have only minimal training and are hired at the end of their primary education as workers without specific qualifications. Many of them learn their tasks mainly by observing the work of their colleagues.

Training at *Toyota* is not limited to production staff; engineers with degrees from Japanese Universities of Technology gradually learn the craft and how to work as a team under the supervision of a mentor, with many courses given by senior engineers. This training develops technical know-how, but also transverse and personnel skills: the first year includes a double training and awareness period of 4 to 5 months for the production and sales staff. An engineer is considered to be experienced after about 9 years of work. It is after this time that he or she can be entrusted with substantial management tasks with additional training to follow (including, once again, mentoring and courses taught by senior managers).

Finally, as with many multinational companies, *Toyota* has its own education system, the Toyota University, which conducts seminars around the world for its executives. The Toyota University does not have a specific building, but rather a curriculum adapted to the specificities of the automotive industry.

10.5 Fordism and Toyotism

Toyotism can be seen as a production model that has introduced corrective actions to Fordism (Figure 10.9). The figure shows that a majority of the corrective actions affect productivity and quality positively. Compared to other quality and research performance approaches, Toyotism adds logical reasons why these actions simultaneously affect the two most important production parameters.

The solutions of Toyotism

● Specific effect on quality
○ Specific effect on productivity
◉ Effects on productivity and quality

Faults of Fordisme	5 S program	Teamwork	Multitasking workers	Production	smooting	JIT-Kanban	SMED	TPM
Prod. tools without improvement						◉		◉
Fragmentized tasks		◉	◉					
Absenteesim		◉	◉					
Overtime			○	◉				
On-call work			○	◉				
Prod. without flexibility			○				◉	
Too much stock					◉			
Irregularities in production				◉				
Disorder and uncleanliness	●							
Extremely specialized engineers		◉						
Costly maintenance								○

Figure 10.9 The solutions of Toyotism.

10.6 TPS, a model?

10.6.1 A model – adopted by the West – in search of meaning

For the past 15 years, Western production systems have hybridized the concepts, methods and tools of Toyotism with those from the Taylorist–Fordist legacy.

The consequence of such hybridization is often ambiguous: we multiply the number of tools (which are as efficient as they are incomplete) for organizing and managing production, without finding a new concept of a production system that is both comprehensive and unifying, nor a champion such as *Ford* or *Ohno* to promote it on a large scale.

In the 1980s, the industrial West was able to study TPS and to appropriate ideas without spending too much time implementing them deeply into the culture of the company and applying them completely. The automotive industry, logically, became interested in the first *Toyota* model. Gradually, other sectors subject to the effects of globalization became interested in TPS or some of its components. If we agree that TPS is a good summary of the various methods in vogue, it becomes an indispensable reference.

TPS also gave birth to *Lean Management*, a summary of Toyotism that can be applied to all forms of business and is often integrated into the tool of Six Sigma total quality management, and is then called Lean Six Sigma. Lean Management and Lean Six Sigma are treated in Chapter 11. However, the productivity and quality of TPS remain largely unmatched.

10.6.2 Evolution and future of TPS

Is it possible for Toyotism to continue to evolve and increase a manufacturers' productivity? According to *Ohno* [10.9], the potential still exists: *"The Toyota system would be nearing its end if it were envisaged as merely a production technique. However, since it continues to change qualitatively in order to become a management system, I think it still has many good years ahead of it."*

Can Toyotism be applied in businesses other than the automobile industry or that of tangible goods? *Ohno*'s answer is: *"TPS is no longer a production system and, insofar as we are in search of a universal management system, what it is called is unimportant. As soon as we apply our system, as well as Autonomation and Just in Time, the name itself becomes universal."* However, in the context of services, autonomation and JIT must be reformulated and reframed.

10.6.3 A bureaucratic risk?

Is it possible that Toyotism, with its strong culture and attention to documentation, could be a form of bureaucracy that in the end resists innovation? This question was investigated in a recent book [10.10] showing that the culture of Toyota also corresponds to a clash of contradictory trends that support the innovation process:

- move slowly, but also make big jumps of innovation;
- cultivate frugality, but also be able to invest large sums;
- operate very efficiently, but accept a redundant mode when necessary;
- cultivate mental stability, while staying alert in an almost paranoiac state;
- respect the bureaucratic hierarchy while allowing the freedom of dissent;
- maintain a simple communication in parallel with a more complex one.

The principles of *genchi genbutsu* and *newamashi*, allied with pragmatism, prevent the system from falling into a bureaucracy that would alienate it from the work floor.

10.6.4 Gray areas at Toyota?

In 2005, the American NGO *Bluewater Network*, a division of *Friends of the Earth*, accused *Toyota* of joining an American lobby to prevent the US from enacting strict rules regarding fuel consumption and CO_2 emissions, but also from producing luxury hybrid vehicles with a fuel consumption very close to conventional vehicles. Nonetheless, *Toyota* has received numerous awards for its policy and its results when it comes to sustainable development, despite the fact that it brings to the market a number of vehicles (those with four-wheel drive, for example) at the same time as its global innovation, a hybrid vehicle with low fuel consumption, the *Prius*.

Other more serious critics, following the establishment of production centers in Europe, focus on the acute stress of employees and the productivity race, favored by a decrease of the *Takt Time* decided unilaterally by management. When the French Toyota factory in Valenciennes opened in 2001, a car was produced every 108 seconds; in 2008, it was every 60 seconds. Sometimes acute stress is induced by the team itself. Those with the poorest performance can be ostracized, accused by their colleagues of impairing productivity and set goals. Absenteeism is also heavily penalized, and *Toyota* employees who are absent are not replaced, inducing additional stress for the team and animosity toward the person who is absent.

Fatigue, stress, depression, professional illness, accidents, phone harassment of employees who are at home sick, pressures and attempts to blackmail the workers in question in order for them not to report accidents, this was the lot of the French workers [10.11], who were physically worn out and morally broken for a monthly salary of between 1100 and 1300 EUR. This is even more serious as Japan does not have a good reputation for decent working conditions (euphemistically stated); employees take only half of the (meager) annual leave allocated to them, and many additional hours are rarely paid (including meetings and discussions of quality circles, which are held outside "official" working hours). It is in Japan that *karoshi* has been identified and conceptualized, i.e. death from exhaustion due to overwork, or *karo-jisatsu*, suicide in relation to work [10.12]. In November 2007, a Japanese court sentenced *Toyota* for driving one of its employees to death by making him work over 106 hours of overtime in a month [10.13]. The dispute arose from the fact that the company refused to compensate the family of the deceased (from 2005 to 2006, over 310 similar requests have been identified by the Japanese Ministry of Labor). Another case, reported in July 2008, concerned a Toyota engineer aged only 45 who died of heart failure after 160 hours of overtime in two months.

This information is to be compared with other sources, from trade unions [10.14], stating that Asian immigrants who are employees of *Toyota* are deprived of their passports on arrival on Japanese soil, work 100 hours per week for wages that may be half those of their Japanese counterparts, have tasks that are dangerous and exhausting, and that those who are injured are simply dismissed without compensation. Moreover, those who remain have no paid leave.

Thus, Toyotism, far from having softened Fordism or decreasing rates, only strengthens the control of body and mind of the worker, causing problems that occupational medicine has difficulties in preventing [10.15]; the situation of temporary workers in this context is very distressing. Although *Toyota* certainly guarantees long-term employment, temporary and even on-demand jobs still represent over one-third of employees, indicating that production smoothing is far from perfect[2], and that the transformation of a fixed term contract to that of a permanent job must generate "eternal" gratitude of an employee towards the group. In times of market contraction, the company outsources permanent workers to suppliers, who then lay off their employees. *A large number of questions thus remain open.*

---

2 This proportion is much larger for *Toyota* suppliers as here it is close to 50%.

10.6.5 Discussion and conclusions

Since the early 20th century, automobile production has revolutionized three times the global industrial methods and landscape (Fordism, management of *GM* with *A. Sloan*, and Toyotism). TPS is often cited as the most successful system for quality and performance. What relationships are there between TPS and the world of Japanese TQC? Is it at its origin or rather a consequence?

The answer to this must consider the following factors:

- the cultural values of Toyota come directly from its founders (autonomation is one such idea formulated early in the company);
- the contributions of Deming and Juran, like the principle of continual improvement and the Deming cycle, were disseminated throughout Japanese industry with the support of the occupation authorities;
- Toyota's production methods (JIT, SMED, TPM) are directly related to the fragility of the Japanese economy after the war, and they have persisted, since they are more competitive than Fordism production methods.

These methods were phased into what would become a single system, TPS. This progressive construction is typical of *kaizen*.

TPS itself has been described an upward spiral, incorporating within it the tools permitting even better achievement of its objectives. Once TPS was established, the system and its tools were distributed to suppliers and distribution services, to finally be broadcast worldwide through the two books by *Ohno* mentioned above.

TPS is not in itself the cause of TQC, but nor is it a simple consequence.

10.7 Illustration: the financial situation proves TPS is right

Since 2008, the automotive industry has been in serious difficulties, but *Toyota* handled the crisis better than its US rivals, the *Big Three*. The market value of *Toyota* fell by 43%, whereas *General Motors*, its rival from Detroit, lost more than 80% of its value in one year. Regarding sales, the situation was even worse: *Toyota* showed a 9.4% decrease in sales for 12 months from August 07 to 08 against a steeper decline of 20% for *GM*.

This gap and the favorable position of *Toyota* were due to better quality and productivity practices. Moreover, in the late 1990s, *GM* and *Toyota* took opposing marketing options. Despite the intense warnings by environmental NGOs on global warming and the finite nature of oil sources, *GM* – relying on the presumed effects of anti-Kyoto lobbying by the *Big Three* – chose to focus on large sports cars, with a certain success at first, but without realizing that customer expectations were changing.

On the contrary, *Toyota* invested more in the design of a hybrid car, while the prospects for this market appeared uncertain. In early 2008, the *Prius* was already a commercial success, and *Toyota* became the world leader while *GM* fought against bankruptcy, with its shares falling to the same value as in the 1950s. *Toyota's* top position was closely followed by *Honda's* cheaper hybrid vehicle. *Honda* was the only car-producing company to conclude the period 2008-2009 with a profit. The European and American models of hybrid cars, on the other hand, remained in limbo.

Sources say that *GM* had even thought of asking *Toyota* to help it reorganize to adapt to the changing automotive market. But alas, it was too late: on June 1 2009, the legacy of *A. Sloan* filed for bankruptcy, as the first step in an announced liquidation (the largest in the

US). Not even the brightest managers from the prestigious Sloan School at MIT succeeded in keeping *GM* afloat. Certainly, it is not (yet) the bankruptcy of American management excellence, but this bitter failure casts real doubts on the true capabilities of American technological universities – even at the highest level and beyond the high-sounding declarations about their programs and intentions – to instill innovation in industrial management and to bring new added value.

By studying the 14 principles of *Toyota*, we see that it is not only the production system but the entire company and marketing that are influenced by the principles of total quality management set by *Toyota* (which is consistent with the vision of *Ishikawa*'s TQC). As such, the first of the 14 principles emphasizes the importance of basing decisions on the long term, which served *Toyota* well 20 years ago and continues to do so today.

10.8 A specific tool of TPS: the A3 Report

A TPS tool for problem-solving makes it possible to become more familiar with the culture of *Toyota*: the A3 report [10.16]. This is a method of problem solving, a communication support, and a tool for process improvement and organizational transformation resulting from the *Deming* cycle.

10.8.1 What is a problem for Toyota?

A problem is:

- failure to satisfy a customer's needs:
  - references,
  - quantity,
  - time,
  - quality,
  - price;
- inability to carry out the work as planned.

Toyota works to bring such problems to the surface and resolve them at the root, not by specialists like *Super Master Black Belts*, but by all without distinction, and cost, and this at all levels [10.17]. The method used is the A3 report, applied both at the top and bottom of *Toyota's* hierarchy. It is a linking and unifying tool of corporate culture.

10.8.2 Toyota's A3 Report

The A3 report is a problem-solving tool that uses an A3-size (420 mm x 297 mm, two A4 sheets) sheet of paper on which a summary of the problem's solution is written and provided to people participating in the action. Toyota has adopted this format because it considers this space to be sufficient: if one is able to communicate to stakeholders the solution of a problem on a sheet of A3 paper, it is a sign that the thinking is concise and structured.

The content is divided into two halves to gather the main results of the basic PDCA cycle: the left part defines the problem and the right offers solutions. A good A3 report tells a story – giving life to a problem and showing the way to a better future!

10.8.3 Staff development at Toyota

The A3 report is used for employee development. It mobilizes the following faculties:

- a logical state of mind capable of separating and determining cause and effect;
- objectivity;
- focus on processes and their outcomes;
- an analytical state of mind, with written and graphic communication skills;
- the search for agreement from stakeholders on the content and proposed solutions;
- consistency, robustness of reasoning, and alignment of the content with the goals and philosophy of the organization;
- system guidance: how the problem and its solutions are connected to the rest of the organizational structure.

During problem solving, the following skills are enhanced:

- perception of problems;
- observation, search for evidence;
- understanding the processes;
- knowledge of equipment;
- communication with other employees;
- imagination and creation of solutions and countermeasures;
- ability to negotiate;
- group work;
- ability to persuade with facts;
- leadership skills.

10.8.4 The steps of A3 for solving problems

The (classical) steps of problem solving are eleven in number and were described in particular by *Durward Sobek* [10.18]:

1. *Identify the problem or need*: if the way the work is carried out does not fit the expectations, or if a goal or objective is not achieved, we have a problem (or, if you will, a need).
2. *Seek to understand the current situation:* before you can correctly identify a problem, it is important to understand the existing situation. To this end, it is recommended that one:
   - observe the work process at the workstation in question and document the results of these observations;
   - draw a diagram that shows how the current work is carried out; any system that formally describes a process in graphic form may be used, but often simple figures of arrows and sticks will do;
   - quantify the importance of the problem and, if possible, represent the data graphically.
3. *Root cause analysis*: when a good understanding of the current process (i.e., the one that needs to be reviewed) has been acquired, it is time to identify the root causes of errors or inefficiencies. To do this, we must first list the main problems. Then, ask the appropriate question "why?" to indentify the root cause. *A good rule of thumb is to assume that the root cause is not identified until the question "why?" has been asked at least five times (and received a satisfactory answer).*

   The succession of "whys?" may prove to be a succession of causes and effects, but may also lead to causes that are independent of each other. The content of the answers to the questions "why?" can spontaneously lead to consequences that may be the cause of the problem, in which case, the answer to the question is followed by another sentence that starts with: consequently, therefore, etc.

A tool commonly used individually or in teams to identify root causes is the *Ishikawa* diagram: indeed, according to the rule, there is not *a single* root cause to a problem, but potentially several.

4. *Develop countermeasures to eradicate the root cause:* when the current situation is completely understood and the root causes of the problem are explicit, the next step is to select countermeasures. These include the changes that must be introduced into the process in question in order for the organization to approach the ideal or make the process more efficient. In general, it is recommended that the countermeasures meet all three rules borrowed from *Steven Spear* and *Kent Bowen* [10.19] (close to SIPOC of Six Sigma)*:*
   - specify the outcome, the content, and the sequence of tasks in the process in question;
   - create clear and direct connections between the applicants and suppliers of goods and services related to the process in question;
   - eliminate loops, unnecessary work, and delays.
5. *Develop the desired situation*: the countermeasures neutralizing the root causes of the problem will change the operating conditions of the process. This step describes how the work will take place after the establishment of countermeasures. In the A3 report, the situation should be represented by a diagram (similar to the one for the current situation) that illustrates the progress of the proposed new process. The specific countermeasures must be rated or listed and the expected improvement must be specified qualitatively and quantitatively.
6. *Draft an implementation plan*: in order to achieve the objective (target conditions), a sound implementation plan must be established. It should include a list of actions that are needed to implement the countermeasures and create the target conditions. It must also include the people responsible for each task and the deadline. Any other relevant factors, such as cost, can also be added.
7. *Prepare a follow-up plan verifying the expected results*: how does one ensure that the measures taken will pay off? Is the described problem really eliminated by the application of countermeasures? The purpose of this step is to prepare a careful plan of analysis and observations that validates the diagnosis and ensures that the A3 project is completed.
8. *Discuss the elements of the A3 report with all involved parties*: it is vital to communicate with all parties concerned by the plan and to reach consensus. This is an important step as it determines the success of the project. Any objections must be taken into account, and it may be necessary to adjust the countermeasures, target conditions or plan. Each stakeholder needs to be aware of any changes and be convinced of the benefit it brings to the organization.
9. *Obtain approval for the implementation:* if the person in charge of the A3 report is not a manager, approval of the hierarchy is required. The line manager verifies that the problem has been adequately studied and that all stakeholders are incorporated into the project. He/she then approves the changes and authorizes their implementation. This step may take the form of coaching, as the managers at *Toyota* assume not only functions of control and supervision but also training. *Hito zukuri* means "create men of talent": the idea is for *Toyota* to consciously use problems to train men on the job.
10. *Implement the plan.*

11. *Evaluate the results*: improvement of the process does not end with the implementation. According to the *Deming* cycle, the actual results of the countermeasures must be evaluated in the long term and compared with those expected. In case of a discrepancy, the A3 process should be repeated until the objective is reached.

Steps 1 through 8 represent *Plan* (where step 5 includes the planning of *Do* and step 6 that of *Check*). Step 10 corresponds to *Do* and step 11 to *Check*. Based on step 11, another problem can be identified and the A3 process can start anew (*Act*).

10.8.5 The content of the A3 report

Steps 1 through 6, as well as 7 and 11 are recorded in the A3 report according to the model in Figure 10.10. The academic equivalent of the A3 report is the flip-charts used in conferences on research topics or the basic structure of scientific articles (e.g., Introduction, Materials and Methods, Results, Discussion, Conclusions).

The A3 report favors graphs:

- Ishikawa diagrams;
- flow diagrams;
- histograms;
- charts;
- scatter graphs;
- pie charts;
- graphs with vertical or horizontal bars.

The model presented is a basic example; other representations exist.

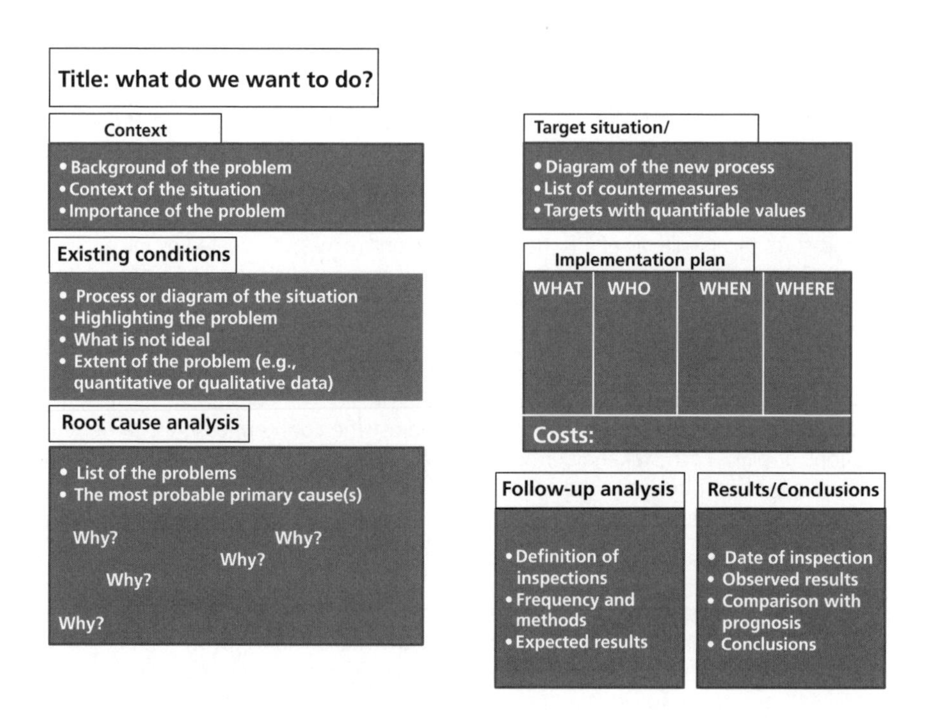

Figure 10.10 Presentation and content of an A3 report.

10.8.6 Advantages of the A3 method for solving problems

- Provides a unique and easy-to-use approach for problem-solving – which can be applied to ANY type of problem: if everyone knows what is means by "do an A3", it allows people to focus on the root causes of problems – without having to reinvent new ways of working together.
- Motivates employees to identify problems, and bring them to attention; this is the opposite of keeping silent about problems or pretending that there are none.
- Stimulates group dialogue and creativity in finding solutions – based on objective observations of indisputable facts; this is the opposite of the person trying to "sell" his subjective opinion and non-accomplished ideas.
- Counteracts the bad habit of humans to determine solutions before they have identified the root causes; and the bad habit of wanting to be the heroic firefighter, instead of preventing fires.
- Enables staff on the front line to share the responsibility to improve their work; this is the opposite of commanding and controlling from above, which does not cause workers to think or be conscientious.
- Encourages managers to manage the means of production by a true understanding of the causes and consequences, seemingly insignificant details; this is in contrast to the approach "just show me the results" of management by objectives, made so popular by many institutes delivering MBAs.
- Develops an organizational culture of the type "learn to learn"; not just do what you are told to do.
- Creates more efficient modes of thinking, working and learning together.
- Prepares for the emergence of new leaders in a scientific and reproducible manner; this actually shows how to manage – and how to transform one's own employees into new leaders.

Finally, the A3 report takes into account two key factors for *Toyota*:

- The employees are the leading resource for the company. Over the last 50 years, the company has progressively gone from an economy of tangible products to an economy relying heavily on services, in which the cost of people is the largest.
- The development of employee competence at all levels is a major factor for the good health of the company. This is in any case the practice of Toyota, who deploys its own and single method of problem solving, the same for the operator and for the CEO: the "A3" method.

10.8.7 Conclusions

To be able to solve a customer's problem or crisis is a key skill for a company and is conferred by the A3 method of *Toyota*. The A3 report is not a technique, it is a way to get people to progress by requiring them to be precise. The method:

- is pragmatic and focused on describing problems, thus facilitating the introduction of simple, economic, and efficient solutions;
- includes the concepts of the Toyota Production System that are relevant when it comes to "seeing the problems";
- renders it possible to describe problems in such a way, and with such precision, that it becomes easier to find solutions;
- is practiced daily by all and allows the appearance of a strong team spirit that develops the strengths of the company;
- gets employees to grow;
- is the basic "brick" of Toyota's culture.

10.9 Glossary of Toyotism

"Words are symbols that postulate a shared memory." J.L. Borges.

Toyotism, like any strong culture, has its language and its expressions. Below are often used phraæses and words in Japanese (or their English translation).

A3 report: More than just a tool, the A3 report is a systematic method to simplify problem solving, a "storyboard" retracing the complete history of an improvement operation on a single sheet of paper measuring 11x17 inches, where the left half describes the problem and the right the solutions.

Andon: Signal or panel that lights up when the operator presses an alert button or pulls on a wire alarm. The *Andon* panel allows the supervisor to immediately identify at what station the problem has occurred and to rush there before the line reaches the next fixed point, where it will stop if the supervisor or team leader has not found a countermeasure and canceled the alert. *Andon* is also used to indicate when there is a need to accelerate replenishment. It constitutes a system of "lights" that indicate the production status; the number of "lights" and their color can vary, but in general: Green = no problem, Yellow = situation requires attention, Red: production is stopped, urgent need of attention.

***CANDO*:** (**C**lean-up, **A**rranging, **N**eatness, **D**iscipline and **O**ngoing improvement.) Compare **5S** (*Seiri, Seiton, Seiso, Seiketsu* and *Shitsuke*) = **5C** (*Clear out, Configure, Clean and check, Conform, and Custom and practice*).

Cells: Implementation of flow machines. The machines perform successive operations on the same part (in the direction of flow) respecting a single piece flow. In general they are "U"-shaped to permit maximum flexibility for the operators.

Chaku-chaku: Term signifying loading-loading in Japanese: takt flow work, which allows one to move parts from one machine to another by loading/unloading. Ideally, the machines unload automatically so that operators can transfer a part directly from one machine to the next without waiting. *Chaku-chaku* may be confusing because the movement of the operator is often done in a direction opposite to the flow of the part: unloading of machine B, then loading it with the new awaiting part, starting of machine A (which is earlier in the flow), unloading, loading, starting up, etcc. There is usually a standard stock of a part between the machines.

Chorobiki: The smooth and regular removal from the *shop stock*. Reducing the batch size is the first step towards reducing inventory, but if the removal frequency does not increase in parallel, stocks will continue to be significant. *Chorobiki* corresponds to the regular removal process from the *shop stock* (every 20 to 40 minutes) starting from the sequencer, in order to continuously fill the staging area of the trucks, rather than arriving all at once when a truck starts.

External operation (see SMED): Any operation that may be carried out safely during the running of the machine.

FIFO: *First In, First out* (supermarket, and streaming). If streaming is not possible, FIFO is attempted; if FIFO is not possible, a supermarket is installed.

5S: This stands for five Japanese words starting with an S, and is used to create a good working environment for added-value operations. *Seiri* means "sort and eliminate" what is needed from what is not; *Seiton* (a dedicated place for everything) means storing at the workstation what is necessary, and placing it in an appropriate location in order to get to work immediately; *Seiso* involves daily cleaning in order to keep the station in good order; *Seiketsu* is the cleaning of workstation elements with a high level of cleanliness and health in order to "Be able to eat off the floor"; and *Shitsuke* refers to the rigor necessary to maintain the other four Ss day after day. 5S is an excellent introduction to the organization of the workplace leading to "work standards" on the one hand and independent teams on the other.

5 Why: The basic method of problem solving. *Ohno* often insists on the need to ask oneself five times the question "why?" to go beyond the symptomatic causes and find the root causes (on which we can then act to eliminate the problem once and for all). For every "why?" one should ask the *gemba*, the person who does the work and who can give a concrete answer rather than something dreamed up by deduction or from behind a desk.

Gemba: The floor (the actual place) where the value creation is actually taking place. It is also referred to as "*genchi genbutsu*": go observe on the floor, in the workshop, how things work and actually talk with the operators of problems when they occur, to gain practical experience of the situation. Originally, *gemba* referred to the workshop.

Gemba kanri: The literal translation of the Japanese expression is "workshop management"; this means managing the process and outcome of production. It is the overall management of a workshop from start to finish, in accordance with *Kaizen* and by applying the 5Ws.

Gemjitsu (reality): Approach reality with real, quantified data.

Genbutsu (actual parts in question; samples, waste): Observe the parts with problems, analyze by comparison with a good part or with the standard.

Genchi genbutsu: Go to the source and onto the floor to "capture" the reality there.

Hanedashi (or *Auto-eject*): Device enabling a machine to automatically unload a part without waiting for the intervention of an operator.

Hansei: Sessions of systematic reflection at the end of each activity. This is not only a learning opportunity but also a moment for sharing experiences between functions. The employment of *hansei* is of utmost importance in *kaizen*, because it is a time for evaluation and the setting of new targets.

Heijunka: Complete smoothing of the diversity and load, or "split-mix", i.e., organizing the queue to optimize the flow of different products using the same resources. This involves, firstly, smoothing the actual demand of the customer (and yes, stocks!) so that the next day's production is as close as possible to the day before and, secondly, to mix production volumes in order to achieve a "single-part flow" philosophy on the production line.

Heinrich principle: A principle of worker safety, the current ratio of incidents and accidents in the factory. It is broken down into serious accidents: minor accident: no accidents = 1: 29: 300. This means that for 1 serious accident, there were previously 29 people who have been through the same incident, but with less severe consequences, and 300 people who were lucky enough to get by unscathed.

Hiyari kyt (kiken-yochi training): Practice that aims to anticipate the occurrence of an accident by detecting and identifying the danger or risk in advance.

Hiyari report: Report made by an employee to his superior to report the existence of a situation that could lead to poor quality or create an accident.

Internal operation (see SMED): Operation that cannot be performed unless the machine is stopped.

Jidoka: Build quality in products and services rather than eliminate waste. This is a set of detection systems of nonconformities that enables stopping of the production, either manually or automatically, in order not to produce bad parts (*muda*). *Jidoka* also makes it possible to instantly focus efforts on problem areas and continuously resolve problems, while giving operators and their supervisors a sense of responsibility.

***Just in time*:** Produce what is needed, when needed and in the desired quantity. JIT is flow where the input of an upstream workstation is defined by the needs of the next one downstream (*standardized work*).

Kaikaku: Drastic change, as opposed to the term *kaizen*, which consists of continuous change of low-amplitude change.

Kaizen: Organization of team discussions to stimulate continual improvement. The goal of *kaizen* is to eliminate *muda* in all its forms. This involves making things simpler and easier to perform. To succeed, *kaizen* starts with "work standards".

Kanban: Decentralized material information systems on the basis of continuous restocking. The term *kanban* corresponds to maps, or sheets, which are a signal to produce only the parts that have been consumed in the order of their consumption. There are several types of *kanban*, mainly: (1) production instructions (triangles or maps), (2) sampling instructions (sampling or supplier *kanban*).

Kosu (= labor time): Time required for the manufacture of a part.

Lead Time: The time from when a part enters a process (a line, a workstation ...) to its output. It is a fundamental concept of lean management and is the time required to produce a product from acceptance of an order to shipment. The lead time is divided into A + B + C:

- A: from receiving the order to the start of the work;
- B: from the early work on materials and components to the finalization of the product (production time + time of non-production)
- C: from the production of the first part in the container to the last (Takt Time of the product multiplied by the number of parts in a container).

Level selling: A system of relationships with a customer, the goal of which is to eliminate peaks in demand caused by the system itself (e.g., due to quarterly or monthly objectives). This system seeks to build a long-term relationship with customers so that future purchases can be anticipated by the production system.

***Manifest*:** Paper corresponding to a vehicle or product giving instructions as to the parts that must be in the truck or attached to the product. This worksheet makes it possible to move both the information and product.

***Mannerika*:** Use of the tools of *kaizen* without the proper spirit – thus reducing it to a mere procedure.

MIFA, M*aterial and* ***I****nformation* ***F****low* ***A****nalysis:* Analysis of material flow and information (also known as *Value Stream Mapping*), the mapping tool that allows *Toyota* to add information flows to conventional maps of the moving of parts, and thus regulate stock to avoid bottlenecks.

Milk run: Loading of parts at several points and delivery to several points in a single loop. This is the opposite of several separate trips.

Morning market: A routine tour of *gemba* to examine the waste (*genbutsu*) of the previous day before the start of the new day, so as to implement countermeasures faster. This is the principle of *gemba genbutsu*. It is done first thing in the morning. The people encountered are the workers on the floor (*gemba*).

Muda: Any activity that consumes resources without adding value to the customer. There are seven main categories of *muda*.

Mura: Interruptions in the workflow. In particular, variability is a form of *mura*: if each person in a workflow follows a repetitive cycle, but one of them has variability in the repetition of his/her operations, this will create waiting times or accumulation in the flow.

Muri: Difficult working conditions for operators or equipment. All work that is mentally or physically difficult for the operator creates *muri* and *muda* in addition to being taxing on the person himself, i.e. a waste of energy.

Nemawashi: Consensual decision-making.

Obeya: Room for carrying out a project where all progress indicators are presented and used.

Poka yoke* or *baka yoke: Anti-error, anti-crazy, anti-moron or fool-proof systems; small practical systems that can immediately identify that the machine produces poor quality or that it does not follow the work standard, either by blocking the following operations or by lighting an LED that indicates a problem. Preferably, *poka-yoke* are designed and created by the operators themselves.

Pool stock: A "lung" that can absorb small variations in quantity and mix from one day to the next in a smooth environment, when the flow of parts goes directly from the *shop stock* at the foot of the cell to the truck preparation area. The actual customer demand is compared to the smooth production program. Two types of maps are introduced into the sequencer: *kanban* maps corresponding to the actual demand of the customer, and some "special" maps if the demand is slightly higher than the average daily consumption. The "normal" maps are drawn from the *shop stock* at the foot of the production cell, while the "special" maps are drawn from another buffer stock: the *pool stock*. If instead, the customer request is below average daily consumption, the change will be absorbed by special maps thus making it possible to replenish the *pool stock*.

Production Program: Production is smoothed over a week so that each day of production is similar to the previous one. On the basis of firm and/or estimated orders during the week, it is possible to establish a *takt time* per part, and therefore a number of parts to be produced every day. The production program can then be smoothed with regard to the mix for each day, so that each team is similar to the previous as well as to the following. The aim is first and foremost to produce all parts every day, then all parts by all teams.

QRQC (*Quick Response to Quality Concern*): Managerial system enabling a quick response to problems of quality (initially), but which applies universally to any type of problem.

San Gen Shugi (*San = Three + Gen = Actual + Shugi = Ideology or Culture = the 3 actual ones):* Posture adopted during analysis of a problem, based on *genjitsu*, *gemba* and *genbutsu*. The keys to make *kaizen* successful are: go to the workshop, work with the on-going products, and establish facts.

Sensei: Highly respected master or instructor carrying out *On the Job Training* or *Learning by Doing*. It is a term used for external consultants who have mastered the skill of TPS.

Shop Stock: The part of the stock of finished or semi-finished products covering the risk process: downtime or defects. The *shop stock* is part of the "supermarket" and is positioned next to the output of the process. The autonomous production team is responsible for it. It allows visual management of the process: during push-driven flow the stock is generated by production programs, whereas for pull-driven flow, the production is organized by the replenishment of the *stock shop* (by kanban maps) based on sampling of the process at subsequent workstations.

Single-piece flow: Part by part in continuous flow.

SMED: *Single Minute Exchange of Die,* method developed by *Shigeo Shingo* in order to reduce the tool change time. This change time is measured from the last good part to first good part. The three essential steps of the method are to:

1. distinguish between internal operations (machine off) and external operations (machine running);
2. transform the maximum number of internal operations into external operations by preparation. For example, the new tool can be kept close to the machine while the latter is still running, rather than going to get it after the machine has stopped;
3. streamline all operations (especially tuning operations once the new tool is mounted). The SMED method is very effective for reducing changing times, with no investment at the beginning (thanks to organization), then by gradually modifying the facilities. It is customary to set goals for reducing levels by 50%. The goal is to go below 10 min, and then in a second step to make the change during a work cycle.
   SMED distinguishes between two time scales:
   - Single-digit setup: preparation and installation of the tool in less than 10 min;
   - OTED: One Touch Setup: change of a tool in one movement (e.g., by pressing a button).

Spaghetti Chart: A graphical aid used in lean manufacturing activities. It is used to detail the actual physical flow and distances involved in a work process. Processes that are not streamlined frequently are poorly laid out with work/product taking a path through the work area that looks like a mass of cooked spaghetti. A spaghetti chart often traces the walking patterns of workers in a process, ranging from manufacturing settings to healthcare.

***Streaming*:** Single piece flow or one piece flow (part by part). This is an important point in lean management because streaming is an effective means of detecting problems in the flow. Detection and prompt resolution of issues is one of the major principles of TPS.

***Supermarket*:** System used to store a certain amount of finished or semi-finished products. It is used when streaming or FIFO are not possible.

Takt Image: Represents the vision of an ideal state where we have eliminated all waste in order to make it possible to perform a *single piece flow* to the rhythm of the *takt time*.

Takt Time: *R*eflects the ideal customer consumption. This is the main tool for smoothing customer orders in production. From the monthly production plan, we fix a production "tempo" corresponding to the daily production time (with 100% equipment during normal working hours, i.e., all day minus breaks), which is divided by the customer demand: working hours/client request. The *takt time* allows for smooth production schedules during a week or two and determines a maximum time to perform each operation.

3 G: "**G**emba, **G**enbutsu, **G**emjitsu":

- Gemba: the actual floor – where the creation of value is actually taking place;
- Genbutsu: what can be actually observed on the floor and, through correlation, can enable to find the causes of the problems;
- Gemjitsu: the actual event that led to something observable on the floor.

***3 K*:** Traditional perception of *gemba*. *Kiken* = dangerous, *Kitanai* = dirty, *Kitshui* = stressful.

3 M: *Muda* (no added value), *Mura* (variability) and *Muri* (overburdening of staff or machines).

TPM, *Total Productive Maintenance:* A form of maintenance that enhances the responsibilities of the production staff, allowing them to gain in-depth knowledge about the machines, and that reduces the downtime of the production equipment for repairs, replacement of parts and controls prescribed by the manufacturer of the equipment.

***Visual Control*:** The placing of tools, parts, and information about the production processes and performance indicators of the production so that they are visible and so that the state/condition of the system can be understood by anyone at a glance (partly carried out with the 5S program).

***Visual management*:** Implementation of physical means in the workshop (*shop stock*, *kanban, andon*, etc.) to ensure at first glance that production operations are proceeding normally – or to quickly identify anomalies. It is an effective management tool to clearly and visibly provide information and *gembutsu* to both operators and managers so that the current state and the goal of *kaizen* are understood by everyone.

VSM (or MIFA) *Value Stream Mapping* (or *Material and Information Flow Analysis*): A mapping tool that allows *Toyota* to add information flows to conventional maps of the movement of parts, and thus consider the stocks to avoid bottleneck issues.

Waste Index (or *Muda Index*): Index measuring the level of waste of an organization primarily for benchmarking between companies in the same industry.

WIP (*Work In Process*): Complete inventory of parts from the beginning to the end of the process.

Work standard (*standardized work):* The sequence of operations to be carried out in order to perform a task without *muda* in a given amount of time (*takt time*). The work standard is the key to *kaizen*. If one fails to perform an operation in the *takt time*, then one should question the differences in practice compared with the standard, and eliminate all causes of divergence. Once all sources of variability have been eradicated, one may find that the production time is sometimes faster than the *takt time* – in this case the standard needs to be altered. A work standard consists of three elements: *takt time*, work sequence and standard stock in the process.

WWWWHW (What, Who, Where, When, How and Why?): The WWWWHW method makes it possible, for all dimensions of a problem, to obtain basic information sufficient to identify its essential aspects. It adopts an analytical approach of constructive criticism based on systematic questioning. The idea is to ask questions in a systematic manner in order not to miss any known information (St. Augustine: *Quis, quid, quando, ubi, cur, quernadmodum, quibus adminiculis?*):

- What do we do?
- With what is it done?
- Who does it? And why this person?
- Where is it done?
- When is it done?
- What quantity? How much does it cost?

- How is it done?
- Why does this problem exist? And why is it done? And why is there ...

To conduct a critical analysis, one should ask "Why?" for each question.

References

[10.1] JAMES P. WOMACK, DANIEL T. JONES and DANIEL ROOS, *The Machine that Changed the World – the Story of Lean Production*, Rawson Associates, 1990, *Chap. 3, The Rise of Lean Production.*

[10.2] SATOSHI HINO, *Inside the Mind of Toyota – Management Principles for Enduring Growth*, Productivity Press, 2006, p. 282.

[10.3] Inspired by Figure 1.1, p. 6, JEFFREY K. LIKEY, *The Toyota Way,* McGraw-Hill, 2004.

[10.4] Ibid., pp. 37-41.

[10.5] TAIICHI OHNO, *Toyota Production System, beyond large scale production* and *Workplace Management*, Productivity Press, 1988.

[10.6] JEFFREY K. LIKER and DAVID MEIER, *The Toyota Way Fieldbook – a Practical guide for implementing Toyota's 4Ps,* McGraw Hill, 2006, p. 64.

[10.7] TAIICHI OHNO, 1988, op. cit., p. 69.

[10.8] JEFFREY K. LIKER, DAVID P. MEIER, *Toyota Talent – Developing Your People the Toyota Way*, McGraw – Hill, 2007.

[10.9] TAIICHI OHNO et SETSUO MITO, *Présent et avenir du Toyotisme*, Masson, Paris, 1992.

[10.10] EMI OSONO, NORIHIKO SHIMIZU, HIROTAKA TAKEUCHI, *Extreme Toyota – Radical Contradictions that Drive Success at the World's Best Manufacturer*, Wiley and Sons, 2008.

[10.11] *Toyota, Cette Machine qui Brise les Hommes*, article in *l'Humanité* from April 11, 2008, the information in which fully tallies the story of S. Kamaka, a Japanese journalist who worked for five months in the factories of Toyota in Nagoya (J), however thirty years earlier. This is proof that the system endures.

[10.12] *Suicides sur les lieux de travail, la faute au toyotisme?*, article by S. Chevalier in *le Monde* on April 22, 2008.

[10.13] *Court rules Toyota employee worked to death*, news from Reuter on November 30, 2007.

[10.14] *International Study: Toyota abuses workers at home and abroad,* published on October 1, 2008, on the website http://www.labournet.net/world/0812/toyota1.html, accessed in July 2009.

[10.15] SATOSHI KAMATA, *Toyota, l'usine du désespoir*, Demopolis, 2008. This story, written in 1972 by a Japanese journalist who worked five months in a factory of the group, focuses on stress, injuries, working 12-hour nights (while production companies generally have three separate teams for the 24 hours, the famous three-shifts), the competition between the plant management to win the distinction of the highest productivity without hesitating to harass the workers, the high staff turnover rate in Japan, the constant supervision of the plant's safety committee, and the French workers of the factory in Valenciennes (see Note 7), but also based on the content of the preface of P. Jobin, which gives newer information on the status of Toyota employees.

[10.16] JEFFREY K. LIKER and DAVID P. MEIER, 2006, op.cit., *Chap. 18. Telling the Story using an A3 Report.*

[10.17] Ibid, p. 310.

[10.18] DURWARD K. SOBEKS and ART SMALLEY, *Understanding A3 Thinking – A Critical Component of Toyota's PDCA Management System*, CRC Press, 2008, p. 20.

[10.19] STEVEN J. SPEAR and H. KENT BOWEN, The DNA of the Toyota Production System, *Harvard Business Review*, Sept-Oct 1999.

Chapter 11

Total Quality and the Six Sigma Approach

Key concepts: Black Belts, box plot, capability, champion, CTQ, Design for Six Sigma, DMADV, DMAIC, Green Belts, Lean, Lean Management, Master Black Belt, median, mode, SIPOC (definition of inputs and outputs), Six Sigma, Six Sigma Council, SMART (definition of goals), standard deviation, tollgates, variance, Yellow Belts.

11.1 Introduction

11.1.1 History

A standard popular all over the world

If ISO 9001 is the most popular European quality standard, the Six Sigma approach takes the prize in the US. It is essential to introduce the Six Sigma approach, which focuses as much on *soft* (not measurable in financial terms) as on *hard* (directly accessible to accounting) profits.

It all started in 1981, when the large American groups began to bear the brunt of the excellence of Japanese companies: *Motorola* was facing aggressive competition that eroded its market share, since the competition was selling similar products of better quality for less. To cope, *Motorola* decided to reduce poor quality by a factor of ten, and began a major program over five years. In 1986, the company developed a method of improving quality, named Six Sigma, with the goal of less than three defects per million parts produced. In 1988, *Motorola* received the Baldridge Award (and was to win it a second time in 2002). According to the awards regulations, *Motorola* had to make the tool available to other US companies, and the name Six Sigma became a trademark of the group. The improvements to *Motorola's* products and processes led to an increase in the group's turnover, remarkable savings, and very good performance of its market share.

Since then, many companies in the US and Asia have introduced the Six Sigma approach; some examples are: *IBM* (1990), *Texas Instruments* (1991), *General Electric* (1996), *Nokia, Toshiba* and *3M* (1998), *Caterpillar* (2001), *Nissan, Société Générale* (2002), *Axa* (2003), *Monsanto, Xerox* (2004), *BNP Paribas, Mastercard Worldwide* (2005), *Orange, France Telecom Group* (2006), etc.

11.1.2 Very visible savings

The improvement in quality and savings that can be accomplished by the implementation of Six Sigma is often impressive. The company *General Electric* announced the following amounts of costs/investment returns (in millions USD) 200/150 in 1996, 400/600 in 1997 and 400/1000 in 1998. In 1999, *Allied/Honeywell* estimated their savings generated by Six Sigma to be 600 million USD. For the period 2000 to 2002, *Ford* estimated savings of 1 billion USD. In 2004, *Dupont* developed no less than 2500 projects and additional revenues were estimated at 1.5 billion USD. Moreover, the US military declared in 2007 that 2 billion USD had been saved in its recruitment process thanks to Six Sigma.

As a final example, *ITT Industries*, a US company of 40 000 employees, was highly successful in implementing "Lean Six Sigma", a variant of Six Sigma. In 2000, *ITT* Chairman, *Lou Giuliano*, former CEO of *ITT*, presented to the Stock Exchange on Wall Street an objective for improving its consolidated earnings within 3 years by 435 million USD. Thanks to Lean Six Sigma, not only was the result achieved, but the share price of *ITT* quadrupled: from 25 USD (March 2000) to 80 USD (April 2004), and then to 125 USD (2007).

This illustrates the characteristics of the Six Sigma approach: it targets highly visible savings, which will be taken into account by the shareholders, but also by the financial authorities of the organization and the board of directors, which determines the remuneration of the president of the organization. Since it is known that many senior executives of a company are partly paid in shares, one can easily understand the purely personal motivation and interest of managers when it comes to Six Sigma. Compared to such success, obtaining the Baldridge Award is simply good to have because, on its own, it has little effect on the board or potential buyers of shares. However, they often go nicely together, such is the lesson of *Motorola*.

Finally, Six Sigma certainly targets significant savings, but its implementation is very expensive (at *ITT*, 1% of the staff was dedicated exclusively to the Six Sigma project, not to mention the involvement of Green Belts and employees). Senior executives must demonstrate to external and internal stakeholders that these amounts have been invested wisely, otherwise they will be penalized by the board and the share price may fall. All these factors explain why, once the decision has been made, managers monitor and lead a Six Sigma project very closely.

In conclusion, Six Sigma owes its success and popularity with managers and presidents of large companies to two features: *quality improvement*, and *substantial associated savings* that boost the performance of the organization. It is (alas?) often the economic incentives that most motivate them to adopt this approach.

Today, Six Sigma is deployed in both service companies (telecommunications, banking, insurance) and in industry.

11.1.3 Six Sigma in brief

Six Sigma is a method, a system for managing and ensuring the success of an organization by focusing on:

- the needs and desires of customers,
- the management and improvement of processes,
- the activity of employees, and
- an appropriate use of facts, data, analysis results and records.

Six Sigma has the following objectives, achieved to varying degrees in companies having implemented the method:

- cost reduction,
- increased productivity,

- increased market share,
- customer loyalty,
- reduced time cycles in the company,
- reduced number of defects,
- change of culture, and
- enhancement of the development of services and products.

Six Sigma makes it possible to achieve goals long regarded as contradictory by American and European companies:

Reduce errors to almost zero	**AND**	*Do things faster*
Motivate staff in its understanding	**AND**	*Respect instructions on how to work*
Improve processes and procedures	**AND**	*Expedite the work to be performed*
Measure and analyze what is done	**AND**	*Develop and implement creative applications to increase performance*
Fully satisfy the customers	**AND**	*Obtain exceptional profitability*

Six Sigma is an eclectic approach that borrows many elements from the contributions of *Deming*, *Juran*, *Taguchi* and common instruments of quality. Six Sigma contains elements common to all previous initiatives, such as:

- continued efforts to obtain stable and predictable processes by reducing variation factors to ensure the success of the organization and its market share;
- assurance that management and manufacturing processes have characteristics that can be measured, analyzed, improved and controlled;
- a way of improving the overall process performance inspired by Deming with two variants:
  - if the process must simply be improved, **DMAIC** (for ***D****efine*, ***M****easure*, ***A****nalyze*, ***I****mprove*, ***C****ontrol*) is used;
  - if the process must be completely rethought and recalibrated, **DMADV** (for ***D****efine*, ***M****easure*, ***A****nalyze*, ***D****esign*, ***V****erify*) is implemented.

What then is its specificity? The key characteristics of Six Sigma are:

- to focus uncompromisingly on clear objectives and measurable (and significant) savings for each Six Sigma project;
- the vital importance of a firm commitment by managers at the highest level for the projects and the Six Sigma approach (in the case of General Electric, its chairman, Jack Welch), but also the direct supervisors who will use the resources;
- an infrastructure for support, staffed by personnel trained specifically for Six Sigma, and divided into several categories of skills (the three main ones are Master Black Belt, Black Belt, Green Belt), for each Six Sigma project;
- an absolute commitment to the decisions made based on facts and studies, with measures and quantifiable data, not opinions.

11.2 Is Six Sigma more efficient than other TQM approaches?

Another factor in the success of Six Sigma is that a number of users claim that this approach is not only more efficient but also that the chances of achieving the goals are much higher

[11.1]. It is true that Toyotism has had a striking success but, unlike Six Sigma, it is not easy to implement in a non-Japanese culture.

Below are some examples of the weaknesses of other TQM approaches that can be avoided by Six Sigma:

1. The concept of TQM training is often either evaded or not clearly defined; this is not the case with Six Sigma, which has specialists with variable skills who are aware and trained in this approach according to appropriate divisions, and this during their first project to implement the approach.
2. The TQM approach is too often seen as an exotic activity, conducted by teams in a context outside that of daily business, and superimposed on the daily activity, which is supposedly more important. By integrating operational managers, and linking for instance their bonuses to achieving Six Sigma goals, the Six Sigma approach avoids this problem.
3. Most TQM methods do not insist enough on the interdepartmental, multi-disciplinary character of performance improvement. Six Sigma uses teams, supported in their reflections by *Green* and *Black Belts* (project managers), whose members must come from many professions, departments, and different hierarchical levels.
4. Quality objectives of the TQM approach are often fuzzy; Six Sigma has a measuring stick: the reduction of defects to a frequency of less than three per million. This value is further modulated by the type of business; it may be even lower in the field of medical instruments and apparatus, for example. Too often, the TQM approach, in its entirety (remember the broad spectrum of business activities covered by the Baldridge Award), ends with a long list of recommendations – without a ranking of priorities. These recommendations will improve the performance of the company incrementally, but are not significant enough for the managers and presidents of a company. Proponents of reengineering blame this approach, while the zealots of TQM condemn it for its devastating effects. Six Sigma, because of its rationality, makes it possible to detect which incremental and exponential changes will have a significant impact on performance.

11.3 What does Six Sigma mean?

Six Sigma refers to variations of a parameter related to a reference value considered decisive for the quality of the product. In most cases, these fluctuations are distributed according to a Gaussian curve characterized by a standard deviation sigma, S_n, which determines the dispersion of the population of n measurements X_i:

$$S_n = \sqrt{\frac{1}{n}\sum_{i=1}^{n}(X_i - \bar{X})^2}$$

where

$$\bar{X} = \frac{1}{n}\sum_{i=1}^{n}(X_i)$$

is the average value for the sample.

A property of Gaussian distribution is that the percentage of values that lie on either side of the mean value is:

- for those in a range of ± 1 sigma: 30.9% of the population[1];
- for those in a range of ± 2 sigma: 69.2% of the population;
- for those in a range of ± 3 sigma: 93.3% of the population;
- for those in a range of ± 4 sigma: 99.4 % of the population;
- for those in a range of ± 5 sigma: 99.98 % of the population;
- for those in a range of ± 6 sigma: 99.9997% of the population.

With a range of ± 6 sigma, only 3 values per million measurements are out of range. It is this quality that is targeted by the Six Sigma approach, which results in a drastic reduction of the dispersion of values found in an existing process. This is because many companies often accepted quality control corresponding to ± 3 or ± 4 sigma.

11.4 Concepts utilized by Six Sigma

The eclectic nature of Six Sigma should not be an excuse to forget that Japan owes its breakthrough in quality notably to the contributions of *Taguchi* and *Ishikawa*, but also to the quality of the TPS system. Here are some key concepts.

11.4.1 Parameters of a distribution

Six Sigma pays attention to the distribution of data around a set-point value and, in addition to the mean and standard deviations, which are characteristics of a normal distribution (but are not found in all cases), Six Sigma uses the following characteristics for all types of distributions [11.2]:

- mode: the most common value in the distribution;
- median: the value below which lie half of the points of the distribution and the other half above; this feature is also used when the distribution includes very extreme values, which usually indicates poor quality;
- to graphically compare two distributions, one often uses the representation below (Figure 11.1 for the standard representation), sometimes called a box plot:
  - draw a line between the maximum and minimum values observed, X_{max} and X_{min};
  - define Q_1, the point below which we find which 25% of the observed data (the first quartile or 25th percentile);
  - define Q_2, the point above which we find 25% of the observed data (the third quartile or 75th percentile);
  - calculate the median *M*.

If $(M - Q_1)$ is significantly different from $(Q_2 - M)$, and the same is true for $(M - X_{min})$ and $(X_{max} - M)$, it is likely that the distribution is not normal. One must verify this by creating a histogram (which leads to a difference between the mean and median). We are then dealing with a Maxwell distribution or a bimodal distribution (e.g., the superposition of two distinct normal distributions). If the distribution is not normal, which can be verified, for example, with a statistical *Anderson Darling* test or *Chi-squared* test, subsequent statistical analyses may give biased results (a statistical test of the significant difference between two means or two standard deviations, for example).

[1] The % takes into account a deviation of 1.5 sigma, totally empirical, reflecting the long-term variability of the process.

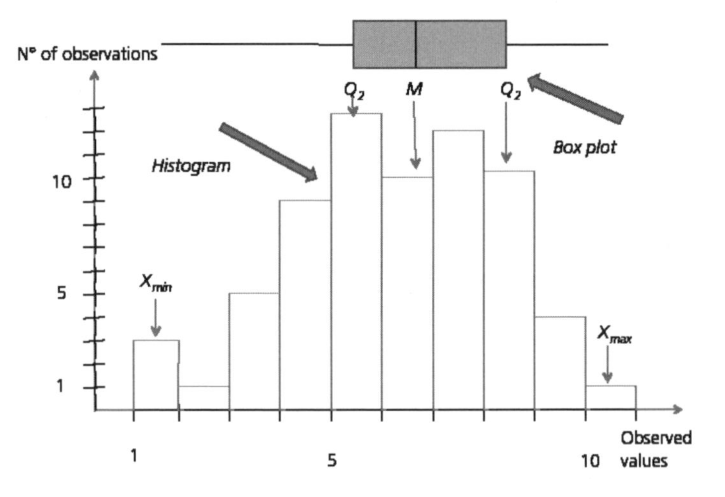

Figure 11.1 Graphic representation of a box plot distribution.

11.4.2 CTQ: critical-to-quality value (SMART)

For Six Sigma, the specification of a consigned variable, which is a critical-to-quality value (CTQ), must be SMART:

- Specific, based on facts that justify the value and ranges;
- Measurable, without dispute regarding the accuracy and credibility of the method;
- Attainable, i.e., not open to discussion and interpretation, but with testing or hard evidence proving its plausibility;
- Realistic, in line with business needs, not resulting in wry nods from the employees;
- Achievable in a determined period of Time, so as not to demotivate employees.

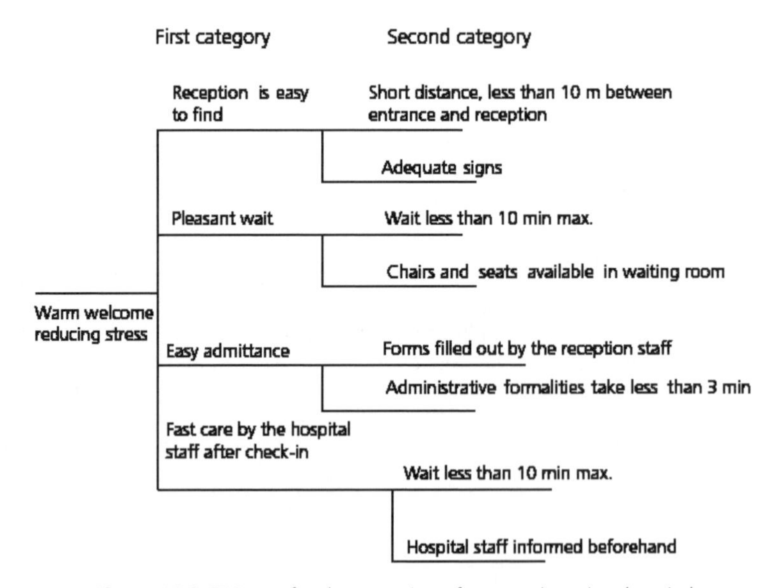

Figure 11.2 CTQ tree for the reception of new patients in a hospital.

A product can have several CTQs, determined by the implicit and explicit expectations of the customer or the user: for example, if each product has 3 CTQs, the opportunities for defects are therefore three per product; consequently, if out of 100 products inspected, there are six non-compliances, *the number of defects per opportunity is* 6/(100 x 3) = 0.02.

A Six Sigma tool to determine the number of CTQs is the CTQ tree, which generally splits the CTQs into two categories. The first represents the direct wishes and expectations of the customer, and the second corresponds to the translation of the first into production control parameters (Figure 11.2). The second category of CTQ criteria are the substituted *Ishikawa* criteria, which can then be integrated into a House of Quality (see Chapter 2).

11.4.3 Capability

The capability of a process is measured by the proportion of products or components manufactured in strict compliance after a first pass in the process. It does not include components or defective products that have undergone adjustment after inspection in order for them to conform (and be reintegrated into the produced batch).

Consequently, the capability of a complex process is measured as the product of the capabilities of all its sub-processes. Six Sigma also recommends measuring the capability of a process in the short and long term, because the variation in the latter case is always greater (personnel changes, changes in external factors, changes in batches of raw materials or components, etc.). This variation between short and long term is the basis for the expression of the sigma deviations (Sect. 11.4).

11.4.4 SIPOC

SIPOC is an acronym for *Supplier – Input – Process – Output – Customers*. This is a tool that follows the life cycle of the design of the product or service used in the tactical deployment of a Six Sigma project.

The SIPOC Six Sigma project is a six-step reflection process that begins with the process itself:

1. *Identify the process to be mapped out,* and define its scope and its interfaces: describe in a few lines what the process does (e.g., by summarizing its action in five to seven steps), what inputs are transformed into outputs and what its interfaces are.
2. *Identify and characterize the outputs*: which products and which services are the result of the process?
3. *Define the customer(s):* name, title, function, the organization or industry, and the company that receives the product or service.
4. *Define the desires and needs of the customers, implicit and explicit:* the construction of a CTQ tree is recommended.
5. *Define the inputs:* human, financial, natural resources, material substrates, information flows.
6. *Identify suppliers:* all persons or entities issuing inputs.

SIPOC (an example is shown in Figure 11.3) provides the foundation for a better ability to listen to:

- the needs of the process itself, the voice of the process;
- the needs of the business, particularly the profit and return on investment (voice of the business) or, for public administration, the optimization of costs or services.

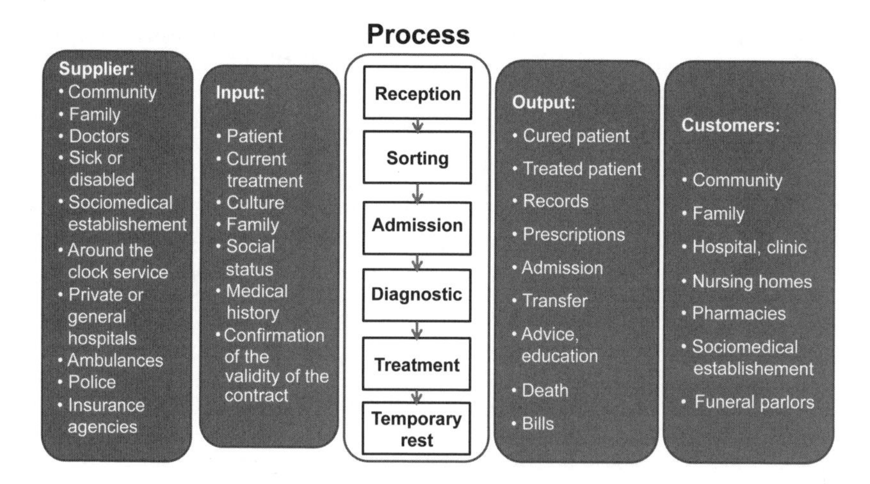

Figure 11.3 An example of SIPOC of an emergency room of a hospital.

Finally, it makes it possible to identify other critical parameters of the type CT:

- CTC: factors that are Critical To Cost,
- CTD: factors that are Critical To Delivery, and
- CTP: factors that are Critical To Price.

11.5 Before starting Six Sigma

11.5.1 Six Sigma, a road of variable width

Many authors recommend that one consider a Six Sigma project like a road (three types: interstate = *business transformation*, highway = *strategic improvement*, provincial road = *problem solving*).

The road ahead is set by the ambition level of the board's objectives:

- ***Business transformation***: this route is chosen when a company faces problems or when significant opportunities present themselves:
  - *emergence of new market segments* (e.g., single women aged 35 with high income);
  - *emergence of a new technology* that will revolutionize the market of an existing product or create a new niche (e.g., Internet, cell phone, etc.);
  - *necessity to change tempo* in business cycles, consistent with a market that has become much more innovative;
  - *perception, due to failures and delays as well as significant losses of market shares*, that the competitive position of the company is falling.

  In this case, the first reflection is accompanied by a data search on the following subjects, which affect the life of the company:
  - distribution mode of the product;
  - effectiveness of sales and processes;
  - product development process;

- financial, marketing and technological intelligence for R&D;
- customer desires, complaints and claims;
- product defects and problems;
- possible reduction of costs on a large scale.

It is this first approach that will touch the largest possible area of the company and the organization of the daily work of the employees, because the Six Sigma project will cover all activities of the company, thus truly embracing a total quality approach. It sometimes involves designing new processes to replace the old. It is this aspect of Six Sigma that brings it closest to *reengineering* and *turn around*, even if the board decides to proceed in stages and at first only selects the two key processes that are most affected. In this case, it is customary to compare several projects with an additional objective for which the earnings add up at the end of the period, which may span several years.

- ***Strategic improvement***, which involves improving a sector or business activity considered to be efficient. It may also be a means of testing the Six Sigma approach on a promising segment, as a first step to a subsequent application of Six Sigma to other processes of the organization. Thus:
  - a biomedical company may want to apply Six Sigma exclusively to the production sector to increase productivity, reduce costs and the rate of product defects, even if this means generalizing this approach to logistics in a second stage;
  - a distributor of consumer goods can focus exclusively on the logistics of distribution of its products from its storage areas to the stores, which carry out retail selling.

  This approach does not fall under total quality, but if its scope is large and it touches the key processes of the organization, the gains may be enormous and the project visible. Its chances of success are thus good. In many cases, the projects focus on significantly improving existing processes.
- ***Problem-solving***, for recurrent poor quality that has not been resolved with common management methods or statistical process controls. The significant contribution of know-how from Black Belts in quantitative methods, and the planning of tests and experiments (Taguchi method) and the methods of coaching by Green Belts can make a difference. In many cases, it is only a question of properly managing existing processes. Project stakeholders are aware of the problem and the perimeters, but lack structure in their resolution process, which Six Sigma generates.
  The low visibility and the restricted level of communication of the project gain very little commitment and adherence of the members of the board. Pressure on staff is lower and some authors consider the risk of failure greater than for the two previous approaches.

11.5.2 Commitment of the Board to Six Sigma

Before deploying the project, the board of the organization needs to perform certain prerequisites:

- *Clarify* and communicate the mission of the company and its key objectives.
- *Identify* the main processes of the organization and discuss them.
- *Select* those processes for which the performance is below that expected and where the Six Sigma approach can have the greatest impact (quality of product/service and productivity, savings).

- *Develop* a logical and convincing argument for all employees: Why has Six Sigma been chosen, what are the reasons and the benefits? What can we expect, what will the impact be on the employee's work, why was this method chosen instead of the EFQM and ISO approaches, etc.? Explanations such as "Wall Street will appreciate our effort", and "this is a request of the board" should be avoided. Management must demonstrate how Six Sigma fits in the improvement of the management system of the company.
- *Actively participate* in the planning and implementation of the project; these two operations cannot be delegated to a consultant (whose experience as a counselor may be appropriate) or a "super" Six Sigma project leader for the following reasons:
  - the members of the board must be the ones to promote and defend it internally;
  - they are the people who can change the orientation of the project as they gain knowledge of new information and needs;
  - they are the best placed to determine the priorities between the daily activities of the company and the effort generated by the Six Sigma project.
- *Create a vision and a marketing plan*: a major weakness of managing change within organizations is the minimal marketing of it to employees; changes may give rise to the disappearance of workstations or modifications of employee habits, for which reasons they provoke resistance, fear, frustration, but also cynical reactions that lead to a loss of motivation and commitment. An internal (and external) marketing plan can mitigate these apprehensions and mobilize resources for innovation. This action is segmented along the following lines:
  - Identify the main theme or vision as a key phrase that is positive, clear, inspiring and easy to remember (a form of slogan), for example:
    - the key motto of *General Electric* was "*Completely Satisfying Client Needs Profitably*";
    - that of *Allied Signal/Honeywell* was "*Creating a Culture of Continuous Renewal*".
  - Develop the marketing plan, by identifying internal and external targets, the corresponding breakdown of the argument and communication channels. The content must be realistic, challenging; wordy phrases should be avoided.
- *Become a strong advocate and supporter* of Six Sigma; this condition is essential according to the experience gained in large corporations. Management must appear strong, energetic and constant in this effort, and this must be visible to all employees, even though in leading this change management, many questions remain unanswered.
- *Define clear objectives (cf. SMART)*; they should be understandable, inspiring, realistic and include a timeframe. This is a focal point of the marketing message.
- *Feel responsible*; this attitude must be generalized for all employees, so that if a Six Sigma project should fail, the blame should not simply be attributed to employees, but management must also ask how it is responsible for the failure, for example:
  - Were the objectives sufficiently quantifiable?
  - Was the project clearly defined?
  - Were the allocated resources in line with the size of the project?
  - Were announced problems listened to?
  - Was the urgency of the project really communicated?

  Linked to this point, we also have the type of encouragement and incentive for the project. How does the employee's compensation vary depending on his/her commitment to the project? Is there a significant difference?

The criteria for obtaining a bonus should also be examined: it is thus useless to reward the marketing department on the number of new products launched per year if the sales department has not received data on their characteristics. Moreover, it does not make much sense to reward managers for making savings if these economies penalize the customer service, etc.

- *Require evidence of tangible and measurable results*; this means to fully exploit the advantages of the Six Sigma approach, notably by mobilizing neutral experts to measure the financial savings achieved. This enables one to:
  - be confident that the gains are real;
  - announce to all the seriousness and confidence of management in the Six Sigma approach and the search for tangible results.
- *Communicate the results with enthusiasm, but also honestly announce the difficulties and failures*: Six Sigma is a proven method and the risk of failure is moderate. The communication of successes and the celebration of them with festive events are good approaches to increase motivation. The communication of difficulties strengthens the credibility of a management team; every employee knows that failures occur in any type of project and that trying to evade them is useless. In this case, the grapevine will undertake to spread the word instead of management, but with false information and magnified by rumor ...

11.5.3 Six Sigma, training of *Black Belts*, Green Belts and other players

One of the first stages of implementing the project is the selection of the roles of project stakeholders and their preparation. Different training for personnel assigned to a Six Sigma project is certainly one of the greatest strengths of this approach. Six Sigma identifies several players (in a descending hierarchical order), and all the players may be present in the project to varying degrees (Figure 11.4).

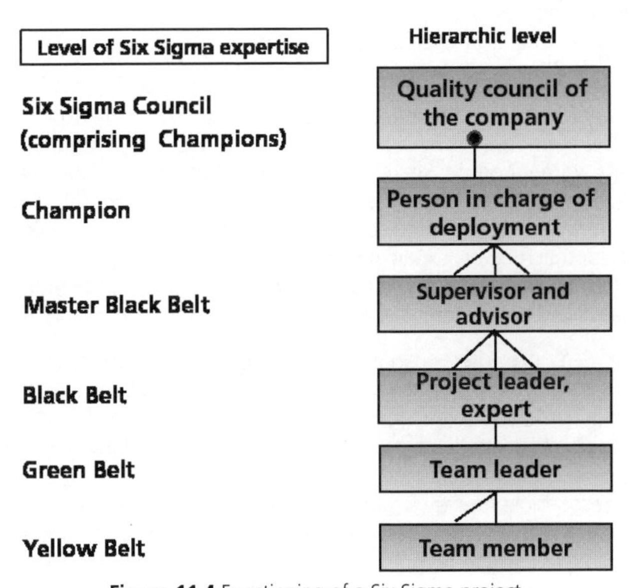

Figure 11.4 Functioning of a Six Sigma project.

- *The Quality Council or Six Sigma Council*, which is the forum where the selected members of management discuss, plan, and guide the Six Sigma initiatives. The activities of this group, who get together monthly or bimonthly, are to maintain pressure and give the appropriate signals to employees, and include:
  - designation of roles and mobilization of necessary infrastructure;
  - selection of specific projects and allocation of resources;
  - periodic review of the progress of projects and, in the case of problems, propositions for help or sound advice;
  - assuming the role of project champion;
  - helping to assess the impact of Six Sigma projects on the financial results of the company;
  - assessing the progress, but also the strengths and weaknesses of efforts, without complacency;
  - transferring good practices in one project to other areas of the organization, even to customers or key suppliers, if appropriate;
  - counteracting the obstacles that the teams have identified in their activities.
- *The Champion or deployment manager* is a role assigned to a senior member of management and involves the following activities:
  - Representing and defending the team at the Quality Council.
  - Setting and maintaining the overall objectives of improvement, determining the logic and the reasons for the project and aligning its goals with the company's priorities.
  - Steering and accepting the changes when it comes to content and objectives during the project, if necessary. This activity is important as teams are reluctant to decide for themselves to change the direction of the project or to concentrate on specific aspects, for fear of disappointing the top executives; a go-ahead from the latter is thus welcome.
  - Finding and negotiating the necessary resources for the project.
  - Working with those responsible for the process to conduct a proper transfer of results at the end of the project.
  - Champion training exists; it lasts 2 to 3 days.
- *The Master Black Belt*, project leader and consultant, who is a top-level engineer or technician, often an accomplished quality controller with wide experience, and trained in statistical methods. His/her task is the control, supervision and coaching of several Black Belts, with whom he shares his experiences and from whom he verifies that the planned and negotiated steps (the tollgates or milestones) are followed and goals met on time. He is also available and can engage in difficult project activities that require the complex approach of an expert (internal consultancy activities). He also intervenes in cases of lack of motivation of the team, despite the efforts of the other *Black Belts*. The Master Black Belt reaches this status after several years of operation as a Black Belt.
- The *Black Belts* or project managers: they are specialists who work full time on Six Sigma projects, usually engineers and quality professionals, with extensive training in basic quality and project management. Their task is central:
  - to inspire, motivate, lead, manage, coach and care for the teams and their members;
  - to be responsible for the constitution of the team, for the development of its motivation and dynamism, for observation, and for the members' participation in training;
  - to be able to carry to completion all activities assigned to the team and to reap the desired results.

Their training in the Six Sigma approach is comprehensive, and includes knowledge of:

- details of the approach and the "philosophy" of Six Sigma;
- description and analysis of DMAIC;
- voice of the customer;
- process management;
- management of teams and work groups;
- complete toolbox of instruments of quality;
- for each step of DMAIC, the most advanced tools and statistical techniques;
- analysis of the measurement system, and capability;
- design of experiments (e.g., according to *Taguchi*).

Their training is the longest; it involves a 20-day course and costs the company about 12000 USD. For some of these quality professionals, there is an overlap between their prior knowledge and the content of training, especially if they have prior training in project management. The function of a *Black Belt* has an average duration of two years and may, depending on the company, be the springboard for subsequent promotion to a senior executive or business manager. In this case, there are good chances that company executives consider the quality function as critical to the proper functioning of the company, creating a supportive culture for management by breakthrough and continual improvement.

- *Green Belts*: These are technicians or engineers, typically executives or managers of production processes with fairly extensive training in the following areas:
  - knowledge of the Six Sigma approach;
  - significant know-how of DMAIC;
  - basic statistical analysis, including elements of statistical process control;
  - analysis of the measurement systems;
  - process capability;
  - introduction to design of experiments.

  Their training typically involves a nine-day course and costs the company about 6000 USD. *Green Belts* distinguish themselves from certain *Black Belts* by their level of training, but especially by the fact that they only work on the project part-time (approximately 25% of their time). *Green Belts* may also lead work groups, which is why team management may be part of their training. *Green Belts* can be considered as agents actually performing physical activities related to the process under scrutiny.

- Yellow Belts: These are employees who are involved in the measurement and analysis activities of a project. They have a basic knowledge of Six Sigma, but are not able to manage projects. Their training is more perfunctory and includes the following topics:
  - basic concepts of Six Sigma, including DMAIC;
  - methods and tools;
  - customers and processes;
  - data collection;
  - data analysis;
  - other simple tools employed in the project.

  The training is three days long and costs the company a few thousand dollars. Typically, *Yellow Belts* assist the *Green Belts* in their tasks and are therefore members of the team.

Many consulting companies have included Six Sigma training in their product selection and have tried to make it a lucrative business, hoping, once the training is accomplished, to also provide consulting services. In this business, the worst seems to be mixed with the best. Still, training requests from boards of directors, but also individuals (in search of better employment) are boosting the market for this type of training.

11.5.4 Employee training in general

In their effort to train teams, *Master Black Belts* and *Black Belts* should [11.3]:

- promote learning in the field (hands-on learning), as it can immediately be capitalized;
- use real examples and link training to the professional world of the team, so that its members are convinced that their new knowledge will fit into their daily practices;
- build knowledge starting with the basics and going from simple to complex, while placing them in the proper context;
- use multiple teaching methods (lectures, exercises, games, etc.).
- use the training in the marketing plan, so as to make it an instrument to promote the approach and Six Sigma projects of the organization;
- place the training in a long-term context; the concepts of Six Sigma must be recalled to anchor them sustainably in the minds of team members.

11.6 How to carry out a Six Sigma project

This section presents the deployment of the Six Sigma approach, the DMAIC process (***D**efine*, ***M**easure*, ***A**nalyze*, ***I**mprove*, ***C**ontrol*). The most obvious distinction between the DMAIC and PDCA lies in the segment formalization of DMAIC: each stage includes objectives that result in observations, reports, and quantifiable data called tollgates (or milestones [11.4]). Their content can be discussed with the *Master Black Belt* or in the Six Sigma Council. The objectives are achieved sequentially (Figure 11.5). is used;

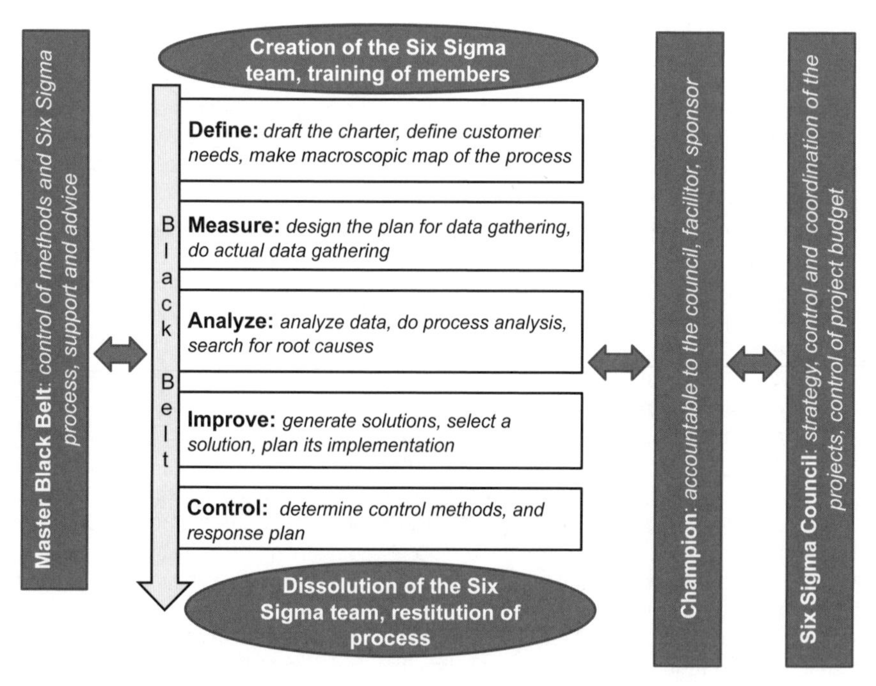

Figure 11.5 The Six Sigma DMAIC process and its tollgates.

11.6.1 *Define*

At first, the Six Sigma approach is only barely noticeable by company employees, because a substantial part of the preparation mobilizes executives, who learned of the process during training for at least one day.

It is in this phase that the team is formed and trained, its leaders appointed, and when it begins to perform work for the Six Sigma project, its activity documented and the necessary means and resources received. The team must deliver the following information and documents (three tollgates):

- *Drafting a charter*: This document summarizes and synthesizes the project, lists the objectives and provides the basic motivation for the team. It contains the following information:
  - *The subject of the study (business case)*: a brief text of a few lines explaining why this project has been chosen, why it has received a higher priority than other projects, and which strategic objectives of the organization will the completion of the project significantly influence;
  - *Definition of the problem*: briefly, this section describes in a neutral and impersonal manner the nature of the problem, its impact, its recurrence, when it first appeared, and what the differences between the current and desired situation are. This section also defines the scope of the investigations, i.e., what the team needs to focus on.
  - *The goals and objectives*: reported here are the team's quantitative goals and objectives that will be achieved 4 to 6 months after the formation of the team.
  - *The steps*: this section lists the dates on which the different phases of DMAIC should be completed.
  - *Roles and responsibilities of the team,* as well as the employees and managers revolving around the project: the names and positions of the members, the *Green Belts*, the *Black Belts*, the *Master Black Belts* and the *Champion*.
- *The needs and desires of the customers*: the team must determine the desires and needs of customers (internal or external) of the first and second categories (criteria substituted) using the CTQ tree. This step is also called "voice of the customers" (VOC). It is at this stage that the product (and/or process) functionalities are determined.
- *A map at the macroscopic level* of the process that is under scrutiny: this document is the result of the corresponding SIPOC.

11.6.2 Measure

The measurements carried out in a Six Sigma project aim to:

- quantify the factors that are vital for the customer and significant for the suppliers;
- determine the levels (and fluctuations) of the influential parameters of the process (time, cost, staff activities, energy use, etc.).

In this step, two tollgates or milestones are listed:

- *The design of a data collection plan*, including:
  - the parameters to be measured, called KPIs (Key Performance Indicators);
  - the type of measurements and the measuring instruments;
  - the type of data to be recorded: discrete (0 or 1, 1, 2 or 3, for example) or continuous? Compliant or non-compliant? etc.;
  - shared and operational definitions of what is measured, so that the team, in a subsequent step, will not be lost in fruitless discussions on the subject;

- the set-point or desired specification, usually stemming from the goals and objectives of the charter;
- forms for recording and processing the data; these will differ depending on whether the measurement results are discrete or continuous:
  - for discrete results, especially defects, they should be defined, each category determined, the time scale for collecting the samples selected, and the record established,
  - for continuous results, one usually employs a histogram;
- sampling should be representative; the method should be specified, and also the number and frequency at which the samples are randomly taken.

• Data collection and the estimation of the number of defects per opportunity: see sections 11.5.1 through 11.5.3 where these methods are described.

11.6.3 *Analyze*

This step is one of the most important in the Six Sigma approach. It focuses on data and process analysis, and seeks to determine the root cause of the malfunction.

It also helps to gauge the performance differential between what is measured and what is desired. It includes three tollgates:

• *Data analysis*, which differs depending on whether one refers to discrete or continuous values:
  - In the case of discrete values, the preferred representation is in the form of a pie chart or a Pareto chart. The latter makes it possible to determine at a glance the major malfunctions and critical factors.
  - In the case of continuous values, the detailed analysis of histograms, but also the kinetics of the observations make it possible to determine which among the six factors affecting a process (Materials, Machines, Methods, Mother Nature, Man power, Measurements) are responsible for deviations (*Ishikawa's* fishbone diagram).
  - Advanced statistical tools are needed to study the relationship between cause and effect according to the type of data (variance analysis, simple and multiple regression, etc.). A *Black Belt* exploits the data with statistical software such as *Minitab* or *Quality Companion*[2] for instance.

• *Process analysis* consists in representing the process in more detail than that defined by SIPOC, in order to examine all the steps, also called sub-processes. The examination of a sub-process involves verifying that all sequences satisfy the following requirements:
  - upstream clients consider this step important;
  - the product in transit undergoes a significant change;
  - the transformation step induces the change at the first pass (no further adjustment after inspection), *Right First Time*.

If any of these criteria fails to be met, then the sub-process or its sequence does not add value, in which case, the team must determine to which of the categories below the sub-process or sequence belongs:

---

2 Commercial software is downloadable from: www.minitab.com.

- internal poor quality: when the sub-process or its sequence must be repeated to obtain the desired effect;
- external poor quality: when the defect is identified by the customer after production;
- delays: queues in the process;
- control/inspection: sorting stations releasing only complying products;
- preparation steps for the next step;
- transporting the product to another location;
- a step which, although not creating value, can be the source of an improvement, due to it being necessary for the organization (e.g., inventory accounts, etc.).

Once this analysis has been performed and represented in tabular form, the team will, by a succession of questions involving "why", initiate the search for the root cause of the dysfunction (at least one, and at the most three for the team to work efficiently).

- Finding the root cause: this step is similar to that of brainstorming. The team searches for all possible causes of the malfunction; good brainstorming activity involves the following conditions:
  - all ideas are documented;
  - the team generates ideas, but it is off limits to discuss, evaluate, or criticize them;
  - each team member must participate by providing several ideas; this can, for example, be done silently by writing each idea on a post-it and sticking it on a wall.

The combination of ideas is most often represented as an *Ishikawa* fishbone diagram.

- The next step is to explain all the factors listed so that they are understood by all participants. This step often makes it possible to restrict their number. Subsequently, each member distributes five votes to the factors that he/she judges paramount for coming to terms with the malfunction. Before the next step, it is possible to retain minority factors that are fiercely defended by one or two team members. This step should reduce the possible factors to seven, for example. One can also use the FMEA, which allows one to quantify the criticality of failures and associated causes and also prioritize corrective actions.
- The last stage is decisive, the most crucial: it consists of testing the hypotheses with the collected data, in generating x/y diagrams with appropriate variables, possibly gathering additional data, or even conducting experiments on or disrupting the process. The work is similar to a mathematical modeling approach of the system to be improved. The fundamental cause is inferred from it and accepted by consensus by the team.

The accumulated data can also be used to *quantify the difference* between the set-point value entered in the chart and the current performance.

11.6.4 *Improve*

Once the root causes of the dysfunction have been found, the improvement phase can take place quickly, unless the process is so under-performing that it must be entirely redesigned. In the latter case, one uses the *Design for Six Sigma* (see section 11.12) and the tactics change from DMAIC to DMADV. The improvement step includes three tollgates, which do not always differ from the search for root causes of the previous step:

- *Generation of solutions*: this consists in finding cures to the root cause by modifying key stages of the process, which leads to the generation of new process maps. The planning of experiments may be necessary if the desired optimum has been subject to no prior action and if compromises are necessary (costs, yields, etc.).

- *Selection of the solution(s)*: choosing the best solution, testing and planning the experiments, full-scale or in a pilot phase (small scale), thus making it possible to focus on a single solution, in parallel with an estimate of costs that may possibly be induced (potential investments, additional human resources, etc.).
- *Planning the implementation of the solution*: this consists in investigating and resolving the practical problems of the process modification, in addition to the deadlines for the implementation.

11.6.5 *Control*

This step includes documentation of the process, its description, procedures and instructions, the control plan, the response plan, the transfer of the project to the person responsible for the process, and project closure, which may include, especially in the USA, a celebration of the success. There are two tollgates:

- determination of the technical method of control, set-point values with their lower and upper limits, and the sampling plan, analytical methods, etc., followed by
- the response plan, a summary document that identifies:
  - mapping of the new process (*"Should-be", Plan*);
  - key data of the new process (e.g., operating conditions, equipment, etc.);
  - target values and specifications of the product/service, as they have been verified and validated by the upstream client;
  - the control method with the appropriate record types;
  - main improvements;
  - alarm thresholds specifying what to do if the input, process and output parameters reach values outside the ranges of the specifications;
  - short-term actions for nonconformities as a preparation for long-term commitments of a Six Sigma project;
  - reflections and ideas for a continual improvement program once the Six Sigma project is completed.

Once the process has been transferred to the person in charge, the leader should perform the following tasks as part of his/her activities:

- update the documentation process;
- monitor and regularly assess the performance, and identify future problems and opportunities;
- support and promote improvement efforts and projects;
- maximize process performance.

11.7 An avatar, *Lean Six Sigma*

11.7.1 What is Lean Six Sigma?

The Six Sigma approach is associated with the concepts of defects and quality, but does not prioritize timeliness, reduction of waste, downtime and the optimization of efficiency. This is where the *lean* approach comes into play (see 11.7.4.), as it has tools to reduce the preparation time of components and products. Since companies must be able to respond quickly to changing customer needs, reducing the chain of deadlines is crucial.

Lean Six Sigma, a structured and complementary approach, involves methods of *Lean management* in the DMAIC cycle. *Lean management* is an American combination of Japanese methods derived from the *Toyota* production system, whose scope covers all organizations and businesses.

11.7.2 The supremacy of TPS (reminder)

The use of the term *lean* is often associated with two researchers at MIT's International Motor Vehicle Program (a program, now international, involving American, European and Asian universities), *James P. Womack* and *John Krafcik*, who in 1987 investigated why the Toyota Production System (TPS) was more efficient than that of US companies. It was because TPS:

- requires less human resources for products and services;
- requires less investment for an equivalent production capacity;
- allows the manufacture of vehicles with much less poor quality;
- needs fewer suppliers;
- has less stock;
- takes less time and human resources to manage the process from concept to commercialization of new products, but also from order to delivery;
- has a much lower number of work accidents[3].

11.7.3 The characteristics of TPS (reminder)

One of the first results of their analysis showed that the Japanese were not more efficient because their costs for wages were lower or because automation was more developed, but rather because *Toyota* [11.5]:

- created a culture where stable and strong values, taught and shared, and had lasted many years;
- constantly reinforced this culture;
- taught its employees to work in groups;
- established interdisciplinary working groups to solve business problems.

In a second step, the review of TPS revealed that the Japanese had generalized the industrial production invented by *Ford* to small – and medium-sized series, which involved the following characteristics:

- ***Heijunka***, a Japanese word for smoothing production by volume and avoiding surges in production due, for example, to changes in orders, or production quotas that need to be met by the end of the month. Smoothing has the following advantages:
  - a more regular production;
  - predictable production schedules;
  - an echoing of this stability among suppliers;
  - a reduction of stress on the production line, and low stocks of components;
  - the ability to work optimally, reproducibly and according to a standard;
  - savings when it comes to overtime;
  - easier management of human resources, on-call not necessary, and no partial unemployment.

Moreover, *Toyota* also sought to smooth its production schedule by producing different vehicles and not working with "packages" of similar vehicles. This was made possible by a fast change of the production tools (less than 10 min per change).

---

[3] For this point, however, see 10.6.4 in Chapter 10 on Toyotism for a much less reassuring version of how *Toyota* executives manage work accidents of their employees.

- ***Jidoka***, a Japanese word related to the concept of prevention, which is broken down into two other concepts:
  1) *Autonomation*, which is the fact that machines on an assembly line are designed to stop when a problem occurs. This stop alerts the operator on the line who then solves the problem. This process of autonomation assisted by people has the following advantages:
     - immediate detection of poor quality;
     - stopping of the line, but also stopping poor quality;
     - immediate repair or correction of the malfunction, and/or search for the cause and triggering of a corrective action;
     - reduction in the number of employees responsible for machines (they manage them only when they are down!).
  2) *JIT* (Just In Time), i.e., producing only what customers need and when they need it. This is achieved by TPS by:
     - adopting a *Single Piece Flow* and not waiting for a significant number (a batch) of parts to be manufactured before providing them to the operator responsible for the step next of production. This procedure enables:
       - poor quality to be detected by autonomation;
       - manufacturing areas to be reduced by removing storage;
       - avoidance of downtime of staff at the next workstation because they are waiting for the batch.
     - *Pull Production*, which signifies producing only what the operators responsible for the downstream workstation need to complete their task; this rule reduces inventory and eliminates production jams where components and products are processed although there is no demand.
     - *Takt Time*, which is the maximum time to perform an operation on a production line. Establishing a *Takt Time* for each step has the following advantages:
       - the product moves on a line, so the bottleneck (the stations where the time required is more than what was planned) are easily identified;
       - the same goes for stations that are not sufficiently reliable (breakdowns, etc.).
       - since the *Takt Time* leaves only a limited time to do the job, employees are motivated to do only the tasks that are strictly necessary; a waste of resources (Japanese: *muda*) is avoided;
       - since all products are linked to a production line, they cannot be placed or stored (or worse, lost!) anywhere in the production facility.

11.7.4 The design of Lean Management

How can we remove ourselves from the cultural (Japanese) and sectorial (vehicle manufacturing) specificity of Toyotism and identify universal guidelines? For *Womack*, the operation was laborious, because the TPS system had largely been established from the bottom-up and was fragmented into multiple practices without an obvious relationship between them. Moreover, apart from a few principles in the Memoirs of the founder of *Toyota*, and in a book by the father of the production of TPS [9.6], summary studies were lacking.

A few years later, the Lean approach was made possible for all businesses and would usefully complement the Six Sigma method. It was based on understanding and improving the flow in the process and removing those activities not adding value (*muda*, the Japanese equivalent of reducing waste). It adhered to the following principles [11.7], while being described as a better method for creating value with less effort:

- *Focusing on the value that the product creates for the customer*: during his visits to partner companies, Womack noted that managers, even though they proclaimed to be primarily concerned with the customer and the CTQ of the product, had a vision clouded by an environment that biased their perception:
  - the Americans were preoccupied with quarterly financial results and escaping lay-offs and hardly knew how their products were manufactured;
  - the Germans were fascinated by technology and engineering, to the point of forgetting that customers do not necessarily share this fascination;
  - the Japanese were primarily focused on not relocating their production outside Japan.

  Womack concluded that despite training, particularly from business schools, the fact of not listening enough to the voice of the customer remained the Achilles heel of most companies and organizations. He felt that this step is critical in the Lean Management approach. In his view, the functionality of a product should be studied in depth with many consumer panels; the design phase is vital.
- *Identifying the value chain: Womack* defined the value chain as all operations and actions that are necessary for the manufacture of an artifact and/or a service. These actions and operations fall into three families:
  - problem-solving tasks, especially during the design phase and the production of the service and/or product;
  - information management tasks, especially those that pass from order to delivery;
  - transformation tasks, covering the production process, raw materials, as well as from components to the finished product.

  For *Womack*, the value chain is rarely analyzed in companies, even though these activities can be systematically evaluated into three types:
  - the steps that create added value (AV);
  - those that create no added value, but which are necessary in light of technology and existing production patterns (inspections, facility maintenance, cleaning, etc.); this type of non-added value is a requirement for the company;
  - those with no added value and whose removal may be immediate.
- *Just-in-time* activities creating added value instead of batches, which is directly borrowed from TPS.
- *Pull Production*, which is to allow, in the sequence of activities that generate added value, the production of only what the person responsible for the subsequent process needs to complete his/her task; this is also directly borrowed from TPS.
- *The quest for perfection and excellence*: Womack notes that once the first four steps above have been implemented, the process leading to savings of means, use of equipment, material and human resources (while delivering a product closer to customer expectations) starts spontaneously. One reason for this is that the first four steps interact with each other in a virtuous cycle, so that:
  - the *teams* in charge of production interact directly with the customers and are able to improve the product functionality while carrying out savings;
  - the acceleration of work through just-in-time activities makes it possible to detect bottlenecks in the process of value creation, and to remedy them.

Finally, implementation of the five stages promotes a ***transparency*** of the processes and activities of the organization. It enables discussions between suppliers and manufacturers, producers and customers and especially between departments of the organization. It is also a motivating factor for the employees, who participate in a more all-inclusive functioning of the company. Part of the alienation of workers prompted by Fordism thus disappears.

11.7.5 Lean Six Sigma

Lean Six Sigma combines the advantages of the two approaches (Figure 11.6):

Lean Management and Six Sigma
The two approaches complement each other

Lean Management highlights:	Six Sigma stresses:
Reduction of downtime and waste, and steps that do not add value	Culture and structure to achieve the objectives
Timeliness and reduced delivery time	Deploying the project by the DMAIC approach and the existence of tollgates
Management tools to facilitate the reduction of delivery time	The essential role played by fluctuations in the process in the generation of poor quality
	A means to identify the critical process steps to satisfy the Voice of the Customer

Figure 11.6 Lean Six Sigma.

Lean Six Sigma uses a reference value to characterize the production efficiency of a service or a tangible object: the ***delivery time***, which obeys Little's law and should be minimized:

$$\textit{Delivery time} = \textit{inventory} / \textit{throughput}$$

Lean Six Sigma has been successfully implemented by *General Electric,* and for services [11.8] by *Lockheed Martin*, but also by public economics institutions such as the *Stanford University Hospital*.

11.8 The current focus on DFSS (*Design For Six Sigma*)

11.8.1 DFSS vs. DMAIC

DFSS (*Design For Six Sigma*) is an emerging methodology for process management, based on Six Sigma. However, it differs by the fact that, in Six Sigma, the method is applied to existing and functioning processes. In DFSS on the other hand, the engineer starts from scratch: instead of improving, he/she creates a product and/or process *ex nihilo*.

Therefore, DFSS is an extension of the DMADV (***D****efine*, ***M****easure*, ***A****nalyze*, ***D****esign*, ***V****erify*) method. It was developed to counteract the decline of American industry with regard to production machinery and capital goods. Indeed, DMAIC tactics are difficult to apply to the development of innovative products:

- statistical and analytical methods available to *Black Belts* are insufficient for the design phase;
- there is either no available data on the processes, or it exists in quantities too small to be used in statistical analysis.

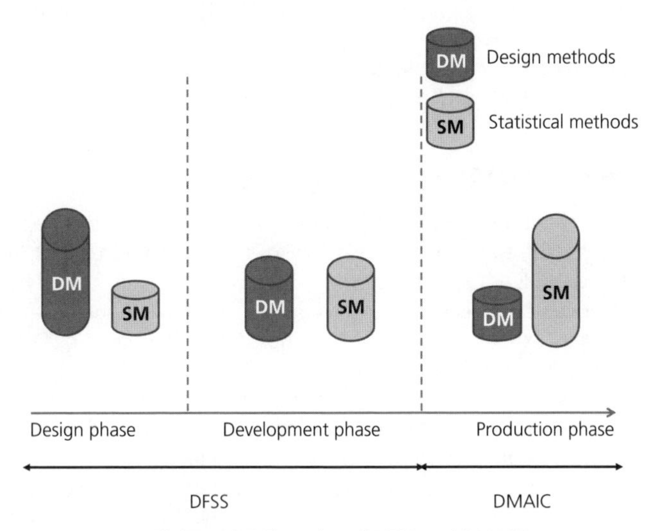

Figure 11.7 The roles of DFSS and DMAIC.

DFSS is intended specifically for the design and development of new products, phases for which the choice and implementation of conceptual tools are vital (Figure 11.7). It addresses the weaknesses often found in a process:

- the product design is vulnerable, particularly in the phase from the determining of functionalities in relation to customer expectations, up to the design of the production process;
- the production process is vulnerable to external fluctuations and changes in operating conditions.

11.8.2 Organizational aspects of DFSS

These have been little discussed in the literature of DFSS. The design from scratch of a product and a process requires more specialized skills than those possessed by *Black Belts*, especially for the design phase. This task is passed on to engineers and technologists who are:

- experienced with similar products and the corresponding market segment;
- able to lead panels of consumers, advisors and clients;
- able to talk to the people in sales, marketing and finance;
- able to deal with company lawyers for legal issues.

In addition to these qualities, the DFSS specialist must master the design tools of quality deployment, methods of creative problem solving, the methodology of surface reaction, etc. Only a few employees in a company are likely to hold this function, which is assigned to a senior research and development engineer of the organization (remember that *Black Belts* represent 1% of the total number of employees). It is in fact such a specialist that is able to determine the extent of the methods to be implemented to achieve the objectives. It is he/she who must negotiate with other interested parties the length of the design and development phase, because marketing generally wants to have the product before the competition, which inevitably shortens the time available for design and development.

Consequently, the organization and project management of DFSS can be distinguished from that of the DMAIC approach, because the skills of *Green* and *Yellow Belts* are only called for at the end of the DFSS project. The potential roles of the Six Sigma Council and *Master Black Belts* should be revised.

Finally, the interface between the project group and the team managing the daily operation requires great care during the transfer.

11.8.3 The stages of DFSS

When it comes to the rugged design of the production process, DFSS involves the statistical contributions of designs of experiments and *Taguchi's* orthogonal matrices, and thus the design of parameters as well as of tolerances.

The merit of the DFSS tactic is essentially to deepen *Taguchi's* phase relating to system design, which is the least formalized part of the approach. Proponents of DFSS define this design phase, known as ICOV, in four steps:

- Identify requirements: This step includes the same tollgates as the Define step of DMAIC, but the design tools used are more sophisticated when determining the CTQ.
- Characterize the design: This step includes three tollgates:
  - the transcription of CTQ parameters and characteristics of the production process (substituted quality criteria);
  - the generation of alternative models for the design of the product;
  - assessment of the proposed models and the selection of one of them.
- Optimize the design: The product development study of the chosen alternative is further elaborated so that the performance of the production process meets the requirements of Six Sigma.
- Validate the design: This includes three tollgates and resembles the Control stage of DMAIC:
  - *Pilot tests and finishing touches*: this step includes testing on a small scale, but in an environment that can be better controlled and with more options for varying the parameters and operation conditions for production.
  - *Validation and process control*: this step determines in a definitive manner the operating conditions, set-points and control methods.
  - *Delivery of the process* to the person in charge with the sole purpose of exploitation. This step also includes the drafting of a response plan.

11.8.4 Discussion

As for Six Sigma, DFSS is an eclectic approach, borrowing many of the tools and methods that are implemented:

- Robust Design by Taguchi for optimization and validation steps of the design;
- the Design practice developed in Germany [11.9];
- the theory of creative problem solving by the Soviet G.S. Altshuller [11.10], etc.

The complexity of the methods used requires a high degree of experience and training, and the duration of the latter needs to be at least a period equivalent to that of *Black Belts* (and starting from that level). This high degree of abstraction makes explaining the approach to less qualified production personnel more laborious, as they are less motivated by theoretical explanations and yet are supposed to understand the ins and outs of the choice of technology, parameters and set-point values, as well as the type of process used. DFSS is applied by many companies who have adopted the Six Sigma approach, including *General Electric* and *Motorola*.

11.9 Conclusions

Six Sigma is, as Toyotism, a proven methodology, and a number of developments have demonstrated its robustness and depth. Six Sigma is a Western variant of total quality, but it is obvious that is has been borrowed from Japanese methods (starting with the names and types of its specialists, and based on the terminology of Japanese martial arts). What is also obvious are the very few academic publications dealing with the subject, particularly those derived from prestigious universities. Moreover, those MIT researchers who have studied the performance of organizations and companies primarily choose careers within consultancy, and not academic ones.

Leading American and English-speaking universities rely on the research and discovery of new technologies, even though the latter, with the exceptions of the Internet, mobile communication and the computer, are dedicated to restricted areas of the economy. They do not therefore greatly increase global productivity and the well-being of the world's population. The savings from Six Sigma, on the contrary, are very sensitive and affect all business segments.

In contrast, Asian universities have professors teaching quality, performance, *Robust Design* and other *Taguchi* methods, as well as the use of standards. Therefore, it may be that the US continues to dominate more in innovation than in efficiency in industry and services (in which, it must be said, the Japanese have yet to excel).

Six Sigma is a method that has proven itself: but what about the long term? One risk arises from the approach itself and involves the level of commitment of the people responsible for processes, and the teams associated with them in terms of continual improvement; once the process has been transferred, the Six Sigma project is terminated and the team dissolves.

The Six Sigma approach by itself does not ensure the success of this crucial step. No monitoring or control of the Six Sigma Council is formally scheduled, despite the fact that Six Sigma is characterized by the rigor of its tactics. It is therefore not certain that Six Sigma enables a culture of sustainable quality once the project is completed. This is why companies perpetuate the gains of project operations by developing staff skills, transferring solutions, and motivation, etc.

Some observers, however, prefer the TPS approach, which, although difficult to implement, provides the company with a commitment to excellence, high performance, continual improvement and waste reduction.

References

[11.1] PETER S. PANDE, ROBERT P. NEUMANN, ROLAND R. CAVANAGH, *The Six Sigma Way: How GE, Motorola, and Other Top Companies are Honing Their Performance*, McGraw Hill, 2000.
[11.2] CRAIG GYGI et al., *Six Sigma for Dummies*, Wiley Publishing, 2005, pp. 85-122.
[11.3] PETER PANDE, LARRY HOLPP, *What is Six Sigma?* McGraw Hill, 2001.
[11.4] GEORGE ECKES, *Six Sigma for Everyone*, John Wiley and Sons, 2003, pp. 27-65.

[11.5] JAMES P. WOMACK, DANIEL T. JONES, DANIEL ROOS, *The Machine that Changed the World: the Story of Lean Production*, Harper Perennial, 1991.
[11.6] OHNO TAIICHI, *Toyota Production System: Beyond large-scale production*, Productivity Press, 1988.
[11.7] JAMES P. WOMACK, DANIEL T. JONES, *Lean Thinking: Banish Waste and Create Wealth in Your Corporation,* Simon & Schuster, 1993, 1996.
[11.8] MICHAEL.L. GEORGE, *Lean Six Sigma for Service: How to Use Lean Speed and Six Sigma Quality to Improve Services and Transactions*, McGraw Hill, 2003.
[11.9] GERHARDT PAHL et al., *Engineering design: A systematic approach*, Springer-Verlag, 1988.
[11.10] KAI YANG, BASEM EL-HAIK. HANG, B. EL-HAIK, *Design for Six Sigma: A Roadmap for Product Development*, McGraw Hill, 2003, pp. 235-306.

Chapter 12

Moral and Ethical Issues

Key concepts: actions, (ethical) altruism, applied ethics, cardinal virtues, categorical imperative, consequentialism (ethical) egoism, ethics, deontological ethics, ethical diamond of the expert citizen, fundamental ethics, (theoretical and practical) intellectual virtues, love of good, love of truth, (ethical) morals, moral virtues, passions, practical ethics, rule-utilitarianism, Sapere Aude!, system of four orders (technico-scientific, juridico-political, moral, ethical), teleological ethics, theoretical ethics, theological virtues, utilitarianism, values, virtues, wills.

12.1 Introduction

Faced with the vagaries of life and the passage of time, we are prey to feelings and emotions, and passions such as sadness, joy, hatred, love, jealousy, pain and suffering. We sometimes have no idea about new situations; we feel penalized by our lack of knowledge, ignorance and the limitations of our reasoning.

So what can we do? How do we think and act justly in order to live in line with personal aspirations and the obligations of family, society and the professional community? This is the field of morality and ethics. This module only gives an outline, providing the minimal foundations to address the moral obligations and social responsibilities of organizations.

12.1.1 The presumption of acting rationally

All morals or ethics presume that human beings are able to act rationally and be responsible. This is also the basis of economic theories. However, the instrument of the economic world – advertising – often uses methods of persuasion (not to say manipulation) that stir the target's passions, and create conditions that lead to an "instinctive" purchase (i.e., contrary to a rational act) – *is this moral?* Passion shapes the life of man, and the practice of virtues steers it.

Any rational act by an individual assumes the permanence of his being, faced with the changing universe of his body (e.g., cell renewal, aging), mind (adolescence, mid-life crisis, etc.) and his environment (society, local economy, employment, socio-political conditions).

What empirical evidence do we have of this permanence? Specialists of the brain and mind have not been able to localize a site in the brain that holds an exclusive function for the permanence of being (the "I" who speaks, communicates, thinks, acts). Buddhism, which now influences the Western approach when it comes to the mind, denies that any permanent "I" exists and refers to five aggregates.

As strange as it may seem, the debate is far from over, but empirical studies [12.1] show that the character of the human being does have constants. One interesting result shows that personality traits [12.2]:

- are present in childhood;
- are independent of ethnicity and race;
- can only be partially influenced by the environment and education;
- are permanently stable from the age of 30;
- show a remarkable consistency in existence;
- move towards a more altruistic perspective in the second part of life, and towards less readiness for change.

Although these results do not provide absolute proof of the permanence of a moral subject, they nevertheless confirm the postulate of morality and ethics.

12.1.2 The past: theoretical and practical ethics

Up to and including the 19th century, morality/ethics was a science studying the laws of human activity. When this approach was developed in a general context, without regard to particular situations, it referred to general or theoretical morals. When, in a second step, it was considered how a principle of moral theory applies in a given situation, this was the field of practical morality/ethics. The first was a science, the second a science and an art. This binary vision would become more complex with the emergence of problems posed by life sciences and technologies developed in the 20th century.

12.1.3 Morals and ethics

Originally, morals and ethics were synonymous. Indeed, their etymology is identical and refers to areas of morality (morality comes from Latin, whereas ethics comes from Greek). The distinction emerged in the 20th century. The meaning of the word ethics is still often confused with that of morals: it is not easy to distinguish one from the other, especially in the institutional communications of companies, where the evocative power of the word ethics outweighs the old-fashioned term morals.

For *Paul Ricoeur* [12.3], morals is the fixed term, the reference plane that has a dual function:

- the principles of the permitted and the forbidden, i.e. the area of standards; and
- the sense of duty as the subjective aspect of the relationship of a subject with respect to standards.

For this same philosopher, ethics has a double meaning:

- ***basic or anterior ethics***, which come before morality; a form of meta-morality that roots standards in life and desire, and which can be summarized in an elegant way as *the aim for a good life (for oneself), with and for others in just institutions*. Basic ethics also encompass minimalist ethics [12.4], ethics of freedom [12.5], ethics of complexity [12.6], etc.;
- ***applied or posterior ethics***, which come after morality, the place of practical wisdom, such as business ethics, judicial ethics or medical ethics. They also include

bioethics, computer ethics (or Netiquette) and environmental ethics. ***Deontology*** is the discipline that develops standards of conduct and actions deduced from applied ethics.

Why do we use the term ethics for these two disciplines? To implement applied ethics, it is generally necessary to start the process in the field of basic ethics. We see possible connections between theoretical morals and basic ethics on the one hand, and practical morals and applied ethics on the other.

12.1.4 Deontological, teleological and consequentialist ethics

Deontological ethics consider foremost the validity of an action or behavior in relation to its compliance with principles or duties. The practice of virtues and morality of the philosopher *Immanuel Kant* are deontological ethics.

Teleological ethics consider as a priority the objectives of a decision or behavior and deem good any result of an action consistent with the original intent. The Deming cycle can be associated with teleological ethics.

When ethics are primarily concerned with the consequences of a decision and chosen behavior, we speak of ***consequentialist ethics***. There are three types of consequentialist ethics:

- ***altruism***, which seeks to maximize benefit to others, regardless of advantages or disadvantages to the originator,
- ***egoism***, which seeks to maximize benefit to the originator, and
- ***utilitarianism***, which seeks the good of a majority of stakeholders, and it is this that is studied in this module.

12.1.5 Values, value systems and axiology

Morality and ethics are part of a complex set, that of values – a ***value*** determines the degree of importance one attaches to an object, its level of preference, esteem, and the concern for ownership in a society or group.

Undoubtedly, the most estimated value in our society is that of social success measured by riches, exterior signs of wealth.

Other values, fortunately, exist: moral, health, and esthetic values, social and political values. In general, the values of a group are developed in the form of a *hierarchical value system*, and ***axiology*** is the discipline that seeks its meaning.

12.1.6 The ethical diamond of the expert citizen

Ethics appear to be the site of analysis, discourse and the development of what to do:

- ***morals*** format these results as absolute standards to be met, and form a prescribed code of conduct for the citizen;
- ***deontology*** clarifies the rules of conduct of experts or technocrats. The expert citizen therefore has a basic tool, the ethical diamond, in which basic ethics are given precedent (Figure 12.1).

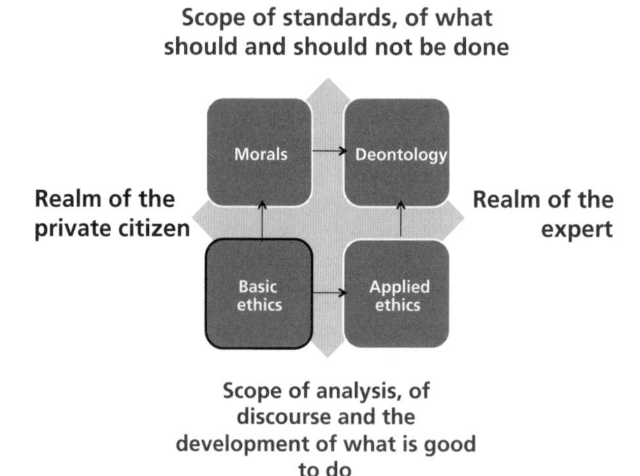

Figure 12.1 The ethical diamond of the expert citizen.

12.1.7 Three adopted moral approaches

Three moral approaches are widely used in Western professions and organizations:

- virtues and their practice;
- morals of the philosopher *Immanuel Kant* and the categorical imperative;
- utilitarianism.

These three approaches are also used for the development of standards for social responsibility of organizations, companies and institutions.

12.2 The practice of virtue

Virtue is a practice which, by repetitive effort, leads man to do good.

Why is this not spontaneous? An answer given by many philosophers is this: Man is naturally subject to passions, prejudices and ignorance that lead him to servitude. The philosopher *Baruch Spinoza* gave an answer to why man is so susceptible to passions: man is a prisoner of his first passion, that of a vague desire to preserve his being. For Jean-*Jacques Rousseau*, who in 1750 published his *Discourse on the Sciences and Arts*, man is born good, but society corrupts him. Psychoanalysis (with the id) and religion (with original sin) have provided other origins.

A very popular theory in the 19th and 20th century is the so-called "veneer" theory, propagated by *Thomas Huxley*, a fervent defender of Darwinism (and called "*Darwin's Bulldog*" for the tenacity of his views). In his book *The Romanes Lecture: Evolution and Ethics* (1893), he says man has an evil nature, compared to a wolf; only education can prepare him to occupy a moral place in society. This perspective is still used to explain the predatory behavior of politicians and top managers, who manifest this fundamental trend, secretly admired because they represent the true nature, the very essence of man to which we must pay tribute, and that only the strongest can manifest. It is a form of justification and subtle absolution of some deviations and diversions in the desires of the powerful.

Studies of the behavior of primates in a Darwinian sense are the most recent (and gradually convincing) contributions to this question [12.7] by showing that moral behavior is probably embedded in the behavior of a species, which is in agreement with *Aristotle*.

12.2.1 Passions and will, definitions

What is passion? Passion comes from the Latin word *patior* meaning *to suffer, feel, endure.* These verbs describe a set of states in which an individual is passive, as opposed to states for which he himself is the cause. In Western philosophy, the word "passion" has had several meanings.

Descartes

Rene Descartes wrote in 1645 (in correspondence with *Elisabeth of Bohemia*) that "we can usually call passions all the thoughts that are [...] excited in the soul without the aid of will, and therefore without any action coming from it, by only impressions that are in the brain, because anything that is not action is passion". In his treatise on the passions (1649), in clause 69, *Blaise Pascal* identifies six basic passions: admiration, love, hate, desire, joy, sadness. The most fundamental passion according to him is (surprisingly) admiration, which triggers other reactions of passion.

This means that, in the sense of *Descartes*, who was thinking perhaps of the Passion of Christ, all passions are not unethical and contrary to reason.

Hume

In 1740, the Scottish philosopher *Hume* wrote "What we commonly mean by passion is a violent and sensitive reaction of the mind to the appearance of good or bad, or an object, which, through the original constitution of our faculties, is likely to excite an appetite" [12.8]. In the mind of *Hume*, the term appetite is synonymous with desire and envy.

Spinoza

Finally, in 1677 *Spinoza* defined in his treatise on *The Ethics III* (general definition of passions) that passion "is a vague idea, mainly abstract and often imaginary, whereby the mind affirms an increase or decrease of its body's strength of existing. For example, pity is a passion since it relies on the vague imagining that a being, such as ourselves, suffers a bad fate, which has the immediate effect of causing sorrow".

The vague wish to perserve one's being is for Spinoza the most fundamental passion from which is derived joy and sadness, then love and hate. Based on a misunderstanding of nature and human nature, passions are incurred rather than marking the fortitude of one who is affected. They are thus passive effects with servitude as a natural consequence.

According to *Spinoza*, passions are opposed to actions, virtues and freedom. It is possible to control them, not by directly challenging them with reasoning and good will, but by opposing them with active feelings and emotions coming from an inner strength, such as firmness (or courage, also called Fortitude, which is one of the cardinal virtues), generosity, and interior acquiescence, knowing that these virtues are for their part born from the joy of understanding the causes of our decisions and particularly our passions.

Will

Will is the ability to do or not do according to one's desire. It is the principle that determines human voluntary action. Let us thus quote the treaty of Descartes in *The passions of the soul* (1649):

"Clause 18: Of the will
Again our wills are of two kinds, because some are actions of the soul, which terminate in the soul itself, as when we love God or generally apply our thoughts to any object that is not material. Others are actions that end in our body, as when we have the will to walk, it follows that our legs move and we walk."

12.2.2 Against slavery and passions: virtues

We usually define a virtue as a way of being resilient, of having a recurring provision of will, acquired by repetition, and which empowers man to act well. Practicing virtues strengthens the mind and helps fight against passions.

We distinguish human virtues from theological virtues that will not be discussed here. There are two types of human virtues: moral and intellectual virtues, which regulate the use of morality and reason. Four virtues, dating back to ancient philosophy, constitute the core of human virtues: Prudence, the mother of all virtues, Fortitude, Temperance and Justice. The list of moral and intellectual virtues is not unified; some authors have listed over 90. The list below is close to that of *Aristotle*, admittedly old, but still relevant.

12.2.2.1 Moral virtues

Moral virtues are powers that empower people to act well in a given activity sphere. They are stimulated by the *Love of Good*. In everyday language, virtue and moral virtue are often synonymous.

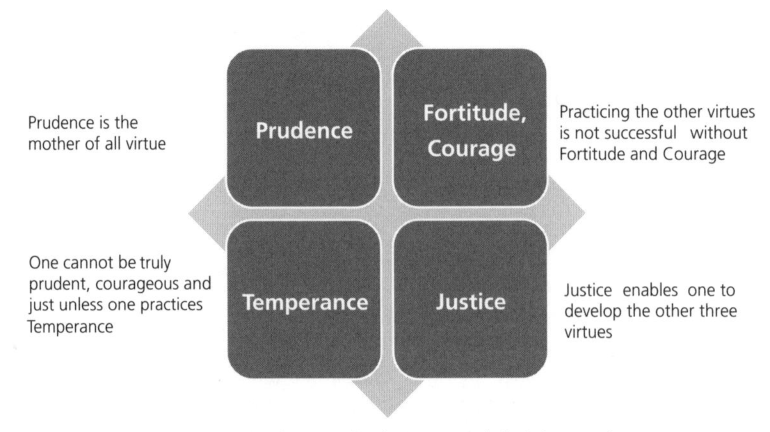

Figure 12.2 The four cardinal virtues and their interactions.

Courage (or ***Fortitude***), as mentioned above, makes it possible to stand firm in pursuing a challenging good, despite all adversities. It is often associated with the Goddess Fortuna, represented in the Middle Ages by the *Wheel of Fortune*. It is associated with the virtue of perseverance.

Temperance enables one to use appropriate means in the enjoyment of delectable goods. It has nothing to do with asceticism, but demands a balance between the sustainability of human life and health, and the satisfaction of needs and desires.

Justice, finally, enables everyone to receive their dues.

Together, *Prudence*, *Temperance*, *Justice* and *Fortitude* form the cardinal virtues since they are civic (Figure 12.2). In Greco-Roman antiquity, each of the three moral virtues that we have named had its basis in the sensitivity of man. *Fortitude* (or courage) regulated the

Mars-inspired combative sensitivity. *Temperance* regulated the Venus-inspired pleasurable sensitivity. *Justice* regulated the Mercurial rational sensitivity.

The moral virtues are in constant interaction forming a system presented in Figure 12.2).

12.2.2.2 Intellectual virtues

The intellectual virtues are powers that enable truth to be achieved in a given sphere. They are stimulated by the *Love of Truth*. They are also based in reason. There are usually five intellectual virtues: Intelligence, Knowledge and Wisdom on the one hand, Art and Prudence on the other. The former are speculative, while the latter two are practical.

Speculative intellectual virtues

Intelligence is that by which we grasp ideas, concepts and principles. For example, what a mouse is.

Knowledge is the way in which we grasp the truth of a conclusion through one of its principles. For example, a mouse is a mammal because it possesses the corresponding characteristics (the mother suckles her young).

Wisdom, made up by intelligence and knowledge, provides information on the concepts and findings that are the most worthy and most difficult. For example, since many species of mammals are endangered, it is urgent to protect them. We should add righteousness to this virtue: rectitude of principles, rectitude of intention, rectitude of mind, and of judgment.

Practical intellectual virtues

Art is a skill in the way things are made. It targets the realization of an asset outside of man. Effectiveness and efficiency are related to Art. The actions we take should reach their goal (***Effectiveness***) and give a maximum of results with implementation using a minimum of time and resources, i.e., effort (***Efficiency***). These two virtues are used daily by all managers.

Prudence (cardinal virtue previously cited) is expertise in the order of exercising freedom and action, the implementation of Wisdom. It aims to build *man himself*. Prudence and audacity are not, in view of this virtue, contradictory. The understanding of the value of this virtue was lost in the 16th century. The virtue of ***Circumspection*** is related to Prudence.

12.2.2.3 Plato's cardinal virtues

It is in the writings of *Plato* that we find the first descriptions of the virtues of *Justice, Temperance, Fortitude* and *Prudence*, still called the cardinal virtues. According to *Plato*, the soul has three parts, each performing a function that can be controlled by a virtue:

- ***Nous***: the principle of reason, is located in the mind (ability to know, judge, reason, to discern right from wrong), which from a moral point of view allows self-control and sensitive desires (*epithymia*). The virtue of *nous* is *Prudence*, the ability to govern and discipline oneself by the use of reason. It connects *nous* to truth and goodness.
- ***Thymos***: the principle of anger and irritability, is very sensitive to upbringing, and located in the heart. When combined with reason (*nous*), it turns into energy, enthusiasm, and emulation. If combined with *epithymia*, it turns into irritation, frustration, and brutality. *Thymos* is related to self-esteem (dignity, self-respect, respect for mankind), and to values we place on ourselves. The higher the moral value we place on ourselves, the more we are frustrated if we do wrong or when we are wronged. The virtue of *thymos* is *Fortitude* or Courage, which consists in firmly

maintaining the commandments of reason by fighting against the internal enemies of *epithymia* and against daily anxieties.

- ***Epithymia***: appetite or sensitive desire, is an inclination to sensual carnal pleasures (needs or desires), such as hunger, sexuality, the desire for survival, and is located in the stomach. The virtue of *epithymia*, harboring passion, is *Temperance* or moderation.

The fourth virtue is *Justice*, the harmonious order and structure of the being, defining the position of each part of the soul, and its function. It integrates the corresponding virtue. The just man is temperate, courageous and wise. But the virtue of *Justice* is also part of *Plato's* ideal society, hierarchically structured according to the three parts of the soul, where it plays a similar role (in the low social class of the peasants, it is the role of the *Schwarzpeter* of passions...!).

The soul wants to ascend to heaven where it can contemplate the eternal ideas of what is Good, True and Beautiful. This is the ultimate function of Reason. But it is diverted from the aspiration to this vision by the passions of *epythemia*. In an allegory, *Plato* compares Man to a team of two horses and the soul to a chariot: reason is the driver, *epythmia* is a black horse, and *thymos* a white horse. This coupling is not initially harmonious: it cannot move in one predetermined direction. The black horse leads the chariot against the will of the white horse, its companion, despite all efforts by the driver to master it. But the driver will eventually master the two horses, and it is at this moment that the chariot becomes winged and the soul is able to rise to contemplate the Truth.

12.2.2.4 The virtues of enlightenment and love of humanity

Two virtues of the Age of Enlightenment are worth quoting:

- *Tolerance*, and
- *Equality*,

which are consistent with the practice of *Charity* (a theological virtue), which combines doing good, the search for public welfare, personal detachment (but not when it comes to obligations of the community and family), and compassion, stimulated by a *Love of Humanity*.

12.2.3 What are the virtues for?

The morality of the virtues [12.9][1] is not a theoretical morality but a practical one. It does not impose an absolute obligation, but aims to raise us, to structure us. The purpose of practicing the virtues ultimately depends on our personal position when it comes to spirituality and death:

- If one believes that life ends with death and that the spirit is a by-product of matter, cultivating the virtues is essentially a way to maximize pleasure and happiness, or even to contribute in a small way to the progress of humanity. These benefits are obtained by cultivating health, family, friends, social circles and constructive activities. Repeated non-virtuous acts ruin friendships, filial and conjugal ties, health, and lead to loneliness, lack of self-esteem, and even to sickness.
- If on the contrary we think that life does not end with death and that honest behavior is related to post-mortem conditions, cultivating the virtues is, in addition to the benefits identified above, at best a passport to the afterlife, but more generally a prerequisite and essential part of the success of spiritual asceticism.

Practice of the virtues is used by deontology to condition professional behavior.

12.2.4 The example of *Benjamin Franklin*

The virtues were highlighted in the 18th century because, according to the philosophy of the Age of Enlightenment, Men are born equal in rights and only their virtual behavior differs, not the privileges of birth. A self-made man, *Benjamin Franklin*, man of the Age of Enlightenment and founding father of the United States, determined the virtues that he wanted to cultivate. He selected thirteen: *temperance, silence, order, determination, frugality, industry, sincerity, justice, moderation, cleanliness, tranquility, chastity and humility*, and for each of which he had a maxim. He practiced self-examination [12.10] each night before bed, noting the times of the day where he had been able to put them into practice and where he had unfortunately transgressed. The list of virtues of *Benjamin Franklin* formed the prototype of those of the middle classes.

Benjamin Franklin was an assiduous practitioner of the virtues, and a fervent opponent of slavery. He was, according to his biographers, a distant father and a poor husband! The practice of virtues has, as any method, its limitations. We will see how, with the contributions of *Kant*, it is possible to complete the practice of virtues.

12.2.5 Abandonment of the practice of the virtues in the 19th and 20th centuries

Practice of the virtues became less important in the second half of the 19th century and in the 20th century. A few reasons are:

- The basis for practice of the virtues is the understanding of their mother *Prudence*, which is an applied *Wisdom*. Understanding of this virtue was already lost in the 17th century, initiating the slow decline of this practice, only partly curbed by the Age of Enlightenment.
- Psychoanalysis, introduced by *Sigmund Freud*, sidelined, apparently excessively, the virtues of introspection and the importance of a conscious mind. Some modern psychological tendencies now make this sidelining relevant, but the human impulses of life and death revealed by *Freud* have undermined the success of the practice of virtues.
- At first, the values of the American Revolutionary War (1775-1783) were based on the theme of returning to the (supposed) moral purity of the early settlers, abandoned over time under pressure from the promiscuous British and their venal exported objects (jewelry, luxury clothing, etc.). Driving out the English was associated with the recapture of a virtuous life based on simplicity and moderation of material assets; this was expected to continue, even with the constant influx of immigrants, by expanding the borders to the west. In order to have a discourse suitable for all religious factions (including the Puritans and the Quakers), the fathers of the American Revolution, such as *Samuel Adams,* shaped in their writings and speeches, the moral idea of *Patriots* and *Minute Men*[1]. They refer to the practice of virtues [12.11] – this from the perspective of the Age of Enlightenment – because practice of the virtues was linked to the democratic life of the city in antiquity. At the end of the war, *Samuel Adams* noted with sadness that his initiative was a failure: the settlers' taste for power, greed, and luxury had increased, contrary to his calculations. Without state or religious compulsion, US citizens abandoned the perspective of a simple life to become Yankees con-

1 Back-up troops from Washington, composed of farmers who could drop their tools within a few minutes and take to their guns.

cerned with material and financial success. The American value system based almost exclusively on comfort, financial, material and social success was almost in place. The practice of virtues can thus not be the only civic discipline and morality of a democracy. At the time, this relative failure fueled the arguments of advocates for spirituality and religion, who believed that man is virtuous only if a spiritual influence reinforces the practice of the cardinal virtues (in the case of Christianity, by adding the practice of the theological virtues[2] and Grace).

- But even more pernicious, *Machiavelli's* book *The Prince* appeared in 1532. With remarks such as "*the Prince should know how to break his promises*", it ruined the effectiveness of the practice of the virtues in political action. Other cynical publications of unequal values followed, such as *the Breviary of Politicians* attributed to *Cardinal Mazarin,* published in 1684 [12.12]. These works argued that the actual practice of the virtues is contrary to major actions and that instead dissimulation, hypocrisy and suspicion are key behaviors to adopt. Who could blame them? This remark is not new, because for many ancient philosophers, the philosopher who loves Wisdom followed a path at odds with the thinking of politicians, and even of men of action.

12.2.6 Renaissance of the practice of virtues

Hedonism, individualism and the desire for transparency in our society, the disenchantment with regard to progress, the development of individual responsibility, the loss of moral, spiritual and religious references, and the success of analytical philosophy mean that the individual practice of the virtues has today, with the rediscovery of ancient philosophy, gained renewed interest. The French philosopher *Pierre Hadot* has put the message of ancient philosophy and its practices back on the agenda [12.13].

12.2.7 A universal taxonomy of virtues?

Is there a universal taxonomy of virtues, a model that could cover all the ethical systems of the civilizations of mankind. Why not? Man has lived in societies for tens of thousands of years and his strength of character should certainly present some invariants. This study was carried out by a team of psychologists [12.14], who classified the virtues into six categories, with a good overlap between Hinduism, Buddhism, Islam, Christianity, Judaism, etc. (Figure 12.3).

The proposed classification is a good summary, and a credible alternative to the one inherited from antiquity.

12.3 Kant, La Bruyère and the categorical imperative

How does one define a general approach to the problem of the definition of ethical or moral behavior? The morality of the Age of Enlightenment has been studied in more detail

[2] The three theological virtues, Faith, Hope and Charity, are cited by the apostle Paul (I Cor 13:13). Their universal (or not) character or the effectiveness of their practice will not be discussed here.

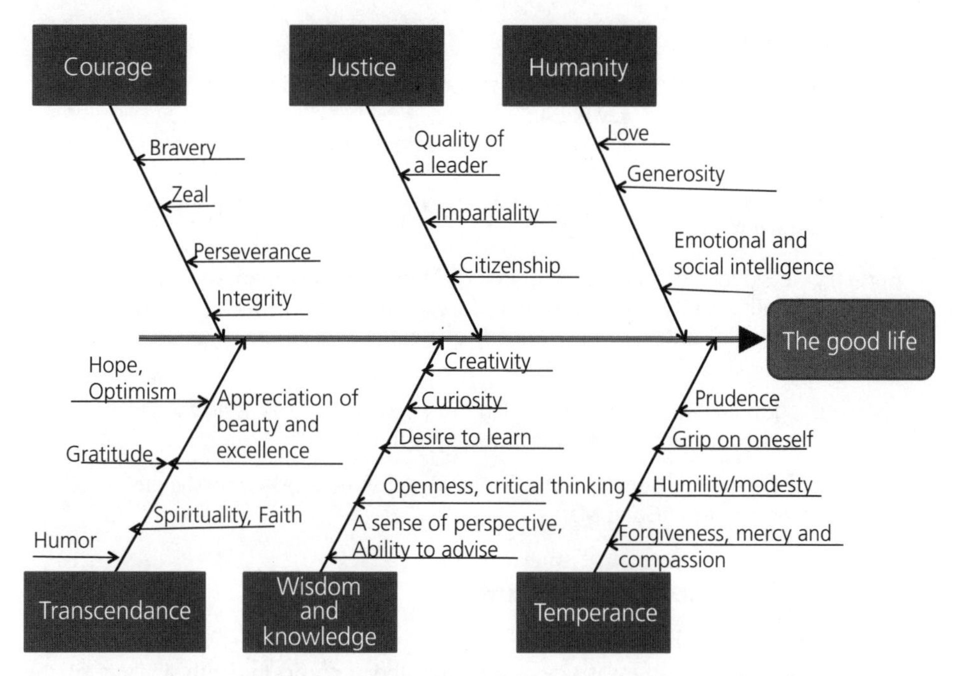

Figure 12.3 The universal taxonomy of the virtues by C. Peterson (2004).

by *Immanuel Kant*. Anyone who would like to summarize the intellectual concerns of all searching men today can refer to the questions he asked on May 4, 1793[3]:

- What may I know?
- What should I do? This question is different from "how to act" or "how to be", the answer to which may arise from practice of the virtues.
- What am I allowed to hope for?
- What Is Man?

We are only interested here in the second and fourth question.

12.3.1 What is Man?

It is amusing to read the book *The Characters* by *Jean de la Bruyère* (1696). *La Bruyère* had a penetrating mind, but a rough exterior. He scrutinized the companies of the Court of France, the City of Paris, and its affluent backgrounds. This book has not aged a bit – human behavior is ever constant. La Bruyère distinguishes between three types of men – in the sense of socioeconomic agents:

- A ***cunning man*** is he who conceals his passions, who understands his well-being, who sacrifices a lot for it, and who has acquired or managed to retain property (55, VII). It is also he to whom the following maxim applies the best:

  "The shortest and best way to make your fortune is to let people see clearly that it is in their interests to promote yours" (42, V).

[3] Letter from May 4, 1793, adressed to *Carl Friedrich Stäudlin*.

And also this one, far less pleasant:

"We open and display every morning to fool our world, and close at night after having deceived for the whole day (42, I).

- An ***honest man*** is one who does not steal, who does not kill, and whose vices are not too outrageous (55, VII).
- A ***good man*** is one who is neither a saint nor a (false) devotee, and who is merely righteous (55, VII).

Certainly we would rather have as a business partner or friend a man who wavers between Honesty and Goodness rather than between Cunning and Honesty. On the other hand, we certainly prefer to face dangers in the company of a man who has a minimum of cunning. *La Bruyère* was very accurate when he added:

"You are a good man, you are not thinking of pleasing or displeasing the favorites, you are only attached to your master and your duty: you are lost" (40, I).

The average person must embrace these three types of men, even if he/she prefers to give priority to the model of the Good Man.

12.3.2 Morality according to Kant

12.3.2.1 Man knows phenomena

Taking human reason as a case study, *Kant* subjected it to an in-depth analysis of the opportunities it gives to man, but also its limits. There is a limit of possible experiences for all men, because man moves into the world of the senses ("phenomena"), and in *Critique of Pure Reason* (1781, 1787) he establishes the impossibility of scientific knowledge of the thing-in-itself (*Noumena*). This is a first answer to the question on knowledge. We can only establish models, although their suitability to describe phenomena is unsettling.

It is also a basis for reflection on racial profiling, this cutting judgment on another human community, from which, on the basis of certain facial features, we deduce hidden defects or character quirks. This error was also noted by *La Bruyère*:

"Facial appearance is not a rule that is given us to judge men, we can use it as conjecture" (31, IV).

12.3.2.2 Liberty and human dignity

Beyond the rather reserved judgment of human reason and, to answer the question of how to do it, *Kant* demonstrates the moral importance to man of freedom, immortality and religious fulfillment. He calls for human freedom and, respecting human dignity, dismisses the awareness of this freedom to the realm of subjectivity, largely treated in *Critique of Practical Reason* (1788). The conclusion of this criticism begins as follows:

"Two things fill the heart with ever new and increasing wonder and awe, the more often and the more seriously reflection concentrates upon them: the starry heaven above me and the moral law within me."

12.3.2.3 The categorical imperative

The moral autonomy of man is the principle of morality. The moral person – that is to say not the empirical man, who is an element of the sensible world, but "humanity present in man" – is an end in itself, and not a means to other ends. Regarding the "how" of the moral approach, *Kant* defines it as a categorical imperative, promoted to great success:

"Act only according to that maxim whereby you can, at the same time, want that it should become a universal law."

We retain this statement as the first formulation of the categorical imperative.

12.3.2.4 The Age of Enlightenment: *Sapere Aude!*

But there's more; linking moral approach to intellectual inquiry, *Kant* sustainably described the Age of Enlightenment of the 18th century. Taking a decisive role in this movement, he gave an appropriate description to his arguments that bore fruit against intolerance, indoctrination, cowardice and laziness. They can still be used today to portray a universal society on solid foundations. Here is his definition of enlightenment:

"Enlightenment is defined as man coming out of the state of minority, where he is maintained through his own fault. The minority is unable to use its own understanding without the guidance of another. It is due to our own fault, as it is not the result of a lack of understanding but a lack of resolution and courage to use it without the guidance of another. Sapere aude! Have the courage to use your own understanding! That is the motto of enlightenment" [12.15].

12.3.2.5 Upbringing and the human person

Cultivating courage and stimulating the desire for knowledge require a definition of upbringing. Its major tasks are fourfold: discipline, cultivate, civilize and moralize. *Kant* did not approve of the inhuman principle of obedience, constant in Prussian history with the lawyer and enlightened monarch, old Fritz (*Frederick II!*), which partly explains why the Nazis went too far. He hoped instead that upbringing and modeling would be governed by a principle of leniency or high standards, respecting and applying, as far as possible, the dignity of human beings. This should be the approach for redefining and consolidating this requirement of "moralization" according to Kantian thought.

12.3.2.6 Morals of action

Let every man in time and space be considered as an end in himself. This is the key idea, thanks to *Kant*, that must guide human conduct. Let us recall the consequences of his concept of a human person:

"Act in such a way that you treat humanity, whether in your own person or in the person of any other, never merely as a means to an end, but always at the same time as an end."

We retain this statement as the second version of the categorical imperative.

or:

"The nature of human beings, constituted by freedom and reason as a moral subject, results in all people being recognized as moral objects, as individuals."

The duty of peace presents itself as absolute and should be applied universally. We can place historical foundation in the categorical imperative, and its impact on activities whose purpose is to make human dignity a reality without reservation, with the establishment of eternal peace as an end (even if this end remains an ideal).

12.3.3 The categorical imperative and the resistance to totalitarianism

Professor Kurt Huber[4] from the German resistance group *The White Rose* praised the categorical imperative in the People's Court in Munich, April 19, 1943, before he was executed by the Gestapo:

> "I had set a goal to raise in student circles, not through an organization but only by the word, not some act of violence but a moral awareness of the serious evils that mark our current political life. The return to clear moral principles, to a state based on law, and trust between men. This is not illegal, it is rather to restore legality. I wonder, in the spirit of *Kant's* categorical imperative, what would happen if this subjective maxim of my action became a universal law. To this question, there can be only one answer: it would be the return, in our political life, of order, security, and confidence in our state.

The categorical imperative is constantly lost by greed and human passions and found by those who wish to become good men. To imagine the categorical imperative to be out of date, would be to think that there is an end to the excitement of passion and human ambition, much like the advent of a classless society allowed the Marxists to solve all problems of society and of humanity.

12.4. Ethics, morality and economy

12.4.1. The ethics and economics of trade

The economy is based on trade: logically, there can be trade only if both parties realize that they have mutual interests. The underlying ethic is one of egoism since, naturally, both parties find it worthwhile despite a lack of concern for the other. The virtue of Justice, which gives everyone his due, is respected, without any human intervention. Everything seems to be for the best in the best of worlds, thanks to an almost divine hand. However, other aspects of the situation of trade as the basis of the economy must be considered:

- Trade only really benefits both parties if the price of the product or service is decided in an atmosphere of free competition, i.e. in a perfect market (no monopoly, for example). These conditions rarely exist.
- Trade is not only carried out in a market, but within specific sociopolitical conditions which often influence the actual conditions of trade. A common example is the plundering of resources by certain mining companies that handsomely pay corrupt elites while expropriating the owners of the territories that they have surveyed.

---

[4] The resistance group *The White Rose* was founded in the spring of 1942 at the University of Munich. Young students refused to accept the totalitarianism in which Germany had sunk, and wanted to safeguard their independence of mind against the Nazis. They talked about the political situation with *Kurt Huber*, a professor at the University of Munich, known for his philosophy classes. He encouraged them to resist and became the mentor of *The White Rose*. Students wrote slogans on the walls of pacifists and anti-fascists, and collected bread for prisoners in concentration camps and took care of their families. The actions of *The White Rose* were taken as examples as of January 1943 by intellectuals in southern Germany and Berlin. In February 1943, students threw hundreds of leaflets into the courtyard of the University of Munich. The caretaker stopped them and turned them over to the Gestapo. They were sentenced to death.

- The practices of business, including relationships of cooperation/competition between companies, and their relations with the state, are full of predatory behavior, which make us doubt that the virtue of Justice is always respected, and the same goes for the categorical imperative.

Finally, managers of large organizations need to address economic problems, including conflicts of values where a misunderstanding could have serious consequences.

12.4.2 Is capitalism moral?

The only survivor of the many previous economic systems, capitalism, the basis of the market economy, now bears the hopes of a just society, given the predominant influence of economics on society (which was not always the case). Is this system able to respond spontaneously to these expectations? The issue is widely discussed, and systems of social responsibility as well as business ethics, introduced recently, cause doubt.

Science and technology are not in themselves moral, so why should the economy be, which is itself a science and a technology? How can the economy be moral, since it is without will or consciousness? Does the fact that economic agents act rationally[5] (however limited) mean that they act virtuously and reasonably[6]?

The philosopher *Comte-Sponville* [12.16], with many economists [12.17], definitely opted for an amoral (but not immoral) capitalism, while stressing that individuals or employers of organizations/companies, if they do so with full awareness, can act in a spirit of justice.

12.4.3 The four orders

Comte-Sponville was against the integration of morality into the economy, and has defined four distinct orders in which what is permitted and forbidden should be studied:

- the techno-scientific order (including economics),
- the juridico-political order,
- the moral order, and
- the ethical order.

It is primarily the juridico-political order that should prescribe what is permitted or prohibited in the techno-scientific order, although other higher orders may have a direct (but still more limited) influence on the functioning of the techno-scientific order. However, it is in ethics that the modes of action and choice of values are made, taking into account the constraints and experience in the techno-scientific domain.

The field of quality and its management systems belong to the techno-scientific order. Nevertheless, it integrates the juridico-political order (product standards, for instance) and deontology as well as applied ethics (Figure 12.4).

12.4.4 A neo-Kantian *homo economicus* to save the economy[7]?

But this position does not appeal to some economists for whom the economy comes to the aid of morality. *Lemennicier* [12.18], a professor at the University of Paris II, had the smart

---

5 In the sense of being in accordance with reason, logic.

6 In the sense of being honest, levelheaded and prudent (a virtue).

7 Economism is an economic reductionism that reduces all social facts in an economic dimension, especially by systematically giving prevalence to the factors of supply and demand.

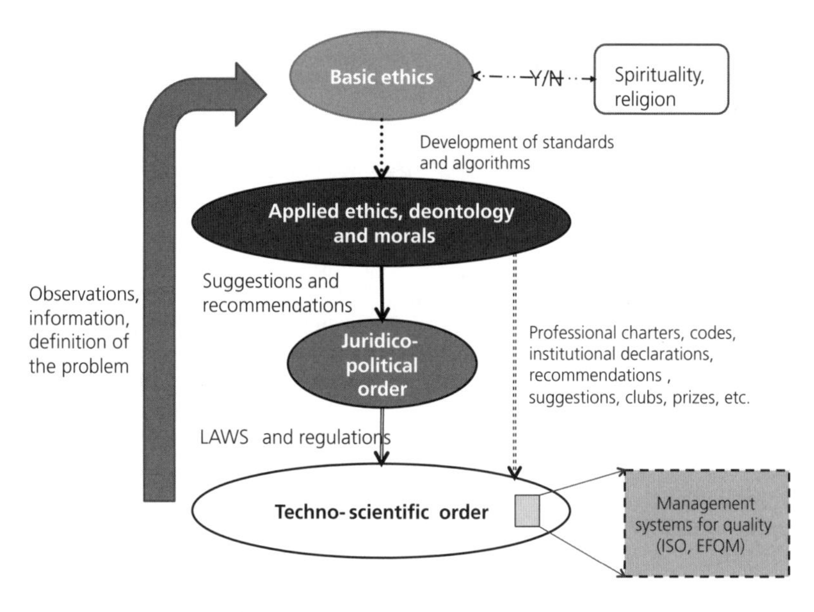

Figure 12.4 The four orders of recommendation and the place of quality.

reflection that Kantian ethics can serve as a basis if they incorporate trade. This amounts to reformulating the categorical imperative as follows:

> "*Act in such a way that you treat humanity, whether in your own person or in the person of any other, always at the same time as an end and never merely as a means without consent* (as part of the autonomy of the will)."

And to this should be added the right of the first occupant or first owner, i.e., for his own body, to use it completely without outside interference.

The remark is sharp and, as *Lemmenicier* indicated, the reader reacts strongly according to his/her own values. The application of economic reasoning shows that many economic decisions based on morals and ethics remind us of the saying: *The road to Hell is paved with good intentions*. The application of this system also allows for physical violence.

Thus, some conclusions are embarrassing:

- private selling of organs;
- complete decriminalization of the sale and consumption of drugs (a fact, indeed, in some jet-set spheres);
- allowing virtually unrestricted abortions;
- privatization of city streets, leading to the formation of mini-cities, where the ruling system (thus the political system…) is determined by the owners (a return to a republic of the powerful elite, even feudalism, etc.);
- acceptance of racial discrimination in employment;
- rejection of the system of political democracy – without any counterproposal – based on the affirmation: *the legal oppression of a minority* (read: wealthy citizens) *by a majority* (read disadvantaged citizens)… *are the brazen laws of any democracy* [12.19].

This mixed result leads us to accept the system of four orders as a model for a manager, businessman and scientist, even though science and technology will actively participate in the resolution of ethical issues, such as the basis of moral rights [12.20].

Moreover the techno-scientific order may set aside a techno-ethical debate: the Age of Enlightenment has certainly contributed, through the rise of human rights, to prohibit slavery, but that status has since then largely disappeared given that the development of energy technologies and machines have replaced slave labor. This is also the case in the biomedical field, in which some ethical problems have been made obsolete, and others induced, by the development of new therapies and healing techniques.

12.5 Utilitarianism

Why address utilitarianism in an introductory module on ethics and morality? Because it is the moral doctrine that contributes most to shaping the answers to legal disputes, or political or economic alternatives. It thus also contributes to the field of applied ethics (euthanasia, abortion, environment, rights of future generations, and animal rights).

12.5.1 Definition

Utilitarianism [12.21] is a moral doctrine belonging to the field of fundamental ethics, whose appearance in the Age of Enlightenment accompanies democratic thought. It coordinates and specifies evaluation and moral action with three recommendations:

- a criterion of good and evil: the happiness, well-being of all stakeholders, including in the long term;
- a moral imperative: maximize this good, give the greatest happiness to as many as possible (in other words, consequentialism);
- a measured evaluation of the action by the above criterion.

The practice of utilitarianism is neither spontaneous nor automatic, but allows for prompt answers that can reconcile ethical altruism and a Darwinian perspective.

Utilitarianism places little emphasis on action: it has been written that the disadvantage of utilitarianism is to confuse impersonality (a-subjectivity) and impartiality (objectivity).

Moreover, utilitarianism faces the subjectivity of the concept of welfare, and the difficulty of quantifying it. It is accused, not without reason, of reducing life to an exclusively material and economic perspective. This is why some strains of utilitarianism have determined a hierarchy of values, in which cultural values are preferred.

12.5.2 Rule-utilitarianism

Utilitarianism does not take into account how good is done. It sets aside Kantian ethics, which means that utilitarianism can be used in a wrong way; an evil action can be accepted in the hope of doing good (the end justifies the means). This is not objectionable in certain circumstances (think of a medical procedure that temporarily causes pain), but can create problems through the use of lies, violence, torture, etc.

This is why the philosophers of the 20th century brought a new form of utilitarianism, that of rule-utilitarianism [12.22]. Implementing rule-utilitarianism involves defining a first algorithm or model of an action. To determine whether this model should be followed, we extrapolate what happens if it is consistently applied in many circumstances. If the application of this rule, when consistently followed, produces more happiness than pain, the rule should be retained and compared with other available models of action that have passed this first test. The rule that is at the top of the list when it comes to maximizing good while minimizing suffering is selected.

12.5.3 Rule-utilitarianism and human rights of 1948

Rule-utilitarianism is a form of ethics, known in the English-speaking world – but often received harshly in French-speaking parts of the globe.

However, rule-utilitarianism cannot provide definitive answers in areas such as torture[8], and the oppression of minorities. Utilitarianism does not seem able to fully take into account the basic adage: *Do unto others as you would have them do unto you*.

This is why the results of rule-utilitarianism must be confronted by, for instance, sections of the Universal Declaration of Human Rights of 1948,[9] which comprises 30 articles and is accepted by all member countries (even if there is some doubt that they all implement them), and in which the preamble states, in its eighth paragraph:

This Universal Declaration of Human Rights... (is)... a common standard of achievement for all peoples and all nations, so that every individual and every organ of society, keeping this Declaration constantly in mind, shall strive by teaching and education to promote respect for these rights and freedoms and by progressive national and international measures promote universal and effective recognition and application, both among the peoples of Member States themselves and among the peoples of territories under their jurisdiction.

When it comes to Europe, it has set up a similar convention in 1950, *The Convention for the Protection of Human Rights and Fundamental Freedoms*[10].

12.6 Summary: an ethical approach to a problem, an educational model

The ethical approach to a problem is a group effort, relying on philosophers, politicians, technocrats[11], citizens and stakeholder representatives. To illustrate the approach, Figure 12.5 shows a hypothetical and pedagogical model of a framework and algorithm.

12.7 A philanthropic application

The wealth differential is growing rapidly and the upper parts of our society [12.23], the "jet set", become nationless, fleeting, volatile, and elusive, moving from one paradise to another, especially if it is a tax paradise. The privileges of birth have not vanished. It is therefore interesting to apply moral theories to philanthropy. Are donations made:

- in a spirit of detachment and charity?
- to have a good image of oneself?
- to use a portion of one's taxes for activities outside the state sphere (this is the case of the US: but is it really philanthropy...)?

---

[8] Of which rightly belong the "mock" executions and drownings obligingly communicated by the press as harsh interrogations, although they were described as torture by the same US authorities when it was the Nazi Gestapo that administered them...

[9] Available at: http://www.un.org/french/aboutun/dudh.htm

[10] Available at: http://conventions.coe.int/Treaty/fr/Treaties/Html/005.htm

[11] Scientists, engineers, managers, health specialists, finance specialists, etc.

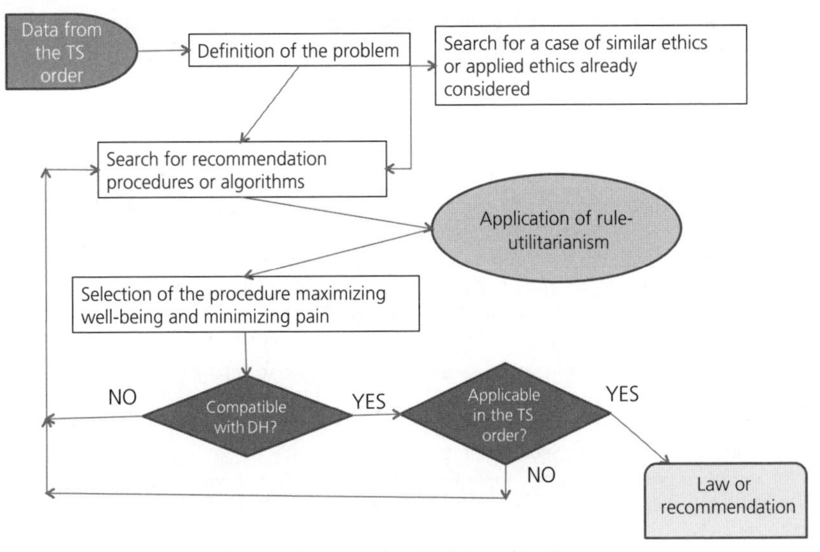

Figure 12.5 Educational model of an ethical approach.

- to avoid taxes?
- in order to benefit, in the latter part of one's life, from additional consideration?
- in order to diversify one's power and expertise by directing the business of the community through public investments in targeted services?

Or more simply, are they made by concern for the reputation of an unfailing integrity in a business, which sometimes really needs it?

"You are great, you are powerful, it is not enough; make me respect you, so that I may be sad to be deprived of your favors, or of not being able to acquire them" (*La Bruyère*, 36, IV).

How can we identify the motivations of a donor? For the recipient, who often makes it a business, should he agree to work with the powerful? It's a risky situation that *La Bruyère* sensed well:

"If it is dangerous to be steeped in a suspected case, it is even more so to find oneself an accomplice of one who is great; he gets away and lets you pay double, for him and for you" (38).

Let us finish with two bitter sentences, with one on insensitivity:

"Health, wealth, taking the experience of evil from men, inspire hardness for their like"(79, VIII).

And the other on networking:

"To evade the Court for a single moment is to give it up; the courtier who sees it in the morning, sees it the evening, and recognizes it the next day so that he himself will be recognized" (4, IV).

The utilitarian version of ethics, which judges mainly the result of actions, accepts the majority of the motivations above. This is not the case for Kantian, teleological and virtuous approaches.

Reference

[12.1] For an introduction to the issue of character traits, see the book by GERALD MATTHEWS and IAN J. DEARY, *Personality Traits*, Cambridge University Press, 1998, pp. 3-35.

[12.2] ROBERT R. McCRAE, PAUL T. COSTA, Jr, *Personality in Adulthood – a five Factor Theory Perspective*, Guilford Press (2003), ISBN 1-57230-827-3, Chap. 4,5,6 and 10.

[12.3] Cf. the ethical article, p. 189, in the *Dictionary of ethics and moral philosophy* (T.1), edited under the direction of MONIQUE CANTO-SPERBER, published by PUF, Coll. Quadrige, 2004.

[12.4] Who limits himself to the need to avoid deliberate harm to others (RUWEN OGIEN, *L'éthique d'aujourd'hui*, Gallimard, 2007.

[12.5] Of which the principle is: act so that others can increase their choices, in EDGARD MORIN, *La méthode: 6. L'éthique,* du Seuil, 2004, p. 132.

[12.6] Of which the principle is: act so that others can increase their choices, *ibid.*, p.132.

[12.7] Read for example the short collection of HENRI ATLAN and FRANS B.M. DE WAAL, *Les Frontières de l'Humain,* Le Pommier, 2007, especially the presence of empathy and concern for justice among hominids. pp.58 et seq.

[12.8] DAVID HUME**,** *Traité de la nature humaine*, Aubier Montaigne, 1992, p. 548.

[12.9] For a brief introduction and simple relationship between morality and virtues, see the book by COLIN McGINN, *Moral Literacy – How to do the right things*, Hacket Publishing Company, 1992.

[12.10] The method developed by *Benjamin Franklin* is still practiced in the USA. See the pamphlet PAULA R. BENNER, *Benjamin Franklin's Art of Virtue Journal*, Infinity Publishing, 2006.

[12.11] DAVID.E. SHI, *The simple life - plain living and high thinking in American culture*, Oxford University Press, 1985, Chap. 3. *Republican simplicity,* and Chap. 4. *Republicanism transformed.*

[12.12] *Bréviaire des politiciens*, Arléa, 1997, prefaced by UMBERTO ECO, whose five precepts are: 1. Simulate, 2. Hide, 3. Do not trust anyone, 4. Speak well of everybody, 5. Get ready for any eventuality.

[12.13] For those who wish to go further: PIERRE HADOT, *La philosophie comme manière de vivre,* Le livre de poche, 2003 — the story of a philosopher, his interest in ancient philosophy and practice; ANDRE COMTE-SPONVILLE, *Petit traité des grandes vertus,* Poche, Points, 2006 – a secular view of the virtues, the essential reference; JOSEF PIEPER, *Four cardinal virtues: Theology*, University of Notre Dame Press, 1990, – a Christian and Catholic vision of virtues, based on the thoughts of St Thomas d'Aquin.

[12.14] CHRISTOPHER PETERSON, MARTIN SELIGMAN, *Character, Strengths and Virtues – A handbook and Classification*, Oxford University Press, 2004, pp. 29 and 30.

[12.15] EMMANUEL KANT, *Qu'est-ce que les Lumières? (1784)*, des Mille et une nuits (2006), pp. 11 et seq.

[12.16] ANDRÉ COMTE-SPONVILLE, *Le capitalisme est-il moral?* Albin Michel, 2004, cf. Chap. 2. *Les Limites*, and Chap. 3. *Le capitalisme est-il moral?*

[12.17] See the Nobel laureate in Economics MILTON FRIEDMAN, *Essays in positive economics*, Chicago University Press, 1953, Chap. *The methodology of positive economics.*

[12.18] BERTRAND LEMENNICIER, *La morale face à l'économie,* d'Organisation, 2006.

[12.19] *Ibid.*, p.226.

[12.20] See the study of "moral" behavior of primates, but also trials such as JEAN-PIERRE CHANGEUX, *Les fondements naturels de l'éthique*, Odile Jacob, 1993.

[12.21] For a short introduction of utilitarianism in the context of ethics, cf. SIMON BLACKBURN, *Being good – a short introduction to ethics,* Oxford University Press, 2001, pp.74-93. For further information, see the book by FRED FELDMAN, *Introductory Ethics*, Prentice Hall Inc., 1978, pp. 16-79. In French, see the pros and cons of this system in the book by SMART JOHN JAMIESON CARSWELL, WILLIAMS BERNARD, *L'utilitarisme- le pour et le contre*, Labor et Fides, 1997.

[12.22] Cf. the article on *utilitariansim* by CATHERINE AUDARD in the *Dictionary of ethics and moral philosophy*, op. cit., 2002 et seq.

[12.23] For a wider view of access to wealth, beyond the glorious success stories of management or finance periodicals - often close to self-celebration - we refer to the caustic book by MICHEL VILLETE and CATHERINE VUILLERMOT, *Portrait de l'homme d'affaires en prédateur*, La Découverte, 2007. Here are two excerpts: "*Marked by signs of ambivalence, the morality of the businessman is rather one of efficiency when it comes to business, and becomes one of generosity when, his fortune made, he tries to redeem his image. However, even at the height of the action, to be effective and achieve fast and dramatic financial results, he cannot ignore that virtue is a capital and that bad reputation comes at a cost" (p. 230) ... "Experts and managers who work alongside businessmen are professionally responsible for producing a morally acceptable discourse about morally reprovable activities and they do it because it is well known that a bad reputation can lead to serious economic difficulties" (pp. 234-235).*

Chapter 13

Deontology of Professions and Functions

Key concepts: accountability, accuracy, constraints, democratic values, deontological charter, deontological code, deontology, diligence, efficiency, ethics, ethics of availability, ethics of efficiency, ethics of responsibility, honesty, impartiality, importance of intention, integrity, intrinsic value of actions, merit-based recruitment, moral values, New Public Management, people values, probity, professional values, public utility, respecting rights, rules, science of duty, selflessness, whistleblowing.

In their work, engineers and scientists act on nature and the community in a significant and visible manner. If their behavior does not comply with certain guidelines, it impairs the confidence of the citizen, user, customer, patient, or representative of the profession, institution, or company employing them. As such, the duties and rules of behavior when exercising a profession can be integrated into the Quality Assurance of an institution, company or professional association. They thus integrate the so-called emotional intelligence, transferable competencies or soft skills.

How does one start this approach? By using the knowledge gained in the previous module, briefly introducing the science of duty, by incorporating some historical yardsticks, and taking into account three types of ethical approaches, of which the first two are of a fundamental nature:

- medicine;
- public administration;
- engineering.

13.1 Deontology

The first definition of ***deontology*** is the science of duty. We are indebted to *Jeremy Bentham* [13.1] for creating this word [13.2]. Although one can find texts in English where this usage persists, the word has changed its meaning twice. It once referred to the empirical study of what to do in a given social situation, but now it is used exclusively in a professional context. For the first sense of this term (the science of duty [13.3]), the French now use the word ***déontologisme*** (deontology).

In deontology [13.4], *the focus is on right before good*. Deontology adopts the following ethical positions:

- *Intrinsic value of actions*: the existence in themselves of obligatory and evil actions that cannot be implemented for charitable purposes, even if they are morally obligatory or morally admirable.
- *Respecting rights*: an immoral act, ethically speaking, is an act that violates the right of others. The *Declaration of Human Rights* is the most complete model. We can exploit these rights in favor of a higher cause (e.g., by the dictatorship of the proletariat, a transitory step towards the ideal classless society… or by any dictatorship, even that of the market…).
- *Rules and constraints*: being more than a vision of the action of good or the good life, deontological ethics emphasize an awareness of the limits of an action. Rules, laws and prohibitions are the main features of deontology.
- *Importance of intention*: what counts in the action is above all the moral intent with respect to ethical principles.

This position has influenced the drafting of charters and codes of conduct, even if an adjustment to the priority of the good over the just can intervene. This is a borderline case that deontology seeks to avoid. Deontology demolishes the saying: *the end justifies the means*.

13.2 Deontology, the science of professional duty

13.2.1 The reason for deontology

Deontology [13.5] *is the set of duties related to exercising a profession*. Why has deontology been developed? The reason is the growing imbalance between a holder of technical or technological competence and users who are uneducated in a subject that falls, at best temporarily, in their power during an activity within a department. The differential of skills has developed asymptotically since the advent of the industrial revolution.

Deontology is available as a set of formal and explicit rules, the transgression of which can give rise to sanctions. This set of rules is called a *deontological code* or *deontological charter*. The first expression is used when deficiencies can lead to serious sanctions. The term *deontological charter* can also be used for institutions and companies. Many charters and codes of conduct prescribe the correct behavior, not only in everyday situations but also in conditions of crisis that a person might encounter in the course of his/her professional activity.

However, protecting the user is not the only reason behind the development of codes of conduct and deontological charters: the differential of skills can be catastrophic for the holder if the user has a substantial advantage when it comes to economic or political power. Thanks to its transparency, the deontological code and the fact of following it protects the provider of a service as far as possible from the vindictiveness of patients or users of shady or powerful character.

13.2.1 First known deontological code

One of the earliest deontological codes that illustrates this danger is probably that… of an astrologer [13.6].

Firmicus, a Roman senator and lawyer born in Sicily in the 4th century, was an astrologer. His reputation and expertise were recognized: he published an astrological analysis of the birth horoscope of the Consul and Prefect of the City of Rome. *Firmicus* wrote eight detailed books on Greco-Roman astrology. At the end of the second book (probably written in 355), he sums up in fifteen items what the life and practice of an astrologer should be.

Firmicus mentions honesty, absence of greed, moderation of desires, transparency and accessibility, probity, moral conduct, a blameless and exemplary life, and the need to maintain a good reputation. He recommends caution and to *stay out of debates, intrigues and political conflicts, to avoid making predictions about the Emperor* (!)[1], *and to cultivate friendship* (the ancestor of networking). He considers it essential to perfect oneself professionally by regularly reading astrology books (continuing education already existed then!).

13.2.3 Deontology: neither morals nor ethics

Deontology is formal, with a content similar to legal thought, and thus distinguishes itself from *professional ethics*. The latter is characterized by a less sharp profile, and involves – beyond maintaining a high level of integrity of practice – research and the defense of individuals confronted by it. Professional ethics develop a broader critical function. Ethical reflection, however, can lead to changes or supplements to a deontological code.

Deontology is not the same as morals, because it can contradict moral requirements, including the observation of confidentiality or privilege (e.g., lawyers have the obligation not to reveal the crimes of their client, even if they have confessed to them). Many deontological charters and codes prescribe behavior in crises or difficult situations and not the moral dilemmas provoked by such situations. However, morality and deontology have the following in common:

1. deontological codes are based on universal values – very often virtues – such as humanity, selflessness, probity, honor, loyalty, and
2. morality and deontology stress the behavior of the individual.

The difference between morality and deontology is the type of constraint:

- the constraint exercised by morals, where noncompliance to these results in feelings of guilt or the more or less strong disapproval of society;
- the constraint of deontology is that authorities are able to impose sanctions ranging from reprimands to exclusion from the profession.

13.2.4 Compliance between a deontological code and virtuous behavior

It follows that a professional following a deontological code does not necessarily seek to behave virtuously. Indeed, his/her search for compliance is partial (based only on certain virtues), and is limited in time (that of the practice of his/her profession). Adhering to a deontological code is therefore not synonymous with citizenship skills.

However, since a virtuous attitude is the result of repeated behavior, adherence to a deontological code is certainly one way that can bring this about…

An eccentric founder

It is interesting to relate deontology to the personality of its founder, *J. Bentham* (1748-1832). As the son and grandson of a lawyer, English and French citizen, world citizen and man of the Enlightenment, he created the word *international*. Founder of utilitarianism, focusing on the

---

[1] Action punishable by death in Imperial China…

rights of all nations, *Bentham* lived, despite his fame, as a hermit. His writings were published through the commitment of the Genevan *Etienne Dumont*.

An eccentric, *Bentham* published at the end of his life a brochure about the corpses of celebrities. Instead of honoring them with a statue, he proposed preserving their bodies as "auto-icons". To this end, *Bentham* set the example: that his own body, embalmed and clad in his own clothes, should appear in a showcase in the library of *University College* in London. Admirers of his thoughts can still hold discussions in his presence.

13.3 A pioneer, an example: the medical profession

The first ethical and moral obligation of a profession is the *Hippocratic Oath*, attributed rightly or wrongly to an ancient Greek citizen of the 5th century BC. Today, we cannot call it a deontological code, since it is an oath made to the gods, and transgressions were punished by them. This moral commitment, however, persists in many countries, even if, for example, French doctors have made a version more suitable for medicine in our society.

13.3.1 A deontological code

Instead of the *Hippocratic Oath*, there is now a detailed and binding deontological code, established for example in Switzerland by the Swiss Medical Association, the FMH. This (*Foederatio Medicorum Helveticorum*) is the umbrella organization of the Swiss medical profession. Over 90% of the 30,000 active practitioners in the country are members. In its capacity as a professional organization, the FMH has the aims of ensuring high quality medical care in Switzerland, assisting physicians in performing their functions, and strengthening the ties between them, by providing them – in particular politically – with the best possible conditions to perform their professional duties.

Its latest version[2], revised in 2003, comprises 49 articles, segmented into the following chapters:

I. Goals.
II. Principles.
III. The doctor and the patient.
IV. The doctor and the community.
V. The doctor and his colleagues.
VI. Exercising the medical profession; attitude toward insurers and other provisions.
VII. Implementing legislation and the execution of the deontological code.

Section 47 lists the sanctions that may be imposed:

a) a reprimand;
b) a fine of up to CHF 50,000;
c) suspension from membership for a specified period;
d) exclusion from the cantonal medical society and the FMH;
e) removal of the FMH title;
f) publication in the organ of the society or ASMAC or in that of the FMH;
g) communication with the Director of Public Health or the health insurance bodies concerned;
h) supervision.

[2] Available (Dec. 2008) from: http://www.fmh.ch/fr/data/pdf/stao_2003_frz.pdf

These sanctions may be cumulative, but they cannot be applied to a physician who is not a member of the FMH. Sanctions are imposed by a cantonal commission. This is in close liaison with the Swiss Council of Ethics, which functions as an appeals body.

An interface with jurisprudence is provided:

- For Article 44: *If a particular question fails to be answered in the deontological code, or in the regulations of the cantonal medical society or ASMAC, or in that of the Swiss Council of Ethics, the provisions of the Federal Law on Administrative Procedure shall apply.*
- For Article 49: *If, in the same case, a formal procedure is initiated by an administrative authority or court, internal proceedings may be suspended or canceled.*

13.3.2 Medical-ethical guidelines

The deontological code also includes 14 guidelines[3] about professional ethics (still called medical-ethical) developed by the SAMS (Swiss Academy of Medical Sciences), including:

- medical guidelines for organ transplantation;
- medical recommendations for sterilization;
- recommendations for the sterilization of mentally deficient persons;
- medical-ethical guidelines for medically assisted procreation;
- medical-ethical guidelines for medical practice among inmates;
- medical-ethical guidelines for patients in intensive care.

The guidelines can be brief, for example a single page.

13.3.3 An ethical think tank within SAMS: the CEC

The Swiss Academy of Medical Sciences (SAMS) was founded in 1943 by the Faculties of Medicine and Veterinary Medicine, as well as by the FMH. The SAMS has set the following lines of action:

- clarification of ethical issues relating to the development of medicine and its impact on society;
- reflections on the future of medicine;
- a commitment to the policy of the Hautes Ecoles, science and training combined with an expert and advisor activity intended for politicians and authorities;
- the promotion of young scientists, particularly in clinical research;
- support of high-quality research in biomedical and clinical research;
- communication between scientific medicine and practice.

Relationships with other countries are widely promoted by memberships of various international organizations, and the participation of SAMS members in international congresses devoted to biomedical ethics and to the future of medicine.

SAMS has a Central Ethical Committee (CEC) that functions as a think tank. It anticipates and discusses ethical issues in medicine. It is an organization monitoring national and international developments, and encourages information exchange and collaboration with related institutions. In this way, the CEC enacts the ethical guidelines destined to support medical practice or biomedical research. In addition, it expresses views on medical-ethical issues in the news or submitted to the SAMS by institutions or by public or private persons.

---

[3] Last accessed (Dec. 2008): http://www.fmh.ch/ww/fr/pub/fmh/code_de_deontologie/annexe1.htm

To conclude, the Swiss medical profession has strict deontology, supplemented by medical-ethical guidelines. These are developed by a Central Ethics Committee within a scientific body, SAMS, founded by Swiss medical academia and the FMH.

This exemplary structure may serve as a model for any profession, because it integrates deontology with professional ethics, which are intimately connected to the practice of a profession, its renewal, its innovations and its integration into the political-economic network.

13.4 Deontology of State functions [13.7]

The state is a pioneer for the deontology of functions within an institution, and has had for several centuries a most complex and complete organization.

Deontology plays a key role in the professions and activities of state officials. Indeed, they must:

- inspire confidence as the representatives of a legally constituted state;
- demonstrate exemplary behavior in accordance with democratic and ethical values of a legally constituted State;
- demonstrate an efficient and parsimonious use of state monies, allocated by taxes from its citizens.

So what are the relationships that the state has woven with deontology?

13.4.1 Deontology of public functions

13.4.1.1 Public administration and its history

Public administration was the first to use deontology in relation with jurisprudence. It later inspired other organizations, institutions and companies to use this deontology and related codes; these bodies were faced with the unitary character that adhesion to the grand principles represented, but also with the pluralistic character (depending on professional sectors) of the deontology that employees must respect to maintain the confidence of principals, customers, and consumers.

This pioneering behavior is due to the size of the governing body of a kingdom and to the trust that its officials should inspire. The situation in France, whose policy of centralization began very early, is exemplary.

The first manifestation of deontology in the civil service in Europe can be found in the ordinance of a government and police reform of the Kingdom of 1254 published by Louis IX (Saint Louis). His son, Philip IV the Fair, developed it in the ordinance of the reformation of the Kingdom of March 18, 1303.

The ordinance already included the three cardinal principles of the functioning of a public administration (probity, impartiality, effectiveness):

- for ***probity***, we read in particular in Article 40: *They will receive neither gold, nor silver, nor gifts of any kind, but only enough to eat and drink*;
- ***impartiality*** is mentioned in Article 38: *They swear they will do justice to large and small and all persons regardless of their conditions;*
- ***effectiveness*** (and timeliness) can be found in Article 13: *Investigations laid before the Court will be handled and judged within two years.*

The French Revolution brought three new elements expressed in the Declaration of Human Rights of 26 August 1789:

- ***recruitment by merit*** (Article 6): *All citizens are equally eligible in all dignity to public positions and occupations according to their capacity and without distinction other than their virtues and talents;*
- ***public interest*** (Article 12): *Public power is instituted for the benefit of all and not for the benefit of those to whom it is entrusted;*
- ***accountability*** (Article 15): *Society has the right to hold to account any public official of its administration – which leads to a sense of responsibility and the requirement of effectiveness.*

At the end of the 19th century the modern profile of the state official or employee appeared. Paul Ferrand [13.8] lists the six collections of virtues that any state agent should have:

1. diligence, accuracy, regularity;
2. probity and selflessness;
3. deference to superior orders;
4. discretion on the subject of the service;
5. obligation of amenity and convenience to the public;
6. decorum, honorable and regular conduct in his/her private life.

Public service agents have their personal morality: they are only requested to be accountable for the practice of their profession, subject to the adage: "*The behavior of an official makes his reputation.*" Deontology can thus be defined as:

Deontology *is the statement as well as the practice of professional duties in concrete situations of a profession, to ensure the efficient performance of duties.*

13.4.1.2 Deontology of public functions: unity and plurality

Using the deontology of public functions, the state may define appropriate behavior and require its officers to be credible. Deontological principles are common to all functions within the state, but professional issues may differ depending on the type of business and mission:

- the deontology of a judge who oversees collegiality and secrecy is not the same as that of
- a teacher, challenged by the diversity of his/her audience, or that of
- hospital staff, faced with the pain, frustration and loneliness of patients.

The deontology of public service is both one and many. But unlike the strict framework of deontology, the deontology of public functions does not make an automaton out of the agent. It does not put in parentheses reflection, questions, discussions, consultations, and research required by the agent to successfully carry out his commitments. This duality will also be found in the deontological charters of professions and companies.

13.4.2 Deontology of public functions: cardinal principles

Probity, impartiality and effectiveness/efficiency are the cardinal principles of public service. "*Probity excludes all injustice, corruption, evil and even bad manners to do good*" (for these last eight words, see the properties of deontological ethics in section 13.1):

- *Probity* (*integrity, honesty, rectitude*) avoids confusion and conflicts of interest, to deal with, to defend or promote ones relatives (nepotism), to fall within the scope of insider trading, to accumulate public and private employments (full time). It prevents the acceptance of gifts that may cast doubt on the integrity of the agent or represent a basis for blackmail, to benefit one party during the submission of a project.

- *Impartiality* (related to *neutrality, fairness, objectivity*) avoids injustice, or not abiding by the principle of proportionality (e.g., length of a given prison sentence unrelated to the severity of the offense), bias, the abuse of authority of an agent with regard to his functions, the submission of interventions and favors. Impartiality has two aspects:
  - a state of consciousness that rises when interests, commitments, feelings and personal relationships are put aside;
  - a moral principle stating that equal consideration should be given to everyone's interests, preferences and dignity.
- *Effectiveness,* in relation to the socio-economic effectiveness, the efficiency of management and the quality of services, makes possible achievements, competence, continuity of services (especially by keeping records), and avoids the illegal switching from public to private employment, lack of mobility and flexibility, and promotes the accountability and determination of the agent. It also implements the principle of general interest.

Impact on the private economy

These three cardinal principles, recognized in public administration from the Middle Ages, are also the ethical core of employees in the private sector, whether in the IT sector, services sector or industry.

13.4.3 Deontology of public administrations and New Public Management

Efficiency, determination and accountability of the agent are the three values that New Public Management[4] has worked to strengthen in public administration.

Could it be that the values of public administration should now coincide with those of the private economy? The answer is nuanced, because a trend is noticeable in two areas:

- certainly, neoliberalism puts more emphasis than before on the principle of subsidiarity of the State and on the introduction of governance practices, leadership and managerial skills developed by the private sector in public service;
- from another point of view, large and multinational companies face social accountability and media exposure that they must consider if they do not want to alienate the consumer and public communities[5]; the deontology that they now require from their employees is not unrelated to that of public administrations.

13.4.4 Values of a public administration

In addition to the cardinal principles, a government develops its own values. Here we consider these values as the ideals that drive public service officers, and not as their motives. Four properties characterize them:

---

4 ***New public management*** is a concept of public management born in the 1970s in neoliberal circles. It calls for the modernization of management of public administrations in order to improve the cost/service ratio. It represents pragmatic management. New public management is based on a sharing of roles between those with political power, which includes taking strategic decisions and setting goals, and the administration that takes operational decisions (this is not new, but has had to be repeated after significant growth of public administrations during the "Thirty Glorious Years"). (Definition according to WIKIPEDIA, accessed in Dec. 2009.)

5 Remember the rigid attitude with regard to the environment of the company SHELL, facing the NGO *Greenpeace* in 1995, which resulted in a huge loss of image and alienated public opinion to such an extent that some of its gas stations were torched.

a. *They are understood*; these are the ideals and projects targeted to best fulfill the unity of the social bond, resistant to all that is ephemeral, to destruction and to chaos.
b. *They have a constitutional basis* (fundamental rights and principles of the Constitution of Switzerland, for example). A surprising example comes straight from the French Republic of Terror, in the Convention of 1793: *The Republic honors loyalty, courage, old age, filial piety, misfortune...*
c. *They are positive affirmations*, not threats or potential sanctions.
d. *They are used in Europe and in the world in terms of public service*.

To remove any doubt concerning the existence of these values, a Canadian study [13.9] has grouped them into four families:

1. *Democratic values* that define the essential function of the agent acting within the framework of a democracy. Thus, in France, the values of the military state in 2005 were those of "A state of mind of sacrifice that could go as far as the supreme sacrifice, and discipline, availability, neutrality and loyalty".
2. *Professional values,* such as excellence, competence, candor, improvement, merit, initiative and creativity.
3. *Moral values*, such as prudence, justice, fairness, altruism and discretion.
4. *Values more specifically related to people,* or those we can show or give to others, such as compassion, forgiveness, good manners, courtesy, openness, reciprocity and discernment.

The three cardinal principles of honesty, impartiality and effectiveness are found naturally in the above values. One advantage of studying these values is that they are easy to detect in conflicts that an agent will face sooner or later:

- compassion as well as indulgence *versus* impartiality;
- transparency against discretion;
- risk as opposed to security, etc.

13.4.5 How do we deal with conflicts of value?

When faced with conflicts of value, *Prudence*, or Applied Wisdom (the cardinal virtue of the Greeks and the mother of all the others, representing a wisdom that was not without courage), encourages a dialogue with peers and superiors. It also leads to analysis of the hierarchy of values at stake, where the dignity of the person remains in the foreground. Certain documents, such as the *European Convention for the Protection of Human Rights and Fundamental Freedoms*[6], can facilitate the untangling of the skein, and lead to possible cases of jurisprudence.

13.4.6 Value conflicts and whistleblowing

A difficult case occurs when an agent must forward confidential information to higher authorities under circumstances where the deontology of public administration would be violated, and especially for situations involving close colleagues. This practice, which is often compared to informing on and denouncing people, is sensitive:

- informing and denunciation in an atmosphere of arbitrariness provide the foundations of an authoritarian regime – or even one based on totalitarian violence and injustice;

---

6 See: http://conventions.coe.int/treaty/Commun/QueVoulezVous.asp?NT=005&CL=FRE

"crows" (those who write anonymous letters of accusation) have had a bad reputation in Europe since World War II;

- conversely, it may be an honor to expose a scandal, a collusion case, a threat to life, or the dignity, integrity or liberty of a third party or even a colleague (sidelining, for instance);
- an example that comes from the top: *the United Nations Convention against Corruption of 2003*[7] states in section 8.4 that *each State... plans to put in place measures and systems to facilitate the reporting by public officials to the competent authorities of corruption that has come to their knowledge in the exercise of their function.* Such a provision is also found in the *Code of Conduct for Public Officials of the Council of Europe (art. 12, 2002)*[8].

The stakes are high. It is primarily up to the authority to establish a transparent procedure and a neutral body able to separate the wheat from the chaff, and restrict or punish any partisan denunciation[9] and not bow to pressure from the employer who requires disclosure of the identity of whistleblowers.

This policy is not completely innocent; it also aims to prevent the dissemination/leakage of this information to the public and the media (a factor that may be critical for institutions and public administrations a few months before an election, for example). It thus prevents the birth of "moles" who can inform the media about difficult situations in institutions, and even identify potential whistleblowers. Let us bet that their career progression would suffer under certain circumstances.

Whistleblowing was introduced in companies in Anglo-Saxon countries. It has spread to companies in Switzerland and in the Swiss federal government. The domain of ETH, for example, which includes two technical universities in Switzerland, has such a body since 2006. It appears on the website of the ETH Board (www.cepf.ch: Reporting body).

13.5 Deontological charters and codes: hazy communications

13.5.1 Codes or charters?

Communications are hazy, especially when it comes to the media, corporate communications or advertising, with regard to the general public: deontological charters, codes and ethical principles are often confused. It is true that the word ethics has a more "upscale" connotation than deontology (or morality...).

Some associations also publish deontological charters without a monitoring mechanism or a catalog of sanctions being specified. The desire to employ good practices to display a standard of quality services, and to introduce a unifying factor within an occupational group is dominant in these cases. Deontology may also be implemented when the profession is on

---

[7] See: http://www.unodc.org/unodc/en/treaties/CAC/index.html (accessed Dec. 2009).

[8] Brochure that can be ordered from the European Council from: http://book.coe.int/FR/ficheouvrage.php?PAGEID=36&=FR&produit_aliasid=785 (accès déc. 2009).

[9] Last we heard, the debate is still ongoing; see for instance the article from Libération of October 2009, *A lawsuit against the crows at work*, available from: http://www.liberation.fr/economie/0101599980-un-proces-contre-les-corbeaux-au-travail

the verge of mutation. We here refer to the deontological code for Swiss librarians established in 1998[10].

Other professions, notoriously unpopular or macabre, also have deontological charters:

- the police: for example, the deontological code of Québec police officers, in close liaison with the Police Act[11];
- funeral parlors: the deontological code for funeral parlors in Belgium[12] *highlights the respect that should be given to the dead regardless of the social origin or religion of the deceased*. It has neither sanctions nor monitoring.

13.5.2 The proliferation of deontological codes: retreat of the State?

Even though a deontological code is supposed to protect the user, closer examination reveals that as a guideline for exercising a profession, it gives rise to self-regulation that, up to a point, can replace legal intervention from the state. For this reason, deontology is sometimes presented as a challenge to the monopoly of state law.

However, this is implausible. Deontological rules are scrupulously observed only when considered in a comprehensive legal system, either by jurisprudence or by legislative recognition. With deontology, the situation is more that of a delegation of State powers, and this delegation can be terminated.

The State may invalidate sanctions imposed by the deontological body of a profession. The proliferation of deontological charters and codes shows the growing participation of occupational groups in the creation of state law. Wise parliamentary lobbying can also help to moderate the action of an authoritarian government with regard to a professional group.

13.6 Deontology of engineering and technology

13.6.1 Ethics and deontology of scientists

The man of science must have morals and ethics [13.10]. These involve the following actions:

- advising and supervising young researchers and PhD students,
- processing and formatting data,
- errors, mistakes and negligence in a research activity,
- violations of standards of conduct,
- the participation of human beings and the use of animals in experiments,
- publication by several authors of research results,
- intellectual property,
- conflicts of interest, and
- the social responsibility of the researcher.

Apart from the last point (we particularly have in mind the use of the atomic bomb), the above-mentioned activities mainly concern the world of science and researchers, although proven scientific fraud can have a harmful media impact on science in general.

---

10 Website accessed in Dec. 2008: http://www.ifla.org/faife/ethics/bbscode_f.htm

11 Website accessed in Dec. 2008: http://www2.publicationsduquebec.gouv.qc.ca/home.php (fee-paying).

12 Website accessed in Dec. 2008: http://www.funebra.be/index.php?option=com_content&task=view&id=49&Itemid=85

13.6.2 Ethics and deontology of the engineer

For the engineering profession, the actions are wider and have a direct impact on society, the economy and even politics and public health. The subject is thus more sensitive and more serious.

The deontological guidelines for engineers follow four axes:

- the virtues inherited from public administration;
- the values from the political system under which these guidelines are developed (in short, for Europe and the US, those of the Age of Enlightenment, while remembering that, for example, the Chinese Neo-Confucianism has a somewhat similar content);
- the virtues associated with exercising the profession;
- the values within the state of mind and obligations of the profession, which in engineering may vary.

National engineer associations, as well as FEANI[13] with its EUR ING members, have very similar codes of conduct and recommend the implementation of the above-mentioned ethical principles.

13.6.3 A historical antecedent – the *Regius Manuscript*

Codes of conduct emerged quite early, especially in the field of construction. As such, one can cite a document of the late Middle Ages, the *Regius Manuscript*[14], which is a charter for workers on construction sites.

As the workers were Christian, it was hardly necessary to state in writing the morality required, because the clergy of the time undertook to recall the basic moral precepts of Christianity in religious ceremonies or by less pleasant means. Remember that adultery was punished in some communities by torture leading to a slow death: hanging upside down (visit the fortifications of Murten Castle in Switzerland and the locations of its gallows for additional information). The nobility seemed to escape this treatment, reserved primarily for peasants.

The *Regius Manuscript*, which deals with the "Royal Art", regulates relationships between an apprentice and master, fixing the duration of apprenticeship to a minimum of 7 years, and delineating the rights and professional duties of the masters: "*Only accept work that one is able to complete, show honesty and excellence in the profession, do not work at night or take the assigned work to another master, begin and end contracts clearly, and when exercising the profession be courteous, polite, honest and discreet with subordinates, peers, customers and authorities.*"

13.6.4 Ethics of availability and effectiveness for engineers

The engineering profession took shape in the second half of the 18th century.

The current activity of an engineer, whether in the service of an employer or a client, has the ethics of loyalty and availability: indeed, an engineer manipulates nature by his/her functions, leading to a success of the technical and social machine that he/she must optimally

---

[13] European Federation of National Engineering Associations: see http://www.feani.org

[14] This manuscript, dating back to 1390, held at the British Museum, can be read in ancient or modern English: http://reunir.free.fr/fm/oldcharges/regius.htm, as a summarized version or translated into French: http://reunir.free.fr/fm/oldcharges/regius-0.htm and http://reunir.free.fr/fm/oldcharges/regius2.htm

lead and assume. An engineer reconciles the two poles of invention and economy of the secondary or tertiary sectors. The power released by technology has introduced ethics of responsibility[15] that must integrate social and political perspectives – perspectives that cannot be entrusted or completely delegated to his/her employer or client.

13.6.5 Evolution of deontological charters for engineers

It was in Britain that the first charter was drafted and accepted in 1910 by the *Institution of Civil Engineers* (ICE). In the United States, the *American Institute of Consulting Engineers* (AICE) pioneered work in 1911, drawing inspiration from the text of the ICE and by involving senior engineers respected by the profession. Other engineering associations published their own charters in the years to follow. All codes insisted mainly on the loyalty of the engineer towards the employer.

It was not until the 1970s that the social responsibility of an engineer with regard to his/her activities was developed, and even later was there responsibility towards the environment. This was mainly the result of working groups including engineers and philosophers, funded by two US agencies, the *National Endowment for the Humanities* and the *National Science Foundation* [13.11].

In 1985, the very powerful ABET (*Accreditation Board for Engineering and Technology*), which accredits all engineering programs in the US, decided that engineering programs must include the ethical foundations of the engineering profession. This point was confirmed in 2000[16]. Such conditions are also found in the 2009 standard for French engineering degrees, which primarily accredits the *Hautes Ecoles* of national engineering[17].

13.7 The FEANI Charter

In Europe, the FEANI[18] Charter of September 29, 2006, has become a standard. Its contents are as follows:

Ethical principles
Decisions and actions of engineers have a major impact on the environment and society. The members of this profession have an obligation to ensure that their work meets public interest, particularly when it comes to health, security and sustainable development.

Framework statement
Individually, engineers have a personal obligation, that of acting with integrity in accordance with public interest and to implement all reasonable care and skill in their profession.

In this implementation, engineers:

- *must maintain their knowledge, expertise and key skills at the required level and only accept tasks for which they are competent;*

---

[15] Solidarity of human beings with their actions, they admit them as theirs; in view of their intention, they assume merit and demerit, which implies consciousness and freedom of the agent.

[16] The standard for accreditation as an engineer can be downloaded from: http://www.abet.org/forms.shtml#For_Engineering_Programs_Only

[17] CTI standard that can be downloaded from: http://www.cti-commission.fr/References-et-Orientations-2009

[18] European Federation of National Engineering Associations, www.feani.org

- *give an exact wording of their professional qualifications or those obtained in the Hautes Ecoles;*
- *provide analysis and impartial judgments to their employers and clients, avoid conflicts of interest and observe their duties when it comes to confidentiality;*
- *perform their duties while preventing danger when it comes to health and safety, and negative impacts when it comes to the environment;*
- *face the responsibilities arising from their work and that of the employees they supervise;*
- *respect the rights of persons with whom they work, and the legal and cultural values of societies in which they carry out their assignments;*
- *be prepared for public debate on issues of science and technology in the field of their expertise, and for which they provide information and take an appropriate stance.*

Codes of conduct

The Pan-European statement above on the conduct and ethics within engineering sciences will be included in the drafting of codes by the National Engineering Associations. These codes can – and many already do – incorporate a list of these objectives in a form that reflects cultures and national circumstances. Additional objectives may be added if they are required by national practice.

Included in this charter are the three key virtues of ethics, integrity/probity, impartiality and effectiveness, inherited from the virtue of tolerance from the Age of Enlightenment, a desire for personal and professional transparency and the development of social responsibility of engineers.

13.8 Charters for software engineers

Finally, let us examine another charter, that of the software engineer[19], developed by the IEEE[20] in 1999.

ABRIDGED VERSION

Preamble

The abridged version of the code sets out the main objectives, whereas the items detailed in the full version of the code provides examples and more information about how these objectives should be reflected in the conduct of a software engineer. Overall, the abbreviated and full versions of the code form a coherent whole. Indeed, without the statement of objectives, details of the code may seem legalistic and boring, and without their detailed development, objectives may seem abstract and meaningless.

Contents

Software engineers must commit themselves to the analysis, specification, design, development, testing and maintenance of a useful and respected profession. In keeping with their commitment to health, safety and public welfare, software engineers shall adhere to the following principles:

---

[19] Version from 2006 available in English from: http://www.ieee.org/web/membership/ethics/code_ethics.html

[20] *Institute of Electrical and Electronic Engineers*, URL: http://www.ieee.org/web/aboutus/history/index.html

1. ***The Public***. *Software engineers shall act in the public interest at all times.*

2. ***The client and the employer***. *Software engineers shall act in a manner that is in the best interests of their clients and their employer, always in the public interest.*

3. ***The product***. *Software engineers must ensure that their products and related modifications meet the highest professional standards.*

4. ***Judgment***. *Software engineers shall maintain integrity and independence in their professional judgment.*

5. ***Management***. *Managers and those responsible for software engineering must adhere to an ethical approach to development management and software maintenance and work to promote it.*

6. ***The profession***. *Software engineers must ensure the integrity and reputation of the profession in accordance with public interest.*

7. ***Colleagues***. *Software engineers should be fair to and supportive of their colleagues.*

8. ***Oneself***. *Software engineers shall participate in lifelong learning and promote an ethical approach when it comes to exercising their profession.*

The three principles of probity/integrity, impartiality and effectiveness are put into perspective, as is also the cardinal virtue of justice. The charter lists the values of health, safety and public welfare. Items 1 and 2 introduce the idea of public good before the interests of the employer or client. This is an open door for whistleblowing.

13.9 Business ethics in other professions

Applied ethics, deontological codes and charters naturally involve other professions, especially [13.12]:

- labor laws and moral rules (e.g., loyalty of superiors);
- deontology of communications and advertising;
- deontology of human resources;
- financial deontology and respect for investors;
- financial deontology and the functions of financial markets, etc.

These provisions are an integral part of the social responsibility of a company and will probably be integrated into future relevant standards (ISO 26000, for example).

References

[13.1] JEREMY BENTHAM, *Deontology and the Science of Morality* (1834), Adamant Media Corporation (2000).

[13.2] For a more detailed analysis of deontology and its history, see the book by STEPHAN DARWALL, *Deontology*, Blackwell Publishing (2003), and especially its introduction, pp.1-9.

[13.3] PAUL JANET, La morale du devoir, dans le langage des utilitaires, *Revue des Deux-Mondes* (1874), p. 105.

[13.4] See the article *Déontologisme* by ANDRÉ BERTEN, in the *Dictionnary of ethics and philosophical morality* (T.1), edited under the direction of MONIQUE CANTO-SPERBER, PUF, Collection Quadrige, 2004.

[13.5] See the article *Déontologie* by DANIÈLE SIROUX, in the *Dictionnary of ethics and philosophical morality* (T.1), *ibid.*

[13.6] JAMES HERSCHEL HOLDEN, *A History of Horoscopic Astrology*, AFA Publications, 2006, pp. 69-72.

[13.7] This subchapter borrows a lot from the magnificent book by CHRISTIAN VIGOUROUX, *Déontologie des fonctions publiques*, Dalloz, 2006.

[13.8] PAUL FERRAND: *Des avantages accordés et des obligations imposées aux fonctionnaires et employés civils*, 1882, pp. 298-299.

[13.9] Workgroup on the values of ethics in the public function, *Rapport de solides assises,* Jan. 2000.

[13.10] See for instance the brochure with a complete bibliograpy, edited by the *Committee on Science, Engineering and Public Policy*, *On being a Scientist – a Guide to Responsible Conduct in Research*, and published by the National Academy of Sciences, National Academy of Engineering and the Institute of Medicine of the National Academies, National Academies Press, 2009.

[13.11] MICHAEL S. PRITSCHARD, *Engineering Ethics* (Chap. 46) from the book by RAYMOND G. FREY and CHRISTOPHER H. WELLMANN, *A Companion to Applied Ethics*, Blackwell Publishing, 2005, pp. 625-632.

[13.12] A good approach can be found in RAYMOND G. FREY and CHRISTOPHER H. WELLMANN, ibid., Chap. 5, 41 and 43.

Chapter 14

Environmental Management and the ISO 14000 family

Key concepts: Environmental aspects, environmental auditing, environmental targets, Coalition for Environmentally Responsible Economics (CERES), life cycle, sustainability development, EMA-Eco-Audit, EMS (Environmental Management System), life cycle assessment, impact assessment, environmental assessment, *crisis management*, environmental management, risk management, Global Compact, Global Reporting Initiative (GRI), ISO 14001, ISO 14004, environmental objectives, environmental performance, environmental policy, precautionary principle, crisis situation, environmental management system (EMS), environmental control.

14.1 Increasing number of environmental problems

In the 1950s, people were worried by nuclear testing in the atmosphere, anxious about the destructive influence on health of radioactive fallout; and pictures of Hiroshima and its survivors had been seen worldwide. Science was no longer considered to be exclusively in the service of progress and the well-being of all – it had a dark side. This was first seen during the first technological war, that of 1914-1918 (gas, etc.), and especially at the end of World War II with the experiments carried out by SS doctors on prisoners in concentration camps (experiments, it is true, carried out by doctors with the same "training" on indigenous peoples thirty years earlier [14.1]).

This represents the decline of popular belief in progress, from the Age of Enlightenment, that was dear to Jules Verne.

14.1.1 The environment becomes a theme

However, growth of the consumer society and its discharge of waste into the environment started to become problematic in the late 1960s. Following the demonstrations in France in May 68, there was a progressive dislocation of Trotskyism and a decline of communist parties. Alternative movements arose to consider a problem other than that of the class struggle: environmental damage.

A new political force emerged, the Green Party, a form of counterculture often conducted by those disappointed by May 68, but also with support from survivors of the hippie movement: their attention was mainly focused on major themes such as nuclear power plants, with their risk of accidents and contamination. These groups became allies with non-governmental organizations (NGOs) (another nascent political force), particularly in the fight against the extermination of animal species (remember the slogan: *Rather naked than wear fur*, which was a great success and angered the bourgeoisie of the time).

At the same time, the Club of Rome was founded on April 8, 1968, on the initiative of *Aurelio Peccei*, an Italian member of the board of FIAT, and *Alexander King*, a Scottish scientist and public servant, former scientific director of the Organization for Economic Cooperation and Development. His first report, still called the *Meadows Report* (1972), has the French title *Halte à la Croissance*? *(Limit growth?)* [14.2]. It was a stunning success, and sold over 12 million copies and was translated into 37 languages. It was followed in 1974 by a second study: *Beyond the age of waste* [14.3].

And what did the reports from the Club of Rome affirm? That if mankind continued to grow exponentially and exploit natural resources at the same rate, there would be a collapse of the economic system. These results, even though they interested the public, only gave rise to shrugs in management circles, where the directors were convinced that it was merely a new version of Malthusianism. It was true that the Club of Rome predicted that the phenomenon would not become large scale until 2010. Time was not running out!

How far ahead that seemed... but the Club of Rome persevered. In 2004, it published the book, *The Limits to Growth*: *The 30-Year Update* [14.4], which confirms its first analyses. One of its conclusions is:

"We... believe that if a profound correction is not made soon, a crash of some sort is certain. And it will occur within the lifetimes of many who are alive today."

14.1.2 The first shock

The birth of this new political sensitivity unleashed hilarity and irritation in established politics and the private economy. All parties, including those of the left, allied themselves against this often hair-raising prediction. Serious events would remind these groups that the rise of ecological awareness was faced with private and public unawareness of serious environmental risks. Five examples are given below (*first shock*):

- *1963*: Japanese authorities recognized that a plant producing acetaldehyde at *Minamata* had been polluting the sea with mercury for more than 15 years, causing serious problems for the population. Warning signs had been seen much earlier in behavioral disorders in the city cats, mockingly called *cat suicide* according to some sources. The cause was already known by the company executives in 1959.
- *1967*: the sinking of the *Torrey Canyon*; 119,000 tons of oil polluted more than 180 km of beaches in France and Britain. Such accidents still occur.
- *1976*: an accident at the ICMESA factory in Seveso (subsidiary of the Swiss company Givaudan) polluted 35 acres with dioxin and injured 200 people. 700 people were evacuated, and 80,000 animals died on the spot or were slaughtered. Tens of thousands of people suffered, immediately or later, with various ailments such as skin lesions, disfiguration, hepatic dysfunction, kidney dysfunction, and thyroid dysfunction. Within a decade, the leukemia rate doubled, and cases of brain tumors tripled. Despite these alarming circumstances, the authorities were not alerted to the seriousness of the situation by the management of ICMESA until five days after the accident.

What really happened? Thirty years later, a book provided some answers [14.5]: the production center of Seveso was the waste bin of the group: inadequately trained personnel, inadequate management, limited facilities, everything was failing. Some installations were used to produce a herbicide but also to manufacture a chemical weapon comprising the herbicide and dioxin, synthesized simultaneously in the same tank. During production, the breakdown of a poorly controlled brewing tank led to overheating of the liquid situated close to the heater, causing an exothermic chain reaction: the liquid exploded and, since no holding tank had been installed for financial reasons, went into the atmosphere. The unbelievable saga of waste from the factory being transported from one European country to another before being properly incinerated further strengthened the scandal of this event, widely covered by the media.

- *1979*: failure of a nuclear reactor at *Three Mile Island* (USA), with 200 000 evacuees.
- *1984*: industrial chemical disaster in *Bhopal*, India, when the *Union Carbide* plant exploded, releasing a hyper-toxic gas (40 tons of methyl isocyanate) and causing nearly 30,000 deaths and 300,000 cases of illness. This is the most serious accident to date. In 2009, 25 years later, although the company has paid about half a billion USD to the Indian government, the people have still not been compensated.

The public had been alerted, and those refusing to take the environment into consideration gradually become discredited. Some actors in the public and private economy now adopt a broader civic consciousness, reckoning that a reputation as champion of environmental protection will give them competitive advantages (political recovery).

14.1.3 Chernobyl, a nuclear remake

The *second shock* was the accident at the Chernobyl nuclear power plant (USSR) in 1986, involving radioactive emissions that contaminated the whole of Europe. The accident left at least 31 dead and 300 injured (figures well below reality), requiring the creation of an exclusion zone of 30 km around the plant and undermining the image of the USSR, which had been somewhat improved by successes in science and space technologies.

The same year, in Basel, 30 tons of mercury were discharged by a factory of the Sandoz group, triggering an outcry from the EU. For such a clean country, the survey revealed many deficiencies in the company's security, sacrificed to optimize profits.

14.1.4 Sustainability, an avatar of the results of the Club of Rome?

The *third shock* was the issue of sustainable development, resource management and global warming, addressed in the early 1990s, and seem to have become acceptable in the early 21st century (the Nobel Peace Prize of 2007 was given to a champion of sustainable development: the American Al Gore).

The problem is more complex than that of environmental protection. The oldest and most prestigious democracy in the world, the United States, has yet to be proactive; it is true that the challenges facing this country in the implementation of programs leading to sustainable development are enormous, given the urbanism and the existing pattern of consumption. However, the US has a reaction force that should not be underestimated.

Development of the Western economy in the 1980s focused on services using fewer natural resources than the secondary and primary sectors. The economic development of Asia has, however, been going in the opposite direction since the 1990s.

14.2 Environmental policy and sustainable development

In the 1970s, the governments of the OECD[1] countries gradually adopted environmental legislation and environmental ministries. The content of toxic compounds in waste, emissions and effluents was regulated, energy policies gave consideration to the environment, and the recycling of obsolete products began. Some universities, such as the Swiss technological university, EPFL, were among the first to create teaching sections and departments of environmental engineering.

Meetings on a global scale were organized to find common ground and conclude the first international agreements. But environmental policy is linked to territory, and is therefore primarily national.

Issues of sustainable development transcends science and technology, borders and continents, and go well beyond the issue of environmental protection. It, therefore, leads to a debate about society and its values: no one knows whether society will be able to control global warming, resource management and transport on a global scale – or whether we will keep the excessively consumerist values of our current society. The debate has only just begun and, given the divergent interests of stakeholders, it is likely to last. Given the many interconnections of the global economy, our society is very fragile and needs to be able to evolve with as few jolts as possible.

14.3 Environmental policy and environmental management

Governments currently use four instruments to promote their environmental policy:

- *direct action*, especially the creation of infrastructure improving environmental protection;
- *regulation*;
- *economic instruments* (tax-based or otherwise);
- *urging volunteers* and advocating the use of standards such as ISO 14001, an instrument of *soft governance*.

14.3.1 What is the role of ISO 14001?

ISO 14001 is primarily based on *soft governance*. It is an environmental management tool that allows a company to implement its own environmental strategy, where the minimum is to respect the constraints given by legislators. An audit of an environmental management system in a company does not replace government controls, but it reduces the frequency of these checks (a form of environmental "self-insurance"). The standard contains no environmental policy, but it requires an organization's management to develop and draft one. Moreover, it

[1] The OECD, the Organisation for Economic Co-operation and Development, brings together the governments of 30 countries committed to the principles of democracy and market economy in order to: (a) support sustainable economic growth, (b) increase employment and raise living standards, (c) maintain financial stability, (d) help other countries develop their economies, and (e) contribute to the growth of world trade. The OECD also shares expertise and exchanges views with more than *100 other countries and economies*, from Brazil, China and Russia to the least developed countries in Africa.

does not set the level of environmental protection to be achieved, or guarantee on its own that the company has the means necessary to comply with legislative requirements.

ISO 14001 is a generic standard but applies to a specific context (determined according to the company's strategy and its significant environmental aspects). Thus, the food industry must primarily manage the treatment of water containing large amounts of organic material (e.g., whey), while chemical industries are mainly concerned with retaining potential pollutants, and the detoxification of effluents. This is why implementation of the standard requires the support of the quality manager of the company in question, and why environmental management is generally entrusted to the environmental (and often safety) specialist.

14.4 Private sector initiatives, and those of the UN and the EU

In widely varying degrees of diligence and commitment, public and private economies have noted the position of a company's business in the environment when:

- Non-renewable natural resources (i.e., minerals, the substratum, including "natural" gas, etc.) are inexorably diminishing;
- natural resources are subject to a renewal rate with a limited flexibility;
- waste decomposition by microorganisms has kinetics well below the pace of their production;
- non-degradable pollutant emissions endanger the habitat (consider Seveso, Bhopal and Chernobyl);
- some emissions seem to be clean but have been shown to endanger the terrestrial ecosystem (carbon dioxide, methane).

The reactions of companies vary from "the polluter pays" to the consciousness of being a player in sustainable development, even if that behavior is not always without ulterior commercial (and economic) motives. Usually a strategy is put in place to reduce the amount of resources used, to utilize natural or recycled resources, to develop low-polluting production processes, and to treat residual waste and effluents by controlled methods and procedures (or store them in appropriate locations). Implementation of this strategy is often done by the ISO 14001 management tool.

14.4.1 CERES

The private sector began to mobilize effectively in the late 1980s, less than a generation after the first emergence of environmental movements.

The **C**oalition for **E**nvironmentally **Res**ponsible Economics (CERES) was born in 1989 in reaction to the grounding of the tanker Exxon Valdez. It comprised three stakeholders: investors wishing to work in more ecological aspects of the economy, companies sharing the same concerns, and advocacy groups for environmental protection. CERES has published a code of environmental ethics, one goal of which is the prevention of pollution. One of the duties of members of CERES is the annual publication of an environmental report.

14.4.2 Business Charter for Sustainable Development

In 1991, the 16 principles of the *Business Charter for Sustainable Development* were published by the World Chamber of Commerce. Some of the 16 principles below are found in the content of ISO 14001 (in the same period, Switzerland established the association ÖBU[2] with the same goals).

1. **Priority** (*of environmental management*) **for the company**
   Make environmental management a top priority of the company and recognize that it is a factor for sustainable development; implement policies, programs and safe practices with regard to the environment.

2. **Integrated management**
   Fully integrate these policies, programs and practices in any company, as essential elements of management in all its aspects.

3. **Improvement process**
   Continue to improve policies, programs and company performance with regard to the environment, taking into account new technical developments, scientific understanding, consumer needs and public expectations, with regulations as the starting point. Apply the same set of international criteria related to the environment.

4. **Training of employees**
   Educate, train and motivate employees to conduct their tasks responsibly with regard to the environment.

5. **Preliminary assessment**
   Assess the impact on the environment before starting a new activity or project and before ending the activity of a facility or leaving a site.

6. **Products and services**
   Design and deliver products and services that will not have an undue impact on the environment and whose intended use is safe and has the best performance with regard to the consumption of energy and natural resources. Moreover, with products, one should be able to recycle, reuse or dispose of them safely.

7. **Tips for consumers**
   Advise and, where relevant, educate customers, distributors and the public about the use, transport, storage and safe disposal of provided products. Similar considerations should be applied to the provision of services.

8. **Facilities and activities**
   Develop, design and operate facilities and conduct activities while considering the efficient use of energy and materials, the sustainable use of renewable resources,

---

[2] *More than 300 Swiss companies of various sizes and orientations have come together to tackle a common task: the continuing evolution of the Swiss economy according to the principles of sustainable development*. The association Öbu (www.oebu.ch) serves as a *reservoir of ideas* for everything concerning environmental and business management. It forges links between companies, governments, politics, environmental organizations, professional associations, the media and the general public. Öbu carries out *projects* for companies and promotes exchange of experience among its members. Affiliated companies consider the ecology and sustainable development to be "the boss's business", and are thus represented by their senior executives.

the minimization of environmental damage and waste generation, and the safe and responsible disposal of residual waste.

9. **Research**
   Conduct or support research on the environmental impact of materials, products, processes, emissions and waste associated with the company's activities, and on means to minimize negative impacts.

10. **Preventive measures**
    Modify the manufacture or use of products or services, or conduct activities based on scientific and technical knowledge, so as to avoid serious or irreversible degradation of the environment.

11. **Subcontractors and suppliers**
    Promote the adoption of these principles by contractors working for the company by encouraging and requesting, where appropriate, an improvement in their practices to match those of the company, and encourage the wider adoption of these principles by suppliers.

12. **Emergency and response plan**
    Where significant hazards exist, develop and maintain preparedness plans for emergencies, in cooperation with emergency services, relevant authorities and the local community, taking into account potential trans-boundary impacts.

13. **Technology transfer**
    Contribute to the transfer of technologies and management methods that respect the environment in all public and private sectors.

14. **Contribute to the common effort**
    Contribute to the development of public policies and initiatives and education programs in the private sector, public sector and intergovernmental bodies designed to raise awareness of the environment and its protection.

15. **Remain open to dialogue**
    Encourage openness and dialogue with employees and the public, anticipating and responding to their concerns about the dangers and potential effects of activities, products, or waste services, including trans-boundary or global ones.

16. **Respect objectives and information**
    Measure outcomes in environmental terms. Regularly conduct environmental audits and assessments of compliance to company objectives, regulations and these principles, and periodically provide appropriate information to the Board of Directors, shareholders, employees, officials and the public.

14.4.3 Global reporting and Global Compact

Another initiative, through a partnership with the United Nations Environment Program, dating back to 1997, is the ***Global Reporting Initiative*** (GRI), which provides guidelines for the production of sustainability reports.

Taking as a basis the human rights and fundamental principles of labor law, on the initiative of the UN Secretary General and with the support of the World Chamber of Commerce, and in the presence of many stakeholders, the ***Global Compact*** was initiated in 2000. It includes nine principles, three of which directly affect the environment:

- *Principle 7*. Companies are asked to apply a precautionary approach to environmental challenges.
- *Principle 8*. Companies are to undertake initiatives to promote greater environmental responsibility.
- *Principle 9*. Companies are to promote the development and diffusion of environmentally friendly technologies.

Global Compact legitimizes the precautionary principle, which is often challenged in the English-speaking business world, where a more open risk culture dominates; this has, for example, led the global economy to the brink with the *subprime* crisis.

The precautionary principle[3] has given rise to many discussions on the scope of its application:

- *On the acceptability of risk*: the choice is not simply between a risky action or cautious inaction, but between two risks: the one related to the action, and that related to inaction; thus penalizing innovation, and creating distortions of competition.
- *On risk assessment*: the subject of the precautionary principle is marked by scientific uncertainty. It is therefore necessary to define criteria for assessing the risk of damage.
- *On the persons or institutions involved*: should the precautionary principle apply to all categories of actors, public authorities, economic operators or private individuals?
- *On the scope of this principle*: should this principle, defined for the environment, be extended, for example to include health?
- *On the duration of the proposed measures*: measures may be temporary or permanent.

On one hand, the precautionary principle could be accompanied by a duty of research to improve knowledge and help review the measures taken. On the other hand, it could justify the permanent ban of a project.

14.4.4 The EU and EMAS – Eco-audit (Smea II)

The Eco-audit regulations *(EMAS: Eco Management and Audit Scheme)* of the European Council came into force in 1995 (and were updated in 2003). They are legislative but not mandatory; companies pledging to respect them must follow the guidelines advocating continual improvement actions for the protection of the environment. EMAS has four strategic components:

- implementation of an environmental management system;
- periodic and complete audits of the system;
- informing the public and stakeholders regarding actions taken to protect the environment and their results;
- staff involvement.

The environmental management system used is generally that of ISO 14001. To become registered, a company must meet a number of criteria, the majority of which can be found in the ISO 14000 family, but it must also provide a list of historical incidents. After verification of

[3] Principle 15 of the Rio Declaration on the Environment and Development, adopted in 1992 during the Earth Summit, states: "*In the case of threats of serious or irreversible damage, lack of full scientific certainty should not be an excuse for postponing the adoption of effective measures to prevent environmental degradation*".

the objectives of the Eco-audit regulations by an accredited body, the company is registered (in a public list). The list of registered companies is published once a year in the *Official Journal of the European Union*, which makes them visible to stakeholders and customers. Although the content is very similar, there are however fundamental differences between the two frames of reference (ISO 14001 and Eco-audit), e.g., territorial, the applicability to each site, the disclosure obligation and the verification procedures.

14.5 Benefits of an environmental management system

Improving environmental performance by introducing a management system has many advantages, particularly in terms of cost reduction:

- cost reduction in the use of energy, water, packaging and raw materials;
- cost reduction in waste and effluent treatment, as well as recycling, usually by a decrease in volume or a change in technology;
- improved recycling (e.g., the reuse of slightly contaminated water in another unit operation) by reducing the volume generated at the source, recycling or recovery (waste in one operation sometimes constitutes the raw materials in another);
- reduction of secondary packaging weight for company products;
- performance optimization or even removal of treatment systems and recycling (wastewater treatment plant, filters);
- reduction in the amount of local or national environmental tax by reducing the volume of effluent, fumes, waste;
- improved risk management has an impact on insurance premiums or other bank ratings.

The management system provides a new tool for management, making it possible to optimize its investment in a facility by taking into account desired environmental performance.

Announcement of this promotes the organization in the eyes of the media, consumers and customers. It is also a pledge to the local community (social image). This last point is important for renewing a lease, purchasing new land, obtaining a building permit or applying for public works.

14.6 Presentation of ISO 14001:2004

ISO 14001:2004, *Environmental management systems – Requirements with guidance for use* [14.6] is brief and adopts the Deming cycle (Figure 14.1). It is similar to the ISO 9000 family. It is supplemented by Annex A that explains its content. The standard focuses on assessing the environmental performance[4] of an organization.

The standard has the following steps, which are sequential in its implementation (the figures in brackets correspond to clauses in the standard; the first three chapters are, as for all ISO standards, 1. Scope, 2. Normative references, 3. Terms and definitions):

4 Environmental performance: measurable results of the management of environmental aspects of an organization (definition from ISO 14000:2004).

Figure 14.1 ISO 14001 and the Deming cycle [14.7].

Plan: *vision, strategy, goals and objectives, metrics and benchmarks, planning*

a) general requirements (4.1) and environmental policy[5] (4.2); it is at this stage that the management of the organization is most involved;
b) environmental aspects[6] (4.3.1);
c) legal and other requirements (4.3.2);
d) environmental objectives[7] and targets[8] (4.3.3);

Do: *implementation, program, process, staff development, traceability and memory, preparedness and emergency management*

e) resources, roles, responsibility and authority (4.4.1);
f) skills, training and awareness (4.4.2);
g) communication (4.4.3);
h) documentation (4.4.4);
i) document control (4.4.5);
j) operational control (4.4.6);
k) prevention and response to emergencies (4.4.7);

---

5 Environmental policy: a formal expression by management at its highest level of its overall intentions and direction of the organization regarding its environmental performance (definition from ISO 14000: 2004).

6 Environmental aspect: element of an organization's activities or products or services that can interact with the environment (definition from ISO 14001:2004).

7 Environmental objective: overall environmental goal, consistent with the environmental policy, that an organization sets (definition from ISO 14001:2004).

8 Environmental target: detailed performance requirement, applicable to the organization or parts thereof, that arises from environmental objectives and that needs to be set and met in order to achieve those objectives (definition from ISO 14001:2004).

Control/check: *monitoring, gathering and processing data, internal audits*

l) monitoring and measurement (4.5.1);
m) assessment of conformity of results to legal standards or others (4.5.2);
n) nonconformities, corrective and preventive action (4.5.3);
o) control of records (4.5.4);
p) verification of the environmental management system, internal audit (4.5.5).

Act: *writing a summary for management for the implementation of remedial actions or as part of continual improvement (not to be confused with linear or incremental improvement)*

q) management review (4.6).

The management review may include new goals for environmental performance. Remember that a policy of small steps can have very limited visible effects according to the principle: *no pain, no gain.*

The implementation of ISO 14001 can be coupled with registration under the Eco-audit regulation by an accredited organization according to Figure 14.2.

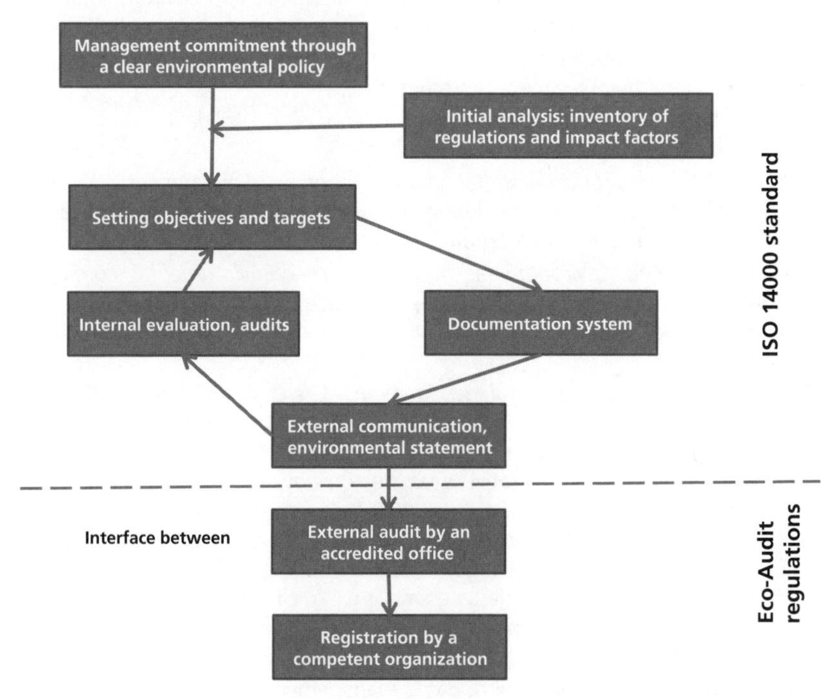

Figure 14.2 Interface between the implementation of ISO 14001 and the Eco-audit regulations.

14.7 Other documents of the ISO 14000 family

Since ISO 14001 is brief, other documents (14 in total) have been prepared: they specify the implementation of the system, environmental performance, the life cycle assessment (LCA) of a product, labeling or environmental labels, internal and external communications, and external quality assurance:

- ISO 14004:2004, *Environmental management systems – General guidelines on principles, systems and support techniques*, states and facilitates the establishment, implementation, maintenance and improvement of an environmental management system (EMS) by addressing in more detail technical sections of ISO 14001 and by giving numerous examples. This standard also shows the links between ISO 14001 and other management systems.

  ISO 14004 specifies that beginning implementation of an EMS must include an initial review of the situation by management, with commitment and leadership from top management, as well as a definition of its scope (the area of the organization for which environmental damage is possible or proven). The management objectives, particularly in terms of waste discharge, must be expressed as measurable data, such as the amount of waste, or the amount (moles) or annual mass of harmful substances discharged. A goal would be, for example, 30% less annual mass of dust released into the atmosphere.
- ISO 14015:2001, *Environmental management – Environmental assessment of sites and organizations (EASO),* also gives guidelines on how to conduct an EASO by applying a systematic process of identifying environmental aspects and environmental concerns and, where applicable, by determining their economic and trade implications. It lists:
  - the functions and responsibilities of stakeholders in the evaluation (sponsors, experts and the assessment representative);
  - the steps of the assessment process (planning, consolidation and validation of information, interpretation and reporting).
- ISO 14031:1999, *Environmental management – Environmental performance evaluation – Guidelines,* for the design and use of an evaluation of the environmental performance within an organization.
- ISO 14063:2006, *Environmental management – Environmental communication – Guidelines and examples*, provides guidance to an organization on general principles, policy, strategy and activities related to internal and external environmental communication.
- ISO 14044:2006, *Environmental management – Life cycle assessment – Requirements and guidelines*, ISO/TS 14048:2002, *Environmental management - Life cycle assessment – Data documentation format*, ISO/TR 14047:2014, *Environmental management – Life cycle assessment – Illustrative examples on how to apply ISO 14044 to impact assessment situations*, and ISO/TR 14049:2014, *Environmental management – Life cycle assessment – Illustrative examples on how to apply ISO 14044 to goal and scope definition and inventory analysis*, specify the methods, formatting, and modes of investigation of the life cycle assessment (LCA) of a product. The LCA involves a procedure, i.e., a series of unit operations and standardized mathematical tools, to characterize the impact of a product on the environment.
- ISO 19011:2011, *Guidelines for auditing management systems,* and ISO 14015:2001 (see above), specify the external quality assurance of ISO 14000 (audit, evaluation mode).
- ISO 14050:2009, *Environmental management – Vocabulary,* contains definitions of basic terms directly related to environmental management.

The ISO 14000 family also includes guidelines for labeling and environmental labels [ISO 14020:2000, *Environmental labels and declarations – General principles*, ISO 14021:1999, *Environmental labels and declarations – Self-declared environmental claims (Type II environmental labeling)*, ISO 14024:1999, *Environmental labels and declarations – Type III environmental declarations – Principles and procedures*, and ISO 14025:2006, *Environmental labels and declarations – Type III environmental declarations – Principles and procedures*], and greenhouse gas emissions (Parts 1 to 3 of ISO 14064:2006), and thus concerns sustainable development and the control of global warming.

The modular advantage of the ISO concept is apparent.

14.8 The heart of ISO 14001: Clause 4

Clause 4 (environmental policy) specifies the requirements of the standard:

- commitment of management,
- development of an environmental policy,
- establishment of an environmental management plan.

14.8.1 Preparing the plan

Several steps are included in the path below:

- Global vision of the management for the environmental policy:
  - what is the pre-existing culture?
  - what role should one take on? Pioneer, champion, leader, or mere compliance with the legislation in place?
- Identification of significant elements of the policy (what?). This dimension concerns:
  - the type of business (pharmaceutical, chemical, steel, food processing, assembly, etc.), so that possible areas of significant contamination can be identified,
  - its size,
  - its direct environment (urban, rural, estuary, by a river, etc.)
  - the degree of excellence in environmental performance that the company or institution wants to achieve, and
  - the fact that there already exists a system or an embryonic management system (environmental or similar to ISO 9001).

  At this stage, it is customary for the management of the company to perform an environmental analysis, including an assessment of risk and loss of control, while also taking into account the impacts under normal circumstances. Factors are identified, followed by measurable critical environmental parameters under normal circumstances (e.g., rate of organic matter in wastewater released from the organization), and significant environmental impacts in the case of non-control: the FMEA approach or another similar method can be used in both cases.
- Identification of regulatory frameworks and other guidelines that management wishes to follow (what settings?). This requirement also means that management:
  - must be up-to-date with regard to legal issues (a legal analysis resulting from detailed work, together with a summary); legislation may differ from one country to another, and some companies know this when they implement polluting production in countries with more tolerant regulations,
  - must develop monitoring methods that give information about new trends and emerging regulations or guidelines, but also
  - must look for good practices when it comes to environmental protection related to the branch, and
  - must gauge published ethical codes and charters of companies or business groups, etc.

- Identification of quantifiable goals (setpoints in terms of annual volumes of discharge and concentrations of pollutants), but also of the level achieved in the context of continual improvement; this impetus is not negligible.
- Identification of an environmental improvement program including the necessary resources (financial, potential infrastructure, human resources, training, communication) and the agenda.

14.8.2 Implementation

The implementation (How?) is outlined below (Figure 14.3):

- planning of programs for raising consciousness and for employee training;
- defining the roles of key players and the structure;
- establishing internal and external communication, especially about the environmental policy that management wants to implement. The aim of internal communication (raising awareness, training) is to inform the staff and anyone working in the organization of:
  - the impact of their activities on the environment;
  - the consequences that deviations from target values have on the environment;
  - their role and responsibility in the implemented system, but also if they are able to act effectively when it comes to prevention or, in the case of exceeded target values, of incidents or of accidents. The procedures for emergency measures must be thoroughly known;
  - the results of the environmental performance of the company;
- developing a documentation plan (including measures for emergency situations), its control and its updating, with emphasis on the traceability of measurements and evaluations, implementation of the program (analysis, organizational structure, record of incidents and emergency measures, etc.). The documentation and record-keeping of measurements and analytical data must be meticulously carried out, because the credibility and efficiency of the management system are largely based on them. ISO 14001 gives importance to record-keeping and processing of *bottom-up* proposals and claims;
- implementation of operational control and emergency measures.

14.8.3 Measurement and monitoring

Clause 4 also states that, prior to a management review (clause 4.7), activities and actions must be adequately documented. The evaluation should include:

- establishment of an internal audit system; the actions needed when it comes to prevention, incidents and accidents, should have the full attention of internal auditors;
- record-keeping (regular measurements of key parameters); in this way, changes of processes or technologies, following a program of corrective action, for example, must also be registered in the list of responsibilities for the treatment of nonconformities, accidents or incidents (according to the guidelines and regulatory level considered);
- monitoring of all actions under the program, taking into account the guidelines and the regulatory level considered.

14.8.4 Identification and management of emergencies

This provision is specific to ISO 14001; there is no equivalent in the ISO 9000 family.

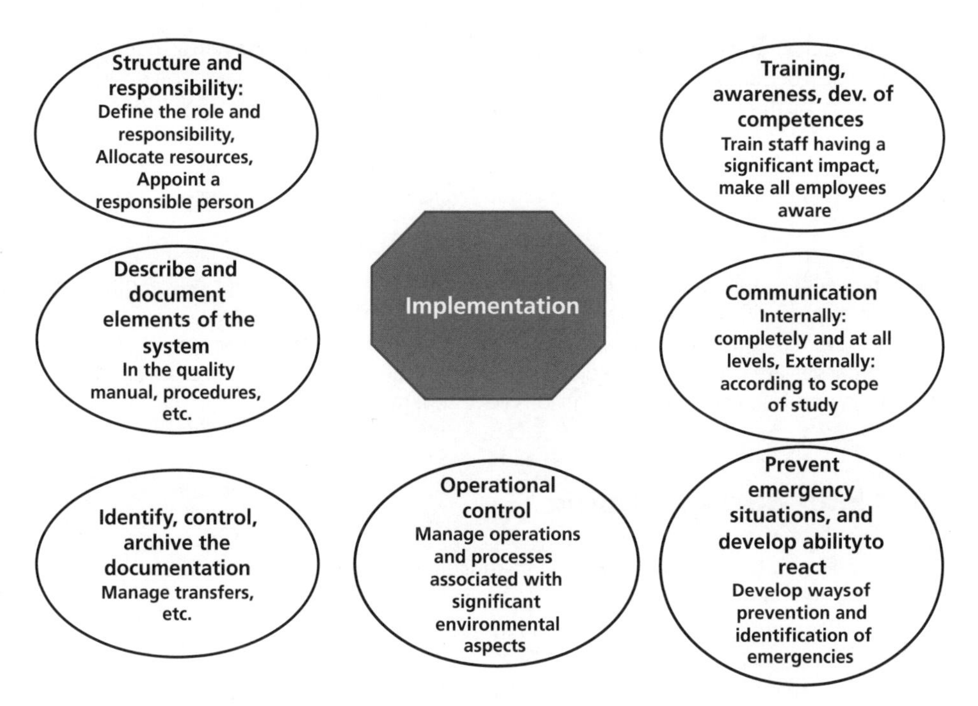

Figure 14.3 Seven key elements of the implementation of ISO 14001.

It is the responsibility of the organization to develop appropriate procedures and a readiness for emergencies (e.g., pollution). In the preparation of crisis management procedures, the organization should consider:

a) the nature of potential accidents at the site (acids, flammable liquids, pressurized gases, organic solvents or carcinogenic pollutants, etc.) and what to do in the case of contamination or discharge; the cause may be technical or due to a human error or inability;
b) the most likely types of emergencies, accidents and their magnitude;
c) the most appropriate response to such accidents or emergencies (crisis management);
d) plans for internal and external communication;
e) actions to be implemented to minimize the deleterious impacts and environmental damage;
f) the corrective and preventive actions to set up after the accident;
g) periodic testing procedures that correspond to emergency situations;
h) staff training when it comes to these procedures;
i) a list of key personnel and emergency services and assistance (hospitals, fire brigade, decontamination service, etc.);
j) assembly points and evacuation routes;
k) the impact of the accident or emergency on the surrounding environment (road, railway or river, etc.);
l) the establishment of a program of mutual assistance with other organizations in the vicinity.

The identification of potential accidents and emergency situations uses a risk management method that resembles the FMEA (see Section D). The organization may also consider the requirements of the British Standard BS 6079-3:2000, *Project Management - Guide to the Management of Business Related Project Risk*. The identification, evaluation and assessment of risks lie in the field of risk management [14.8], and are included in the Austrian Standards of the family on risk management ONR 49000 (which is a point of reference in Switzerland on the subject).

The impact of natural phenomena is not mentioned in the strict sense in the ISO 14000 family, however, natural phenomena such as hurricanes, storms and floods can lead to major crises if critical facilities are affected. Crisis management integrated in the ISO 14000 family has some overlap with the security management of the organization. For this reason, these two approaches are often integrated and the head of the EMS can also have the function of safety manager (even if these two functions have separate tasks).

14.9 EMS documentation

The EMS documentation broadly takes on the documentary structure of ISO 9001. ISO 14001 states in clause 4.4.4 that it must include:

a) environmental policy, objectives and targets;
b) description of the scope of the EMS;
c) description of the main elements of the EMS and their interactions, in addition to reference to related documents;
d) the documents, including records, required by this international standard;
e) the documents, including records, determined as necessary by the organization to ensure the planning, operation and effective control of the processes that relate to its significant environmental aspects.

These documents, which are often used by the staff, should be precise, simple and easy to understand and implement.

Document control (clause 4.4.5 of ISO 14001) generally follows the same principles and the same cycle as those recommended by ISO 9001.

Although the standard does not specify whether an EMS manual is necessary, the documentary structures of many EMSs follow the pyramidal structure recommended by ISO 9001. However, the documentation for an EMS should be more simple and brief:

a) *EMS manual*: this is a short summary document that does not include any confidential data, illustrating how the EMS is managed and used. It often includes the items described in Chapter 6, or clauses 4.1 through 4.6 of the standard. The manual is used in external communication as a "map of the organization," for which reason its contents should be explicit and easily understood. The manual refers to:
b) *The process*, which usually comprises the following:
    1. phases critical for the environment;
    2. measurement systems and controls in order to analyze the performance and make improvements;
    3. methods for reducing impacts;
    4. legal limits to follow;
    5. operational responsibilities;
    6. relevant documents;

7. mode of action in the event of an incapacity (who informs who, who becomes involved when, where, how, and what to record);
8. training requirements;
9. communication criteria;
10. procedures for maintenance of the equipment concerned.

c) *Procedures*, that represent the operational part of the EMS, internal documents intended exclusively for the personnel.
d) *Instructions* that include well defined procedures and checklists for line staff.
e) *Forms* that, once filled out with the *measurement and control data*, become records.

14.10 Moral obligations in ISO 14004?

ISO 14001 [14.9] alone enables the introduction of a minimalist management system, short on regulatory requirements, which is why ISO 14004:2004, *Environmental management systems – General guidelines on principles, systems and support techniques*, specifies that management must first seek to implement ISO 14001 in accordance with current regulations, with the objective of preventing pollution. Management should then specify in its policy if it intends to introduce the following points, while keeping the perspective of continual improvement:

- solely the management of compliance (analogy: the clever man, who does nothing against the law); some service sectors, with only low emissions, adopt this policy;
- a leadership management, with a more caring attitude with regard to the environment, and which goes hand-in-hand with active communications (analogy: the honest man, who includes ethical or moral considerations in his behavior, even if this choice is not just selfless, but also economic), and if so, how?
- a pioneering environmental management, in which all aspects of environmental management are taken into account, with a view to sustainable development and resource management on a global scale (analogy: a good man, who wishes to adopt a posture resolutely ethical and moral with regard to his behavior) (e.g., Eco-Design and Sustainable Management).

Thanks to a growing awareness, the aim is now for a company to climb these three levels of commitment for sustainable (economic, social and environmental) development.

In order to promote investment in relation to the environment, ISO 14004 specifies in clause 4.3 that the company must "develop procedures to calculate the profits, such as the costs of activities, related to the environment, such as the costs of limiting pollution, waste."

The treatment of nonconformities (Figure 14.4) is an important part of the ISO 14000 family. A nonconformity may result from an emergency situation, measurement results, or an internal audit. In all cases, treatment of effects and cause analysis must be carried out. For the latter, corrective and preventive actions must be implemented.

14.11 Certification, recognition and the effects of ISO 14001

Companies usually prefer to have their system recognized and certified by an independent office, accredited by a state body - in Switzerland this is SAS-METAS (Federal Office of Metrology and Accreditation) - which holds regular external audits and issues a certificate, rather than self-proclamation or assessment by a client.

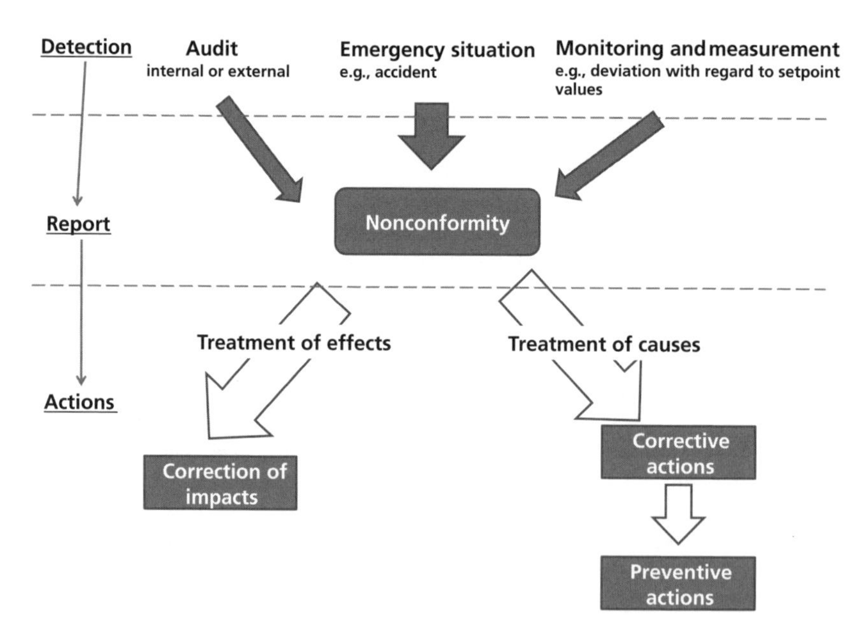

Figure 14.4 Treatment of nonconformities according to the ISO 14000 family.

Procedures for certification, including the preliminary steps of the organization and the certification office, constitution of an expert committee, audit, report issued by experts in order for the organization's management to make a decision, and the certification decision of the office, follow the same process as that of ISO 9001, while keeping in mind that the scope of an EMS can be well below that of a QMS.

What does certification of an environmental management system represent with regard to the law? Is it legal proof of compliance in the event of a dispute with a third party? The answer is unclear, but certification provides conclusive evidence of reasonable conscientiousness (in the sense of preventive measures/precautions that are implemented) of the company's environmental protection. Certification, in conjunction with the results of controls or external audits performed by state administrations, makes it harder to prosecute a company.

14.12 Environmental audits and ISO 19011:2002

ISO 14001 certification implements the ISO standard dealing with audits: ISO 19011 has already been presented in the context of audits associated with ISO 9001. What are the requirements for the verification of an ISO 14001 system? A convenient definition of environmental auditing is that it is systematic documented verification, which occurs periodically and is an objective evaluation by an accredited or regulated group, involving operations and practices on sites, designed to meet regulatory requirements.

ISO 14001 only requires an audit of an environmental management system, but many companies in the initial review (Chapter 2) include an *environmental assessment*, a systematic process where one attempts to measure (or better, qualitatively weigh) and characterize the level of pollution existing on the site and its exposure to pollution. During the management review (clause 4.7 of ISO 19011), numerous companies also systematically include an *impact study* (that attempts to assess the environmental consequences of a project) for each project in order to minimize the negative effects of its implementation on the environment and society.

14.13 Post-certification

In analogy with ISO 9001, the effort for environmental protection does not end after certification, although monitoring may be less and recommendations reduced and directly related to the environmental performance aimed at by management. Nevertheless, it is important to plan for post-certification, as some companies have failed the renewal audit due to their not having kept up pressure on the system. However, the rate of occurrence of renewal audits varies from one country to another.

The mobilization of company management is improved if one introduces the EMS performance chart in the annual progress report, for example. Another success factor is the publication of an annual environmental report and other communication operations that maintain the interest of employees and external stakeholders.

The annual reports of internal audits are drafted by specialists of the company, who are often more capable than outside experts of discovering the potential for improvement, but also the risks of failures of the system. They go hand-in-hand with the management review, which includes:

- the actual environmental performance;
- the result of internal audits;
- failures, incidents and potential accidents;
- preventive and corrective measures;
- communication;
- the future strategy of the company.

Targeted communication of the minutes of that meeting also serves to underscore the importance of the EMS by management and to convey this impression to the staff.

14.14 Environmental communication and ISO 14063:2006

ISO 14063 distinguishes between internal and external communication; it gives more freedom for the latter.

14.14.1 ISO 14063 and internal communications

ISO 14063:2006, *Environmental management – Environmental communication – Guidelines and examples,* states "*with regard to its environmental aspects and its environmental management system, the organization shall establish and maintain procedures for internal communication between different levels and different functions of the organization*" (clause 4.4.3, Communication).

Generally, such communication uses the following channels (depending on the company's size):

- e-mail and e-newsletters;
- posters;
- newsletters;
- management, who patrol the periphery areas and facilities that are key to environmental performance;
- special meetings, celebrations on the occasion of certification or the inauguration of a new facility;
- information sessions (including new staff) to raise awareness;
- training sessions;
- executive meetings;
- annual progress reports by the person responsible for the EMS;
- annual activity report of the company;
- internal valuation of the external communication.

Internal communication should involve:

- policy and commitment of management;
- environmental concerns related to products and services;
- objectives and continual improvement of environmental performance;
- the environmental management system;
- current values of the environmental performance, new developments and ongoing projects;
- results of the follow-up, of the audit and of the management review of the environmental system. The latter communication is generally restricted to top management, executives and employees directly involved in the EMS.

The challenge is to connect each channel to a target audience and to integrate necessary and sufficient communications, and also to introduce redundancy on the basis of McLuhan's[9] communication principle: "the medium is the message" without causing fatigue. This long-term work should not be underestimated because it drives awareness of issues of environmental performance.

Finally, even though internal communication is primarily "top down", meetings and direct contact should always be used to gather – openly, and even proactively – any information, complaint, or issue of the "bottom up" type, etc.

14.14.2 ISO 14063 and external communication

The standard is not very demanding here; it specifies simply that the management's policy must be accessible by the public. The second requirement is that there must be a procedure that gathers demands, relevant questions and complaints, and which documents the responses to external transmitters. Finally, if there is active external communication, the procedure must be described and the actions documented.

14.14.3 A proactive external communication for the Eco-audit

Registration to the Eco-audit of the EU is subject to a requirement related to external communication: the company must produce an audited environmental report for the public.

[9] Herbert Marshall McLuhan (1911- 1980) was a Canadian communications theorist.

In practice, a company with an EMS, especially if it is certified, develops external communication, particularly if it involves leadership management or a pioneering ecology management. Some possible actions when it comes to certification are:

- introduce certification and environmental performance in the marketing strategy;
- inform the sales department so that it informs customers, buyers and sponsors either in writing or orally;
- inform shareholders, insurance agencies, creditors and suppliers by letter, and inquire about their level of environmental performance;
- affix the certification logo on the company's paper or electronic documents (including the web);
- use networks to convey not only the certification, but also prospects for continual improvement and environmental performance, possibly with comparisons with other areas or demanding standards;
- include environmental performance in the annual report;
- directly inform the authorities, local media and external stakeholders, inviting them to a ceremony or celebration;
- inform and seek the views of NGOs.

Finally, the effect of these actions must be gauged. What is the actual impact? Does it correspond to what management was looking for? Are there unanticipated boomerang effects, smirks, ironic smiles, during get-togethers? Have we managed to communicate the quality of the environmental performance of the organization and not just a certificate of conformity to a management system?

This assessment requires careful monitoring, not only by the communication department of the organization, but also by managers and executives. This is sometimes underestimated in companies.

14.15 Conclusions: ISO 14001, advantages and ambiguities

The implementation of an environmental management system is now standard for OECD companies, even if it is not always certified. For service companies, this implementation is easy, whereas for heavy industries, it is an expensive investment, but can be quickly amortized by the savings it generates and the effect on the company's image.

ISO 14001 has been successfully established, although the implementation of such a standard does not guarantee courageous behavior with respect to environmental protection, or a commitment to sustainable (environmental, societal and economic) development. Some lobbies may use this private commitment of relative transparency to counter the initiatives of governments when it comes to the scrutiny of good practice for environmental protection.

ISO 14001 certification can be granted to organizations in the same sector that are not on the same level of environmental performance. To an uninformed public, this may suggest that the problems of waste and pollution are totally controlled by the management of an organization; this may be false. Certainly, the standard allows certification authorities, or the public if the agency is registered with the Eco-auditing body, to quickly measure the degree of commitment and sincerity: this information can be found in clauses 4.1, 4.2 and 4.6.6 of the standard, and is included in the EMS manual.

But reading this material requires an expert's eye when it comes to environmental science and engineering, i.e., by someone well aware of regulatory requirements and best practices of the audited industry. This is not true of all certification agencies in the EU. A reputation for probity and competence, and international visibility of the latter, is essential.

References

[14.1] Reference to genocide of the Herero of south-western Africa in 1904 by the Germans, herded into camps. They underwent genetic crossbreeding experiments planned by Prof. E. Fischer, who became rector at the University of Berlin and one of whose pupils was Dr. Mengele, whose medical experiments at Auschwitz have been widely reported (in MAHMOOD MANDANI *La CIA et la fabrique du terrorisme islamique*, ed. Demopolis, 2007, p. 11).

[14.2] DONELLA H: MEADOWS, *Limits to Growth*, ed. Signet, 1972.

[14.3] EDNNIS GABOR, *Beyond the Age of Waste,* ed. Elsevier, 1974, 2nd ed., 1981.

[14.4] DONELLA H: MEADOWS et al. *The Limits to Growth: The 30-Year Update*, Chelsea Green Publishing Company, 2004.

[14.5] JORG SAMBETH, *Incident à Seveso*, ed. Héloïse d'Ormesson, 2006.

[14.6] For a detailed presentation, see the book by PAOLO BARACHINI, *Guide à la mise en place du management environnemental en entreprise selon ISO 14001*, PPUR, 2007.

[14.6] According to the book by CORINNE GENDRO, *La gestion environnementale et la norme ISO 14001*, Presses de l'Université de Montréal, 2004, p. 82.

[14.7] JACQUES CHARBONNIER, *Le risk management - méthodologie et pratiques*, ed. L'argus de l'assurance,2007, chap.1 L'identification des risques, chap. 2 L'évaluation des risques, chap. 3 Les méthodes combinant l'identification et l'évaluation des risques, chap. 4 Le bilan des risques, pp. 45-98.

[14.8] See KEN WHITELAW *ISO 14001 – Environmental Systems Handbook*, Elsevier, 2004, pp. 1-23.

Chapter 15

Social Responsibility of Organizations and ISO 26000

Key concepts: architecture of social responsibility, sustainable development, corporate governance, ISO 23299, ISO 26000, ISO 27001, OHSAS 18000, social responsibility, SA8000, integrated management system of social responsibility, three pillars of social responsibility.

15.1 Introduction

Transparency is of the spirit of the time. "Say what you do, do what you say"; this is the motto of a twenty-first century led by a globalization that brings together more and more distant partners for trade and communication. The sophistication of consumer products creates an imbalance between the manufacturers' knowledge, the opacity of some of their practices – products are often made far from the countries where they are sold – and the consumers' more restricted field of knowledge.

In a highly virtual universe (think of the extremely limited time a person is in contact with the rest of nature), where the properties of tangible products are juxtaposed with their image, the quality standards for products are no longer sufficient. It is necessary to introduce procedures that account for the actions of a company in relation to democratic, ethical (e.g., Universal Declaration of Human Rights, 1948) or religious (Muslim *Halal* and Jewish *Kosher*) values.

Institutions – in particular, companies – acquire ethical charters, often referred to as ethical principles, especially if they are multinational.

15.2 Corporate social responsibility (SR) and the Triple Bottom Line (TBL)

15.2.1 Corporate Social Responsibility and Sustainable Development

The concept of sustainable development was introduced in 1987 by the World Commission on Environment and Development:

Sustainable development *meets the needs of the present without compromising the ability of future generations to meet theirs.*

Two concepts are inherent in this notion:

- the concept of "*needs*" and, more particularly, the *essential needs* of the poor, who should be given the highest priority, and
- the *idea of limitations* imposed on the environment's ability to meet present and future needs by the state of our technology and our social organization.

This concept was discussed again at the World Summit on Sustainable Development in Johannesburg[1] in 2002, attended by 180 company representatives. It was not until years later that a transcript of sustainable development appeared in corporate life: social responsibility [15.1].

In 1998, *Elkington* [15.2] made a statement that forms the basis of corporate social responsibility. He suggested that the success of a company or organization cannot be measured only by financial indicators of profit or return on investment. He introduced the concept of *Triple Bottom Line* or *3P* (Planet, People, Profit). Here, we prefer the term social responsibility.

In a next step, *Andrew Sawitz* [15.3] used a striking example in the introduction to his book: whaling [notorious thanks to the story of *Moby Dick* by *Herman Melville* (1851)], an industry that occupied up to 70,000 people in the mid-nineteenth century. Whales became threatened with extinction, which led to the near disappearance of this occupation and also of the capital that was invested in it. The three parts of the 3P were, therefore, destroyed.

According to *Sawitz*, all companies must now avoid the fate of the whaling industry. This is part of sustainable development (we will use the term "social responsibility"), i.e., "creating profit for shareholders, while protecting the environment and improving the lives of those with whom the company interacts."

15.2.2 Ambiguous from the start?

It is unclear whether the reasoning used by *Sawitz* goes beyond concerns for a robust and resilient economic performance. Indeed, he is not appalled by the slaughter of whales, nor does he question the killing of a species that clearly appears to have a brain function similar to ours. So there is something amiss from the start of the process. Indeed, one can cynically imagine:

- that the concern for environmental protection was linked to the economic risk posed by poor company image and the high costs of insurance, and
- that the desire to improve the human living conditions in question was only aimed at increased consumption by generating revenue, and by avoiding social upheaval and perturbations within the company and stock markets, which would lead to heavy losses among investors.It is, nonetheless, true that transparent accountability started in the twenty-first century. An organization must now consider its impact over a wider area, that of the overall economy, and the environment and society in which the company operates. Resource management is no longer considered as exclusively financial: environmental and social resources (employees, their time and expertise, the infrastructure provided by the government, etc.) must also be taken into account. The three pillars of social responsibility must be positive.

[1] Web-site: http://www.sommetjohannesburg.org/

15.2.3. The three pillars of social responsibility

Sawitz envisaged a structure with three pillars (Figure 15.1).

This model does not take into account the fact that social responsibility is not just a form of administrative management in the form of a *scorecard*[2], but rather clever entrepreneurial management that can maximize financial gain by conciliating parties, or by using their support with honesty and integrity.

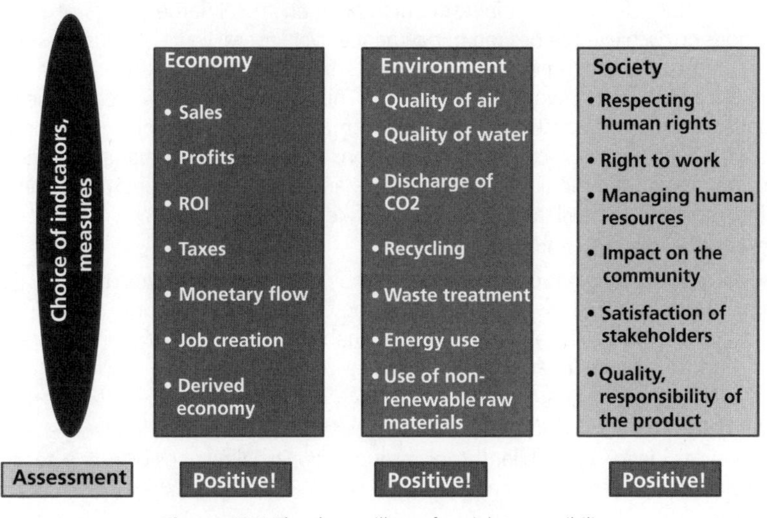

Figure 15.1 The three pillars of social responsibility.

15.2.4 Critics of social responsibility

The implementation of corporate social responsibility has its detractors:

- The opponents, who are recruited from NGOs and progressive circles, see it as a smokescreen or a media initiative of industry and finance to minimize or destabilize the role of the state. We cannot, however, blame company managers for developing instruments to contribute to sustainable development as it was defined in 1987.
- Opponents from the business community rely on the neoliberalism of *Milton Friedman*, who stated that company managers must above all maximize company profits for shareholders [15.4]. *Adam Smith* showed that human behavior is primarily governed by personal interest and not by altruism, and it is thus unrealistic to integrate another perspective into the management of a company. However, when reading the works of *Adam Smith* (the father of the market economy), one can see that his position was more subtle. Finally, the societal impact of business economics on society in the twenty-first century cannot be compared to that of the British economy… of the eighteenth century.

---

[2] In 1992, *Robert S. Kaplan* and *David Norton* launched the Balanced Scorecard (BSC) method to measure a company's activities with regard to four main perspectives: learning, processes, customers and finances. First, the vision, values and mission of the entity should be explained, to give managers a comprehensive understanding of their organization. The new element focuses not only on financial results but also on the human issues that bring results, so that organizations concentrate on the future and act in their best long-term interest (text source: WIKIPEDIA).

Moreover, there is no official directive that forces business managers to maximize profit, even if its optimization is sought: economic practices ruining social and environmental atmospheres have a negative impact on the performance of the company in the short and medium term as well as on its capital.

The reason behind profit maximization cannot be sought in the remuneration of loyal shareholders, to whom one might explain the reasons for optimization but not maximization of profit. Shareholders have been replaced by investors who are without qualms to seek profit maximization of short-term investments and then to sell large shares of stock if the company does not achieve the desired performance over several years. As business results are not always predictable over time, there is a great temptation for company managements and boards to give a single watchword: maximization. This trend is also reinforced by the incentive of top management to increase the revenue and margins of the company and the growing influence of the board in executive compensation. A career as top manager can be short (and dangerous), and the medium-term prospects are often given second priority in favor of maximizing an action that will allow the CEO to take a tempting *cash-out*.

But some criticisms are more relevant:

- Social responsibility has often been evoked by executive boards to counter acquisitions and mergers [15.5], which should in no way be its function, but since social responsibility has as its objective the sustainability of the organization, we see how far off track this concept can sometimes be.
- The concept of social responsibility has been around since at least 1919 [15.6], when *Henry Ford* tried to justify investments to expand his plants (by depriving shareholders of their dividend) with philanthropic arguments. Social responsibility is a concept that surfaces whenever a company is in crisis.

The social responsibility policy of a company depends on the definition of the scope and limits of responsibility. The landscape and perimeters depend on the type of business and the direct social and natural environment of the company or organization.

15.2.5 An unfortunate omission of TBL: Corporate governance

But the volatilization of invested capital and all shareholder dividends is alas not only due to poor management of the three pillars of social responsibility as defined by *Sawitz*. *Corporate governance and respect for good management practices* are ignored; is this due to nostalgia for Fordism? In the 2000s, however, there occurred a succession of scandals that showed that these elements are largely hidden, not without reason.

15.2.5.1 Questionable practices

In 2001, the fraudulent management of *ENRON* and the blindness of its board of directors caused the bankruptcy of the company. The value of shares was divided by 350, resulting in the loss of capital for the staff pension fund – which had an equity portfolio of 60% *ENRON* shares – and leading to the end of *Arthur Andersen*, the consulting firm that had certified the accounts. *ENRON* recorded that year a loss of over a billion dollars.

Also in 2001, in January, the share value of *Hewlett Packard* collapsed to less than 10 USD, after having a maximum value of 60 USD in 1999. "Creative" financial practice had increased turnover by 3 billion dollars and profits by 1.5 billion from 1995 to 1999, while the company's growth was in reality near zero during the 1990s. The executives sold shares for 90 million dollars while the value was at its peak. In 2002, Hewlett Packard admitted an accounting error and paid 10 million dollars in fines. Unfortunately, this money did not come from the pockets of the directors, but those of the shareholders, because the money was taken from the accounts of the company.

In 2002, similar circumstances led to the bankruptcy of *WorldCom*, the second largest communications provider in the US. The debts were estimated at 41 billion dollars.

Also in 2002, *IBM* was accused of financial trickery in order to present an annual growth of over 10% while the average real growth did not exceed 5%.

Many observers were convinced that these scandals were only the observable part of a set of questionable practices, benefitting from a lack of curiosity, foresight and the will to control boards of directors, but also due to the difficulty in obtaining key information regarding the operation of a company.

In 2008, the *subprime* crisis almost derailed the global economy. Governments of all countries were forced to invest huge sums to compensate aberrant management practices of executives and directors of banks and their "bonus" management. The crisis plunged these companies into an unprecedented slump and marked the end of the dictates of neoliberalism.

15.2.5.2 The shipwrecking of boards of directors [15.7]

Too often, the Board was chaired by the CEO of the company, who saw the members as advisors at his own discretion. Also too often, members of the Board were chosen exclusively by the CEO of the company, selected only from his or her address book, for their complaisance (the *Yes Men*), their visibility that flattered the board, or for their ability to fall asleep while sitting, or exclusively for their network which enabled the CEO to climb the social ladder. The CEO was granted financial privileges or services (apartments, jets, company cars) by the board, of which he was president. Curiously, the financial press, which often wrote of misuse by politicians in this area, was hardly inquisitive about this conflict of interest before the scandals of the 2000s.This suggests that if staff or unions participate in a Board of Directors (as is the case in Germany and the Scandinavian countries), the actual skills they bring to the table could be more useful than those of the showier members. When we imagine the pressure on middle management to account for and to achieve annual goals, or even quarterly objectives, this inadequacy suggests that the revolt of the elite described by *Lasch* [15.8][8] is more than a polemicist vision.

15.2.6 Quality of corporate governance, OECD principles, the *Sarbanes-Oxley* act and *COSO2*

In response, corporate governance and their boards are one of the subjects of the OECD *Principle of Corporate Governance (1999, 2004).* It is useful to know the definition*:*

> *"Corporate governance is one of the main factors for improving efficiency and economic growth and for strengthening investor confidence. Corporate governance refers to the relationship between company management, its board, its shareholders and other stakeholders. It also provides the structure used to define the company's objectives as well as the means to achieve them and ensure the monitoring of obtained results. Corporate governance should provide proper incentives for the board and management to pursue objectives in the interests of the company and its shareholders and should facilitate effective monitoring of obtained results."*

The board must be ethically committed, which contradicts neoliberal thinking, which sees the mission of corporate governance as exclusively the maximization of shareholder interests:

> *"The board plays a key role in defining the policies of the company when it comes to ethics, not only through its own actions but also in appointing and overseeing key*

executives and, consequently, management in general. It is in the long-term interest of the company to maintain high ethical standards in order to establish credibility and reliability not only in daily activities, but also as part of its long-term commitments. In order for the aims of the board to be clear and achievable, many companies have found it useful to develop their own codes of conduct, based in particular on professional standards, and sometimes more general codes. These may include voluntary commitment of the company (and its subsidiaries) to follow the OECD guidelines for multinational companies[3]*, which include the four principles contained in the Declaration on the Rights at Work*[4] *by ILO."*

The US has set up stricter regulations, with the publication in 2002 of the *Sarbanes-Oxley* (OX) act, and the introduction of a control framework, COSO2[5], the objectives of which correspond to the expectations of investors, and are as follows:

- implementation and optimization of operations,
- reliability of financial information, and
- compliance with laws and regulations.

The implementation of COSO2 requires the intervention of a new executive, the risk manager, and gives more responsibility to the board of directors, including criminal risk. The proclamation of the *Sarbanes-Oxley* act had an impact on many OECD countries, where similar measures were taken. However, few countries have laws as restrictive as the US.

Subsequently, numerous works of corporate governance have been published by consultancy firms. Their contents, often upsettingly simple, particularly with regard to the recommendations [15.9], reveal much about the moribund nature of the governance of many companies before the *Sarbanes-Oxley* act.

For its part, the European Union has published its eighth Directive, called *EUROSOX*, effective from July 1, 2008. In Switzerland, a risk management based on the US initiative

---

[3] **Eleven principles that exclude banana republics**. Companies should be fully aware of policies established in the countries where they operate and should consider the views of others. In this regard, companies should: **1.** Contribute to economic, social and environmental progress to achieve sustainable development. **2.** Respect the human rights of those affected by their activities, consistent with the international obligations and commitments of the government of the host country. **3.** Encourage the creation of local capacities by working closely with the local community, including local business communities, while developing the company's activities in both the domestic and foreign markets in a manner consistent with sound commercial practices. **4.** Encourage the creation of human capital, in particular by creating employment opportunities and facilitating the training of employees. **5.** Refrain from seeking or accepting exemptions not allowed in the statutory or regulatory framework related to the environment, health, safety, labor, taxation, financial incentives or other areas. **6.** Support and uphold the principles of good corporate governance and develop and apply good corporate governance practices. **7.** Develop and implement self-regulatory practices and effective management systems that foster a relationship of mutual trust between companies and the communities in which they operate. **8.** Ensure that their employees are fully aware of company policies and comply with them, disseminating them as appropriate, especially through training programs. **9.** Refrain from discriminatory or disciplinary action against employees who make bona fide reports to management or, where appropriate, the competent public authorities, on practices that contravene the law, the guidelines or company policies. **10.** Encourage, where possible, their business partners, including suppliers and subcontractors, to apply principles of corporate conduct compatible with the guidelines. **11.** Abstain from any improper involvement in local political activities.

[4] See: http://www.ilo.org/declaration/thedeclaration/lang–fr/index.htm

[5] Committee of Sponsoring Organizations of the Treadway Commission.

appeared with the Basel II standards, adopted by the EU and implemented by Directives 2006/48/EC and 2006/49/EC.

In view of the causes of the unprecedented financial crisis of 2008 and the abysmal situation of the Union Bank of Switzerland (UBS), one wonders if the risk management included in these guidelines was sufficiently restrictive. There are voices saying that risk management should be unified and strictly formalized within banking institutions and that its control should rely on the state. The problem remains just as acute.

15.2.7 Governance and social responsibility

The three pillars are replaced by a structure (Figure 15.2), which includes:

- governance of administrative boards and boards of directors (corporate governance);
- the morality of the management;
- good financial practices and business ethics;
- the implementation of management science.

Finally, the performance of the company depends on its base, its political framework, as well as the economic and social environment in which the company operates and with which it interacts. The structure of social responsibility describes a new form of quality, not total, but extended.

Figure 15.2 Social responsibility of organizations.

15.3 Corporate social responsibility (SR): what to choose?

15.3.1 A wealth of initiatives

The definition and control of the SR policy of a company (and its implementation) require the choice of one or more key charters of indicators and partner organizations dedicated to this goal. The list below is not exhaustive, but it is based on measures implemented to gauge the financial performance of companies [15.10]:

1. frames of reference for international commitments (UN, OECD, EU); these include guidelines, ethical charters, lists of recommendations[6];
2. frames of reference for reporting, mostly frameworks for formatting data, and proposals for standardized communication, enabling comparisons between organizations;
3. frames of reference for certification, standards and guidelines for organizations, approaching the ISO model;
4. frames of reference for evaluation established by rating agencies[7], who, on the basis of data and questionnaires sent to organizations, establish a classification of organizations that can be used by insurance companies, banks, public communities, government authorities, investors, shareholders, etc.

15.3.2 Calling for meta-standards?

There are numerous current initiatives, covering the various objectives of social responsibility, and they are often the prompt, and perhaps volatile, result of unbearable pressure from stakeholders.

While it seems possible to reach a global consensus on sustainable "environmental" development, the definition of a common social policy seems difficult (many economic partners are far from adopting similar policies, despite having signed the UN charter). A policy of corporate social responsibility can therefore not avoid the use of "meta-standards", for example the Universal Declaration of Human Rights, and basic charters of international organizations such as UNESCO, and the International Labour Organization (ILO), etc.

This results in an unclear situation because social responsibility (SR) does not yet have a comprehensive framework that would facilitate its implementation. To achieve this goal, an organization should:

- have a policy according to the Triple Bottom Line model, clearly defining the underlying motivation for such a policy: SR is deduced either from economic imperatives only (which?), or from the use of ethical or moral principles and, if so, which and why?
- define a strategy and objectives for SR: which model and why?
- establish a system of quality management for SR and for reporting it;
- implement a system of internal audits and, if desired, external ones;
- implement a system of corrective and remedial actions;
- develop continual improvement, and
- make management (or the board?) perform periodic policy reviews.

---

[6] For example *Global Compact* of the UN, comprising 10 principles, see: www.un.org/french/globalcompact/principles.shtml

[7] We have seen an emergence in the field of sustainable development of agencies rating companies according to social and environmental criteria. These agencies determine the level of corporate responsibility, corporate social responsibility with regard to sustainable development based on different criteria. Initially, these specialized rating agencies worked primarily for investors in the field of socially responsible investment (SRI). This is a growing sector that includes 30 exclusive structures, of which the most important in Europe are: *Oekom* in Germany, *Triodos* in the Netherlands, *Avanzi* in Italy, *BMJ Ratings* and *Vigeo* in France.

15.4 A pioneer, the French Standard AFNOR SD 21000

15.4.1 A standard and an application document

In continental Europe, France has rarely – compared with Germany – presented itself as a pioneering player when it comes to environmental protection and sustainable development. Yet it is France, through its standards agency AFNOR, which published in 2003 the first frame of reference for sustainable development and social responsibility[8]: the *SD 21000 standard*. This is intended for all companies regardless of sector or size. The standard explains the interests of sustainable development for companies and proposes a management system to implement them. ISO 26000 is based on SD 21000 (15.5).

Figure 15.3 Social responsibility according to SD 21000.

The AFNOR document has two parts:

- *the standard itself,* "Sustainable Development – Corporate Social Responsibility", whose objective is not certification of an organization, but to give guidance on understanding the stakes of sustainable development in the strategy and management of a company (SD 21000; FD X 30-021; May 2003);
- *an application document*, i.e., guidance on the identification and prioritization of sustainability issues, and a method for the diagnosis of sustainable development for an organization. It also contains recommendations for the development and implementation of an action plan for continual improvement of the organization's performance with regard to the issues identified (FD X 30-023, May 2003).

---

[8] Sustainable development, from strategy to operation, AFNOR (2007).

15.4.2 Strategic issues

The principle of sustainable development is conceived as a learning curve for a company, but also as a global project for economic development, in which each actor is the author. This apprenticeship begins with a strategic analysis and an inventory of the markets/companies who make their leaders ask themselves the following strategic questions, and the definition of a business plan over a much larger timeframe than usual (Figure 15.3):

- What are the reasons that now necessitate consideration of new factors or actors?
- Why behave differently towards them?
- Who are they?
- Who are the stakeholders? Why do their influences on the company's performance grow significantly?
- What are these influences (positive or negative) on the company's performance? Now? In the short and medium term? Issues over time? What areas will be particularly concerned?
- Where is the organization positioned on this path compared to its competitors and its market segment?
- If the company decides to comply with the principles of sustainable development, what priority actions should be taken? What resources should be mobilized, and how soon in order to avoid jeopardizing development?
- How can the existing policies and programs be extended and altered while introducing the necessary breakthroughs when permanent continual improvement is not enough?
- How can the objectives and business processes be adapted to take into account the principle of sustainable development?

These questions require the coordinated management of three long disconnected domains:

- the area of regulations and laws,
- the management of stakeholders, and
- the management of significant environmental and social aspects.

SD 21000 therefore describes the strategic approach and emphasizes the operational aspects to be integrated into the management and implementation of means that lead to the achievement of objectives (Figure 15.4).

15.4.3 The strategic approach

The strategic approach includes the following actions:

- An adjustment of corporate governance, given the future implementation of a multi-year plan for sustainable development.
- Openness to stakeholders, in order to know and understand them completely, to accurately determine their needs, and to assess their degree of influence on the company's performance. Also those who are willing to collaborate and prepare for the development of a metric that can communicate in an honest, sincere, transparent and verifiable way the performance results of the organization should be identified. Recognize that they have the full right to know for example:
  - universal principles (human rights, children's rights, the ILO conventions);
  - the application of principles derived directly from the decision to respect the principle of sustainable development (prevention, care, equity of access to resources, participation…);

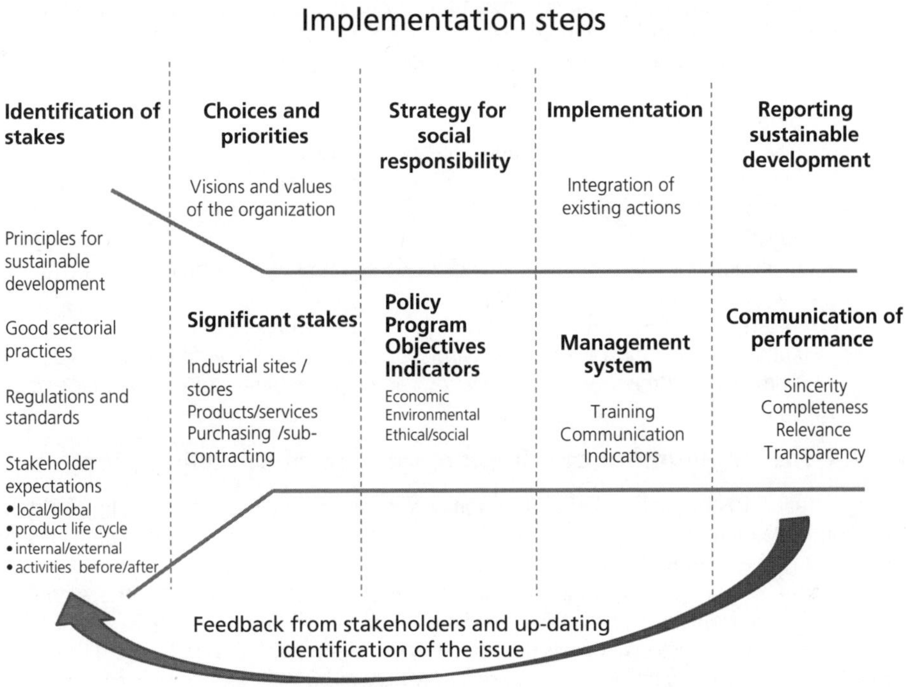

Implementation steps

Identification of stakes
Principles for sustainable development
Good sectorial practices
Regulations and standards
Stakeholder expectations
• local/global
• product life cycle
• internal/external
• activities before/after

Choices and priorities
Visions and values of the organization
Significant stakes
Industrial sites / stores
Products/services
Purchasing /sub-contracting

Strategy for social responsibility
Policy
Program
Objectives
Indicators
Economic
Environmental
Ethical/social

Implementation
Integration of existing actions
Management system
Training
Communication
Indicators

Reporting sustainable development
Communication of performance
Sincerity
Completeness
Relevance
Transparency

Feedback from stakeholders and up-dating identification of the issue

Figure 15.4 The global approach of SD 21000.

- considerations relating to fair trade and fair compensation for work and resources;
- weak stakeholder interests (poorly represented minorities, future generations, animal or plant species), etc.

- An inventory that specifies:
  - identification of the characteristics, expectations and market trends of the business;
  - an assessment of the resources and skills, strengths and weaknesses of the company.
- A risk assessment and identification of significant issues. The above steps can reveal a high number of potential issues, however the company must in the short-term focus only on those judged to be paramount.
- These are taken into account when developing strategies, policies and an action plan of priorities, and are then translated into measurable objectives: quantitative or qualitative indicators are associated with them, and other issues are integrated into the policy in the medium to long term.
- The taking into account frames of reference such as ISO 9001, ISO 14001, SA 8000 and the management system models ISO 9004 and ISO 14004 already in place.
- Once an analysis of risks and opportunities is completed, management is able to determine its new vision for the company, and update its strategy, its policy and its objectives to develop a multi-year program, a real "business plan" for sustainable development. From this, management can conclude its action plan of priorities, involv-

ing different disciplines and across all functions of the company, and can carry out the plan and categorize it on an annual basis, in collaboration with the company.

- The multi-year program may involve such diverse areas as:
  - product design;
  - production technologies;
  - procurement policies and changes in specifications of raw materials and components;
  - subcontracting policies;
  - logistics;
  - management of infrastructure and heavy equipment, and policies for acquiring them;
  - information and communication: this affects employees, the public, political authorities and all stakeholders, with a categorization for each target. Guidelines from ISO 14004 about communication can help companies formalize their efforts.

15.4.4 The deployment and implementation of an action plan

Implementation of the plan follows the traditional rules for any project by relying on data for preliminary planning, allocation of powers and resources, and implementation of key processes to achieve the set goals in the allotted time.

Of course, indicators and monitoring systems must also be put in place. The emphasis should be on sustainable development indicators. Some standards, such as the Global Reporting Initiative or sectorial reports developed by some companies, together with the UNEP (United Nations Environment Program) and WBCSD (World Business Council for Sustainable Development), can guide organizations in their choice of relevant indicators.

The implementation includes the introduction of a *Deming* cycle, and thus a regular review of management and a program of continual improvement.

15.4.5 Self-assessment in the light of sustainable development (SD)

SD 21000 details the mode of self-assessment for an organization regarding sustainable development issues. This should enable the organization to:

- identify and assess its performance with regard to each of the challenges of the sustainable development proposed;
- revise the magnitudes of various issues relating to its stakeholders;
- prioritize issues according to performance and their importance for the company and its stakeholders.

The self-assessment involves three steps:

- a study of sustainable development issues;
- a study of stakeholders;
- prioritization of the issues.

Definition and positioning issues

The analysis is based on a list of 34 issues split into five categories:

- *Cross-category SD issues*: 1. Products/eco design. 2. Purchasing policy. 3. Management and risk prevention. 4. Storage. 5. Territorial integration of the organization and management of external issues. 6. Transport of employees, site accessibility.
- *Social issues of SD*: 7. Work: general conditions and atmosphere. 8. Fair trade, etc. 9. Employment, skills, training. 15. Health, safety and hygiene.

- *Environmental/ecological issues of SD*: 16. Water: consumption management. 17. Water: pollution. 18. Energy: consumption. 19. Air: pollution and greenhouse gas emissions. 20. Waste. 21. Soils: management and pollution. 22. Biodiversity. 23. Noise and odors. 24. Transport and logistics.
- *Governance and management practices of SD*: 25. Management commitment. 26. Strategy, policy and objectives. 27. Management system. 28. Organization and responsibilities. 29. Participation, involvement and motivation of staff. 30. Internal communication. 31 External communication. 32. Monitoring of regulations. 33. Taking into account other factors. 34. Identification of stakeholders and links between the expectations of stakeholders and the policy.

It appears that the economic wheel, one of the three pillars of TBL (corporate financial results, changes in turnover, etc.), appears only to a modest degree in these issues. For each of the 34 issues, the organization needs to take a position with regard to:

- its performance; assign a rating of 1 (simple awareness of the issue, no action) to 5 (high level of excellence and innovation in the state of the art brought into play by the organization);
- the importance of this issue; assign a score of 1 (no priority, issue deemed unimportant) to 5 (essential to the life and sustainability of the organization);
- the selection of stakeholders who, from their point of view, are affected by this issue: for each issue, rate the influence of stakeholders from 1 (no risk of harm to the organization) to 5 (the stakeholder may challenge the very existence of the organization).

In a second step, the organization proceeds with:

- a rebalancing of the importance of stakeholders according to the number of issues that concern them;
- a rebalancing of issues depending on the number of stakeholders who are involved in each issue.

The corrected data is entered in a table that determines the criticality of issues and actions to be implemented (see Figure 15.5).

15.5 A promising start: ISO 26000

15.5.1 Foundations and functions

Among recent initiatives, ISO 26000:2010, *Guidance on social responsibility*, is the most prominent, because it guarantees a degree of continuity to the process.

ISO launched the project for ISO 26000 [15.11] on social responsibility[9] in 2006. The standard is not intended for certification. Its first edition dates from November 1, 2010. Here is the basic text, the moral basis underlying its creation:

"The world is still not fair[10], but overall we still have the possibility of eliminating poverty, inhuman situations and environmental degradation that are still far too common. As individuals we all have the opportunity to take responsibility and it is no longer possible

---

9 Social: relating to society, its values, its institutions (Le Nouveau Robert méthodique 2003), the adjective "social" has a wider content.

10 Fair: having fairness, equity: virtue that consists in regulating one's conduct in the natural sense of right and wrong (Le nouveau Robert méthodique, 2003).

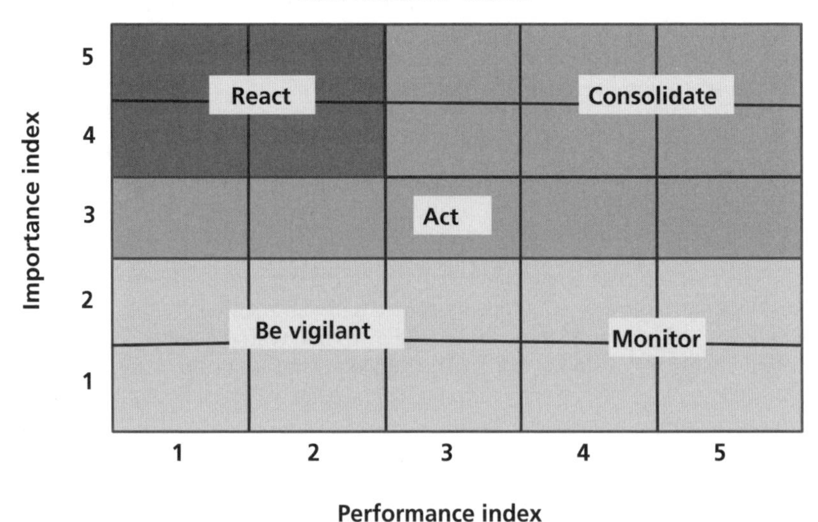

Figure 15.5 Table of priorities of issues.

to say that it is up someone else to solve global problems… The world is still not fair but together we still have the option to change this situation, if we can reach consensus and take action… Sustainable development, a possible future. "

The foundations of the work in hand:
[…] *The foundations of the work of ISO on social responsibility rely on the general recognition that the social responsibility of organizations is essential for the sustainable development of an organization. This recognition is based on both the Earth Summit in Rio in 1992 on the environment, and the World Summit on Sustainable Development (WSSD) in Johannesburg in South Africa in 2002.*

And the function of ISO 26000:
[…] *Any organization that wishes to improve its social responsibility should regularly review its processes and its role in society. All sectors of society can contribute more to sustainable development and there is a need for tools that help the practitioner state and put into practice sustainable development goals. ISO 26000 addresses one need* […]

15.5.2 Comparisons with other basic texts

To study the evolution of this debate, it is useful to compare the above texts with other humanitarian or egalitarian texts, in particular with:

- The last sentences of the Communist Manifesto of Marx and Engels (1848):
  Finally, the Communists work for the union and agreement of the democratic parties of all countries. The Communists disdain to conceal their views and aims. They openly declare that their ends can be attained only by the forcible overthrow of all existing

social conditions. Let the ruling classes tremble at the idea of a communist revolution! The proletarians have nothing to lose but their chains. They have a world to win. Workers of all countries, unite.

- Paragraph 1.1 of the Social Contract of Rousseau (1762):
  Man is born free and everywhere he is in chains. Those who think themselves the masters of others are indeed greater slaves than they. How did this change come about? I do not know. What can make it legitimate? That question I think I can answer.

  If I took into account only force, and the effects derived from it, I should say: "As long as a people is compelled to obey and obeys, it does well; as soon as it can shake off the yoke, and shakes it off, it does still better: for, regaining its liberty by the same rights as took it away, either it is justified in resuming it, or there was no justification for those who took it away." But the social order is a sacred right which is the basis of all rights.

- The preamble to the Universal Declaration of Human Rights (1948)[11]:
  Whereas recognition of the inherent dignity and of the equal and inalienable rights of all members of the human family is the foundation of freedom, justice and peace in the world.
  Whereas disregard and contempt for human rights have resulted in barbarous acts which have outraged the conscience of mankind, and the advent of a world in which human beings shall enjoy freedom of speech and belief and freedom from fear and want has been proclaimed as the highest aspiration of the common people.
  Whereas it is essential, if man is not to be compelled to have recourse, as a last resort, to rebellion against tyranny and oppression, that human rights should be protected by the rule of law, [...]

15.5.3 Scope of ISO 26000

The scope of ISO 26000 can be summarized to:

- help an organization to support and develop its social responsibilities.
- provide guidelines to:
  - implement social responsibility;
  - identify and engage with stakeholders;
  - improve the credibility of reports and statements about SR.
- put the results and improvements into perspective.
- increase customer satisfaction and customer confidence.
- promote common terminology in the field of SR.
- obtain consistency[12] and non-contradiction with documents, existing treaties and conventions and other ISO standards.

---

[11] Comprising 30 articles available at: http://www.un.org/fr/documents/udhr/

[12] ISO and the International Labour Organization (ILO) have signed a Memorandum of Understanding to ensure the consistency of ISO 26000 with the ILO Conventions.

15.5.4 Areas concerned

The areas covered are based on the TBL types:

- environment;
- human rights and good practices in the workplace;
- organizational governance and fair operational (managerial) practices;
- problems with regard to consumers, community involvement, development of society.

ISO 26000 integrates the views of the following families of standards: ISO 9000, ISO 14000 and ILO-OSH 2001 (management tools developed by the International Labour Organization, ILO). ILO-OSH 2001, entitled *Guidelines for management systems for health and safety,* is the only international framework agreed upon on a tripartite basis – government, employer, worker. Collaboration with the *Global Compact* has also been established.

15.5.5 Type of standard

This standard includes only recommendations, and is not intended for certification by accredited agencies or offices. Its major contribution is to provide a reference for vocabulary, concepts, and best practices.

Companies are solicited by many external stakeholders, with sometimes conflicting interests and often using disparate standards, so they struggle to meet the expectations of stakeholders and NGOs, as there is no basic consensual document.

ISO 26000 is the common denominator of expectations when it comes to social responsibility for government offices, companies and organizations and their internal and external stakeholders. It is complex, detailed and well researched. Its implementation is a long-term task, especially in large organizations with a significant portfolio of tangible products or services.

15.5.6 Structure and contents of ISO 26000[13]

Its structure consists of numerous clauses and subclauses. Its content is rich, since the length of the standard is 118 pages, whereas ISO 9001:2008 for example only has 36. It includes the following clauses (the first three items are common to all ISO standards):

0. *Introduction:* information about the content of the standard and the reasons that led to its development.
1. *Scope*: Definition of the subject of the standard, its scope and application limits.
2. *Terms and definitions*: this is an important part of the standard for all the key concepts that revolve around social responsibility, since it allows partners to understand one another, using concepts that have the same meaning for all. Social responsibility is defined as the responsibility of an organization for the impacts of its decisions and activities on society and the environment, through transparent and ethical[14] behavior that:

---

[13] Validated by the ad hoc Working Group in September 2005 in Bangkok.

[14] Behavior that is in accordance with accepted principles of right or good conduct in the context of a particular situation and is consistent with international norms of behavior (def. 2.7 of ISO 26000). These are expectations of socially responsible organizational behavior derived from customary international law, generally accepted principles of international law or intergovernmental agreements that are universally or nearly universally recognized (def. 2.11 of ISO 26000).

a. contributes to sustainable development, including health and the welfare of society;
b. takes into account the expectations of stakeholders;
c. is in compliance with applicable law and consistent with international norms of behavior;
d. is integrated throughout the organization and practiced in its relationships.

It is important that sustainable development is a key objective of any organization engaged in a socially responsible approach.

3. *Understanding social responsibility*. This clause describes important factors and conditions that have affected the development of social responsibility and which continue to influence its evolution and practice (global issues related to the environment and health, etc.). It also describes the concept of social responsibility itself – what it means and how it applies to organizations, and includes, in particular, equality between men and women. Finally, it contains guidelines on the implementation of the standard for small and medium-sized companies, emphasizing flexibility and structure, in short, decisive advantages in the implementation of the standard.

4. *Principles of social responsibility*. This clause details and defines the seven principles of SR that are to be precisely followed in the implementation process:
   a. *Accountability*: the organization must be answerable for its impacts on society, the economy and the environment and, consequently, it must agree to be subjected to appropriate examinations (such as assessment by external auditors on the basis of this standard) and the corresponding duty to answer the question: how will future control of the implementation of ISO 26000:2010 be performed? Certainly not by a certification, but by evaluation by a committee of external auditors, for example on the basis of a self-assessment prepared by the organization (and here, given the complexity of the area to be scanned, we have to rely on the assessment practices of academic institutions or organizations with excellent knowledge...).
   b. *Transparency:* the organization must ensure the transparency of its decisions and its activities when they have an impact on society and the environment.
   c. *Ethical behavior*, i.e., based on the virtues of honesty, fairness and integrity, which means that the organization cares for third parties (stakeholders), animals, and the environment.
   d. *Respect for stakeholder interests*.
   e. *Respect for the rule of law*.
   f. *Respect for international norms of behavior*, and avoidance of complicity[15].
   g. *Respect for human rights*.

5. *Recognizing social responsibility and engaging stakeholders.* This clause presents two practices to identify the field of social responsibility of an organization, and to facilitate the establishment of the sphere of influence of the organization[16]:
   a. evaluation activities and decisions of the organization on the one hand,

[15] Defined as performing an act, or not acting, with a significant impact on a wrongdoing, such as a crime, and this, knowingly or with the intention of contributing to this illegal act.

[16] Sphere of influence means the range/extent of political, contractual, economic or other relationships through which an organization has the ability to affect the decisions or activities of individuals or organizations (def. 2.19 of ISO 26000).

b. identifying the expectations of stakeholders on the other hand, identifying them and establishing a dialogue with them.

6. *Guidance on SR core subjects* (see Figure 15.6): This clause, which is basic to the approach, outlines areas of responsibility that should be processed and placed under the control of the organization, and areas or core issues for which it will be called to account:
   a. organizational governance: the quality of the governance enables the successful implementation of controls in the following areas, and is thus critical, despite the fact that after this the prescriptive content of the standard is surprisingly brief;
   b. human rights;
   c. labor relations and working conditions;
   d. environment;
   e. fair operating practices;
   f. consumer issues;
   g. community involvement and development.

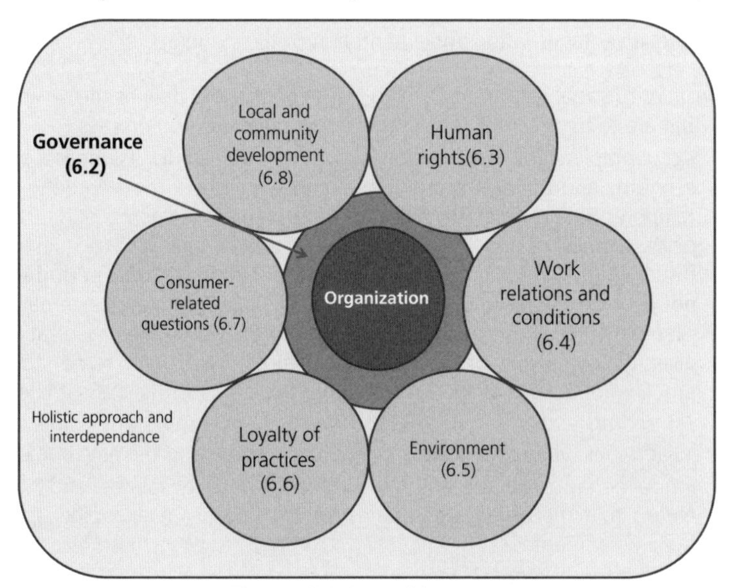

Figure 15.6 Core subjects.

These seven areas must be addressed within an overall approach that is holistic and takes into account their constant interdependence. Each subject area will, on the one hand, be presented in general terms with regard to the organization and the requirements of SR and, on the other hand, presented with regard to specific principles of the central issue under scrutiny.

Finally, each area includes several fields of action that should also be:

- presented in general terms,
- presented with regard to possible adjustments to the organization and a census of societal expectations, as well as those of stakeholders when it comes to the fields of action. (See the table below.)

CORE SUBJECTS AND FIELDS OF ACTION	Clause of the standard
1. Governance	6.2
2. Human rights	6.3
2.1: Due vigilance	6.3.3
2.2: Human rights risk situations	6.3.4
2.3: Avoidance of complicity	6.3.5
2.4: Resolving grievances	6.3.6
2.5: Discrimination and vulnerable groups	6.3.7
2.6: Civil and political rights	6.3.8
2.7: Economical, social and cultural rights	6.3.9
2.8: Fundamental principles and rights at work	6.3.10
3. Work relations and conditions	6.4
3.1: Employment and employer/employee relationships	6.4.3
3.2: Work conditions and social protection	6.4.4
3.3: Social dialogue	6.4.5
3.4: Health and safety at work	6.4.6
3.5: Human development and training in the workplace	6.4.7
4. Environment	6.5
4.1: Pollution prevention	6.5.3
4.2: Using sustainable resources	6.5.4
4.3: Climate change mitigation and adaptation	6.5.5
4.4: Protection of the environment, biodiversity and restoration of natural habitats	6.5.6
5. Fair operating practices	6.6
5.1: Fight against corruption	6.6.3
5.2: Responsible political involvement	6.6.4
5.3: Fair competition	6.6.5
5.4: Promoting social responsibility in the value chain	6.6.6
5.5: Respect for property rights	6.6.7
6. Consumer issues	6.7
6.1: Fair practices in marketing, information and contracts	6.7.3
6.2: Protecting consumers' health and safety	6.7.4
6.3: Sustainable consumption	6.7.5
6.4: Consumer service, support, and complaint and dispute resolution	6.7.6
6.5: Consumer data protection and privacy	6.7.7
6.6: Access to essential services	6.7.8
6.7: Education and awareness	6.7.9

7. **Community involvement and development**	6.8
7.1: Community involvement	6.8.3
7.2: Education and culture	6.8.4
7.3: Job creation and skills development	6.8.5
7.4: Technology development and access	6.8.6
7.5: Creating wealth and income	6.8.7
7.6: Health	6.8.8
7.7: Investment in society	6.8.9

Careful examination of the central issues and policy areas reveals that ISO 26000 places particular emphasis on the corporate and environmental pillars of TBL.

ISO 26000 also includes in its core issues the governance of an organization, but its content is meager compared to the cardinal issues and serious malfunctions identified under 15.2.5, since its length does not exceed 2 pages, compared with 50 pages devoted to clause 6 of the standard. The corresponding references are also sparse, although the bibliography includes OECD guidelines from 2004.

7. *Guidelines for integrating social responsibility throughout an organization*: these cover understanding of the SR of an organization, its integration throughout the organization, communications in connection with SR, improvement of the credibility of the organization with regard to SR, reviewing progress, and improving the performance and assessing the value for voluntary initiatives for SR. The dynamics of the standard are shown in Figure 15.7.

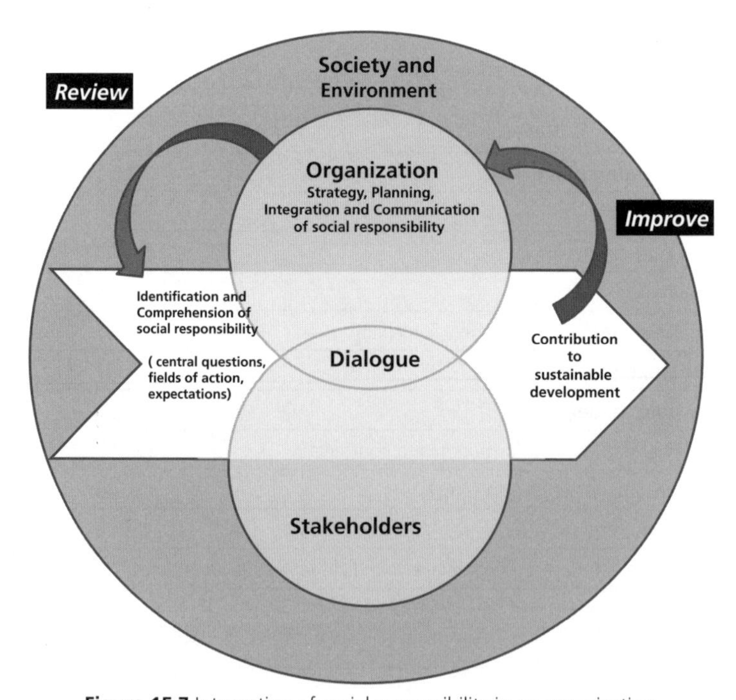

Figure 15.7 Integration of social responsibility in an organization.

Annex A (informative): Examples of voluntary initiatives and tools for social responsibility.
Annex B (informative): Abbreviations.

The standard also includes an extensive *Bibliography*, which is rare in the field of ISO standards and guidelines. It includes over a hundred references, related to SR and used by the drafters of the standard.

15.5.7 ISO 26000 in a nutshell

Figure 15.8 shows a schematic view of the standard and its clauses. It also makes it easier to understand. The drafters of ISO 26000 recommend:

- To begin the process with clause 3, *Understanding social responsibility*, and reviewing or assimilating *The principles of social responsibility* listed under clause 4, since these principles must be scrupulously respected for implementation of the standard.

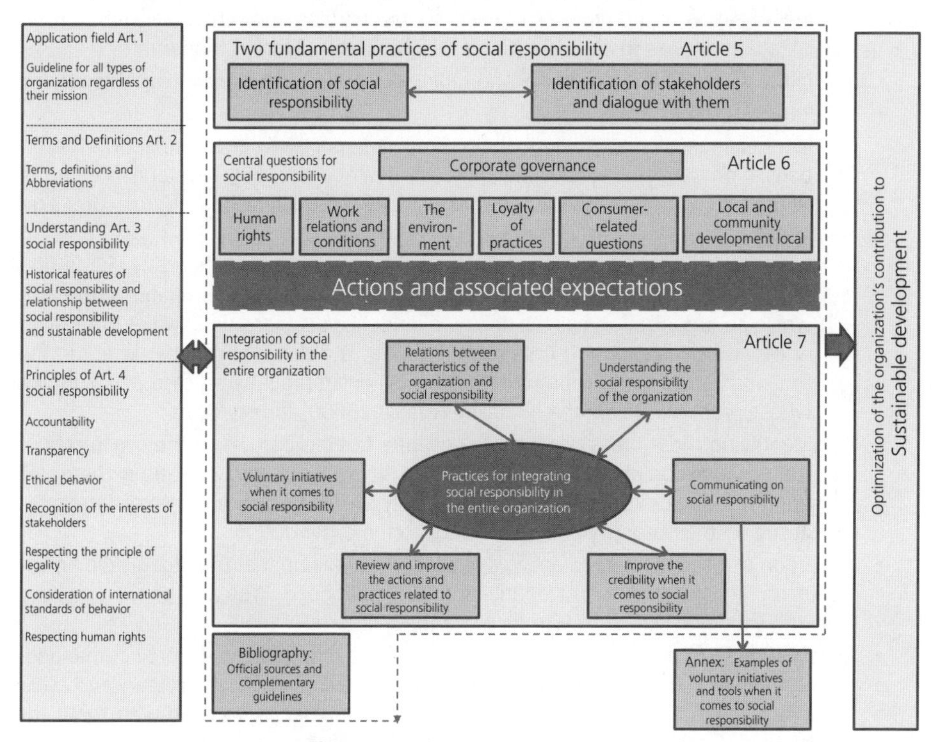

Figure 15.8 ISO 26000, an overview.

- The next step is the study of two fundamental practices of SR: identification of the sphere of influence, identification of stakeholders, and dialogue with them (clause 5).
- Then, it is recommended to analyze the issues and fields of action of SR, as well as individual actions and associated expectations (clause 6).
- By applying the guidelines in clause 7, the organization seeks to integrate SR into all its decisions and activities.
- To facilitate this implementation, the organization should refer to the guidelines listed in the Bibliography, as well as in Annex A, which provides examples of voluntary initiatives in social responsibility.

Implementation of the standard follows the order of the clauses: once the step described in one clause is completed, one should refer to the next.

15.5.8 Key steps in implementing ISO 26000

Aubrun et al. [15.12] from AFNOR advocate an implementation in six steps, which are summarized below:

- *Scope*, that is to say:
  - Determination of the limits of the approach: for example, integration of the entire company or leaving out certain units that are not involved much?
  - Motivation of management: here, an analysis is performed similar to that recommended by ISO 14004 (militant or pioneering state of mind, work under stress, improve the image).
  - Identification of the project's key success factors (company characteristics that facilitate deployment) and risks (those which, on the contrary, hinder it.
  - Initial review of organizational and financial requirements to carry out the project.
- *Inventory* and, specifically:
  - Identification of stakeholders, their strategic importance, their impact on the organization: we refer here to the approach described in 15.4, which deals with the AFNOR SD 21000 standard and makes it possible to limit the number of stakeholders.
  - Evaluation of the sphere of influence of the organization, defined by ISO 26000 as the scope/extent of political, contractual, economic or other relations, through which an organization has the ability to influence decisions or activities of persons or other organizations. This step is important as it determines what leverage the organization can use in promoting SR in its environment (the case of a supplier whose practices violate the guidelines of ISO 26000, for example).
  - Identifying those areas of SR that constitute the basic issues of the organization: the organization should indeed focus on these areas, and mastering them will bring more value to SR. The areas of activity should be put into several categories based on their urgency and importance to the organization.
- *Selection of priorities*: careful analysis of the fields of action that the organization must focus on, according to the categorization outlined above. We can, for example, use the grid shown in Figure 15.9 to facilitate these choices:
- *Establishment of action plans*: this step follows the requirements of project management outlined in 4.1.4 of this book, i.e., careful evaluation of the necessary and available resources. One should also consider the stakeholders associated with the fields of action in question, as they can provide technical assistance or, even better, sponsorship, but also pay attention to activities that are likely to have the greatest impact on the seven principles cited in clause 4 of the standard.
- *Implementation and performance measurement:* the goal is to modify the existing management systems and organization towards greater social responsibility and conformity to the seven principles, which means:
  - Reassess the organization's values, policy and strategy, and readjust them; this step can be difficult because it addresses the "fundamentals" of the organization (especially values, which are sometimes implicit).
  - Determine the objectives and categorize them within the organization, associating performance indicators to them, which will be essential during reviews.

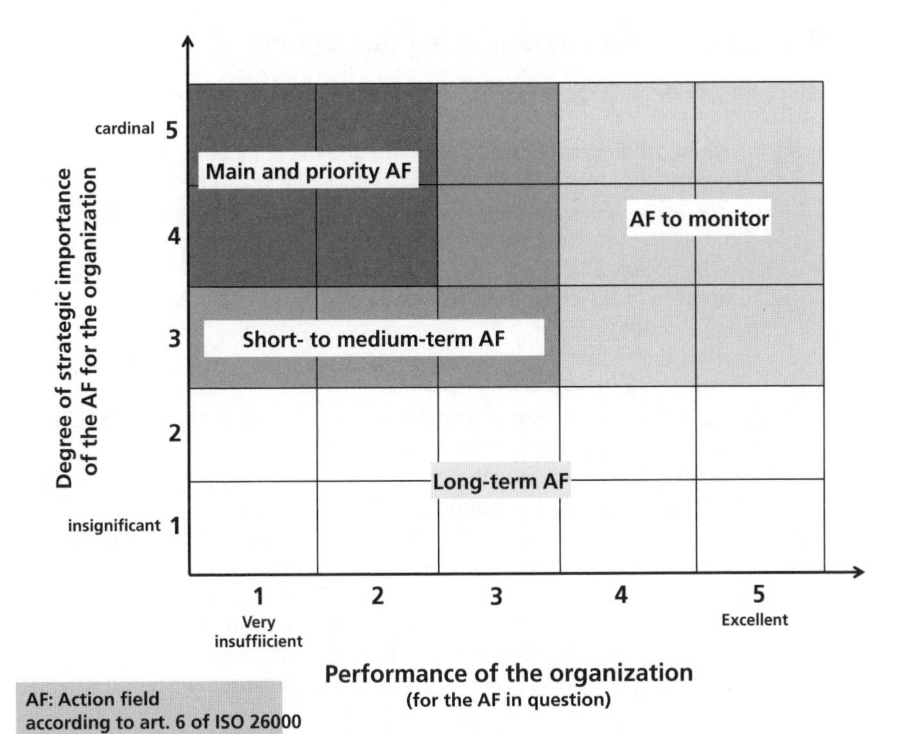

Figure 15.9 Analysis grid for prioritizing fields of action.

- Integrate SR into the organizational process, which requires:
  - awareness at all levels;
  - the development of skills as determined by the organization;
  - relying on existing management systems, while reassessing them.
- *Reviews and improvements:* this is to implement clause 7.7 of the standard with a view, common in management systems, to continual improvement by relying on a model similar to that recommended in ISO 9001 or ISO 14001, particularly when it comes to analysis of the development of indicators, related to, for instance, annual management reviews.
- *Reporting to stakeholders:* this is important because it plays a key role in the dialogue with stakeholders; it also demonstrates the organization's compliance with the principles of accountability and transparency contained in clause 4. This communication should be done:
  - internally and externally, and include the progress of the project of implementing ISO 26000;
  - externally, as ongoing daily communication, and address emergencies that impact on social responsibility, the labeling of products or service descriptions, and public statements to promote a particular aspect of SR;
  - externally again, regarding the implementation, periodically or annually, as a more succinct document, such as a Corporate Responsibility Report.

The guidelines of ISO 26000 may be supplemented by drawing an analogy to the content of ISO 14063:2006 (see 12.14).

15.5.9 Advantages of implementing ISO 26000

The standard includes a list of potential benefits that the implementation of ISO 26000 can bring:

- more informed decision-making, based on an improved understanding of the risks of society, opportunities related to social responsibility (including better management of legal risks) and risks related to the failure of fulfilling one's social responsibility.
- improving the organization's risk management practices.
- improving the organization's reputation and public trust in it.
- social acceptance of the organization's activities.
- developing innovations.
- improving the organization's competitiveness, including access to funding and the status of preferred partners.
- improving its relationships with stakeholders, hence opening up new prospects and contacts with a diverse range of stakeholders.
- increasing staff loyalty, involvement, participation and psychological well-being.
- improved health and safety of employees and collaborators.
- a positive impact on the organization's ability to recruit, motivate and retain staff.
- cost savings related to improved productivity and greater resource efficiency, lower consumption of energy and water, waste reduction and recycling of production scrap.
- improved reliability and fairness of transactions though responsible political commitment, fair and undistorted competition, and lack of corruption.
- prevention and/or reduction of potential conflicts with consumers about products or services.

No improvement of corporate governance is mentioned in this list.

> Although governance is listed as one of the six core subjects and the study of the other five issues depends on it, ISO 26000 hardly addresses this point, which is fundamental to social responsibility and sustainable development according to the concept of TBL (see 15.2.5, where the difficulties of having adequate governance are treated). This question is also the only one not to have fields of action.

15.5.10 Towards integration?

> To some extent, ISO 26000 surpasses the other management systems existing in the organization. It is regrettable that the content of ISO 26000 does not devote a section to this issue, including the interaction with ISO 9004, which focuses on the sustainable performance of an organization, and that makes all management systems proposed by the ISO a meta-system, a form of "Swiss army knife", with many interactions among its parts.

Figure 15.10 provides a plausible model of a management system for social responsibility, integrating standards and other ISO frames of reference – or similar structures – described briefly below; some of these are presented elsewhere in the book:

- **ISO/IEC 27001:2005,** *Information technology – Security techniques – Information security management systems – Requirements,* specifies an Information Security Management System (ISMS) that respects the principle of the *Deming* cycle. It helps

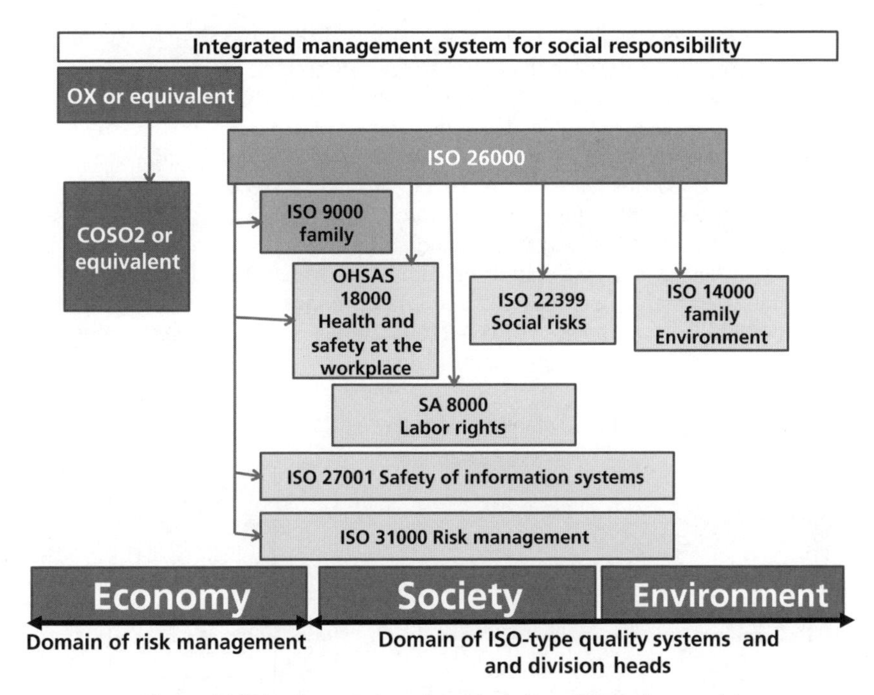

Figure 15.10 Implementation of standards in an SR structure model.

identify security measures to be implemented. It includes an annex as a frame of reference bringing together 39 security objectives, broken down into 133 measures in 11 fields (political security, personnel security, access control…).

- **OHSAS 18001** and **OHSAS 18002** (Occupational Health and Safety Assessment Series): these British standards, created from the failure to establish a corresponding ISO 18000 standard, have a structure similar to that of ISO 14001. They address the implementation of a ***health and safety policy***. They are certification standards, and their contents can be completed by another British Standard (BS 8800:1996) and the frame of reference ILO-OSH (2001), *Guidelines for management systems of health and safety*, from the International Labour Organization (ILO).

- **SA8000 (1997),** a standard based on the requirements of the ***labor law***, can be used for auditing and certification, in accordance with the ISO model. This standard – rarely used in Switzerland, where labor rights are recognized and child labor prohibited – is recommended for EU companies with subsidiaries outside the EU (or in regions belonging to the new members of the EU). Many companies seek to ensure the respect of fundamental workers' rights in their subsidiaries, and especially in the supply chain. SA8000 takes into account the Universal Declaration of Human Rights, the UN Convention on Children's Rights, and eight agreements signed by the ILO. SA8000 has eight chapters (child labor, forced labor, health and safety, freedom of association and right to collective bargaining, discrimination, disciplinary practices, working hours, pay) and a section on the implementation of a management system.

- **ISO/PAS 22399:2007,** *Societal security guideline for incident preparedness and continuity management.* Natural disasters, terrorist actions, technological accidents and environmental incidents have shown that crisis situations, caused intentionally or involuntarily, can affect both the public and private sectors. The PAS (Publicly Available Specification) informs organizations on how to prepare for unexpected and potentially devastating incidents, and how to respond. It describes a holistic management process that identifies potential impacts that threaten an organization and provides a framework to reduce their effects to a minimum. Similar standards in use in continental Europe are the ONR 49000 family. Some private agencies in Switzerland, such as the SQS[17], use a risk management standard (not specifically financial) developed by Germany, Austria and Switzerland and compatible with ISO 9001 and ISO 18000, and the standards mentioned above: it is the ***risk management system of the family ONR 49000***[18]. It was replaced by **ISO 31000:2009,** *Risk management – Principles and guidelines for implementation* (Figure 15.11).

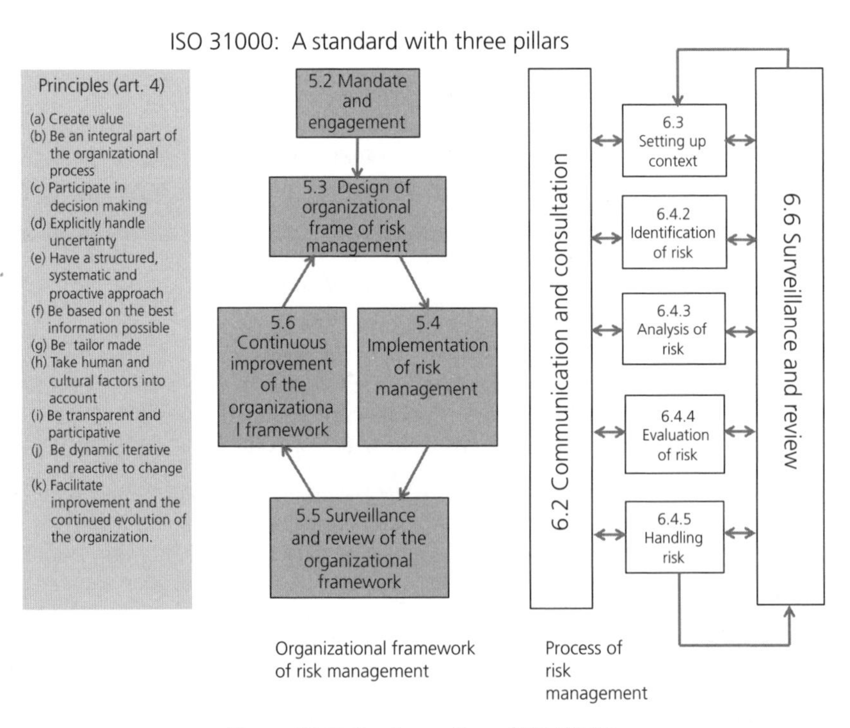

Figure 15.11 The three pillars of ISO 31000.

ISO 31000 is based on three pillars: the principles, the organizational framework and the process for managing risks. Its implementation brings many benefits to an organization: promotion of a proactive rather than reactive management, the need to identify risks in all areas of the organization, easier identification of opportunities and threats, improvement of corporate governance, increase in stakeholders' assurance, the establishment of a sound basis for decision-making and planning, and the effective allocation and use of resources for the treatment of risk, etc.

17 SQS: Swiss Association for Quality Management Systems, URL: http://www.sqs.ch/fr

18 Basic information available at: http://www.sqs.ch/fr/507.pdf#46

15.6 Corporate charters on ethical principles – a marketing strategy?

15.6.1 Charter of ethics or social responsibility?

Many companies have stated their ethical principles, describing the scope of social responsibility that they implement. Such communications emphasize the subjects treated, but do not generally describe the management and control systems used.

This is not without commercial interest; the goal is to create a climate of trust by communicating to customers and users the best practices that the company implements and the social responsibilities (child labor, environmental protection, etc.) it wishes to assume (image of the brand).

But the goal is also to allow the consumer or customer to view the company's commitment to the values associated with its products, which will thus be implicitly communicated to relatives, neighbors and coworkers. A product associated with a charter of ethics indicates, by its purchase, a differentiation of the roles and status of the social actors who bought it. It is a form of mask in the word's original meaning[19].

The sustainability of the implementation of a charter of ethics varies, depending on whether the charter refers to the fundamental values of the company or only to a range of products. In the latter case, the ethical values promoted are more volatile and may disappear when removing the product from the collection. This is logical because the charter of ethics is then associated with the marketing mix[20] of the product. Fundamental ethical values persist because their abandonment very often means a change in culture, which is very difficult in large companies. Their selection should thus be carefully considered.

Monitoring compliance with the provisions of the charter is sometimes carried out by the organization itself, which may cast doubt on its rigor. The solution is, whenever possible, to incorporate the terms of the charter in the *Quality Assurance of the institution*. ISO 10001:2007, *Quality management – Customer satisfaction – Guidelines for codes of conduct for organizations,* tracks the stages of drafting such a charter (code of conduct). Other companies rely on external control organizations or NGOs.

15.6.2 Three case studies: *The Body Shop, L'Oréal* and *Nivea*

The Body Shop

The Body Shop, founded by Anita Perella-Roddick in England in 1976, has its own distribution channels, and is a cosmetics company that has developed rapidly thanks to a charter of ethics adapted to a feminine (and somewhat feminist) niche marketing.

The Body Shop's[21] charter of ethics is broken down into five mottos:

- *Respect for the animal kingdom*: no testing of products on animals.
- *Fair trade*: with disadvantaged communities worldwide.
- *Development of esteem and self-acceptance*: no advertising with "anorexic" or very young models, no claim to eternal youth for the customers, emphasis on comfort and well-being.

---

19 In ancient Greece, masks were used to indicate the roles and emotions of the characters played.

20 Marketing is all technical and market studies that aim to predict, observe, create, renew or enhance consumers' needs and continuously adapt the productive and commercial device to the determined needs. This is done through four elements, the so-called marketing mix or four Ps of operational marketing: product, price, place, promotion.

21 www.thebodyshop.com, accessed in Jan. 2013.

- *Defense of human rights*: working conditions that respect the rights of the employee, especially in emerging and developing countries.
- *Protection of the planet*: use of renewable raw materials, respect for forests and the environment, collaboration with "militant" NGOs.

Although criticized in the 1990s for the discrepancy between the actual practices of the company and its ethical statements, *The Body Shop* was taken over by *L'Oréal* in 2006 for 662 million pounds.

L'Oréal

The commitment (also known as ethical principles) of *L'Oréal*[22] is another positioning that integrates Quality Assurance. Its mottos are:

- *Corporate citizenship*: profit-sharing for employees, women and science, solidarity, education.
- *Employee safety*: achieved by measures that drastically lower the number of accidents.
- *Quality*: ISO-certified company.
- *Product safety*: achieved by significant investments in research and development.
- *Protection of the environment*: use of natural gas (!) for the power stations in the factories, high percentage of recycled waste.
- *Commitment*: patronage, long-term projects, collaborations with UNESCO.

The values of *L'Oréal* correspond to the profile of a customer who is responsible, educated, mature, from the middle class and above, rational, believer in social and scientific progress, and who accepts market globalization. These values are therefore distinct from the more environmentally friendly, globalist, emotional, and hippie activist views of the customers of *The Body Shop*. The parallel with a niche marketing approach is striking, especially when one realizes that both companies are under the same corporate roof. Conversely, the company *Nivea*[23], founded in 1911 and also in the field of cosmetics, makes no mention of ethical principles, but merely communicates its history and affirms the quality of its products (without ever really describing what lies behind). This company also focuses on beauty, but cautiously stresses that this definition varies across cultures.

15.6.3 A skillful and responsible policy, that of Nestlé

Founded in 1866 in Vevey, Switzerland, Nestlé is the world's largest food and beverage company. The company employs around 280,000 people and has factories or operations in almost every country in the world.

More cautious, but closer to business activities, the multinational company Nestlé, often associated with L'Oréal, prefers the term Nestlé Corporate Business Principles (last edition June 2010), based on legislation, human rights, water management, anti-corruption, etc. In a 15-page document, prefaced by the highest governing body of Nestlé, it is difficult to find the words "ethical" "moral" in this carefully polished text, while its commitments are as strict as those of L'Oreal and The Body Shop. In short, Nestlé's *ten principles of business operations* are:

---

[22] www.loreal.fr, accessed in Jan. 2013.

[23] www.nivea.ch, accessed in Jan. 2009.

Consumers

1 *Nutrition, Health and Wellness*: Our core aim is to enhance the quality of consumers' lives every day, everywhere, by offering tastier and healthier food and beverage choices and encouraging a healthy lifestyle. We express this via our corporate proposition *Good Food, Good Life.*

2 *Quality assurance and product safety*: Everywhere in the world, the Nestlé name represents a promise to the consumer that the product is safe and of a high standard.

3 *Consumer communication:* We are committed to responsible, reliable consumer communication that empowers consumers to exercise their right to informed choice and promotes healthier diets. We respect consumer privacy.

Human rights and labor practices

4 *Human rights in our business activities*: We fully support the United Nations Global Compact's (UNGC) guiding principles on human rights and labor, and aim to provide an example of good human rights and labor practices throughout our business activities.

Our people

5 *Leadership and personal responsibility*: Our success is based on our staff. We treat each other with respect and dignity and expect everyone to promote a sense of personal responsibility. We recruit competent and motivated people who respect our values, provide equal opportunities for their development and advancement, protect their privacy and do not tolerate any form of harassment or discrimination.

6 *Safety and health at work*: We are committed to preventing accidents, injuries and illness related to work, and to protect employees, contractors and others involved along the value chain.

Suppliers and customers

7 *Supplier and customer relationships:* We require our suppliers, agents, subcontractors and their employees to demonstrate honesty, integrity and fairness, and to adhere to our non-negotiable standards. In the same way, we are committed to our own customers.

8 *Agriculture and rural development:* We contribute to improvements in agricultural production, the social and economic status of farmers and rural communities, and in production systems to make them more environmentally *sustainable.*

The environment

9 *Environmental sustainability:* We commit ourselves to environmentally sustainable business practices. At all stages of the product life cycle, we strive to use natural resources efficiently, favor the use of sustainably managed renewable resources, and target zero waste.

10 *Water:* We are committed to the sustainable use of water and continual improvement in water management. We recognize that the world faces a growing water challenge and that responsible management of the world's resources by all water users is an absolute necessity.

Nestlé states that its principles are put into work in all its markets, and that they are subject to regular internal audits, during which the assessment protocol is certified by a consulting firm and KPMG inspections. The company also states that the results are submitted by management to its board of directors.

Core dimensions of Nestlé's approach to business

Furthermore, the elements that make up the core dimensions of Nestlé's approach to business are *compliance*, *environmental sustainability* and *creating shared value.*

Compliance

The company is built on a strong base of compliance – national laws, relevant conventions, as well as its own regulations. The Nestlé Corporate Business Principles, updated in June 2010, have been made available to each of the Company's 280,000 employees, in over 50 languages.

Sustainability

Nestlé applies a life-cycle approach, involving its partners from farmer to consumer, to improve the environmental impacts of its operations. At all stages of the cycle, the company aims to use natural resources efficiently, promote the use of sustainably managed renewable resources and achieve zero waste.

Creating Shared Value

"Creating Shared Value," a concept developed by Professor Michael Porter from Harvard University, is at the heart of Nestlé's business. Creating Shared Value encourages each organization to create economic and social value simultaneously by focusing on the social issues that each is uniquely capable of addressing. For a business to be successful in the long term, Nestlé believes that there should be a clear economic or business benefit and clear social benefit. For maximum impact, Nestlé has focused its Creating Shared Value efforts and investments on three areas – *nutrition*, *water* and *rural development*.

- **Nutrition:** because food and nutrition are the basis of health and of its business – it's the reason why the Company exists.
- **Water:** because the ongoing quality and availability of it is critical to life, the production of food and to the Company's operations.
- **Rural development:** because the overall well-being of the farmers, rural communities, workers, small entrepreneurs and suppliers are intrinsic to the Company's ability to continue to do business in the future.

The Company's actions to become environmentally and socially responsible have been globally recognized. In 2011, Nestlé was the first infant formula manufacturer to be included in FTSE4Good, FTSE's responsible investment index. This index is designed to help investors identify companies that meet globally recognized corporate responsibility standards, including standards on human rights, supply chain and the marketing of breast-milk substitutes.

To address some of the challenges it faces, Nestlé engages with a wide range of international stakeholders to share insights, identify ways to overcome challenges and improve its decision-making and accountability. To combat unacceptable labor practices and improve working conditions in the agricultural supply chain, the Company is collaborating with the Fair Labor Association (FLA), a non-profit multi-stakeholder initiative. Through its partnership with The Forest Trust (TFT), Nestlé has developed Responsible Sourcing Guidelines with the aim of eliminating deforestation within its supply chain. The Company also engages with the World Economic Forum's World Resources Group as well as the UN Global Compact CEO

Water Mandate to help formulate strategies aimed at addressing the water 'overdraft' and foster innovative thinking.

Through these actions, the Company aims to bring the best and most relevant products to people, wherever they are, whatever their needs, throughout their lives.

15.6.4 A paragon of ethics and sustainability (www.switcher.com)

Founded in 1981 by *Robin Cornelius*, Switcher, a clothing brand with a small whale as its logo, has promoted ethical and ecological approaches in its management and production (in India), for over 20 years. Their unique philosophy is sustained by an active foundation in the fields of sustainable development, education and training. Indeed, the company, based in Mont-sur-Lausanne in Switzerland, has as its soul values grouped in the form of an ethical code that reflects its commitment especially in terms of environmental protection and social standards.

Strict guidelines govern the application of the code down to the smallest detail. In this way, all products are fully traceable (labeled Respect-code: www.respect-code.org). The headquarters, called "Switcherland", uses solar panels, recycled cardboard and hybrid vehicles. The focus of *Robin Cornelius* is respect for Man and Nature. His vision of performance is consistency over time: *You are effective when people "vote" for you because they believe in what you do. The image of a brand becomes more important than the brand itself: our values of respect are liked and attract interest, both internally and from customers.*

15.7 Limits and opportunities of the market for virtue

15.7.1 The real impact is weak

What is the real impact of the market for virtue? It appears to have reached a plateau, because the results are variable. The consumer is often willing to buy responsibly a product or "virtuous" service if the price difference is not significant. According to *D. Vogel* (professor at the University of California at *Berkeley*), author of a study that reported on the potential and limits of social responsibility [15.13], the initiatives of many companies are still too erratic and subject to the will of top management (little information is collected, often hampered by Boards that should instead be an essential driving force).

A recent study[24] (October 2007) reinforces this impression: BT and the *Economist Intelligence Unit* interviewed 1,200 senior executives from all continents, of whom 29 % were CEOs:

- for the majority, sustainable development is not a full-fledged strategy;
- 45 % of the companies do not publish a report on social responsibility;
- 23 % of the companies do not have a person officially responsible;
- 60 % of the people interviewed felt that management is the most difficult group to convince regarding a sustainable development program;
- 72 % report that they have no professional goal set in this area;
- 66 % have never been asked by their company to consider how their activities could interact with sustainable development;
- 46 % of them recognize that initiatives in this area contribute to improving the brand image, but only 20 % of them believe that this can increase profitability;

[24] Available at: www.globalservices.bt.com/static/assets/insights_and_ideas/sustainability/pdf/EIU

- 30 % of the people interviewed admit that the efforts of their company when it comes to sustainability focus more on communication than on actual changes.

15.7.2 Enhance collaboration between governments and organizations

The next step should be to study more intense collaboration between governments and organizations. This bridge-building is difficult because, although the *lobbying* of political authorities by large groups of manufacturers is known, these actions do not generally involve new regulations, because the strengthening of legislation is seen as contrary to the business and organizations do not want to strengthen the power of the state. These initiatives can also lead to conflicts between groups of manufacturers.

Vogel gives examples that illustrate the limits of social responsibility in its current paradigm:

- It would be good for public health if the US government prohibited the use of antibiotics for cattle (this might circumvent a rapid drop in their effectiveness) but, thanks to *McDonald's*, which does not use such treated cattle, many US citizens steer clear of absorbing these drugs.
- It would be better if the Indonesian government reinforced legislation responsible for environmental protection but, through the initiative of *Chevron Texaco*, part of the fragile ecosystem of New Guinea has been better preserved.
- It would be more effective if China raised its standards when it comes to labor laws but, thanks to *Nike*, some Chinese workers have bearable working conditions.

These examples clearly show that, too often, social responsibility is a second choice. Let us study the following examples cited by Vogel:

- If *Ford*, which has invested heavily in developing low-consumption vehicles, wants to market such products, why does the company not more actively support government campaigns to this effect?
- If *Home Depot* wants to improve forest management, why not support US and Canadian legislation that would seek to improve forestry practices?
- If *Interface* believes that the production of carpets is contrary to good environmental practices and finds that the market reacts too weakly to the launch of a new product in line with sustainable development, why does the company not support government initiatives for recycling this type of product?

This shows that companies are still far from accepting that social responsibility (private initiative) and government legislation can contribute to a world based on sustainable development and TBL.

By refusing to take this step, companies lend more weight to the arguments of critics who claim that the implementation of policies of social responsibility is an initiative of the economy to limit, or even reduce, the role of the state in major issues in our society.

15.7.3 The economy at a turning point?

This is certainly the conviction of *John Elkington*, the promoter of corporate social responsibility, who, in a study in 2001 [15.14], imagined companies based on a capitalism of stakeholders led by citizen directors.

But what are the factors that would enable this transformation? The very ones that guide companies in the direction of TBL: the population explosion, the increasing number of potential consumers, the scarcity of raw materials, environmental degradation and global warming, but also technological advances.

Companies will survive only if they are able to develop a *business model* using ultra clean technologies, strategically employing renewable materials and energy at the forefront of innovation, with a foolproof business ethic and able to form alliances with all stakeholders.

Ten years after the publication of this book, this type of business is still too rare. *John Elkington* is moderately optimistic; it is not known whether he will win his bet. If he doesn't, he predicts an implosion of the capitalist system, which, for a revered business guru supported by the *World Economic Forum*, may seem like a sweeping statement, but based on the current global crisis, may not be far off.

References

[15.1] For more on the ambiguity of the word "social" in this context, see the work by CAPRON and FRANCOISE QUAIREL-LANOIZELEE, *La responsabilité sociale d'entreprise* ed. La Découverte, 2007), p. 28.

[15.2] JOHN ELKINGTON, *Cannibals with Forks: The Triple Bottom Line of 21st Century Business,* New Society Publishers, 1998.

[15.3] ANDREW W. SAWITZ, *The Triple Bottom Line,* How Today's Best-Run Companies Are Achieving Economic, Social and Environmental Success – and How You Can Too, ed. Jossey-Bass, 2006.

[15.4] *Social responsibility is to increase profits, statement of "minimum" ethics,* by MILTON FRIEDMANN, reported *Levin* on the same page: *Just like in a good* (!) *war* (was he at the Battle of the Somme...?), a *manager must fight with courage and bravery but certainly not morals (*but courage and bravery are moral virtues, so...?). in SAMUEL MERCIER, *L'éthique dans les entreprises,* ed. La Découverte, 2004, p. 59; and this tasty quote by *T.*

[15.5] WILLIAM A. DIMMA, *Excellence in the Boardroom: Best Practices in Corporate Directorship*, ed. Wiley, 2002, chap. 11, *Corporate responsibility*, p. 165.

[15.6] Ibid., p. 164.

[15.7] See the book by PAUL W. MAC-AVOY et al., *The recurrent crisis in corporate governance*, Pagrave Mac Millan editions, 2003, p. 6 et seq.

[15.8] CHRISTOPHER LASCH, *La révolte des élites et la trahison de la démocratie*. ed. Flammarion, coll. Champs, 1996.

[15.9] See especially the subsections *Board competencies and Board skills inventory* in the book by DAVID A. NADLER (Mercer Delta Consulting), *Building better boards,* Wiley, 2006, pp. 30-31 and compare the sections in question with the ISO 19011:2002 standard.

[15.10] See for instance the pamphlet by JOHN.R. NOSFINGER, KENNETH A. KIM, *Corporate governance,* ed. Pearson Education, 2004.

[15.11] See the brochure *Participer à la future Norme internationale ISO 26000 sur la Responsabilité sociétale* (éditions de l'ISO, juillet 2006*),* from which these paragraphs have been taken.

[15.12] MÉRYLLE AUBRUN, et al. (collectif), ISO 26000, Responsabilité sociétale, comprendre, déployer, évaluer, AFNOR (2010), Partie II. Mise en œuvre, pp. 71-104.

[15.13] DAVID VOGEL, *The Market for Virtue – the potential and limits of corporate social responsibility*, Brooking Institution press, 2006, pp. 162-173.

[15.14] JOHN. ELKINGTON, *The Chrysalis Economy – how citizen CEOs and corporations can fuse values and value creation*, Capston Publishing Ltd., 2001.